Compendium of Plant Genomes

Series editor

Chittaranjan Kole
Raja Ramanna Fellow, Department of Atomic Energy, Government of India,
Kalyani, India

Whole-genome sequencing is at the cutting edge of life sciences in the new millennium. Since the first genome sequencing of the model plant Arabidopsis thaliana in 2000, whole genomes of about 70 plant species have been sequenced and genome sequences of several other plants are in the pipeline. Research publications on these genome initiatives are scattered on dedicated web sites and in journals with all too brief descriptions. The individual volumes elucidate the background history of the national and international genome initiatives; public and private partners involved; strategies and genomic resources and tools utilized; enumeration on the sequences and their assembly; repetitive sequences; gene annotation and genome duplication. In addition, synteny with other sequences, comparison of gene families and most importantly potential of the genome sequence information for gene pool characterization and genetic improvement of crop plants are described.

Interested in editing a volume on a crop or model plant? Please contact Dr. Kole, Series Editor, at ckole2012@gmail.com

More information about this series at http://www.springer.com/series/11805

Marcelino Pérez de la Vega
Marta Santalla · Frédéric Marsolais
Editors

The Common Bean Genome

Springer

Editors
Marcelino Pérez de la Vega
Facultad de Biología
Universidad de León
León
Spain

Marta Santalla
Misión Biológica de Galicia, Grupo de Biología de Agrosistemas
Consejo Superior de Investigaciones Científicas
Pontevedra
Spain

Frédéric Marsolais
Genomics and Biotechnology
London Research and Development Centre, Agriculture and Agri-Food Canada
London, ON
Canada

ISSN 2199-4781 ISSN 2199-479X (electronic)
Compendium of Plant Genomes
ISBN 978-3-319-63524-8 ISBN 978-3-319-63526-2 (eBook)
https://doi.org/10.1007/978-3-319-63526-2

Library of Congress Control Number: 2017956330

Printed on acid-free paper

This Springer imprint is published by Springer Nature
The registered company is Springer International Publishing AG
The registered company address is: Gewerbestrasse 11, 6330 Cham, Switzerland

This book series is dedicated to
my wife Phullara, and our children
Sourav, and Devleena

Chittaranjan Kole

Preface to the Series

Genome sequencing has emerged as the leading discipline in the plant sciences coinciding with the start of the new century. For much of the twentieth century, plant geneticists were only successful in delineating putative chromosomal location, function, and changes in genes indirectly through the use of a number of 'markers' physically linked to them. These included visible or morphological, cytological, protein, and molecular or DNA markers. Among them, the first DNA marker, the RFLPs, introduced a revolutionary change in plant genetics and breeding in the mid-1980s, mainly because of their infinite number and thus potential to cover maximum chromosomal regions, phenotypic neutrality, absence of epistasis, and codominant nature. An array of other hybridization-based markers PCR-based markers, and markers based on both facilitated construction of genetic linkage maps, mapping of genes controlling simply inherited traits and even gene clusters (QTLs) controlling polygenic traits in a large number of model and crop plants. During this period a number of new mapping populations beyond F2 were utilized and a number of computer programs were developed for map construction, mapping of genes, and for mapping of polygenic clusters or QTLs. Molecular markers were also used in studies of evolution and phylogenetic relationship, genetic diversity, DNA-fingerprinting and map-based cloning. Markers tightly linked to the genes were used in crop improvement employing the so-called marker-assisted selection. These strategies of molecular genetic mapping and molecular breeding made a spectacular impact during the last one and a half decades of the twentieth century. But still they remained 'indirect' approaches for elucidation and utilization of plant genomes since much of the chromosomes remained unknown and the complete chemical depiction of them was yet to be unraveled.

Physical mapping of genomes was the obvious consequence that facilitated development of the 'genomic resources' including BAC and YAC libraries to develop physical maps in some plant genomes. Subsequently, integrated genetic-physical maps were also developed in many plants. This led to the concept of structural genomics. Later on, emphasis was laid on EST and transcriptome analysis to decipher the function of the active gene sequences leading to another concept defined as functional genomics. The advent of techniques of bacteriophage gene and DNA sequencing in the 1970s was extended to facilitate sequencing of these genomic resources in the last decade of the twentieth century.

As expected, sequencing of chromosomal regions would have led to too much data to store, characterize, and utilize with the-then available computer software could handle. But development of information technology made the life of biologists easier by leading to a swift and sweet marriage of biology and informatics and a new subject was born—bioinformatics.

Thus, evolution of the concepts, strategies and tools of sequencing and bioinformatics reinforced the subject of genomics—structural and functional. Today, genome sequencing has traveled much beyond biology and involves biophysics, biochemistry and bioinformatics!

Thanks to the efforts of both public and private agencies, genome sequencing strategies are evolving very fast, leading to cheaper, quicker and automated techniques right from clone-by-clone and whole-genome shotgun approaches to a succession of second generation sequencing methods. Development of software of different generations facilitated this genome sequencing. At the same time, newer concepts and strategies were emerging to handle sequencing of the complex genomes, particularly the polyploids.

It became a reality to chemically—and so directly—define plant genomes, popularly called whole-genome sequencing or simply genome sequencing.

The history of plant genome sequencing will always cite the sequencing of the genome of the model plant *Arabidopsis thaliana* in 2000 that was followed by sequencing the genome of the crop and model plant rice in 2002. Since then, the number of sequenced genomes of higher plants has been increasing exponentially, mainly due to the development of cheaper and quicker genomic techniques and, most importantly, development of collaborative platforms such as national and international consortia involving partners from public and/or private agencies.

As I write this preface for the first volume of the new series "Compendium of Plant Genomes", a net search tells me that complete or nearly-complete whole-genome sequencing of 45 crop plants, eight crop and model plants, eight model plants, 15 crop progenitors and relatives, and three basal plants are accomplished, the majority of which are in the public domain. This means that we nowadays know many of our model and crop plants chemically, i.e., directly, and we may depict them and utilize them precisely better than ever. Genome sequencing has covered all groups of crop plants. Hence, information on the precise depiction of plant genomes and the scope of their utilization is growing rapidly every day. However, the information is scattered in research articles and review papers in journals and dedicated web pages of the consortia and databases. There is no compilation of plant genomes and the opportunity of using the information in sequence-assisted breeding or further genomic studies. This is the underlying rationale for starting this book series, with each volume dedicated to a particular plant.

Plant genome science has emerged as an important subject in academia, and the present compendium of plant genomes will be highly useful both to students and teaching faculties. Most importantly, research scientists involved in genomics research will have access to systematic deliberations on the plant genomes of their interest. Elucidation of plant genomes is not onlyof interest for the geneticists and breeders, but also for practitioners of an array of plant science disciplines, such as taxonomy, evolution, cytology, physiology, pathology, entomology, nematology, crop production, bio-chemistry, and obviously bioinformatics. It must be mentioned that information regarding each plant genome is ever-growing. The contents of the volumes of this compendium are therefore focusing on the basic aspects of the genomes and their utility. They include information on the academic and/ or economic importance of the plants, description of their genomes from a molecular genetic and cytogenetic point of view, and the genomic resources developed. Detailed deliberations focus on the background history of the national and international genome initiatives, public and private partners involved, strategies and genomic resources and tools utilized, enumeration on the sequences and their assembly, repetitive sequences, gene annotation, and genome duplication. In addition, synteny with other sequences, comparison of gene families, and, most importantly, potential of the genome sequence information for gene pool characterization through genotyping by sequencing (GBS) and genetic improvement of crop plants have been described. As expected, there is a lot of variation of these topics in the volumes based on the information available on the crop, model, or reference plants.

I must confess that as the series editor it has been a daunting task for me to work on such a huge and broad knowledge base that spans so many diverse plant species. However, pioneering scientists with life-time experience and expertise on the particular crops did excellent jobs editing the respective volumes. I myself have been a small science worker on plant genomes since the mid-1980s and that provided me the opportunity to personally know several stalwarts of plant genomics from all over the globe. Most, if not all, of the volume editors are my longtime friends and colleagues. It has been highly comfortable and enriching for me to work with them on this book series. To be honest, while working on this series I have been and will remain a student first, a science worker second, and a series editor last. And I must express my gratitude to the volume editors and the chapter authors for providing me the opportunity to work with them on this compendium.

I also wish to mention here my thanks and gratitude to the Springer staff, Dr. Christina Eckey and Dr. Jutta Lindenborn in particular, for all their constant and cordial support right from the inception of the idea.

I always had to set aside additional hours to edit books besides my professional and personal commitments—hours I could and should have given to my wife, Phullara, and our kids, Sourav, and Devleena. I must mention that they not only allowed me the freedom to take away those hours from them but also offered their support in the editing job itself. I am really not sure whether my dedication of this compendium to them will suffice to do justice to their sacrifices for the interest of science and the science community.

Kalyani, India Chittaranjan Kole

Preface

The relevance of common bean (*Phaseolus vulgaris* L.) is determined by the fact that it is the most important grain legume for food consumption worldwide and has a role in sustainable agriculture owing to its ability to fix atmospheric nitrogen. It is a staple crop with a major societal importance as main source of proteins and nutrients in developing regions such as Africa and Latin America. Due to its domestication, it can be grown in a wide range of environments, from near sea level up to 3000 m above sea level, in arid, semiarid and tropical environments, and under high or moderate temperatures. Its main consumption is for its edible dry seed (dry beans) or unripe fruit (green beans). This double use determines that modern cultivars have been bred for different fruit characteristics such as those that influence the seed (shape, color, cooking ability, etc.) and the pod (length, tenderness, stringless, etc.) apart from common features as disease resistance, yield, precocity and plant architecture.

Common bean was domesticated in two geographically isolated and differentiated areas, according to multiple lines of evidence, generating two distinguishable gene pools, Mesoamerican and Andean, within a single species. The domestication of two independent lineages implicated a parallel evolution and introgression events with important implications for modern common bean crop improvement. In fact, common bean was likely domesticated concurrently with maize as part of the 'milpa' cropping system (featuring common bean along with maize and squash), which was adopted throughout the Americas. Common bean was introduced in Europe soon in the XVI century through Spanish and Portuguese trade routes, and independently introduced into Africa. Large variation of common bean evolved in Europe as a result of adaptation to new ecological and manmade conditions. In consequence, common bean has four well-identified centers of diversification, two in America, where it was domesticated, one in Europe, and one in Africa, where it was introduced after the discovery of America.

Common bean is also an important species from the scientific point of view. It has been widely used in research and breeding. Both Darwin and Mendel used this species in their research, and it was widely used as material for genetic experimentation and breeding during the early years of the development of genetics, at the beginning of the twentieth century. For example, the demonstration by Johansen that the phenotype is the result of the interaction between the genotype and the environment; or the direct relationship between a quantitative and a qualitative character, seed size and seed coloration of the common bean seed, by Sax. A large corpus of genetic data on common bean has been generated throughout the twentieth century, now widely complemented and surpassed by the genomic information.

The recent release of the Andean and Mesoamerican common bean genomes is enabling a new wave of cutting-edge research, including epigenomics and translatome analyses, in a crop that has fed billions of people for more than 5000 years. Moreover, the comparison of the Andean and the recently released Mesoamerican genome has initially revealed interesting differences. For example, the size and the number of genes in the Mesoamerican genome is smaller than the Andean genome. The availability of the genome sequence has completely changed the paradigm of the species' genetics and genomics.

Genomic information on the other domesticated or wild *Phaseolus* species range from scarce to null. However, the already sequenced genomes of common bean will help in sequencing and assembling the genomes of other species of the genus. In fact, common bean is one of the five domesticated species of the genus *Phaseolus*, a genus formed by a relatively large number of species with broader or narrower geographical distribution and some of them adapted to particular environments. The other four domesticated species are tepary bean (*P. acutifolius* A. Gray), runner bean (*P. coccineus* L.), lima bean (*P. lunatus* L.) and year-long bean (*P. dumosus* Macfad.), all four originated and domesticated in America. The genomic information from the other *Phaseolus* species will give more complete information on the evolutionary processes associated to domestication and on the adaptive processes to particular environments with special relevance for breeding strategies.

Genomic resources include not only data from the nuclear genome but also from organelle genomes. In particular, mitochondrial and chloroplast genomes gives complementary information on the evolutionary processes within common bean and between species within the genus *Phaseolus*. Genomics data allow knowing the whole set of genes implicated in a metabolic route, including the genes coding for the transcription factors implicated in their expression control, helping in the transcriptomic analyses of their expression under different environments or in different tissues. Likewise, they allow the analysis of the relationships between different metabolic routes and biological processes. A particular set of genes of interest in a crop mainly exploited by its seeds is the set of seed storage proteins, which determine a great part of the nutritive value of the main crop product. New tools will be likely developed from these data such as complete set of molecular markers, microarrays, etc., which will speed and facilitate assisted selection processes.

The present book summarizes data on the economic and scientific relevance of common bean, its relation with other species of the genus *Phaseolus*, and insists on the importance of the domestication events, parallel evolutionary history and further expansion of the cultivated forms. It also describes the numerous genetic and genomic resources available, the identified genes and quantitative trait loci (QTL) identified, as well examples of the study of functionally related genes and future prospects. This book shows that these are exciting times for common bean research in a field with the potential to reduce the threat of food insecurity by releasing crops tolerant to biotic and abiotic stresses, increasing yields and enhancing the nutritional quality of beans.

León, Spain — Marcelino Pérez de la Vega
Pontevedra, Spain — Marta Santalla
London, Canada — Frédéric Marsolais

Contents

Abbreviations

ADP	Andean diversity panel
AFLP	Amplified fragment length polymorphism
AM	Association mapping
BAC	Bacterial artificial chromosome
BBH	Best-bidirectional blast hit
CDS	Coding sequence (Coding DNA sequence)
CGIAR	Consultative Group for International Agricultural Research
CIAT	International Center for Tropical Agriculture (Centro Internacional para Agricultura Tropical)
CMA	Chromomycyn A3
cpSSR	Chloroplast simple sequences repeat
DAF	Days after flowering
DAPI	4′-6-diamidino-2-phenylindole
DDP	Durango Diversity Panel
DRK	Denaturation-renaturation kinetics
ER	Endoplasmic reticulum
EST	Expressed sequence tag
FISH	Fluorescent *in situ* hybridization
GBS	Genotyping by sequencing
GO	Gene ontology
GWAS	Genome wide association study
IGS	Intergenic spacer
IR	Inverted repeat (plastid genome)
ITS	Internal transcribed spacer
KEEG	Kyoto encyclopedia of genes and genomes
KOG	Eukaryotic orthologous groups
LD	Linkage disequilibrium
LG	Linkage group
lncRNA	Long non-coding RNA
LSD	Large single sequence (plastid genome)
MAS	Marker assisted selection
MDP	Mesoamerican Diversity Panel
NCF	Nested chromosome fusion
NGS	Next generation sequencing
P450	Cytochrome P450

PA	Phytic acid
PANTHER	Protein analysis through evolutionary relationships
PBPI	Potential biparental plastid inheritance
PCC	Pearson correlation coefficient
PCR	Polymerase chain reaction
Pfam	Protein families database
PGC	Protein coding genes
QTL	Quantitative trait locus/loci
RAPD	Random amplified polymorphic DNA
RFLP	Restriction fragment length polymorphism
RGA	Resistance gene analog
RIL	Recombinant inbred line
RNAseq	RNA sequencing
RPKM	Reads per kilobase per million mapped reads
SCAR	Sequence characterized amplified region
SNP	Single nucleotide polymorphism
SSC	Short single sequence (plastid genome)
SSR	Simple sequence repeat
TEC	Tissue expression complementarity
UniProt	Protein database
WGD	Whole genome duplication
WGS	Whole genome sequencing

1 Common Bean: Economic Importance and Relevance to Biological Science Research

James R. Myers and Ken Kmiecik

Abstract

Common bean (*Phaseolus vulgaris* L.) is broadly adapted to environments with moderate growing temperatures, about 400 mm of precipitation and a growing season of 60–120 days. The popularity of the crop originates from the fact that it is relatively easy to produce, it is flavorful and versatile, and it is a good source of nutrition. The two major types of common bean are dry edible beans and snap or garden beans. Precise economic valuation of the common bean crop is difficult to obtain on a global scale because other species are often included in the statistical data collected in different countries, but with production of 18.9 million T for all types, it is the most widely produced grain legume and ranked third after soybean and groundnut for oilseed and grain legumes combined. Common bean is produced in both developed and developing countries and is an important source of protein, carbohydrates, some vitamins, and micronutrients. Common bean first became known to the scientific world with the Columbian exchange beginning in 1493, but little is known about the genetic diversity of the early introductions to Europe. Systematic breeding of common bean began in the nineteenth century in Europe and the USA. Common bean was the subject of Mendel's genetics research, was used by Johannsen to investigate quantitative inheritance, and has the distinction of being the first plant species where a quantitative trait locus was identified. Contemporary research on common bean in the recent past has been conducted in about 21 academic disciplines with plant physiology, medicine, microbiology, and food science, resulting in the most publications. Plant breeding, genetics, plant pathology, and

J. R. Myers (✉)
Department of Horticulture, Oregon State University, Corvallis, OR 97331, USA
e-mail: james.myers@oregonstate.edu

K. Kmiecik
7202 Portage Rd, De Forest, WI 53532, USA
e-mail: kakmiecik@sbcglobal.net

M. Pérez de la Vega et al. (eds.), *The Common Bean Genome*,
Compendium of Plant Genomes, https://doi.org/10.1007/978-3-319-63526-2_1

genomics and bioinformatics are generally less well represented, but this may change as more genomics studies are conducted. The special traits of common bean that have driven most research are the seed storage proteins, the symbiotic relationship with rhizobium species, the history of plant domestication, and the architecture of genetic diversity within the species.

Keywords
Grain legume · Dry bean · Snap bean · Production
Consumption · Genetics · Citation analysis

1.1 Introduction

Originating in the New World and widely dispersed because of its broad adaptation, the common bean (*P. vulgaris* L.) has become an important grain and vegetable legume on a global scale. It performs best in moderate growing temperatures (>10 °C and <30 °C) with about 400 mm of precipitation during the growing season. Common bean is found throughout temperate growing regions where the season permits 60–120 days of frost-free growth as well as in the tropical highlands with growing temperatures <30 °C. The popularity of the crop originates from the fact that it is relatively easy to produce, it is flavorful and versatile in its preparation, and it is a good source of nutrition. Uncooked dry bean contains approximately 22% protein, several micronutrients (Ca, Fe, Mg, P, and K), complex carbohydrates (62%), soluble fiber (15%) and is a significant source of folate (USDA 2015). As a vegetable, the immature pods have high moisture content, with raw pods containing about 1.9% protein and 7% carbohydrate, and significant quantities of vitamin C, carotenoids, and vitamin K, which dry beans lack. Common bean is particularly important to the developing world in providing a source of protein, calories, and trace nutrients to individuals who cannot afford more expensive sources of nutrition.

1.1.1 Kinds and Uses of Common Bean

The term bean is widely applied to many different species of large seeded legumes (and some non-legumes), but those English terms that apply specifically to common bean include dry, kidney, French, navy, pinto, garden, snap, green, wax, and string. These names often refer to specific types or uses of common bean (Table 1.1 and Fig. 1.1). The two major types of contemporary agriculture are dry edible beans where the mature seed is harvested and prepared in main dishes and garden beans—the vegetable form where immature pods are cooked and eaten (Table 1.2).

Considerable evidence from several disciplines points toward the independent domestication of common bean occurring in at least two regions (Gepts et al. 1986; Koenig and Gepts 1989; Gepts and Debouck 1991), which are generally termed the Mesoamerican and Andean centers of domestication. Within each of these centers, distinctive races and market classes are found (Table 1.3). The races of common bean (Singh et al. 1991) are generally defined by certain morphological and biochemical characteristics as well as by traditional regions of use. Beans of the Mesoamerican center are classified into three races, and those of the Andean center of domestication are also subdivided into three races (Table 1.3). Within races, dry beans are

Table 1.1 Some of the terms used for different types of *Phaseolus vulgaris* in the scientific literature and statistical databases

Name	Usage/definition	FAO classification
Dry grain		
Common	Term refers generally to all forms of *Phaseolus vulgaris* and is a direct derivative of the Latin name	None
Dry	Mature grain where the seed is boiled to prepare for consumption	Dry bean; in some countries, FAO data may refer to other species as dry bean (including other *Phaseolus* beans such as *P. coccineus, P. lunatus* and *P. acutifolius* as well as *Vigna angularis, V. aureus, V. calcaratus, V. mungo, V. radiata, and V. aconitifolia*, but excluding cowpea or *V. unguiculata*)
Kidney	Name appears in the scientific literature as a general reference to common bean, but specifically refers to a market class of dry bean, usually large seeded, reniform in shape, and red or white in color	None
French	A term commonly applied to *P. vulgaris* beans in Europe and in the scientific literature. The term is used for both dry and snap beans	None
Navy	Market class of dry bean, small seeded and white in color	None
Popping or Nuña	Dry bean that is cooked using dry heat and where the cotyledons expand and soften	None
Vegetable		
Shell or Fresh	Seeds are consumed when mature but still high moisture	Green bean and may include *Vigna* species in some countries
Snap (green, wax) or garden	Green pods at various stages of maturity are cooked and consumed. Snap beans lack pod wall fiber and suture fiber or strings	String bean and may include *Vigna* species in some countries
String bean	Similar to snap beans but pod suture string is present and must be removed by hand	No distinction is made between types with or without pod suture fiber

further separated into market classes, which are generally defined by seed characteristics, but have superimposed upon them their genetic heritage and its influence in adaptation and various developmental characteristics. A classification of dry beans for North America is shown in Table 1.3. It should be noted that the relationships among center of domestication, race, and market class are based on traditional information, but contemporary plant breeding has blurred the boundaries among groups, and what have once been considered good indicators of centers of domestication (e.g., phaseolin seed storage protein) may no longer show absolute associations.

Dry beans (Fig. 1.1a and b) are usually prepared by soaking in water to imbibe the seed followed by cooking in a water-based broth either boiling or using a pressure cooker to shorten preparation time. They may be eaten directly in this fashion, but are more often cooked or combined with other more savory ingredients. They may be reprocessed (such as with refried beans) into new forms. Plant type for dry edible beans ranges from determinate bush to indeterminate upright types and to indeterminate vining non-climbers to indeterminate climbers.

Another type of dry bean is nuña or popping bean (Fig. 1.1e). The popping trait is limited to a

Fig. 1.1 Examples of various forms of common bean used for food around the world. **a** Kablanketi type of dry bean used in eastern and southern Africa; **b** large red bean typical of that consumed worldwide; **c** fresh snap bean pods consumed as a vegetable; **d** snap bean seeds showing distinctive cylindrical shape; **e** Nuña beans after being popped for consumption; **f** dry bean leaves picked for use as a vegetable in southern Tanzania

Table 1.2 Pod and seed characteristics that influence preparation and consumption of the major common bean types

Type	Pod wall fiber	Pod wall thickness	Pod suture fiber (strings)	Method of cooking	Seed shape
Dry	Present	Thin	Present	Moist heat	Variable
Popping	Present	Thin	Present	Dry heat	Round
Shell	Present	Thin	Present	Moist heat	Variable
Romano	Absent	Thin	Generally absent	Moist heat	Variable
String	Absent	Thick	Present	Moist heat	Variable
Snap	Absent	Thick	Absent	Moist heat	Cylindrical

Types are described in Table 1.1

Table 1.3 Races and selected market classes of common bean [Modified from (Singh et al. 1991)]

Race	Seed size	Phaseolin	Market classes
Mesoamerican domestication center			
Mesoamerica	Small	S	Navy, Black, Small Red
Durango	Medium	S	Pinto, Bayo, Red Mexican, Great Northern, Pink, Blue Lake type snap beans
Jalisco	Medium	S	Flor de Mayo, Apetito, Flor de Junio
Andean domestication center			
Nueva Granada	Medium–large	T	Light Red Kidney, Dark Red Kidney, White Kidney, Yellow-eye, Peruano, Tendergreen type snap beans
Chile	Medium–large	C, H	Cranberry, Romano type snap beans
Peru	Medium–large	T, C, H	Nuña, Poroto

Examples of representative market classes of North America, including Mexico

narrow group of dry edible beans originating in the Andean areas of South America. At high altitudes, cooking dry beans by boiling becomes problematic because of the lowered boiling point of water and the extra time required to cook the food. Popping beans are heated directly and require less time to expand or puff into a soft edible state, thereby requiring less cooking fuel.

Vegetable use of common beans includes immature pods (Fig. 1.1c), high moisture seed, and leaves (Fig. 1.1f) as greens. Use of immature pods is common in North America, Europe, Middle East, Africa, and throughout Asia including India, Indonesia, Thailand, Philippines, China, and Japan. Preparation includes boiling, steaming, and frying. Immature pods may have suture strings or be stringless; some wall fiber or none is present depending on use such as processing, market garden, or home garden. The plant type encompassed in garden beans includes indeterminate climbing (pole) types or half runner prostrate types, and fully determinate (bush) plant habits.

The snap bean (Fig. 1.1c and d) is mainly consumed in developed countries. Snap beans have been selected for reduced fiber in the green pod. In its more ancestral form (typified by Romano beans), fiber is reduced in the pod walls only; pod walls are thin, and suture strings are present (Table 1.2). String beans in addition to having low pod fiber have a second trait for fleshy pod walls that give the pod an oval to round cross-sectional shape. However, suture strings may still be present. In the modern forms of snap beans, the pod suture string is also lacking (Tables 1.1 and 1.2).

Use of high moisture seed shelled in the green stage is less common. It includes the French Flageolet types, large Horticultural or Cranberry types in the USA, and similar types in Italy such as Borlotto. Frequently, large seeded types are also shelled when used in high altitudes in the Andes, rather than used dry. Many large seeded dry edible types grown in parts of Asia and Africa may be used as immature pods or shelled, in addition to dry seed.

The development of beans as vegetable is unclear, but it is evident that Andean and Mesoamerican germplasm introduced into Europe is part of the basis of modern garden bean cultivars. The presence of S, T, and C phaseolin types (Gepts and Bliss 1986; Angioi et al. 2010) in pre-1950s' snap bean lines suggests a complex history.

1.2 Value, Production, and Consumption

Exact valuation of the common bean crop is difficult to obtain on a global level. This is because the main source of worldwide agricultural statistics (FAOSTAT 2015) does not distinguish among species of grain legumes when collecting statistical information and more than one species of bean may be aggregated (Table 1.1). This is especially true of India and China where *Vigna* spp. may account for 93 and 56%, respectively, of dry bean use in these regions (Akibode and Maredia 2011). Estimates for the amount of dry beans as classified by FAO that are actually *P. vulgaris* for Asia range from 17% for 1998 data (Singh 1999) to 9% for 2006–2008 data (Akibode and Maredia 2011). According to FAO, the estimates exclude cowpea (Table 1.1). Similarly, green or shell bean and string or snap bean categories in FAOSTAT may include various *Vigna* species (Table 1.1). For the latter categories, even less is known about the percentages that are actually common bean. Here, we report figures adjusted using the more recent 9% estimate of Akibode and Maredia (2011). We applied this adjustment to both dry and shell beans, but not snap beans. Our rationale was that many of the same species are used as dual purpose dry and as shell-outs and that proportional usage should be similar. With snap beans, the other main edible pod bean besides common bean is yard-long bean or asparagus bean (*Vigna unguiculata*), which is the vegetable form of cowpea and generally of minor importance in the countries included in this analysis. Additional caveats to the data are that these numbers do not account for non-commercial production in small holder enterprises and that in many parts of the tropics, common bean is intercropped, which may lead to an overestimation of area and an underestimation of yield (Akibode and Maredia 2011).

Worldwide, approximately 18.9 million T (all types combined) are produced (Table 1.4 and Fig. 1.2). At 13.9 million T, dry bean production is considerably greater than shell (3.5 million T) and snap (1.6 million T) production. Hectares for dry beans show an upward trend over time, while that for shell and snap remain essentially unchanged (Table 1.4 and Fig. 1.3). Yields for dry and snap bean have remained essentially unchanged from 1992 to 2012, but those for shell beans have increased sharply. The increase in area for dry beans and the increase in yield for shell beans have increased production in these types over the past two decades, whereas snap bean production has remained essentially unchanged. On a megaregional basis, Africa and the Americas are the largest producers of dry beans, accounting for 66% of production (Fig. 1.2).

1.3 Importance of Common Bean in Science

1.3.1 Common Bean Becomes Known to Science

Common bean became known to science following the Columbian exchange. What was most likely common bean (and not a related species) was described from the first Columbian expedition (Markham 2010) and were introduced into Europe soon thereafter. What we know of the

Table 1.4 World area (Ha), yield (T ha^{-1}), and production (T) of dry, shell, and snap beans from 1992 to 2012

Bean type	Year				
	1992	1997	2002	2007	2012
Area harvested (1000 ha)					
Dry bean (adjusted)	13,510	13,888	15,360	15,156	16,230
Shell (adjusted)	350	384	395	394	377
Snap	217	217	223	239	182
Yield (T ha^{-1})					
Dry bean	0.65	0.64	0.72	0.73	0.82
Shell	6.51	6.92	8.08	12.34	13.51
Snap	6.82	7.82	8.56	8.59	8.47
Production (1000 T)					
Dry bean (adjusted)	9413	10,070	12,342	12,135	13,904
Shell (adjusted)	2269	2546	2850	3472	3484
Snap	1478	1695	1914	2052	1575

Data source FAOSTAT (2015) with dry and shell bean area and production adjusted to remove non-common bean data from the figures

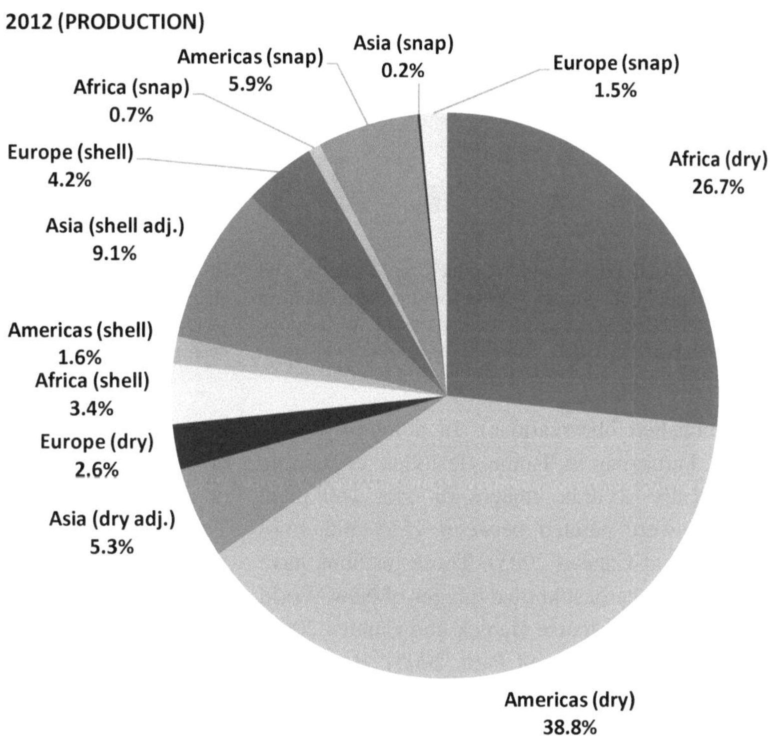

Fig. 1.2 Production of dry, shell, and snap beans in 2012 (FAOSTAT 2015)

early history of beans comes from paintings and illustrations in buildings, prayer books, and herbals. The oldest image of what is thought to be a bean plant is found in Livre d'Heures d'Anne de Bretagne which was illustrated between 1505 and 1508. Camus (1894) cataloged the plants in the prayer book and concluded that one represented *P. vulgaris*, but more recent examination of the image in question raises doubt as the image is not morphologically accurate

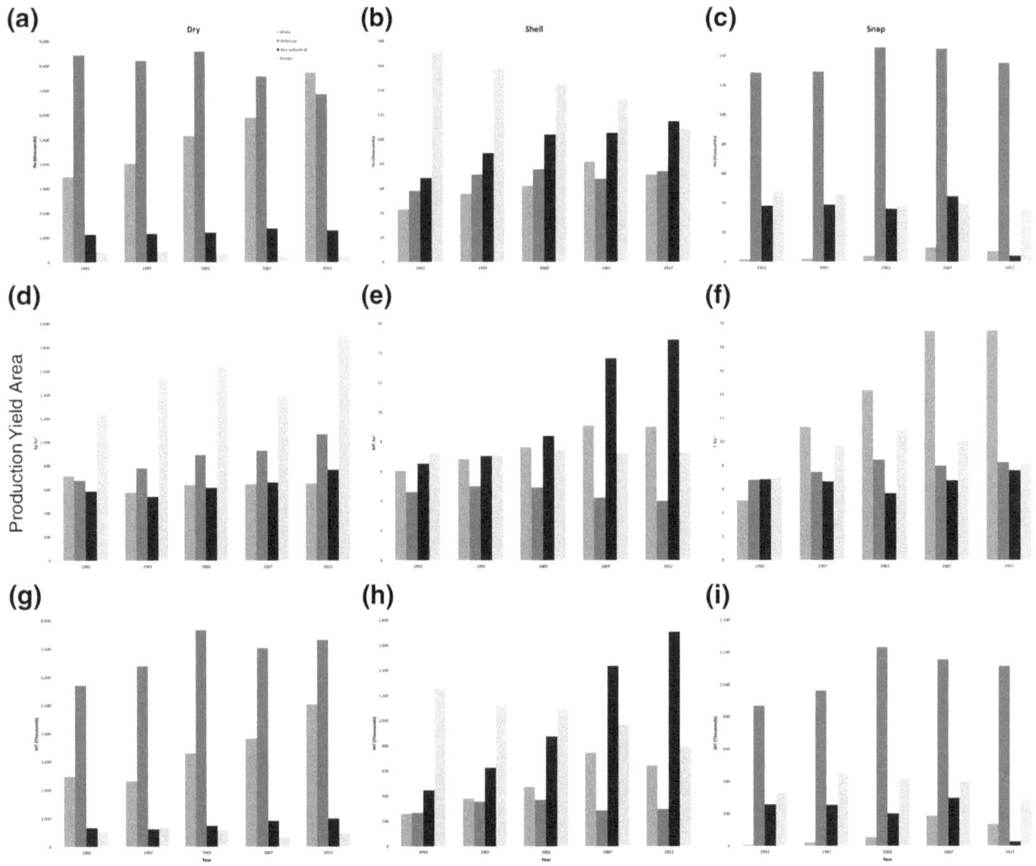

Fig. 1.3 Area (Ha), yield (Kg ha^{-1} or T ha^{-1}), and production (T) of common bean in four megaregions from 1992 to 2012. Data for Asia adjusted to remove non-common bean data from these figures. Data from FAOSTAT (Jan, 2015). **a** Dry bean area; **b** shell bean area; **c** snap bean area; **d** dry bean yield; **e** shell bean yield; **f** snap bean yield; **g** dry bean production; **h** shell bean production; and **i** snap bean production

(unpublished observations). In a loggia of the Villa Farnesina in Rome, festoons containing botanically accurate images of over 160 plant species were painted between 1515 and 1518 (Janick and Caneva 2005). These festoons have some of the earliest known images of New World crops including maize (Janick and Caneva 2005) and squash (Janick and Paris 2005). Bean pods are also illustrated and have been reported to be those of *P. vulgaris* (Caneva 1992). It is not until Fuchs' (1542) Di Historias Stirpium that we have unequivocal images of common bean. Fuchs herbal was followed by others (reviewed in Hedrick 1919) with similar images of Welsh (foreign) bean. These images provide a glimpse into what was being grown in Europe, but they do not show the diversity of germplasm present or any changes that may have been happening. The early science concerning common bean was focused on plant taxonomy, and little is known about the diversification of the crop in Europe during that time period. Various forms of evidence point toward the derivation of snap beans during the sixteenth–eighteenth centuries. Contemporary Native American varieties show little evidence that they were selected for snap bean traits, and the bulk of the evidence suggests that they were derived in Europe, before being disseminated back to the USA and elsewhere in the world (Myers and Baggett 1999).

1.3.2 Early Plant Breeding and Classical Genetics in Europe/USA

Efforts to systematically breed common bean are documented beginning in the nineteenth century in Europe and the Americas (Wade 1937), but scientific research did not begin in earnest until the beginning of the twentieth century. *Phaseolus* beans were the basis for several important studies of classical genetics. Mendel observed partial confirmation of his results in pea for three traits in intraspecific *P. vulgaris* crosses, but found complexities in the inheritance of flower color from an interspecific cross to *Phaseolus coccineus* (Olby 1985). Following the rediscovery of Mendel's work, researchers applied qualitative genetic analysis to the study of various traits in common bean. One of the earliest works was by Emerson (Emerson 1904) where he examined inheritance of traits ranging from plant habit to flower and seed color to pod characteristics. A good review of the bean genetics literature from the early twenteith century can be found in Kooiman (1931). A plant geneticist who contributed immensely to the qualitative genetics of common bean on many different traits was Herbert Lamprecht. Much of what we know about the genetics of seed coat colors comes from his work (Lamprecht 1932a, b, 1933, 1934a, b, 1935, 1936, 1939, 1940a, b, c, 1951, 1952, 1955, 1960, 1961, 1964) along with those of Dr. Prakken (1934, 1940, 1972a, b, 1974) and more recently by Bassett (2007). A list of genes is maintained by the Bean Improvement Cooperative Genetics Committee (BIC 2015) and includes a comprehensive set of references to qualitatively inherited genes.

Common bean has the distinction of being one of the species by which the principles of quantitative genetics were first elucidated. Johannsen (1903) investigated inheritance of several quantitative traits, including seed weight. He found that progeny of lines maintained by self-pollination over several generations tended to more closely resemble ancestral types of that lineage and were distinctly different from one another. He also argued that differences within a pure line were the result of environmental variation and not due to genetic causes (Carlson 2004). Based on these results, Johannsen formulated the concept of genotype versus phenotype.

The first report of linkage between qualitative and quantitative traits was that of Sax (1923) where he established the methodology for contemporary approaches to mapping quantitative trait loci (QTL). At the time of its discovery, it was regarded as a mysterious phenomenon whereby seed size (a quantitative trait) was found to be associated with seed color (a qualitative trait). Researchers were aware of the concept of linkage among qualitatively inherited genes and were constructing some of the first linkage maps, but until Fisher's landmark paper (Fisher 1918), the genetic control of quantitative traits was thought to be fundamentally different from qualitative traits. Modern genetic approaches have further elucidated the QTL that Sax discovered. The seed color gene in question is *P* (Emerson 1909), which controls expression of flavonoids in the seed testa and other plant parts. A gene underlying the QTL for seed size may be the phaseolin (*Phs*) seed storage protein (Johnson et al. 1996) where seeds with S phaseolin tend to be smaller than those with T or C phaseolin (Johnson et al. 1996). It is not known whether *Phs* is directly responsible for seed size differences, or whether other tightly linked loci might be involved. While *Phs* has been sequenced, the definitive experiment to re-engineer and transform back into the species is difficult to perform in common bean. *P* has been mapped to about 10 cM from *Phs* (Johnson et al. 1996). In preliminary work, a candidate gene for *P* has been identified (Mafi-Moghaddam et al. 2014).

1.3.3 Contemporary Research with Common Bean

To better understand where common bean has featured most prominently in contemporary research, a citation analysis was conducted in Google Scholar (GS) and Web of Science (WoS) databases using the search terms common, kidney, dry, French, green, and snap with bean and separately *P. vulgaris* where these words were found in the title or abstract of the paper. The two databases provide different aspects of the data. While WoS focuses on a set of accredited, refereed journals, GS counts citations from journals not accredited by WoS, non-refereed journals, and other gray literature. As a result, GS's citation counts are almost always higher than WoS, and GS can find important and highly cited works that were not published in refereed journals. Citations were reviewed and were classified by academic discipline and research topic. In addition to citation counts, we also calculated average citations per year as means of identifying those areas where current research is quite active, but may not yet have high total numbers of citations (particularly true for the field of genomics). WoS will also identify highly cited papers, and these were obtained for the search terms common bean and *P. vulgaris*. One caveat about using citation databases is as follows: citation counts are constantly increasing, and as such, the numbers reported in this chapter represent a snapshot in time as of March 2015.

Papers obtained from Google Scholar searches were ranked by number of citations per paper, and we chose an arbitrary cutoff of 50 citations or more in Google Scholar which resulted in 397 papers being retained. These were placed into 21 academic disciplines, some with many papers and others with only one or two (Table 1.5). Plant physiology was the largest category followed by plant genetics and plant pathology. Papers involving medicine (neurology, immunology, hematology, clinical nutrition, gastroenterology, and oncology), microbiology, and food science fell into the middle range, whereas plant breeding, molecular biology, and biotechnology comprised the low end. Considerable variation for most highly cited papers by academic discipline was observed, with over a thousand citations for a paper in neurology (Table 1.5). Other fields with relatively high citation rates included plant genetics, plant physiology, immunology, plant pathology, microbiology, and plant breeding.

The 397 papers were further reduced to a set of 46 (Table 1.6) by choosing the top three by citation count within an academic discipline and research area. From these data, it is apparent that seed storage proteins of common bean have played a significant role in research in several unrelated disciplines. These include the biotechnology (bruchid resistance), the medical sciences (neurology, oncology, immunology, and clinical nutrition), entomology (bruchid resistance), plant genetics (especially plant domestication and genetic diversity), and plant physiology. The most highly cited paper from the Annual Report of the Bean Improvement Cooperative (52 citations) concerns standardization of linkage groups and chromosomes for common bean (Table 1.7). The papers identified in WoS as those representing areas of increased activity are a diverse set, ranging from studies on isotopic diversity in Mexico, to genetic diversity studies, to analysis of bean growth exposed to radiation (Table 1.7). Interestingly, while genomics and bioinformatics of common bean is an extremely active area of research, WoS has not yet identified this as an active area of inquiry.

1.3.4 Important Programs and Centers Supporting Common Bean Research and Genetics Predating the Genomics Era

Several programs which have had a profound impact on common bean genetics and breeding are summarized below.

The International Center for Tropical Agriculture (CIAT) based in Cali, Colombia, has been a catalyst for both basic and applied research on common bean. The center has the most extensive germplasm collection of *Phaseolus* species, and researchers at the institution have helped establish much of what we know of species relationships within the genus, as well as providing the foundation for understanding the domestication and dissemination of common bean. Researchers at CIAT helped establish some

Table 1.5 Disciplines with papers focusing on common bean research, total number of papers by discipline with more than 50 citations per paper, and number of citations for the most highly cited paper, by discipline

Research area	No. papers	No. citations for most highly cited paper
Biotechnology	9	318
Clinical nutrition	3	121
Crop ecology	1	290
Entomology	5	181
Food science	20	293
Gastroenterology	3	91
Genomics	4	147
Geography	1	227
Hematology	1	111
Immunology	4	502
Microbiology	25	488
Molecular biology	10	179
Neurology	38	1093
Oncology	2	185
Plant anatomy	5	174
Plant breeding	11	445
Plant genetics	61	688
Plant pathology	57	489
Plant physiology	134	575
Soil science	2	108
Weed science	1	103
Total no. papers	397	–
Mean no. citations	–	324

of the first molecular genetics efforts in bean and developed an extensive library of simple sequence repeat (SSR) markers that provided the backbone for efforts to understand genetic diversity, create linkage maps, and map QTL. CIAT researchers have examined host–pathogen relationships for several diseases of common bean that are especially problematic in the tropics. A program to develop germplasm with enhanced micronutrient content had led to a better understanding of genetic control. CIAT researchers have developed and released germplasm and cultivars that are used globally, especially in the tropics either directly by farmers or by national and regional breeding programs.

United Stated Agency for International Development (USAID) has funded common bean research for three decades, originally through the Bean/Cowpea Collaborative Research Support Program (Bean/Cowpea CRSP) and subsequently via Feed the Future Legume Innovation Lab (FtF LIL). The basic model has been pairing of US universities and research institutions with national programs and universities in developing countries in South and Central America, Caribbean, and sub-Saharan Africa to provide a multidisciplinary approach to increasing bean productivity in the developing world. Funds have not only supported training, infrastructure development, and crop improvement activities in developing countries but have had a profound impact on breeding and genetics programs in the USA. Many researchers from developing countries have acquired graduate degrees from US universities while working in bean breeding programs in the USA. This has facilitated research on a number of problems of significance to bean production in developing

Table 1.6 Most frequently cited papers on common bean listed by research area and ranked by Google Scholar

Article title	No. of citations[a]		Research topic[b]	Reference
	GS[c]	WoS		
Biotechnology				
Transgenic pea seeds expressing the α-amylase inhibitor of the common bean are resistant to bruchid beetles	318	215	Transformation (bruchids)	(Shade et al. 1994)
Regeneration in *Phaseolus vulgaris* L.: High-frequency induction of direct shoot formation in intact seedlings by N6-benzylaminopurine and thidiazuron	216	152	Tissue culture (regeneration)	(Malik and Saxena 1992)
Inheritance of foreign genes in transgenic bean (*Phaseolus vulgaris* L.) co-transformed via particle bombardment	177	97	Transformation (virus)	(Aragão et al. 1996)
Clinical nutrition				
A dietary supplement containing standardized *Phaseolus vulgaris* extract influences body composition of overweight men and women	121	–	Phytohemagglutinin (obesity)	(Celleno et al. 2007)
Crop ecology				
Studies on mixtures of maize and beans (*Phaseolus vulgaris*) with particular reference to plant population	290	98	Intercropping	(Willey and Osiru 1972)
Entomology				
Fumigant toxic activity and reproductive inhibition induced by monoterpenes on *Acanthoscelides obtectus* (Say) (Coleoptera), a bruchid of kidney bean (*Phaseolus vulgaris* L.)	181	125	Bruchid resistance (monoterpenes)	(Regnault-et al. 1995)
Growth Inhibitory effects of an α-amylase inhibitor from the kidney bean, *Phaseolus vulgaris* (L.), on three species of bruchids (Coleoptera: Bruchidae)	142	108	Bruchid resistance (Seed storage protein)	(Ishimoto and Kitamura 1989)
Resistance to the bean weevil and the Mexican bean weevil (Coleoptera: Bruchidae) in non-cultivated common bean accessions	124	58	Bruchid resistance (Seed storage protein)	(Schoonhoven et al. 1983)
Food science				
Functional properties of the great northern bean (*Phaseolus vulgaris* L.) proteins: emulsion, foaming, viscosity, and gelation properties	293	205	Functional properties (starch)	(Sathe and Salunkhe 1981)
Chemical composition, dietary fiber and resistant starch contents of raw and cooked pea, common bean, chickpea, and lentil legumes	200	94	Functional properties (starch)	(de Almeida Costa et al. 2006)
Hard-to-cook phenomenon in common beans—a review	184	120	Functional properties (hard-to-cook)	(Reyes-Moreno and Paredes-López 1993)
Genomics				
Sequencing and analysis of common bean ESTs. Building a foundation for functional genomics	132	98	Biochemical markers	(Ramírez et al. 2005)
Legumes as a model plant family. Genomics for food and feed report of the cross-legume advances through genomics conference	147	90	Review	(Gepts et al. 2005)

(continued)

Table 1.6 (continued)

Article title	No. of citations[a]		Research topic[b]	Reference
	GS[c]	WoS		
Geography				
Atlas of common bean (*Phaseolus vulgaris* L.) production in Africa	227	–	Production (statistics)	(Kirkby et al. 1998)
Hematology				
A method for the rapid separation of leukocytes and nucleated erythrocytes from blood or marrow with a phytohemagglutinin from red beans (*Phaseolus vulgaris*)	111	144	Phytohemagglutinin	(Li and Osgood 1949)
Immunology				
Characterization of the structural determinants required for the high-affinity interaction of asparagine-linked oligosaccharides with immobilized Phaseolus vulgaris leukoagglutinating and erythroagglutinating lectins	502	467	Phytohemagglutinin	(Cummings and Kornfeld 1982)
A mouse lymphoma cell line resistant to the leukoagglutinating lectin from *Phaseolus vulgaris* is deficient in UDP-GlcNAc: alpha-D-mannoside beta 1, 6 N-acetylglucosaminyltransferase	166	167	Phytohemagglutinin	(Cummings et al. 1982)
Mitogenic leukoagglutinin from *Phaseolus vulgaris* binds to a pentasaccharide unit in N-acetyllactosamine-type glycoprotein glycans	122	146	Phytohemagglutinin	(Hammarström et al. 1982)
Microbiology				
Rhizobium tropici, a novel species, nodulating *Phaseolus vulgaris* L. beans and *Leucaena* sp. trees	488	400	Host-microbe interactions (BNF)	(Martínez-Romero et al. 1991)
Biological nitrogen fixation by common beans (*Phaseolus vulgaris* L.) increases with bio-char additions	367	202	Host-microbe interactions (BNF)	(Rondon et al. 2007)
Classification of rhizobia based on nodC and nifH gene analysis reveals a close phylogenetic relationship among *Phaseolus vulgaris* symbionts	323	263	Host-microbe interactions (BNF)	(Laguerre et al. 2001)
Molecular biology				
Regulation of beta-glucuronidase expression in transgenic tobacco plants by an A/T-rich, cis-acting sequence found upstream of a French bean beta-phaseolin gene	259	179	Seed storage protein (phaseolin)	(Bustos et al. 1989)
The glycosylated seed storage proteins of *Glycine max* and *Phaseolus vulgaris*. Structural homologies of genes and proteins	224	160	Seed storage protein (phaseolin)	(Doyle et al. 1986)
Complete nucleotide sequence of a French bean storage protein gene: Phaseolin	210	170	Seed storage protein (phaseolin)	(Slightom et al. 1983)
Neurology				
Topographical organization of the efferent projections of the medial prefrontal cortex in the rat: an anterograde tract-tracing study with *Phaseolus vulgaris* leucoagglutinin	1093	990	Anterograde tracing	(Sesack et al. 1989)

(continued)

Table 1.6 (continued)

Article title	No. of citations[a]		Research topic[b]	Reference
	GS[c]	WoS		
An anterograde neuroanatomical tracing method that shows the detailed morphology of neurons, their axons, and terminals—immunohistochemical localization of an axonally transported plant lectin, *Phaseolus vulgaris* leukoagglutinin (pha-l)	939	1015	Anterograde tracing	(Gerfen and Sawchenko 1984)
Organization of the projections from the subiculum to the ventral striatum in the rat. A study using anterograde transport of *Phaseolus vulgaris* leucoagglutinin	630	587	Anterograde tracing	(Groenewegen et al. 1987)
Oncology				
The binding of kidney bean phytohemagglutinin by Ehrlich ascites carcinoma	185	228	Phytohemagglutinin	(Steck and Wallach 1965)
Consumption of black beans and navy beans (*Phaseolus vulgaris*) reduced azoxymethane-induced colon cancer in rats	148	64	Diet (cancer)	(Bennink 2002)
Plant anatomy				
Chromium VI induced structural and ultrastructural changes in bush bean plants (*Phaseolus vulgaris* L.)	174	99	Abiotic stress (Cr)	(Vazquez, Poschenrieder, and Barcelo 1987)
Structural and ultrastructural disorders in cadmium-treated bush bean plants (*Phaseolus vulgaris* L.)	169	118	Abiotic stress (Cd)	(Barcelo et al. 1988)
Leaf surface and histological perturbations of leaves of *Phaseolus vulgaris* and *Helianthus annuus* after exposure to simulated acid rain	122	106	Abiotic stress (pH)	(Evans et al. 1977)
Plant breeding				
Basis of yield component compensation in crop plants with special reference to the field bean, *Phaseolus vulgaris*	445	324	Yield (yield components)	(Adams 1967)
Broadening the genetic base of common bean cultivars	266	143	Genetic diversity	(Singh 2001)
Common bean breeding for resistance against biotic and abiotic stresses: from classical to MAS breeding	257	149	Abiotic/biotic stress (biochemical markers)	(Miklas et al. 2006)
Plant genetics				
The association of size differences with seed coat pattern and pigmentation in *Phaseolus vulgaris*	688	–	Quantitative genetics (QTL)	(Sax 1923)
Races of common bean (*Phaseolus vulgaris*, Fabaceae)	574	375	Domestication (genetic diversity)	(Singh et al. 1991)
Phaseolin-protein variability in wild forms and landraces of the common bean (*Phaseolus vulgaris*): evidence for multiple centers of domestication	410	303	Domestication (seed storage protein)	(Gepts et al. 1986)
Plant pathology				
Volatile products of the lipoxygenase pathway evolved from *Phaseolus vulgaris* (L.) leaves inoculated with *Pseudomonas syringae* pv *phaseolicola*	489	376	Plant defense (*Pseudomonas*)	(Croft et al. 1993)

(continued)

Table 1.6 (continued)

Article title	No. of citations[a]		Research topic[b]	Reference
	GS[c]	WoS		
Gene cluster of *Pseudomonas syringae* pv. *phaseolicola* controls pathogenicity of bean plants and hypersensitivity of non-host plants	462	326	Pathogenicity (*Pseudomonas*)	(Lindgren et al. 1986)
Ethylene-regulated gene expression: molecular cloning of the genes encoding an endochitinase from *Phaseolus vulgaris*	325	274	Plant defense (Ethylene pathway)	(Broglie et al.1986)
Plant physiology				
Cadmium and zinc induction of lipid peroxidation and effects on antioxidant enzyme activities in bean (*Phaseolus vulgaris* L.)	575	442	Oxidative stress (Cd, Zn)	(Chaoui et al. 1997)
Phytotoxicity of cadmium ions on germinating seedlings of mung bean (*Phaseolus vulgaris*): involvement of lipid peroxides in chlorophyll degradation	483	352	Oxidative stress (Cd)	(Somashekaraiah et al. 1992)
Effects of salt stress on the growth, ion content, stomatal behavior, and photosynthetic capacity of a salt-sensitive species, *Phaseolus vulgaris* L.	444	338	Abiotic stress (salinity)	(Seemann and Critchley 1985)
Soil science				
Effects of supplied nitrogen form on growth and water uptake of French bean (*Phaseolus vulgaris* L.) plants	108	85	Plant nutrition	(Guo et al. 2002)
Weed science				
Influence of common ragweed (*Ambrosia artemisiifolia*) time of emergence and density on white bean (*Phaseolus vulgaris*)	103	76	Crop ecology	(Chikoye et al. 1995)

A minimum requirement to be included in this list was a citation count of 100 or more in Google Scholar. For research areas with many papers that meet this criterion (see Table 1.7), a maximum of three papers are shown
[a]GS: Google Scholar search engine; WoS: Web of Science database. [b]Abbreviations—BNF: biological nitrogen fixation; Cd: cadmium; Cr: chromium; QTL: quantitative trait locus; Zn: zinc

countries. Another profound impact of the B/C CRSP and FtF LIL programs has been to facilitate multidisciplinary approaches to problems that take into account social and gender issues.

The Bean Improvement Cooperative (BIC) was founded in the USA in 1957 by William Tex Frazier from Oregon State University and William Zaumeyer from USDA. The organization initially had 60 members but currently stands at about 300 with representation from about 40 countries. The group holds biennial meetings and publishes the Annual Report of the Bean Improvement Cooperative. The non-peer refereed journal limits articles to two pages and serves as a forum for preliminary reports on research to facilitate rapid exchange of ideas in the bean community. The BIC also serves as a forum and means of coordinating bean genetics and pathology information. The BIC genetics committee curates a list of gene symbols and coordinates rules of nomenclature (BIC 2015). The plant genetics committee also facilitates efforts to establish and coordinate race structure for pathogens such as anthracnose and rust. The Web site posts standardized protocols for evaluating diseases of bean and publishes letters on the GMO and gluten-free status of common bean.

Table 1.7 Most highly cited articles (not in Table 1.6) identified by Web of Science as the top one percent in their field, and most frequently cited paper in Google Scholar from the Annual Report of the Bean Improvement Cooperative

Discipline	Article title	Total citations[a]	Z9[b]	GS[c]	Reference
Archeology	Maize, beans, and the floral isotopic diversity of highland Oaxaca, Mexico	11	11	19	(Warinner et al. 2013)
Plant domestication	Structure of genetic diversity in the two major gene pools of common bean (*Phaseolus vulgaris* L., Fabaceae)	78	83	118	(Kwak and Gepts 2009)
Plant physiology	Growth alteration and leaf biochemical responses in *Phaseolus vulgaris* exposed to different doses of ionizing radiation	5	5	6	(Arena et al. 2014)
Plant genetics	Microsatellite marker diversity in common bean (*Phaseolus vulgaris* L.)	108	115	162	(Blair et al. 2006)
Plant genetics	Standard nomenclature for common bean chromosomes and linkage groups	–	–	52	(Pedrosa-Harand et al. 2007)

[a]Web of Science index. [b]Total times cited in Web of Science indexed databases (Web of Science Core Collection, BIOSIS Citation Index, Chinese Science Citation Database, and SciELO). [c]Google Scholar search engine

1.4 Conclusions

Common bean will likely remain one of the most important grain legume and vegetable crops in the twenty-first century. Its broad adaptation, ease of production, and consumer preference will likely allow it to keep an edge over other grain legumes. Bean consumption is likely to remain steady in the near term, but may rise in the long term as plant-based sources of protein provide alternatives to animal-based protein.

In terms of research focus, common bean is unusual in being the subject in several unrelated disciplines. The common theme in all cases involves the unique set of seed storage proteins that bean possess. These are involved in fields as disparate as medicine, entomology, and plant domestication. While research has historically been focused on plant genetics, physiology, and pathology, there are relatively few works on genomics. As such, we are just entering an era where genomics and bioinformatics studies of common bean will begin to proliferate.

References

Adams MW (1967) Basis of yield component compensation in crop plants with special reference to the field bean, *Phaseolus vulgaris*. Crop Sci 7:505–510

Akibode S, Maredia M (2011) Global and regional trends in production, trade and consumption of food legume crops. Standing Panel on Impact Assessment (SPIA), CGIAR Independent Science and Partnership Council (ISPC). http://impact.cgiar.org/sites/default/files/images/Legumetrendsv2.pdf

Angioi SA, Rau D, Attene G, Nanni L, Bellucci E, Logozzo G, Negri V, Zeuli PLS, Papa R (2010) Beans in Europe: origin and structure of the European landraces of *Phaseolus vulgaris* L. Theor Appl Genet 121:829–843

Aragão FJL, Barros LMG, Brasileiro ACM, Ribeiro SG, Smith FD, Sanford JC, Faria JC, Rech EL (1996) Inheritance of foreign genes in transgenic bean (*Phaseolus vulgaris* L.) co-transformed via particle bombardment. Theor Appl Genet 93:142–150

Arena C, De Micco V, De Maio A (2014) Growth alteration and leaf biochemical responses in *Phaseolus vulgaris* exposed to different doses of ionising radiation. Plant Biol 16 Suppl 1 (January):194–202

Barcelo J, Vazquez MD, Poschenrieder CH (1988) Structural and ultrastructural disorders in cadmium-treated bush bean plants (*Phaseolus vulgaris* L.). New Phytol 108:37–49

Bassett MJ (2007) Genetics of seed coat color and pattern in common bean. Plant Breeding Rev 28:239–315. John Wiley & Sons

Bennink MR (2002) Consumption of black beans and navy beans (*Phaseolus vulgaris*) reduced azoxymethane-induced colon cancer in rats. Nutr Cancer 44:60–65

BIC (2015) BIC Genetics Committee. Bean improvement cooperative. List of genes. http://bic.css.msu.edu/_pdf/Bean_Genes_List_2014.pdf

Blair MW, Giraldo MC, Buendia HF, Tovar E, Duque MC, Beebe SE (2006) Microsatellite marker diversity in common bean (*Phaseolus vulgaris* L.). Theor Appl Genet 113:100–109

Broglie KE, Gaynor JJ, Broglie RM (1986) Ethylene-regulated gene expression: molecular cloning of the genes encoding an endochitinase from *Phaseolus vulgaris*. Proc Natl Acad Sci USA 83:6820–6824

Bustos MM, Guiltinan MJ, Jordano J, Begum D, Kalkan FA, Hall TC (1989) Regulation of beta-glucuronidase expression in transgenic tobacco plants by an A/T-rich, cis-acting sequence found upstream of a French bean beta-phaseolin gene. Plant Cell 1:839–853

Camus J (1894) Les noms des plantes du Livre d'Heures d'Anne de Bretagne. J Botanique 8(19–23):325–336, 345-352, 366–375, 396–401

Caneva G (1992) La Loggia Di Psiche: Una delle prime testimonianze dell'introduzioni di piante Americane in Europa. Atti Della Accademia Nazionale Dei Lincei 9:163–172

Carlson EA (2004) Mendel's legacy. Cold Spring Harbor Laboratory Press, Cold Spring Harbor, NY

Celleno L, Vittoria Tolaini M, D'Amore A, Perricone NV, Preuss HG (2007) A dietary supplement containing standardized *Phaseolus vulgaris* extract influences body composition of overweight men and women. Int J Med Sci 4:45–52

Chaoui A, Mazhoudi S, Ghorbal MH, El Ferjani E (1997) Cadmium and Zinc induction of lipid peroxidation and effects on antioxidant enzyme activities in bean (*Phaseolus vulgaris* L.). Plant Sci 127:139–147

Chikoye D, Weise SF, Swanton CJ (1995) Influence of common ragweed (*Ambrosia artemisiifolia*) time of emergence and density on white bean (*Phaseolus vulgaris*). Weed Sci 43:375–380

Croft KPC, Juttner F, Slusarenko AJ (1993) Volatile products of the lipoxygenase pathway evolved from *Phaseolus vulgaris* (L.) leaves inoculated with *Pseudomonas syringae* pv. *phaseolicola*. Plant Physiol 101:13–24

Cummings RD, Trowbridge IS, Kornfeld S (1982) A mouse lymphoma cell line resistant to the leukoagglutinating lectin from *Phaseolus vulgaris* is deficient in UDP-GlcNAc: alpha-D-mannoside beta 1, 6 N-acetylglucosaminyl transferase. J Biol Chem 257:13421–13427

Cummings RD, Kornfeld S (1982) Characterization of the structural determinants required for the high affinity interaction of asparagine-linked oligosaccharides with immobilized *Phaseolus vulgaris* leukoagglutinating and erythroagglutinating lectins. J Biol Chem 257:11230–11234

de Almeida Costa G., Queiroz-Monici K da S, Reis SMPM, Costa de Oliveira A (2006) Chemical composition, dietary fibre and resistant starch contents of raw and cooked pea, common bean, chickpea and lentil legumes. Food Chem 94:327–330

Doyle JJ, Schuler MA, Godette WD, Zenger V, Beachy RN, Slightom JL (1986) The glycosylated seed storage proteins of *Glycine max* and *Phaseolus vulgaris*. Structural homologies of genes and proteins. J Biol Chem 261:9228–9238

Emerson RA (1909) Factors for mottling in beans. Am Breeders' Assoc 5:368–376

Emerson RA (1904) Heredity in bean hybrids. Ann Rep Nebraska Agric Exp Stat 17:33–78

Evans LS, Gmur NF, da Costa F (1977) Leaf surface and histological perturbations of leaves of *Phaseolus vulgaris* and *Helianthus annuus* after exposure to simulated acid rain. Am J Bot 64:903–913

FAOSTAT (2015) Food and Agriculture Organization of the United Nations Statistics Division. http://faostat3.fao.org/home/E

Fisher R (1918) The correlation between relatives on the supposition of Mendelian inheritance. Phil Trans Royal Soc Edinburgh 52:399–433

Fuchs L (1542) De Historia Stirpium Commentarii Insignes. Basel, Switzerland: In Officina Isingriniana

Gepts P, Bliss FA (1986) Phaseolin variability among wild and cultivated common beans (*Phaseolus vulgaris*) from Colombia. Econ Bot 40:469–478

Gepts P, Osborn TC, Rashka K, Bliss FA (1986) Phaseolin-protein variability in wild forms and landraces of the common bean (*Phaseolus vulgaris*): evidence for multiple centers of domestication. Econ Bot 40:451–468

Gepts P, Beavis WD, Brummer EC, Shoemaker RC, Stalker HT, Weeden NF, Young ND (2005) Legumes as a model plant family. Genomics for food and feed report of the cross-legume advances through genomics conference. Plant Physiol 137:1228–1235

Gepts P, Debouck D (1991) Origin, domestication, and evolution of the common bean (*Phaseolus vulgaris* L.). In: van Schoonhoven A, Voysest O (eds) Common beans: research for crop improvement. Colombia, CIAT, Cali, pp 7–53

Gerfen CR, Sawchenko PE (1984) An anterograde neuroanatomical tracing method that shows the detailed morphology of neurons, their axons and terminals: Immunohistochemical localization of an axonally transported plant lectin, *Phaseolus vulgaris* leucoagglutinin (PHA-L). Brain Res 290:219–238

Groenewegen HJ, Vermeulen-Van Te, der Zee E, Te Kortschot A, Witter MP (1987) Organization of the projections from the subiculum to the ventral striatum in the rat. A study using anterograde transport of *Phaseolus vulgaris* leucoagglutinin. Neuroscience 23:103–120

Guo S, Brück H, Sattelmacher B (2002) Effects of supplied nitrogen form on growth and water uptake of French bean (*Phaseolus vulgaris* L.) plants. Plant Soil 239:267–275

Hammarström S, Hammarström ML, Sundblad G, Arnarp J, Lönngren J (1982) Mitogenic leukoagglutinin from *Phaseolus vulgaris* binds to a pentasaccharide unit in N-acetyllactosamine-type glycoprotein glycans. Proc Natl Acad Sci USA 79:1611–1615

Hedrick UP (1919) Sturtevant's edible plants of the world. J.B. Lyon Co, Albany

Ishimoto M, Kitamura K (1989) Growth inhibitory effects of an α-amylase inhibitor from the kidney bean,

Phaseolus vulgaris (L.) on three species of bruchids (Coleoptera: Bruchidae). Appl Entomol 24:281–286

Janick J, Caneva G (2005) The first images of maize in Europe. Maydica 50:71–80

Janick J, Paris HS (2005) The cucurbit images (1515–1518) of the Villa Farnesina, Rome. Ann Bot 97: 165–176

Johannsen WL (1903) Über Erblichkeit in Populationen Und Reinen Linien. Eine Beitrag Zur Beleuchtung Schwebender Selektionsfragen [On Heredity in Pure Lines and Populations. A Contribution to Pending Questions of Selection]. Gustav Fischer. Jena

Johnson WC, Menendez C, Nodari R, Koinange EMK, Magnusson S, Singh SP, Gepts P (1996) Association of a seed weight factor with the phaseolin seed storage protein locus across genotypes, environments, and genomes in *Phaseolus-Vigna* spp.: Sax (1923) revisited. Journal Agric Genomics 2: Paper 5

Kirkby RA, Aledu C, Allen DJ (1998) Atlas of common bean (*Phaseolus vulgaris* L.) Production in Africa. CIAT, Cali

Koenig R, Gepts P (1989) Allozyme diversity in wild *Phaseolus vulgaris*: further evidence for two major centers of genetic diversity. Theor Appl Genet 78:809–817

Kooiman HN (1931) Monograph on the genetics of *Phaseolus*, (especially *P. vulgaris* and *P. multiflorus*). Bibliographica Genet 8:296–413

Kwak M, Gepts P (2009) Structure of genetic diversity in the two major gene pools of common bean (*Phaseolus vulgaris* L., Fabaceae). Theor Appl Genet 118: 979–992

Laguerre G, Nour SM, Macheret V, Sanjuan J, Drouin P, Amarger N (2001) Classification of rhizobia based on *nodC* and *nifH* gene analysis reveals a close phylogenetic relationship among *Phaseolus vulgaris* symbionts. Microbiology 147:981–993

Lamprecht H (1932a) Beiträge zur Genetik von *Phaseolus vulgaris*: zur Vererbung der Testafarbe [Contributions to the Genetics of *Phaseolus vulgaris*: The inheritance of testa color]. Hereditas 16:169–211

Lamprecht H (1932b) Zur Genetik von *Phaseolus vulgaris*: III. Zweiter Beitrag zur Vererbung der Testafarbe [The Genetics of *Phaseolus vulgaris*: III. Second contribution to the inheritance of testa color]. Hereditas 17:1–20

Lamprecht H (1933) Zur Genetik von *Phaseolus vulgaris*: VI. Vierter Beitrag zur Vererbung der Testafarbe [The Genetics of *Phaseolus vulgaris*: VI. Fourth contribution to the inheritance of testa color]. Hereditas 17:249–316

Lamprecht H (1934a) Zur Genetik von *Phaseolus vulgaris*: VII. Zwei Weitere Gene für Sameneigenschaften, *Cor* und *Fast* [The Genetics of *Phaseolus vulgaris*: VII. Two more genes for seed characteristics, *Cor* and *Fast*]. Hereditas 19:163–176

Lamprecht H (1934b) Zur Genetik von *Phaseolus vulgaris*: VIII. Über Farbenverteilung und Vererbung der Teilfarbigkeit der Testa [The Genetics of *Phaseolus vulgaris*: VIII. Of color distribution and inheritance of partial testa coloring]. Hereditas 19:177–222

Lamprecht H (1935) Zur Genetik von *Phaseolus vulgaris*: IX. Über der Einfluss des Genpaares R-r Auf die Testafarbe [The Genetics of *Phaseolus vulgaris*: IX. On the influence of the gene pair *R-r* on testa color]. Hereditas 20:32–46

Lamprecht H (1936) Zur Genetik von *Phaseolus vulgaris*: XIII. Ein Neues Grundgen für Testafarben, Ein Weiteres Testafarbgen Sowie Etwas Über Blütenfarben [The Genetics of *Phaseolus vulgaris*: XIII. A new basis for testa colors with another testa color as well as flower colors]. Hereditas 22:241–268

Lamprecht H (1939) Zur Genetik von *Phaseolus vulgaris*: XIV. Über die Wirkung der Gene *P, C, J, Ins, Can, G, B, V, Vir, Och* und *Flav* [The Genetics of *Phaseolus vulgaris*: XIV On the effect of the Genes *P, C, J, Ins, Can, G, B, V, Vir, Och* and *Flav*]. Hereditas 25:255–288

Lamprecht H (1940a) Zur Genetik von *Phaseolus vulgaris*: XV. Über die Vererbung der Mehrfarbigkeit der Testa [The Genetics of *Phaseolus vulgaris*: XV. On inheritance of testa variegation]. Hereditas 26: 65–99

Lamprecht H (1940b) Zur Genetik von *Phaseolus vulgaris*: XVI. Weitere Beitrage zur Vererbung der Teilferbigkeit [The Genetics of *Phaseolus vulgaris*: XVI. Another contribution to the inheritance of partially colored testa]. Hereditas 26:277–291

Lamprecht H (1940c) Zur Genetik von *Phaseolus vulgaris*: XVII–XVIII Zwei Neue Gene für Abzeichen auf der Testa, *Punc* und *Mip*, Sowie Über Die Wirkung von *V* und *Inh*; Und Ober Matte Samenschale und Ihre Vererbung [The Genetics of *Phaseolus vulgaris*: XVII–XVIII Two new genes for partially colored testa, *Punc* and *Mip*, and on the effect of *V* and *Inh*; and exterior matte seed coat and its inheritance]. Hereditas 26:292–304

Lamprecht H (1951) Die Vererbung der Testafarbe Bei *Phaseolus vulgaris* L [The Inheritance of testa color in *Phaseolus vulgaris* L]. Agri Hort. Genet 9:18–83

Lamprecht H (1952) Ein Gen Für Truncata-Samen Bei *Phaseolus* [A gene for truncated-seed in *Phaseolus*]. Agri Hort Genet 10:105–112

Lamprecht H (1955) Die Vererbung der Caruncula-Warze bei *Phaseolus vulgaris* und die Koppelungsgruppe *Sur-Y-Cav-Te-Miv-P* [The inheritance of caruncula-shape in *Phaseolus vulgaris* and the linkage group *Sur-Y-Cav-Te Miv-P*]. Agri Hort Genet 13:143–153

Lamprecht H (1960) The synonymy of the genes *Sh* and *D* with *J* and *B* for the seed coat colour of *Phaseolus vulgaris*. Agri Hort Genet 18:205–208

Lamprecht H (1961) Die Vererbung der Rezessiv Roten Testafarbe von *Phaseolus*-Sowie Bemerkungen zur Manifestation und Symbolik von Testafarbgenen [The inheritance of recessive red testa color of *Phaseolus* and comments on the expression and gene symbols for testa color genes]. Agri Hort Genet 19:344–359

Lamprecht H (1964) Die Vererbung Eines Neuen Types von Marmorierung der Samen von *Phaseolus vulgaris* L [The inheritance of a new type of marbling in seeds of *Phaseolus vulgaris* L]. Agri Hort Genet 22: 256–271

Li JG, Osgood EE (1949) A method for the rapid separation of leukocytes and nucleated erythrocytes from blood or marrow with a phytohemagglutinin from red beans (*Phaseolus vulgaris*). Blood 4: 670–675

Lindgren PB, Peet RC, Panopoulos NJ (1986) Gene cluster of *Pseudomonas syringae* pv. *phaseolicola* controls pathogenicity of bean plants and hypersensitivity of non host plants. J Bacteriol 168:512–522

Mafi-Moghaddam S, Stonehouse R, Lee R, Mamidi S, Bello M, Miklas P, McClean P, Bett K (2014) Molecular genetic analysis of the *Phaseolus vulgaris* *P* locus. In San Diego, CA: Plant and animal genome. https://pag.confex.com/pag/xxii/webprogram/Paper10921.html

Malik KA, Saxena PK (1992) Regeneration in *Phaseolus vulgaris* L.: high-frequency induction of direct shoot formation in intact seedlings by N6-benzylaminopurine and thidiazuron. Planta 186:384–389

Markham CR (ed) (2010) Journal of Christopher Columbus (during His First Voyage, 1492–93): and documents relating to the voyages of John Cabot and Gaspar Corte Real. Ashgate Publishing Ltd. ProQuest ebrary. Hakluyt Society, First Series, vol 86, Farnham, Surrey, GBR

Martínez-Romero E, Segovia L, Mercante FM, Franco AA, Graham P, Pardo MA (1991) *Rhizobium tropici*, a novel species nodulating *Phaseolus vulgaris* L. beans and *Leucaena* sp. trees. Int J Syst Bacteriol 41:417–426

Miklas PN, Kelly JD, Beebe SE, Blair MW (2006) Common bean breeding for resistance against biotic and abiotic stresses: from classical to MAS breeding. Euphytica 147:105–131

Myers JR, Baggett JR (1999) Improvement of snap bean. In: Singh SP (ed) Common bean improvement in the twenty-first century. Kluwer, Dordrecht, pp 289–329

Olby RC (1985) Origins of Mendelism, 2nd edn. University of Chicago Press, Chicago

Pedrosa-Harand A, Porch T, Gepts P (2007) Standard nomenclature for common bean chromosomes and linkage groups. Ann Rep Bean Improv Coop 51:106–107

Prakken R (1934) Inheritance of colours and pod characters in *Phaseolus vulgaris* L. Genetica 16:177–296

Prakken R (1940) Inheritance of colours in *Phaseolus vulgaris* L. Genetica 22:331–408

PrakkenR (1972a) Inheritance of colours in *Phaseolus vulgaris* L. III. On genes for red seedcoat colour and a general synthesis. Mededelingen van de Landbouwhogeschool Te Wageningen 29:1–82

Prakken R (1972) Seedcoat colour in *Phaseolus vulgaris* L.: attempt to a general synthesis. Ann Rep Bean Improv Coop 15:4–79

Prakken R (1974) Inheritance of colours in *Phaseolus vulgaris* L.: IV Recombination within the 'complex locus *C*'. Mededelingen van de Landbouwhogeschool Te Wageningen 24:1–36

Ramírez M, Graham MA, Blanco-López L, Silvente S, Medrano-Soto A, Blair MW, Hernández G, Vance CP, Lara M (2005) Sequencing and analysis of common bean ESTs. Building a foundation for functional genomics. Plant Physiol 137:1211–1227

Regnault-Roger C, Hamraoui A (1995) Fumigant toxic activity and reproductive inhibition induced by monoterpenes on *Acanthoscelides obtectus* (Say) (Coleoptera), a bruchid of kidney bean (*Phaseolus vulgaris* L.). J Stored Products Res 31:291–299

Reyes-Moreno C, Paredes-López O (1993) Hard-to-cook phenomenon in common beans—a review. Crit Rev Food Sci Nutr 33:227–286

Rondon MA, Lehmann J, Ramírez J, Hurtado M (2007) Biological nitrogen fixation by common beans (*Phaseolus vulgaris* L.) increases with bio-char additions. Biol Fert Soils 43:699–708

Sathe SK, Salunkhe DK (1981) Functional properties of the great northern bean (*Phaseolus vulgaris* L.) proteins: emulsion, foaming, viscosity, and gelation properties. J Food Sci 46:71–81

Sax K (1923) The association of size differences with seed-coat pattern and pigmentation in *Phaseolus vulgaris*. Genetics 8:552–560

Seemann JR, Critchley C (1985) Effects of salt stress on the growth, ion content, stomatal behaviour and photosynthetic capacity of a salt-sensitive species, *Phaseolus vulgaris* L. Planta 164:151–162

Sesack SR, Deutch AY, Roth RH, Bunney BS (1989) Topographical organization of the efferent projections of the medial prefrontal cortex in the rat: an anterograde tract-tracing study with *Phaseolus vulgaris* leucoagglutinin. J Comp Neurol 290:213–242

Shade RE, Schroeder HE, Pueyo JJ, Tabe LM, Murdock LL, Higgins TJV, Chrispeels MJ (1994) Transgenic pea seeds expressing the α-amylase inhibitor of the common bean are resistant to bruchid beetles. Nat Biotechnol 12:793–796

Singh SP (1999) Common bean improvement in the twenty-first century. Kluwer Academic Publishers

Singh SP (2001) Broadening the genetic base of common bean cultivars. Crop Sci 41:1659–1675

Singh SP, Gepts P, Debouck DG (1991) Races of common bean (*Phaseolus vulgaris*, Fabaceae). Econ Bot 45:379–396

Slightom JL, Sun SM, Hall TC (1983) Complete nucleotide sequence of a french bean storage protein gene: phaseolin. Proc Natl Acad Sci USA 80: 1897–1901

Somashekaraiah BV, Padmaja K, Prasad ARK (1992) Phytotoxicity of cadmium ions on germinating seedlings of mung nean (*Phaseolus vulgaris*): Involvement of lipid peroxides in chlorophyll degradation. Physiol Plant 85:85–89

Steck TL, Wallach DFH (1965) The binding of kidney-bean phytohemagglutinin by Ehrlich ascites

carcinoma. Biochim Biophy Acta-General Subjects 97:510–522
van Schoonhoven A, Cardona C, Valor J (1983) Resistance to the bean weevil and the Mexican bean weevil (Coleoptera: Bruchidae) in noncultivated common bean accessions. J Econ Entomol 76:1255–1259
USDA (2015) Nutrient data : USDA national nutrient database for standard reference release 27. http://www.ars.usda.gov/Services/docs.htm?docid=8964
Vazquez M, Poschenrieder CH, Barcelo J (1987) Chromium VI induced structural and ultrastructural changes in bush bean plants (*Phaseolus vulgaris* L.). Ann Bot 59:427–438
Wade BL (1937) Breeding and improvement of peas and beans. In Yearbook of Agriculture, 251–82. United States Government Printing Office
Warinner C, Garcia NR, Tuross N (2013) Maize, beans and the floral isotopic diversity of Highland Oaxaca, Mexico. J Archaeol Sci 40:868–873
Willey RW, Osiru DSO (1972) Studies on mixtures of maize and beans (*Phaseolus vulgaris*) with particular reference to plant population. J Agric Sci 79:517–529

Domestication and Crop History

2

Valerio Di Vittori, Elisa Bellucci, Elena Bitocchi,
Domenico Rau, Monica Rodriguez,
Maria Leonarda Murgia, Laura Nanni, Giovanna Attene
and Roberto Papa

Abstract

A new era has begun for *Phaseolus vulgaris*, and other *Phaseolus* spp., with the release of the reference genomes of both the Mesoamerican and Andean genotypes. Exploiting the new genome sequences information and the derived tools, important insight in the common bean genomics can be achieved. A major breakthrough in common bean will be the identification of the molecular basis of the domestication syndrome, representing a main step towards our understanding of the evolutionary processes and a fundamental support for researchers and breeders involved in crop improvement. *P. vulgaris*, along with the other *Phaseolus* species, represents a unique model to study evolution, domestication and environmental adaptation, focusing on the major phenotypic changes occurring during its evolutionary histories and trajectories and unveiling the molecular mechanisms and genetic basis responsible for the observed changes.This chapter offers an overview of the current knowledge of the evolutionary history of common bean, and of the outcomes relating to the genetic bases of important domestication and adaptation traits. We provide an analysis of the process of domestication, with the focus on convergent phenotypic evolution, and a survey of the studies on common bean that have been specifically carried out on genes related to the domestication syndrome; with particular focus on studies that have compared wild and domesticated forms, highlighting findings on the genetic control of the domestication syndrome and on the genetic architecture of environmental and agronomic important traits.

V. Di Vittori · E. Bellucci (✉) · E. Bitocchi · L. Nanni · R. Papa
Department of Agricultural, Food and Environmental Sciences, Università Politecnica delle Marche, Via Brecce Bianche, 60131 Ancona, Italy
e-mail: e.bellucci@univpm.it

D. Rau · M. Rodriguez · M. L. Murgia · G. Attene
Department of Agriculture, Università degli Studi di Sassari, Via de Nicola, 07100 Sassari, Italy

M. Pérez de la Vega et al. (eds.), *The Common Bean Genome*,
Compendium of Plant Genomes, https://doi.org/10.1007/978-3-319-63526-2_2

Keywords
Adaptation · Convergent evolution · Domestication syndrome
Evolution · Selection

2.1 Introduction

The common bean, *Phaseolus vulgaris* L. ($2n = 2x = 22$), is a crop species that in addition to the importance of its societal, nutritional and environmental sustainability has a very interesting and distinctive evolutionary history. Indeed, it appears to represent a unique model for domestication and evolutionary studies. This also arises through the domestication of five *Phaseolus* species and the occurrence of multiple independent domestications in Mesoamerica and the Andes for both *P. vulgaris* and *Phaseolus lunatus* (for review, see Bitocchi et al. 2017).

The common bean wild forms originated in Mesoamerica about 165,000 years ago and spread southwards towards the Andes (Bitocchi et al. 2012; Schmutz et al. 2014). At least two of its independent domestication events determined the formation of two distinct domesticated gene pools that evolved under isolation, one in Mesoamerica and one in the Andes. These gene pools underwent parallel evolution that was associated with partial reproductive incompatibility (i.e. low hybrid fertility due to gene conditioning hybrid weakness and breakdown (Johnson and Gepts 1988; Koinange and Gepts 1992; Singh and Molina 1996)), and they spread further through the development of landraces with distinct characteristics and specific adaptations. Such distinct and replicated domestication events that occurred for the same species (or the same genus) that led to morphological and functional changes represent an almost unique experimental feature for evolutionary studies. This is different from other examples of multiple domestications events (Meyer et al. 2012) that were not independent due to the lack of reproductive isolation (Bitocchi et al. 2017). However, some similarities can be seen in rice with the *indica* and *japonica* subspecies (Vitte et al. 2004; Londo et al. 2006; but see also Molina et al. 2011; Choi et al. 2017).

For these reasons, the common bean is an ideal model to study domestication and evolution, and the present review aims to cover the current knowledge of its evolutionary history. This provides an analysis of the process of domestication, with the focus on convergent phenotypic evolution. It also highlights current knowledge of the genetic control of the domestication syndrome from the perspective of the new era that is associated with the release of both the Mesoamerican (Vlasova et al. 2016) and Andean (Schmutz et al. 2014) reference genome sequences.

2.2 Origins of *Phaseolus vulgaris*

P. vulgaris originated in America, and specifically in Mesoamerica, in the state of what is now Mexico (Bitocchi et al. 2012). The wild form remains widely distributed from northern Mexico to north-western Argentina (Toro et al. 1990), and it is characterised by three eco-geographical gene pools. The Mesoamerican and Andean gene pools are the main ones, and they show parallel wild and domesticated geographical distributions, as has been reported in several studies based on different datasets, which included plant morphology, seed proteins, allozymes, many different molecular markers and sequence data (Gepts et al. 1986; Gepts and Bliss 1985; Koenig and Gepts 1989; Singh et al. 1991; Becerra-Velásquez and Gepts 1994; Freyre et al. 1996; Papa and Gepts 2003; Rossi et al. 2009; Kwak and Gepts 2009; Bitocchi et al. 2012, 2013, 2016; Bellucci et al. 2014b; Rodriguez et al. 2016). The third gene pool was discovered in the 1980s and described in 1993 (Debouck et al. 1993). This gene pool comprised wild populations from northern Peru and Ecuador, which was suggested to be the region of origin of the wild form of the common bean (Kami et al. 1995). However, further data (Bitocchi et al. 2012) indicated that the common bean originated in Mesoamerica and that the other two wild gene pools originated from two independent migration events. The Mesoamerican origin was also supported by whole-genome sequencing analysis (Schmutz et al. 2014) that also estimated the divergence between the Mesoamerican and Andean gene pools at some 165,000 years ago.

Bitocchi et al. (2012) clearly defined the Mesoamerican wild population structure. Although previous studies had highlighted the presence of population structure in the Mesoamerican gene pool (e.g. Papa and Gepts 2003), they had not demonstrated any clear subdivisions into different sub-populations. On the other hand, by using sequence data, Bitocchi and collaborators (2012) demonstrated the presence of four distinct genetic groups, two of which were more related to the Andean (i.e. Mesoamerican B3) and the northern Peru and Ecuador (i.e. Mesoamerican B4) populations. From the Mesoamerican centre of origin, different groups migrated from central Mexico to South America, which led to the formation of the two South American gene pools, as Andean and from northern Peru and Ecuador.

In a recent paper, Rendón-Anaya et al. (2017) confirmed that the populations occurring in North Peru and Ecuador represent a distinct population that migrate in South America much earlier than the Andean gene pool. Moreover, these authors, analysing 29 accessions from 12 *Phaseolus* species, based on nuclear and chloroplast genome sequences and on metabolomics data, suggested that this third gene pool should be considered a sister species of *P. vulgaris* (*Phaseolus pseudovulgaris*, Rendón-Anaya et al. 2017). However, a larger sample of Mesoamerican accessions and further analysis, including hybridisation experiments, are needed to confirm the hypothesis of a new true species.

2.3 Domestication

Domestication is a complex process that starts from a wild plant population or several populations, and through adaptation and the shaping of the natural environment leads to a crop plant that is modelled on human needs and agricultural practices. The domestication process involves several morphological and physiological changes that result in genetic, structural and functional modifications that are shared among most crop species (i.e. domestication syndrome). These processes make the developing crop genetically different from its wild relatives and confer better adaptation to different agro-ecosystems (Gepts and Papa 2002; Bellucci et al. 2014b).

In the *Phaseolus* species, the main differences between the wild and domesticated forms are related to gigantism (e.g. small *vs* large seeds and pods), growth habit (e.g. more compact in the domesticated form, occurrence of bush structures and no climbing types), seed dormancy (i.e. present *vs* absent), photoperiod sensitivity (i.e. short-day *vs* complete or partial insensitivity), shape and colour of the plant and its harvested parts (e.g. seeds and pods) and the dissemination mechanisms (e.g. high shattering *vs* low shattering or non-shattering pods).

For the *Phaseolus* genus, domestication occurred as at least seven independent events (five species, and *P. vulgaris* and *P. lunatus* with two gene pools), with the convergent phenotypic evolution of very similar set of traits. The *Phaseolus* species have different breeding systems (i.e. autogamous, allogamous) and life history traits (i.e. annual, perennial), thus making *Phaseolus* an extremely interesting and unique model to study domestication dynamics and evolution under domestication (Bitocchi et al. 2017).

At the genome level, the main consequence of domestication, which is common to most crop species, is a reduction in their genetic diversity when compared to their wild ancestors. At the origins of agriculture, farmers selected a reduced number of individuals, and consequently they collected only a small portion of the diversity of the entire wild gene pool and populations, which is known as the 'founder effect' (Glémin and Battaillon 2009).

For the common bean, the process of domestication has been rather widely studied, and in particular detail. Major domestication traits have been mapped (Koinange et al. 1996), and some genes associated with domestication have been characterised (Kwak et al. 2008; Repinski et al. 2012). Compared to the wild gene pool, it is well known that strong reductions in the genetic diversity in the Mesoamerican and Andean domesticated populations of the common bean have occurred (Papa et al. 2005, 2007; Kwak and Gepts 2009; Rossi et al. 2009; Nanni et al. 2011; Bitocchi et al. 2012, 2016; Desiderio et al. 2013; Schmutz et al. 2014; Bellucci et al. 2014a; Rodriguez et al. 2016). Two independent domestication events have been reported in several studies, one in Mesoamerica and one in the Andes, where the two major domesticated gene pools originated (see Bellucci et al. 2014b), plus single domestication events that occurred within each gene pool (Nanni et al. 2011; Bitocchi et al. 2013). The two domesticated gene pools were differentiated according to their morpho-agronomic traits and biochemical patterns, and at the molecular level (Gepts et al. 1986; Koenig et al. 1990; Singh et al. 1991; Papa et al. 2006; Acosta-Gallegos et al. 2007; Bitocchi et al. 2013; Schmutz et al. 2014). When the domesticated forms are compared to the wild forms, the bottleneck due to domestication was three-fold greater in the Mesoamerican than in the Andean. Indeed, the Andean populations underwent a bottleneck before domestication that initially impoverished the genetic diversity of the Andean wild germplasm, which resulted in minor effects of the subsequent domestication bottleneck (i.e. sequential bottleneck; Bitocchi et al. 2013).

Together with the sequencing of the first reference genome of the common bean, Schmutz et al. (2014) reported the analysis of 60 wild genotypes and 100 landraces from Mesoamerica and the Andes, which confirmed the occurrence of two independent domestications from wild gene pools that had diverged before humans arrived in America. They also found diversity reduction in the Mesoamerican landraces, while for their Andean samples they proposed the occurrence of admixture events with Mesoamerican accessions and the emergence of new mutations because the landraces were more diverse than the wild populations (Schmutz et al. 2014).

The severe reduction in genetic diversity for the Mesoamerican accessions was also observed at the nucleotide level (Bellucci et al. 2014a) when the transcriptomes of wild and domesticated common bean accessions were analysed. By exploiting the RNA-seq technique and de novo assembly, Bellucci et al. (2014a) showed a reduction at the phenotypic level (i.e. gene expression) in the domesticated form when compared to the wild form. For the first time in a crop species, they showed that the reductions highlighted at the transcriptomic level decreased the phenotypic diversity at the gene expression level by about 18%. Moreover, they reported that, in comparisons between wild and domesticated forms, the majority (74%) of the contigs identified as differentially expressed were down-regulated in the domesticated forms. This suggested that the occurrence of loss-of-function mutations (which are relatively frequent compared to gain-of-function changes) was a common source of variation. This also supports selection during rapid environmental changes (Olson 1999), as they occur for the adaptation to an agro-ecosystem from the wild environment. These results also support the observation that most of the traits of the domestication syndrome are recessive.

At the genome-wide level, lower gene expression was found for the domesticated compared to the wild transcripts, as if there had been an accumulation of deleterious mutations due to hitchhiking, as mostly loss-of-function, or with reduced expression (Bellucci et al. 2014a); they referred to this as the 'cost of domestication'. As suggested for rice (Lu et al. 2006), the accumulation of loss-of-function mutations, or reduced expression mutations, might also have been due to reduced effective recombination, which would result in increases in the frequency of deleterious mutations in the domesticated pool and would negatively influence the fitness.

In Mesoamerica and the Andes, the wild and domesticated forms grow under different levels of sympatry, and the effects on the population structure and gene flow were analysed by Papa and Gepts (2003) using Mexican populations. They demonstrated that the wild and domesticated common bean were not genetically isolated, with moderate gene flow detected, which was three-fold higher from domesticated to wild, than in the opposite way. In the presence of such levels of asymmetric gene flow, the high phenotypic differences between the two forms were maintained due to selection against domesticated alleles in the wild environment and against wild alleles in the cultivated agro-ecosystems (Papa and Gepts 2003). The presence of asymmetric introgression was also recently reported by Rendón-Anaya et al. (2017), who also confirmed, in agreement with Papa et al. (2005), that the level of introgression is higher in the genome area not involved in the genetic control of the domestication syndrome. One of the most important factors causing such asymmetry is the recessive nature of domestication traits that make the F_1 hybrids, from the crosses between domesticated and wild individuals, more similar to the wild forms, favouring the conscious and unconscious selection against wild alleles in the domesticated environment compared to the selection against domesticated alleles in the wild environment.

The work of Bellucci et al. (2014a) also allowed the analysis of the selection signature due to domestication and the size of the genome affected. Indeed, the knowledge of the genes and the genome regions involved in the process of domestication is crucial for any successful breeding and to unravel the genetic diversity carried by the wild forms (Tanskley and McCouch 1997; McCouch 2004). Papa et al. (2005) highlighted the presence of domestication genes in genome regions of high divergence between the wild and domesticated forms, and the highest diversity of the wild common bean was observed in genome regions linked to the domestication loci, which was probably the least exploited by farmers and breeders. In another study, Papa et al. (2007) used amplified fragment length polymorphism (AFLP) markers to reveal that about 16% of the genome of the common bean was under the effects of selection due to domestication. Bellucci et al. (2014a) using simulated demographic dynamics during domestication and RNA-seq data found that about 10% of the contigs analysed were affected by selection during domestication or were physically linked to selected genes. In most cases, results showed a reduced diversity in the domesticated forms compared to the wild, as expected following positive selection due to domestication. Transcripts analysed by RNA-seq showed further reduction in the diversity of gene expression (by 26%) and a five-fold enrichment of the differentially expressed genes. Likewise, Bellucci et al. (2014a) conducted a detailed survey of the functions of the contigs that showed effects of selection due to domestication, and this will be discussed later below.

A very interesting example of diversifying selection that acts on the domesticated forms, whereby domestication increases the level of functional diversity, was observed for 2.8% of the transcripts that showed effects of selection due to domestication (Bellucci et al. 2014a). Here, no diversity was observed in the wild forms, while diversity was detected in the domesticated. Among these transcripts, the analysis of the gene functions highlighted the example of the drought-related and growth-related *KUP6* (K^+ uptake transporter-6) gene (Osakabe et al. 2013). *KUP6* was significantly overexpressed in the domesticated form compared to the wild, as if domestication had also increased the functional diversity of the selected gene in addition to the increased nucleotide diversity. Bellucci et al. (2014a) suggested that for further crop improvements, a key aspect was not only depletion of the wild germplasm diversity, but also the diversity contained in the domesticated pool (e.g. in the traditional landraces). This was originated by the fixing of useful mutations after domestication, and it needs further consideration.

Schmutz et al. (2014) identified candidate genes that were associated with domestication by a comparison of wild and landrace populations across 10-kb/2-kb sliding windows, where they examined the empirical distribution of the diversity ratios and population differentiation statistics. They found 1835 Mesoamerican and Andean candidates with negative Tajima's D values, which indicated positive selection. They then investigated the functions of the genes identified and applied a genome-wide association study (GWAS) approach and defined a set of genes that are linked with flowering time, leaf and seed size and seed weight (see details below).

Another important aspect related to domestication of the common bean was the identification of the geographical centres where the process took place. Kwak and Gepts (2009) proposed the Lerma–Santiago basin as the domestication site for the Mesoamerican gene pool, while Chacón et al. (2007) proposed southern Peru as the Andean domestication site. More recently, Bitocchi et al. (2013) suggested Oaxaca Valley in Mesoamerica and southern Bolivia and northern Argentina in South America as the areas of *P. vulgaris* domestications.

To unravel the respective roles of the Mesoamerican and Andean areas in common bean domestication, Rodriguez et al. (2016) integrated the spatial, phenotypic and molecular data with those from different disciplines, including archaeological and glotto-chronological data. For Mesoamerica, the data of Rodriguez et al. (2016) confirmed Oaxaca Valley as the putative region where domestication of the common bean took place. The genetic diversity data were supported by previous studies that detected archaeological sites in this area with common bean macroremains that were dated from 2300 to 2100 years BP (Kaplan and Lynch 1999). This area also included the homeland sites of the Zapotecan, Mixtec-Cuicatec and Popolocan protolanguages, for which ancient bean words can be reconstructed from 3149 to 3036 years BP (Brown et al. 2014).

In the Andes, Rodriguez et al. (2016) proposed the region of northern Argentina and southern Bolivia as the putative Andean domestication area, as the wild accessions from Argentina–Bolivia that were analysed were genetically more similar to the Andean domesticated forms, and showed lower 100-seed weight when compared with other Andean accessions. Their data were consistent with those from previous genetic (Beebe et al. 2001; Bitocchi et al. 2013), archaeological (Tarrago 1980) and glotto-chronological (Brown et al. 2014) studies.

The common bean is the most cultivated crop worldwide among the *Phaseolus* species, as after its domestication it underwent dissemination and evolution out of these American centres of origin and domestication. The pathways of distribution of *P. vulgaris* were complex and involved several introductions from the New World, combined with exchanges between continents, and among different countries within continents.

In the Old World of Europe, both of the common bean domesticated pools were introduced after the travels of Columbus and were then rapidly disseminated to many different European areas that were characterised by varied environmental conditions and agronomic practices. The levels of diversity for the domesticated common bean in Europe as determined using molecular markers are comparable to that observed in the Americas, without any detectable genetic bottleneck effects (Angioi et al. 2010; Gioia et al. 2013a). Moreover, due to the breakdown of the spatial isolation between the two gene pools in Europe, hybridisation and introgression occurred between the Andean and Mesoamerican gene pools, which led to the hybrid development of 40% of the European landraces (Angioi et al. 2010; Gioia et al. 2013a). Many studies have indicated that Europe was the secondary centre of diversity for the common bean (Santalla et al. 2002; Angioi et al. 2010, 2011; Gioia et al. 2013a). The high level of hybridisation that would not be expected for an autogamous species was most likely the result of selection for adaptation to these new environmental conditions. This will have exploited hybridisation and recombination between the two different gene pools to create novel genotypic combinations when compared to those of their centres of origin. Other continents and countries have also been proposed as secondary centres of diversification for the common bean, including Brazil (Burle et al. 2010), central, eastern and southern Africa (Martin and Adams 1987a, 1987b; Asfaw et al. 2009; Blair et al. 2010), and China (Zhang et al. 2008).

Recent reviews (Bellucci et al. 2014b; Bitocchi et al. 2017) afford a detailed analysis of the dissemination and evolution of the common bean and the other *Phaseolus* crop species outside their centres of origin.

2.4 Convergent Evolution

As mentioned above, domestication offers numerous examples of convergent phenotypic evolution that were associated with adaptation to human needs and novel agro-ecosystems. For instance, favourite plants showed adaptive features related to different climatic and environmental conditions, such as cold or drought tolerance. Most domesticated animals were selected to maximise the yield of useful products (i.e. meat, milk, wool) and for their docile behaviours, while crops were selected for the size of the plant organ used by humans (i.e. seeds, fruit) and for reduced, or lack of, seed dispersal. Indeed, during domestication, similar sets of phenotypic characteristics were selected (i.e. traits of the domestication syndrome), which has provided the opportunity to study convergent phenotypic evolution for many responses to selection pressures.

An interesting question thus arises in terms of whether convergent responses due to selection pressures under domestication are limited to the observed effects, or whether they are also related to the molecular mechanisms that control the phenotypic traits, thus acting on the same genomic regions or set of genes responsible for the same trait. Schmutz et al. (2014) were the first to investigate the convergent evolution between the two main gene pools of the common bean. By comparing wild and landrace populations across 10-kb/2-kb sliding windows and analysing the empirical distribution of the diversity statistics ratios and population differentiation statistics, they were able to compare the effects of selection that occurred within gene pools. These were subjected to independent domestications events, and therefore Schmutz et al. (2014) tried to determine whether in order to obtain the same convergent phenotypes, selection had acted on the same genomic regions, or on a completely different set of genes that coded for the same phenotype. They showed that <10% of the 74 Mb of genome sequences that were putatively involved in selection during domestication was shared between the Mesoamerican and Andean gene pools, thus suggesting different genetic routes to domestication. However, Schmutz et al. (2014) did not use explicit demographic modelling to generate an expectation of the number of potential false-positive regions. Thus, an alternative explanation of their data is that there were high levels of false positives (i.e. regions of the genome with reduced diversity due to stochastic effects of domestication bottlenecks), which would lead to a lack of shared genome regions and genes that would be predicted to be involved in domestication between the two gene pools.

To better understand this interesting phenomenon, Bitocchi et al. (2016) further investigated common bean domestication in the Mesoamerican gene pool by sequencing 49 gene fragments from a sample of 45 wild and domesticated accessions and compared the candidate genes they identified for selection during domestication with those from other studies (Bellucci et al. 2014a; Schmutz et al. 2014; Rodriguez et al. 2016). In doing this, they tried to understand whether the sexually compatible Mesoamerican and Andean lineages with similar morphologies and life cycles underwent independent selection based upon distinct sets of genes or not. They found that two genes out of the four strong Mesoamerican candidate genes identified were also detected as outliers by Schmutz et al. (2014) only during Andean domestication. This suggested that more studies and evidence are needed to understand the convergent responses due to selection pressures under domestication.

2.5 Domestication Traits

The domestication process induced several changes in the common bean plants for major traits associated with adaptation and cultivation and to address human needs. At present, the domesticated species can be clearly distinguished from their wild progenitors by a set of traits, which is known as the 'domestication syndrome'. These changes in the domesticated individuals have guaranteed higher productivity in cultivated environments, although at the same time they have reduced the adaptation to erratic environment variations, where wild traits show much greater fitness over domesticated traits.

In recent years, many studies have been conducted in common bean through different approaches in the search of an understanding of the genetic control of these traits, among which the molecular linkage mapping approach has been widely adopted. Recently, GWAS analysis has provided a powerful tool to search quantitative trait loci (QTLs) through the use of markers such as single-nucleotide polymorphisms (SNPs), which show wide distribution across the genome. The common bean genome sequence (Schmutz et al. 2014) has also allowed establishing a link between genetic and physical maps, which thus facilitates the identification of candidate genes for domestication traits in genomic regions where significant QTLs are found. Furthermore, the availability of the entire reference genome can facilitate the identification of regions where the effects of selection are more evident, and differentiation between wild and domesticated populations is greater; these regions might contain genes that were involved in the domestication process.

There are several traits that have been linked to domestication in the common bean, which are related to vegetative growth, phenological features, size, colour and shape of the harvested parts. In particular, these include two main target traits that were selected for during the domestication process: the seed dispersal mechanism and the seed dormancy. Here, we provide a survey of the studies on common bean that have been specifically carried out on genes related to the domestication syndrome (Table 2.1), with particular focus on studies that have compared wild and domesticated forms. Studies where no direct comparisons have been made for the target traits between wild and domesticated forms were not covered in the present survey.

2.5.1 Seed Dispersal Mechanism

Many wild plants are characterised by seed shattering, which represents a strategy that ensures seed dispersal at maturity. Indeed, this trait is fundamental for propagation of progeny in wild individuals. Seed dispersal occurs through various mechanisms in species that have different kind of fruit.

Wild common bean is characterised by a dry strongly dehiscent legume fruit (Gepts and Debouck 1991) that opens at maturity along the ventral suture to ensure seed release. This trait has been a target of selection, and now many domesticated varieties have totally or partially lost the ability to disperse their seeds after ripening. Indeed, while dry beans have dehiscent pods, snap beans are completely indehiscent (i.e. stringless varieties; Gepts and Debouck 1991), as they no longer have the fibres in the pod sutures (i.e. string) and walls (Prakken 1934; Koinange et al. 1996) (Fig. 2.1). In the common bean, the seed dispersal mechanism is associated with the content and location of the fibres in the pods (Prakken 1934; Murgia et al. 2017), with strict positive correlation between shattering ability and increased carbon and lignin content, as was recently highlighted (Murgia et al. 2017). A QTL analysis carried out by Koinange et al. (1996) on a recombinant inbred line (RIL) population derived from a cross between Midas (an Andean domesticated accession) and G12873 (a wild Mesoamerican genotype) mapped the locus *St* to chromosome Pv02. This locus controls the presence or absence of pod suture fibres and co-segregates with the trait of lack of pod wall fibres (Koinange et al. 1996; Freyre et al. 1998).

Identification of the genetic controls of seed shattering in common bean has also been carried out using candidate gene approaches that focussed on the homologues to the *A. thaliana*

Table 2.1 Genes and QTLs identified as related to the domestication syndrome in the common bean

Chromosome	Domestication trait category	Locus/QTL/gene name	Trait/function	Approach and population	Literature
Pv02	Seed dispersal	*St*	Seed dispersal: pod suture fibres	Linkage mapping—RILs	Koinange et al. (1996)
Pv02	Seed dispersal	*St*	Seed dispersal: pod wall fibres	Linkage mapping—RILs	Koinange et al. (1996)
Pv06	Seed dispersal	*PvSHP1*	Seed dispersal[d]	Candidate gene approach[a]/linkage mapping—RILs	Nanni et al. (2011)
Pv02	Seed dispersal	*PvIND*	Seed dispersal[e]	Candidate gene approach[a]/linkage mapping—RILs	Gioia et al. (2013b)
Pv02	Seed dormancy	*DO*	Germination	Linkage mapping—RILs	Koinange et al. (1996)
Pv03	Seed dormancy	*DO*	Germination	Linkage mapping—RILs	Koinange et al. (1996)
Pv04	Seed dormancy	*DO*	Germination	Linkage mapping—RILs	Koinange et al. (1996)
Pv03	Seed dormancy	*DO*	Germination	Linkage mapping—RILs	Koinange et al. (1996)
Pv01	Growth habit	*fin*	Determinacy	Linkage mapping—RILs	Koinange et al. (1996)
Pv09	Growth habit	*GH*	Determinacy	Linkage mapping—$F_{2:4}$ pop	Tar'an et al. (2002)
Pv01	Growth habit	*PvTFL1y*	Determinacy[f]	Candidate gene approach[a]/linkage mapping—RILs	Kwak et al. (2008)
Pv01	Growth habit	*PvTFL1y*	Determinacy[f]	Candidate gene approach[a]/linkage mapping—RILs	Repinsky et al. (2012)
Pv01	Growth habit	NA	Determinacy	GWAS—domesticated	Moghaddam et al. (2016)
Pv04	Growth habit	NA	Determinacy	GWAS—domesticated	Moghaddam et al. (2016)
Pv06	Growth habit	NA	Determinacy	GWAS—domesticated	Moghaddam et al. (2016)

(continued)

Table 2.1 (continued)

Chromosome	Domestication trait category	Locus/QTL/gene name	Trait/function	Approach and population	Literature
Pv07	Growth habit	NA	Determinacy	GWAS—domesticated	Moghaddam et al. (2016)
Pv11	Growth habit	NA	Determinacy	GWAS—domesticated	Moghaddam et al. (2016)
Pv11	Growth habit	NA	Determinacy	GWAS—domesticated	Moghaddam et al. (2016)
Pv01	Growth habit	*fin*	Twining	Linkage mapping—RILs	Koinange et al. (1996)
Pv04	Growth habit	*Cab1-1*	Climbing ability	Linkage mapping—RILs	Checa and Blair (2008)
Pv04	Growth habit	*Cab1-2*	Climbing ability	Linkage mapping—RILs	Checa and Blair (2008)
Pv04	Growth habit	*Cab2-1*	Climbing ability	Linkage mapping—RILs	Checa and Blair (2008)
Pv05	Growth habit	*Cab1-3*	Climbing ability	Linkage mapping—RILs	Checa and Blair (2008)
Pv07	Growth habit	*Cab1-4*	Climbing ability	Linkage mapping—RILs	Checa and Blair (2008)
Pv10	Growth habit	*Cab1-5*	Climbing ability	Linkage mapping—RILs	Checa and Blair (2008)
Pv11	Growth habit	*Cab1-6*	Climbing ability	Linkage mapping—RILs	Checa and Blair (2008)
Pv01	Growth habit	*NM (fin)*	Number of nodes on the main stem	Linkage mapping—RILs	Koinange et al. (1996)
Pv01	Growth habit	*NM*	Number of nodes on the main stem	Linkage mapping—RILs	Koinange et al. (1996)
Pv08	Growth habit	*NM*	Number of nodes on the main stem	Linkage mapping—RILs	Koinange et al. (1996)
Pv10	Growth habit	*TN*	Number of nodes on the main stem	Linkage mapping—$F_{2:4}$ pop	Tar'an et al. (2002)

(continued)

Table 2.1 (continued)

Chromosome	Domestication trait category	Locus/QTL/gene name	Trait/function	Approach and population	Literature
Pv04	Growth habit	*TB*	Number of branch on the main stem	Linkage mapping—$F_{2:4}$ pop	Tar'an et al. (2002)
Pv04	Growth habit	*Brn1*	Number of branch on the main stem	Linkage mapping—RILs	Checa and Blair (2008)
Pv01	Growth habit	*NP (fin)*	Number of pods	Linkage mapping—RILs	Koinange et al. (1996)
Pv08	Growth habit	*NP*	Number of pods	Linkage mapping—RILs	Koinange et al. (1996)
Pv04	Growth habit	*NP*	Number of pods	Linkage mapping—RILs	Koinange et al. (1996)
Pv04	Growth habit	*PPP*	Number of pods	Linkage mapping—$F_{2:4}$ pop	Tar'an et al. (2002)
Pv07	Growth habit	*Pp7.2*	Number of pods	Linkage mapping—BC pop	Blair et al. (2006)
Pv09	Growth habit	*Pp9.2*	Number of pods	Linkage mapping—BC pop	Blair et al. (2006)
Pv11	Growth habit	*Pp11.3*	Number of pods	Linkage mapping—BC pop	Blair et al. (2006)
Pv05	Growth habit	ss715649615	Number of pods	GWAS—domesticated	Kamfwa et al. (2015)
Pv07	Growth habit	ss715647649	Number of pods	GWAS—domesticated	Kamfwa et al. (2015)
Pv01	Growth habit	*L5*	Internode length	Linkage mapping—RILs	Koinange et al. (1996)
Pv04	Growth habit	*Int2*	Internode length	Linkage mapping—RILs	Checa and Blair (2008)
Pv04	Growth habit	*Int3*	Internode length	Linkage mapping—RILs	Checa and Blair (2008)
Pv04	Growth habit	*Int4*	Internode length	Linkage mapping—RILs	Checa and Blair (2008)
Pv03	Growth habit	*Int1*	Internode length	Linkage mapping—RILs	Checa and Blair (2008)

(continued)

Table 2.1 (continued)

Chromosome	Domestication trait category	Locus/QTL/gene name	Trait/function	Approach and population	Literature
Pv03	Growth habit	*PH*	Plant height	Linkage mapping—$F_{2:4}$ pop	Tar'an et al. (2002)
Pv07	Growth habit	*PH*	Plant height	Linkage mapping—$F_{2:4}$ pop	Tar'an et al. (2002)
Pv01	Growth habit	*Ph1.1*	Plant height	Linkage mapping—BC pop	Blair et al. (2006)
Pv06	Growth habit	*Ph6.1*	Plant height	Linkage mapping—BC pop	Blair et al. (2006)
Pv06	Growth habit	*Ph6.2*	Plant height	Linkage mapping—BC pop	Blair et al. (2006)
Pv07	Growth habit	*Ph7.1*	Plant height	Linkage mapping—BC pop	Blair et al. (2006)
Pv04	Growth habit	*Plh1-2*	Plant height	Linkage mapping—RILs	Checa and Blair (2008)
Pv03	Growth habit	*Plh1-1*	Plant height	Linkage mapping—RILs	Checa and Blair (2008)
Pv04	Growth habit	*Plh1-3*	Plant height	Linkage mapping—RILs	Checa and Blair (2008)
Pv08	Growth habit	*Plh1-4*	Plant height	Linkage mapping—RILs	Checa and Blair (2008)
Pv04	Growth habit	*Plh2-1*	Plant height	Linkage mapping—RILs	Checa and Blair (2008)
Pv11	Growth habit	*Plh2-3*	Plant height	Linkage mapping—RILs	Checa and Blair (2008)
Pv04	Growth habit	*Plh2-2*	Plant height	Linkage mapping—RILs	Checa and Blair (2008)
Pv06	Growth habit	*Pw6.1*	Plant width	Linkage mapping—BC pop	Blair et al. (2006)
Pv06	Growth habit	*Pw6.2*	Plant width	Linkage mapping—BC pop	Blair et al. (2006)
Pv07	Growth habit	*Pw7.1*	Plant width	Linkage mapping—BC pop	Blair et al. (2006)
Pv01	Gigantism	*PL*	Pod length	Linkage mapping—RILs	Koinange et al. (1996)
Pv02	Gigantism	*PL*	Pod length	Linkage mapping—RILs	Koinange et al. (1996)

(continued)

Table 2.1 (continued)

Chromosome	Domestication trait category	Locus/QTL/gene name	Trait/function	Approach and population	Literature
Pv07	Gigantism	*PL*	Pod length	Linkage mapping—RILs	Koinange et al. (1996)
Pv08	Gigantism	ss715639408	Pod weight	GWAS—domesticated	Kamfwa et al. (2015)
Pv08	Gigantism	ss715649359	Pod weight	GWAS—domesticated	Kamfwa et al. (2015)
Pv08	Gigantism	ss715647392	Pod weight	GWAS—domesticated	Kamfwa et al. (2015)
Pv01	Gigantism	*SW*	100-seed weight	Linkage mapping—RILs	Koinange et al. (1996)
Pv07	Gigantism	*SW*	100-seed weight	Linkage mapping—RILs	Koinange et al. (1996)
Pv07	Gigantism	*SW*	100-seed weight	Linkage mapping—RILs	Koinange et al. (1996)
Pv11	Gigantism	*SW*	100-seed weight	Linkage mapping—RILs	Koinange et al. (1996)
Pv04	Gigantism	*SW*	100-seed weight	Linkage mapping—$F_{2:4}$ pop	Tar'an et al. (2002)
Pv11	Gigantism	*SW*	100-seed weight	Linkage mapping—$F_{2:4}$ pop	Tar'an et al. (2002)
Pv02	Gigantism	*Sw2.1*	100-seed weight	Linkage mapping—BC pop	Blair et al. (2006)
Pv02	Gigantism	*Sw2.2*	100-seed weight	Linkage mapping—BC pop	Blair et al. (2006)
Pv03	Gigantism	*Sw3.1*	100-seed weight	Linkage mapping—BC pop	Blair et al. (2006)
Pv06	Gigantism	*Sw6.1*	100-seed weight	Linkage mapping—BC pop	Blair et al. (2006)
Pv07	Gigantism	*Sw7.1*	100-seed weight	Linkage mapping—BC pop	Blair et al. (2006)
Pv08	Gigantism	*Sw8.1*	100-seed weight	Linkage mapping—BC pop	Blair et al. (2006)
Pv08	Gigantism	*Sw8.2*	100-seed weight	Linkage mapping—BC pop	Blair et al. (2006)
Pv09	Gigantism	*Sw9.1*	100-seed weight	Linkage mapping—BC pop	Blair et al. (2006)
Pv10	Gigantism	*Sw10.1*	100-seed weight	Linkage mapping—BC pop	Blair et al. (2006)

(continued)

Table 2.1 (continued)

Chromosome	Domestication trait category	Locus/QTL/gene name	Trait/function	Approach and population	Literature
Pv11	Gigantism	*Sw11.1*	100-seed weight	Linkage mapping—BC pop	Blair et al. (2006)
Pv06	Gigantism	*SW6*	100-seed weight	Linkage mapping—RILs	Peréz-Vega et al. (2010)
Pv08	Gigantism	*SW8.1*	100-seed weight	Linkage mapping—RILs	Peréz-Vega et al. (2010)
Pv08	Gigantism	*SW8.2*	100-seed weight	Linkage mapping—RILs	Peréz-Vega et al. (2010)
Pv03	Gigantism	Ref_259_comp6493[g]	Fruit size	Candidate gene approach[c]	Bellucci et al. (2014a, b)
Pv03	Gigantism	Ref_220_comp2070[g]	Fruit size	Candidate gene approach[c]	Bellucci et al. (2014a, b)
Pv03	Gigantism	Ref_25_comp3527[g]	Fruit size	Candidate gene approach[c]	Bellucci et al. (2014a, b)
Pv03	Gigantism	Ref_259_comp4515[g]	Fruit size	Candidate gene approach[c]	Bellucci et al. (2014a, b)
Pv08	Gigantism	*Phvul.008G168000* (nitrate reductase)	100-seed weight/plant growth	Candidate gene approach[b]	Schmutz et al. (2014)
Pv08	Gigantism	*Phvul.008G168000* (nitrate reductase)	100-seed weight/plant growth	GWAS—domesticated	Schmutz et al. (2014)
Pv02	Gigantism	*SL2*	Seed length	Linkage mapping—RILs	Peréz-Vega et al. (2010)
Pv03	Gigantism	*SL3*	Seed length	Linkage mapping—RILs	Peréz-Vega et al. (2010)
Pv06	Gigantism	*SL6*	Seed length	Linkage mapping—RILs	Peréz-Vega et al. (2010)
Pv08	Gigantism	*SL8*	Seed length	Linkage mapping—RILs	Peréz-Vega et al. (2010)
Pv10	Gigantism	*SL10*	Seed length	Linkage mapping—RILs	Peréz-Vega et al. (2010)

(continued)

Table 2.1 (continued)

Chromosome	Domestication trait category	Locus/QTL/gene name	Trait/function	Approach and population	Literature
Pv06	Gigantism	*SH6*	Seed height	Linkage mapping—RILs	Peréz-Vega et al. (2010)
Pv08	Gigantism	*SH8*	Seed height	Linkage mapping—RILs	Peréz-Vega et al. (2010)
Pv03	Gigantism	*WI3*	Seed width	Linkage mapping—RILs	Peréz-Vega et al. (2010)
Pv06	Gigantism	*WI6*	Seed width	Linkage mapping—RILs	Peréz-Vega et al. (2010)
Pv07	Gigantism	*WI7*	Seed width	Linkage mapping—RILs	Peréz-Vega et al. (2010)
Pv01	Earliness	*DF (fin)*	Number of days to flowering	Linkage mapping—RILs	Koinange et al. (1996)
Pv01	Earliness	*DF*	Number of days to flowering	Linkage mapping—RILs	Koinange et al. (1996)
Pv08	Earliness	*DF*	Number of days to flowering	Linkage mapping—RILs	Koinange et al. (1996)
Pv09	Earliness	*DF*	Number of days to flowering	Linkage mapping—$F_{2:4}$ pop	Tar'an et al. (2002)
Pv01	Earliness	*df1.1*	Number of days to flowering	Linkage mapping—BC pop	Blair et al. (2006)
Pv02	Earliness	*df2.1*	Number of days to flowering	Linkage mapping—BC pop	Blair et al. (2006)
Pv06	Earliness	*df6.1*	Number of days to flowering	Linkage mapping—BC pop	Blair et al. (2006)
Pv06	Earliness	*df6.2*	Number of days to flowering	Linkage mapping—BC pop	Blair et al. (2006)
Pv09	Earliness	*df9.1*	Number of days to flowering	Linkage mapping—BC pop	Blair et al. (2006)
Pv09	Earliness	*df9.2*	Number of days to flowering	Linkage mapping—BC pop	Blair et al. (2006)

(continued)

Table 2.1 (continued)

Chromosome	Domestication trait category	Locus/QTL/gene name	Trait/function	Approach and population	Literature
Pv11	Earliness	*df11.1*	Number of days to flowering	Linkage mapping—BC pop	Blair et al. (2006)
Pv01	Earliness	*DF1*	Number of days to flowering	Linkage mapping—RILs	Peréz-Vega et al. (2010)
Pv02	Earliness	*DF2*	Number of days to flowering	Linkage mapping—RILs	Peréz-Vega et al. (2010)
Pv08	Earliness	*DF8*	Number of days to flowering	Linkage mapping—RILs	Peréz-Vega et al. (2010)
Pv01	Earliness	ss715646578	Number of days to flowering	GWAS—domesticated	Kamfwa et al. (2015)
Pv08	Earliness	ss715646088	Number of days to flowering	GWAS–domesticated	Kamfwa et al. (2015)
Pv01	Earliness	NA	Number of days to flowering	GWAS—domesticated	Moghaddam et al. (2016)
Pv01	Earliness	*DM*	Number of days to maturity	Linkage mapping—RILs	Koinange et al. (1996)
Pv01	Earliness	*DM (fin)*	Number of days to maturity	Linkage mapping—RILs	Koinange et al. (1996)
Pv08	Earliness	*DM*	Number of days to maturity	Linkage mapping—RILs	Koinange et al. (1996)
Pv09	Earliness	*DM*	Number of days to maturity	Linkage mapping—$F_{2:4}$ pop	Tar'an et al. (2002)
Pv10	Earliness	*DM*	Number of days to maturity	Linkage mapping—$F_{2:4}$ pop	Tar'an et al. (2002)
Pv05	Earliness	*Dm5.1*	Number of days to maturity	Linkage mapping—BC pop	Blair et al. (2006)
Pv07	Earliness	*Dm7.1*	Number of days to maturity	Linkage mapping—BC pop	Blair et al. (2006)
Pv01	Earliness	*DM1*	Number of days to maturity	Linkage mapping—RILs	Peréz-Vega et al. (2010)
Pv02	Earliness	*DM2.1*	Number of days to maturity	Linkage mapping—RILs	Peréz-Vega et al. (2010)

(continued)

Table 2.1 (continued)

Chromosome	Domestication trait category	Locus/QTL/gene name	Trait/function	Approach and population	Literature
Pv02	Earliness	*DM2.2*	Number of days to maturity	Linkage mapping—RILs	Peréz-Vega et al. (2010)
Pv06	Earliness	*DM6.1*	Number of days to maturity	Linkage mapping—RILs	Peréz-Vega et al. (2010)
Pv06	Earliness	*DM6.2*	Number of days to maturity	Linkage mapping—RILs	Peréz-Vega et al. (2010)
Pv01	Earliness	ss715646578	Number of days to maturity	GWAS—domesticated	Kamfwa et al. (2015)
Pv05	Earliness	Ref_259_comp19102_c0	Vernalisation and flowering time[h]	Candidate gene approach[c]	Bellucci et al. (2014a, b)
Pv03	Earliness	*Phvul.003G033400*	Vernalisation and flowering time[h]	Candidate gene approach[b]	Schmutz et al. (2014
Pv02	Earliness	*Phvul.002G000500*	Vernalisation and flowering time[i]	Candidate gene approach[b]	Schmutz et al. 2014)
Pv01	Photoperiod sensitivity	*Ppd*	Delay in flowering under 16 h days	Linkage mapping—RILs	Koinange et al. (1996)
Pv11	Photoperiod sensitivity	*PD*	Delay in flowering under 16 h days	Linkage mapping—RILs	Koinange et al. (1996)
Pv04	Photoperiod sensitivity	Ref_25_comp11990_c0[j]	Photoperiod response	Candidate gene approach[c]	Bellucci et al. (2014a, b)
Pv05	Productivity	*Y*	Seed yield (g/m)	Linkage mapping—$F_{2:4}$ pop	Tar'an et al. (2002)
Pv09	Productivity	*Y*	Seed yield (g/m)	Linkage mapping—$F_{2:4}$ pop	Tar'an et al. (2002)
Pv10	Productivity	*Y*	Seed yield (g/m)	Linkage mapping—$F_{2:4}$ pop	Tar'an et al. (2002)
Pv02	Productivity	*yld2.1*	Seed yield (kg/ha)	Linkage mapping—BC pop	Blair et al. (2006)
Pv03	Productivity	*yld3.1*	Seed yield (kg/ha)	Linkage mapping—BC pop	Blair et al. (2006)
Pv03	Productivity	*yld3.2*	Seed yield (kg/ha)	Linkage mapping—BC pop	Blair et al. (2006)

(continued)

Table 2.1 (continued)

Chromosome	Domestication trait category	Locus/QTL/gene name	Trait/function	Approach and population	Literature
Pv04	Productivity	*yld4.1*	Seed yield (kg/ha)	Linkage mapping—BC pop	Blair et al. (2006)
Pv04	Productivity	*yld4.2*	Seed yield (kg/ha)	Linkage mapping—BC pop	Blair et al. (2006)
Pv04	Productivity	*yld4.3*	Seed yield (kg/ha)	Linkage mapping—BC pop	Blair et al. (2006)
Pv04	Productivity	*yld4.4*	Seed yield (kg/ha)	Linkage mapping—BC pop	Blair et al. (2006)
Pv09	Productivity	*yld9.1*	Seed yield (kg/ha)	Linkage mapping—BC pop	Blair et al. (2006)
Pv09	Productivity	*yld9.2*	Seed yield (kg/ha)	Linkage mapping—BC pop	Blair et al. (2006)
Pv03	Productivity	ss715648538	Seed yield (kg/ha)	GWAS—domesticated	Kamfwa et al. (2015)
Pv09	Productivity	ss715646178	Seed yield (kg/ha)	GWAS—domesticated	Kamfwa et al. (2015)
Pv06	Productivity	*sp6.1*	Seeds per plant	Linkage mapping—BC pop	Blair et al. (2006)
Pv07	Productivity	*sp7.1*	Seeds per plant	Linkage mapping—BC pop	Blair et al. (2006)
Pv07	Productivity	*sp7.2*	Seeds per plant	Linkage mapping—BC pop	Blair et al. (2006)
Pv03	Productivity	ss715639901	Seeds per plant	GWAS—domesticated	Kamfwa et al. (2015)
Pv05	Productivity	ss715650235	Seeds per plant	GWAS—domesticated	Kamfwa et al. (2015)
Pv08	Productivity	ss715639408	Yield per plant	GWAS—domesticated	Kamfwa et al. (2015)
Pv08	Productivity	ss715649359	Yield per plant	GWAS—domesticated	Kamfwa et al. (2015)
Pv09	Productivity	ss715647002	Yield per plant	GWAS—domesticated	Kamfwa et al. (2015)
Pv02	Productivity	ss715647433	Biomass	GWAS—domesticated	Kamfwa et al. (2015)

(continued)

Table 2.1 (continued)

Chromosome	Domestication trait category	Locus/QTL/gene name	Trait/function	Approach and population	Literature
Pv08	Productivity	ss715639408	Biomass	GWAS—domesticated	Kamfwa et al. (2015)
Pv01	Productivity	*HI*	Harvest index	Linkage mapping—RILs	Koinange et al. (1996)
Pv08	Productivity	*HI*	Harvest index	Linkage mapping—RILs	Koinange et al. (1996)
Pv06	Productivity	*H*	Harvest index	Linkage mapping—$F_{2:4}$ pop	Tar'an et al. (2002)
Pv03	Productivity	ss715639243	Harvest index	GWAS—domesticated	Kamfwa et al. (2015)
Pv03	Productivity	ss715641141	Harvest index	GWAS—domesticated	Kamfwa et al. (2015)
Pv04	Productivity	ss715648677	Pod harvest index	GWAS—domesticated	Kamfwa et al. (2015)
Pv07	Seed and pod pigmentation	*P*	Colour of seed coat	Linkage mapping—RILs	Koinange et al. (1996)
Pv04	Seed and pod pigmentation	*G*	Colour of seed coat	Linkage mapping—RILs	McClean et al. (2002)
Pv06	Seed and pod pigmentation	*V*	Colour of seed coat	Linkage mapping—RILs	McClean et al. (2002)
Pv08	Seed and pod pigmentation	*C*	Colour of seed coat	Linkage mapping—RILs	McClean et al. (2002)
Pv08	Seed and pod pigmentation	*Gy*	Colour of seed coat	Linkage mapping—RILs	McClean et al. (2002)
Pv09	Seed and pod pigmentation	*T*	Pattern of seed coat	Linkage mapping—RILs	McClean et al. (2002)
Pv10	Seed and pod pigmentation	*Bip*	Pattern of seed coat	Linkage mapping—RILs	McClean et al. (2002)
Pv10	Seed and pod pigmentation	*Ana*	Pattern of seed coat	Linkage mapping—RILs	McClean et al. (2002)

(continued)

Table 2.1 (continued)

Chromosome	Domestication trait category	Locus/QTL/gene name	Trait/function	Approach and population	Literature
Pv10	Seed and pod pigmentation	*J*	Pattern of seed coat	Linkage mapping—RILs	McClean et al. (2002)
Pv03	Seed and pod pigmentation	*Z*	Pattern of seed coat	Linkage mapping—RILs	McClean et al. (2002)
Pv02	Seed and pod pigmentation	*y*	Pod colour	Linkage mapping—RILs	Koinange et al. (1996)

[a]Candidate gene approach searching for homologues gene with a known function in *A. thaliana*
[b]Candidate gene approach searching for genomic region under selection between wild and domesticated common bean accessions
[c]Candidate gene approach searching homologues genes for transcripts (RNA-seq analysis) that were under selection between wild and domesticated Mesoamerican accessions
[d]Homologues to *AtSHP* (*SHATTERPROOF*), a gene involved in the seed dispersal in *A. thaliana*
[e]Homologues to *AtIND* (*INDEHISCENT*), a gene involved in the seed dispersal in *A. thaliana*
[f]Homologues to *TFL1* (*TERMINAL FLOWER 1*) that controls determinacy in *A. thaliana*
[g]Homologues to a YABBY-like transcription factor (*FASCIATED*) that controls number of locules and fruit size in tomato
[h]Homologues to *VRN1* (*VERNALISATION 1*), a gene involved in the vernalisation pathway and flowering time
[i]Homologues to *VRN2* (*VERNALISATION 2*), a gene involved in the vernalisation pathway and flowering time
[j]Homologues to *GI* (*GIGANTEA*), involved in the photoperiod response and flowering time

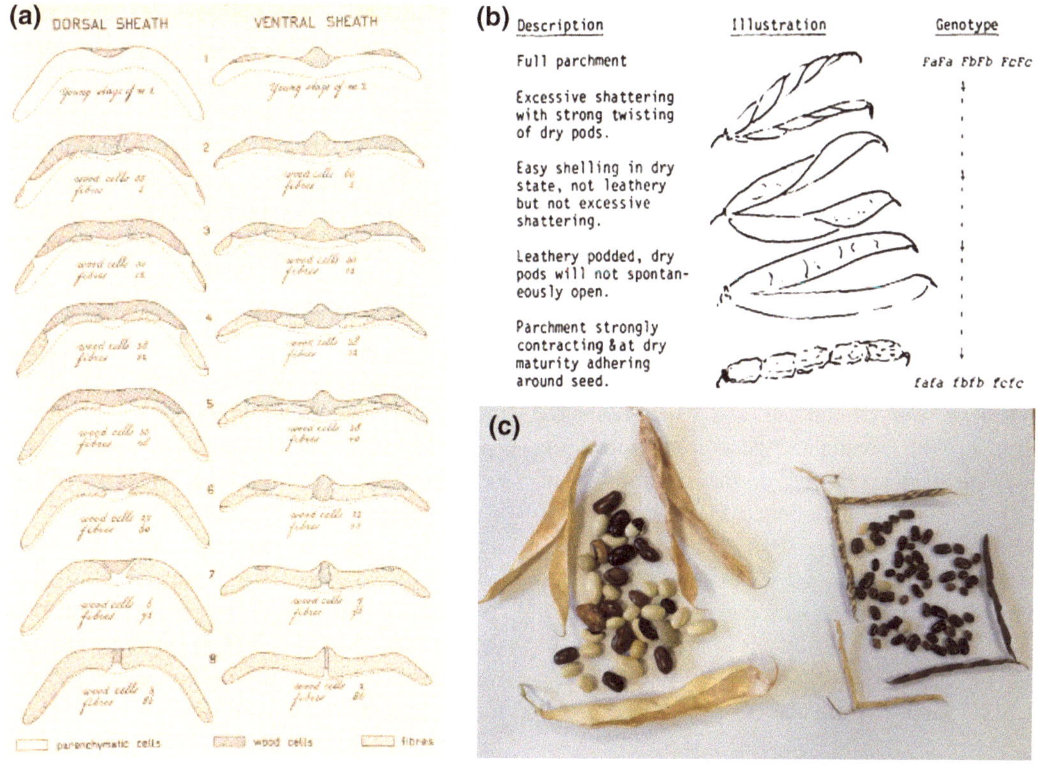

Fig. 2.1 Seed dispersal mechanism of the common bean. **a** From Prakken (1934): location and content of fibres, parenchymatic and wood cells in the dorsal and ventral sheets of the pod valves in stringy (dehiscent, bottom), stringless (indehiscent, above) and intermediate varieties (in the middle); **b** from Lamprecht (1932): schematisation of the hypothesis of Lamprecht (1932) on the genetic control of pod shattering in common bean; **c** domesticated pods and seeds, on the left, and twisted pods and seeds from a wild common bean, on the right

SHATTERPROOF-1 (*SHP1*; Nanni et al. 2011) and *INDEHISCENT* (*IND*; Gioia et al. 2013b) genes. These represent two genes that are directly involved in seed shattering and the primary factors that are required for silique shattering in this species (Liljegren et al. 2000, 2004). Nanni et al. (2011) identified, characterised and mapped a sequence (*PvSHP1*) in the common bean using the RIL mapping populations BAT93 × Jalo EEP558 (Freyre et al. 1998) and Midas × G12873 (MG RIL population; Koinange et al. 1996). They mapped *PvSHP1* to chromosome Pv06, in proximity to the *V* gene that controls flower colour in the common bean (Nodari et al. 1993; McClean et al. 2002), although on a different chromosome from that of the *St* locus. Similarly, Gioia et al. (2013b) identified and mapped the *PvIND* (GenBank KC192374) sequence using the same mapping populations as Nanni et al. (2011). *PvIND* was located on chromosome Pv02 near to the *St* locus, although complete segregation between these two loci was not observed. Moreover, when Gioia et al. (2013b) examined the association between polymorphisms in the *PvIND* sequence and the dehiscent/indehiscent phenotype in 105 wild and domesticated lines, they did not identify any SNPs that were significantly associated with pod shattering.

Studies have indicated that a convergent phenotypic response to selection due to domestication for a specific trait might not be produced by the same molecular mechanism, also in related species (Nanni et al. 2011; Doust et al. 2004).

2.5.2 Seed Dormancy

Seed dormancy is a plant strategy that allows the delay of seed germination to avoid seedling growth under unfavourable environmental conditions. Loss of seed dormancy is considered as a main trait in the domestication syndrome in common bean and other crop species because this is crucial for cultivation (Koinange et al. 1996). The reduction of dormancy in domesticated beans has ensured more rapid and simultaneous germination. Very few studies on the seed dormancy trait have been carried out, and when Koinange et al. (1996) investigated genetic control of the domestication syndrome in the common bean, they identified four unlinked QTLs on chromosomes Pv02, Pv03 and Pv04. These four QTLs cumulatively explained 69% of the total phenotypic variation for the seed dormancy trait.

2.5.3 Growth Habit

The bush growth type is a common feature that characterises the compact growth habit of the domesticated common bean, which includes no twining branches, few vegetative nodes and long internodes. Among the traits that contribute to this growth habit, the main descriptors of the plant architecture include the climbing ability, twining, number of nodes on the main stem, number of branches on the main stem, number of pods, internode length and plant height and width. Several differences between wild and domesticated common bean can be observed, and there is also variability between the domesticated genotypes. In addition, these traits have an impact on the agriculture practices and on the yield and they can also be correlated to each other and subjected to the effects of the environment. Indeed, the length of the main stem (i.e. the plant height) is related to the numbers and lengths of the internodes, while the internode length (which also changes along the main stem) depends mainly on the environmental effects and the growth stage of the plant (Debouck et al. 1986).

The common bean plant can be either determinate or indeterminate for its growth habit, which is defined by the characteristics of the terminal part of the stem and branches (Fernández et al. 1986). The common bean shows a wide range of phenotypic variability for this determinate/indeterminate trait that can be summarised as five main types (Evans 1973; Singh 1982; Debouck et al. 1986; Fernández et al. 1986; Leakey 1988):

Type I: determinate bush growth habit that shows reproductive terminal buds both on the main stem and on the branches (Singh 1982). After flowering, apical vegetative growth stops.

Type II: indeterminate bush growth habit with a vegetative bud on the apical part of both the main stem and the branches. After flowering, new leaves and vegetative nodes are produced.

Type III: indeterminate climber growth habit that is characterised by open branches and semi-prostrate habit. In some cases, these have moderate climbing ability.

Type IV: indeterminate climber growth habit with very long branches that show strong climbing and twining.

Type V: determinate climber growth habit that shows climbing and twining.

Wild common bean is indeterminate, and selection for the more compact growth habit (e.g. the bush habit) was one of the aims of the domestication process and breeding to promote earlier flowering and maturation (e.g. determinacy) in the crops compared to the wild ancestors. The inheritance of determinacy was investigated using a linkage mapping approach and segregation analysis in the RIL population of Midas × G12873 (Koinange et al. 1996), and a single locus (*fin*) that controlled determinacy was identified on chromosome Pv01.

Using a candidate gene approach on two RIL of the BAT93 × Jalo EEP 558 (Freyre et al. 1998) and Midas × G12873 (Koinange et al. 1996) mapping populations, Kwak et al. (2008) mapped Pv*TLF1y* to chromosome Pv01, which is a sequence homologous to *Terminal Flower 1* (*TLF1*) of *A. thaliana* (Shannon and Meeks-Wagner 1991). This locus is responsible for the development of the terminal flower, and it acts as a repressor of flowering, with a role in the inflorescence meristem identity, and co-segregates with the *fin* locus (i.e. determinacy). Foucher et al. (2003) identified *PsTFL1a*, a pea homologue of *TFL1*, as the gene controlling the determinacy phenotype in pea (*Pisum sativum* L.). More recently, Repinsky et al. (2012) confirmed the co-segregation of *PvTLF1* and *fin* and validated the function of *PvTFL1y*, whereby they confirmed it as the functional homologue of *TFL1*. Indeed, Repinsky et al. (2012) detected a 32–133-fold decrease in expression between the indeterminate haplotype BAT93 and the determinate haplotype CDRK, and the reduction in the expression was 20–91-fold for the determinate G00750 compared to BAT93. This finding confirmed the function of *PvTFL1y* as a flowering repressor, as it prevents the transition of the vegetative apical bud into a terminal flower. *PvTFL1y* sequences for some important reference lines of the common bean have been deposited with GenBank (Repinsky et al. 2012).

Moghaddam et al. (2016) used GWAS analysis on a panel of 280 domesticated common bean genotypes and confirmed the co-localisation on chromosome Pv01 of *PvTFL1y* (*Phvul.001G189200*) and the *fin* locus (determinacy) (Koinange et al. 1996; Kwak et al. 2008; Repinsky et al. 2012). When performing GWAS analysis with the exclusion of the determinate genotypes, they also detected significant associations with other QTLs on chromosomes Pv04, Pv06, Pv07 and Pv11, which were not detected using the entire collection.

Among the five growth habit classes, three include common beans that can climb and also have erect growth if supports are provided. In a traditional intercropping system known as 'milpa', the common bean is also cultivated in association with maize (*Zea mays* L.), which provides the support, and also maybe including squash (*Cucurbita* spp.). Moreover, climbing is related to twining, another important feature that characterises the main stem of the common bean.

For the climbing ability, using the RIL population G2333 × G19839, Checa and Blair (2008) identified seven QTLs. G2333 is an indeterminate climbing (Type IV) Mesoamerican landrace, while G19839 is an indeterminate bush (Type II) Andean landrace. Among these seven QTLs, one was located on chromosome Pv04 (*Cab 1-1*) in two field trials at 45 days after planting, in the same genomic region in which a further QTL (*Cab 2-1*) was associated with the climbing ability at two sowing locations and at 75 days after planting. The other five QTLs were mapped by Checa and Blair (2008) to chromosomes Pv04 (*Cab 1-2*), in the same genomic region where QTLs for plant height and internode length were also located, Pv05 (*Cab 1-3*), Pv07 (*Cab 1-4*), near the *Phs* locus (i.e. the phaseolin gene), Pv10 (*Cab 1-5*), and Pv11 (*Cab 1-6*).

In their investigations into the genetic control of the twining predisposition using the Midas (non-twining) × G12873 (twining) RIL population, Koinange et al. (1996) mapped this trait (*Tor*) in the same region as *fin* (i.e. chromosome Pv01), the locus for determinacy. As a possible explanation, they proposed that the *fin* locus might have a pleiotropic effect on both determinacy and twining, or that the genes responsible for these two traits are strictly associated in the MG RIL population.

Among the growth habit traits, the number of nodes and the number of branches on the main stem are considered to be reliable descriptors for the common bean phenotypic architecture. Domestication caused a reduction in the number of nodes in the main stem, and for this trait, three QTLs were defined by Koinange et al. (1996). One of these, on chromosome Pv01, was linked to the *fin* locus for determinacy in the same genomic region where QTLs related to earliness and number of pods were mapped. Similarly, for the two other QTLs, which were mapped in the same regions where QTLs for earliness traits, yield components and plant architecture were identified, one was located on chromosome Pv01, tightly linked to the *Ppd* locus for the photoperiod sensitivity, and the other on chromosome Pv08.

Tar'an et al. (2002) studied 142 $F_{2:4}$ individuals derived from a cross between two inbred lines: OAC Seaforth (determinate bush; Type I) and OAC 95-4 (indeterminate bush; Type II). They mapped one QTL for the number of nodes of the main stem on chromosome Pv10. They also identified a QTL on chromosome Pv04, for the number of branches on the main stem, as did Checa and Blair (2008), who mapped a QTL for the number of branches (*Brn1*) in the same region. Moreover, *Brn1* was located close to other QTLs for plant architecture traits, such as climbing ability, internode length and plant height (Checa and Blair 2008).

For the common bean, the domesticated growth habit is generally characterised by a lower number of pods per plant, on the main stem and branches. The parental lines of the MG RIL population (Koinange et al. 1996) are representative for this trait. Indeed, the domesticated parent Midas produced a mean of 13.9 pods per plant, while the wild parent G12873 gave 43.2 pods per plant. Using the molecular linkage mapping approach for the number of pods, six QTLs were mapped in two different RIL populations derived from a cross between wild and domesticated parental lines (Koinange et al. 1996; Blair et al. 2006, using a backcross $BC_2F_{3:5}$ population derived from a cross between ICA Cerinza, cultivated, Type I and G24404, wild, Type IV). Two QTLs were mapped, on chromosome Pv01 linked to the *fin* locus, and on chromosome Pv08 in the same region where QTLs for the number of nodes on the main stem and the earliness traits were identified (Koinange et al. 1996). The other four QTLs were mapped on Pv04 (D14 on the Midas × G12873 linkage map; Koinange et al. 1996) and on chromosomes Pv07 (*Pp7.2*), Pv09 (*Pp9.2*) and Pv11 (*Pp11.3*) (Blair et al. 2006). Using the same approach on domesticated individuals, Tar'an et al. (2002) identified a further QTL for the number of pods (*PPP*) on chromosome Pv04, in a region where QTLs for numbers of branches (Tar'an et al. 2002; Checa and Blair 2008), climbing ability, plant height and internode length (Checa and Blair 2008) have been identified. More recently, Kamfwa et al. (2015) using a GWAS approach with the Illumina BARCBean6K_3 BeadChip genotyped 237 Andean domesticated individuals (i.e. varieties, elite lines, landraces). They searched for genomic regions that were associated with important agronomic traits that are also related to the domestication process, and they identified two significant QTLs for the number of pods on chromosomes Pv05 and Pv07.

As indicated above, internode length is a growth habit trait that can show high phenotypic variability due to both the different growth stage of the plant and environmental effects. In general, domestication selected individuals with lower numbers of vegetative nodes on the main stem, although with longer internodes, which correspond to the stem portion between neighbouring nodes. Using the molecular linkage mapping approach for the internode length, a QTL (*L5*) was mapped to chromosome Pv01 by Koinange et al. (1996), while Checa and Blair (2008) worked on the G2333 × G19839 RIL population and mapped four QTLs, one on chromosome Pv03 (*Int1*) and three on chromosome Pv04 (*Int2*, *Int3*, *Int4*). Interestingly, most of the QTLs associated with plant architecture were mapped to chromosome Pv01 by Koinange et al. (1996), Checa and Blair (2008), using domesticated individuals, observed the co-localisation of QTLs for plant architecture traits on chromosome Pv04. To dissect out the plant architecture components, Blair et al. (2006) searched for QTLs for plant height identifying four QTLs, one on chromosome Pv01, two on chromosome Pv06 and one on chromosome Pv07. Working on a RIL population derived from a cross between wild and domesticated lines, Blair et al. (2006) also mapped three QTLs for plant width to chromosomes Pv06 (two QTLs) and Pv07. Interestingly, the QTL for plant height on chromosome Pv07 (*ph7.1*) was close to the *Ph* (phaseolin) locus, and it fell in the same region where Checa and Blair (2008) mapped the QTL *Cab 1-4* for climbing ability. Using the same approaches on domesticated materials, a QTL for plant height was mapped to chromosome Pv07 by Tar'an et al. (2002); this QTL mapped near to *PvTFL1z* (Kwak et al. 2008), another homologue of *Terminal flower1* that controls the indeterminate phenotype in *A. thaliana*.

2.5.4 Gigantism

During the domestication process, humans propagated the individuals that showed appreciable features for consumption, and in general the selection was related to all of the usable parts of the plants, such as the edible parts, which can differ on different species. In legumes such as the common bean, artificial selection favoured large pods and seeds. Indeed, one of the most obvious differences between wild and domesticated beans is the different of shapes, sizes and weights of the fruits, which together constitute the trait known as 'gigantism'.

Pod length and pod weight are two traits that are considerably important in terms of yield. Three QTLs related to pod length were mapped by Koinange et al. (1996) using linkage mapping approaches, on chromosomes Pv02, Pv07 and Pv11 (as D1b, in Freyre et al. 1998). The QTL on chromosome Pv11 was the most significant, as this explained 23% of the total phenotypic variance, while the three QTLs together explained 37% of the total phenotypic variance for pod length.

More recently, three QTLs that mapped on chromosome Pv08 were associated with pod weight, using GWAS analysis on Andean domesticated individuals (Kamfwa et al. 2015). The SNP showing the highest association with pod weight (i.e. ss715639408; $P = 4.3 \times 10^{-8}$; position 5150618) was also associated with plant biomass (as g/plant). Indeed, the trait of pod weight contributes to the biomass, and the significant genetic association between these two traits might depend on the high phenotypic correlation between them ($r = 0.87$; $P = 0.001$; Kamfwa et al. 2015). Moreover, and of particular interest, two significant SNPs on chromosome Pv08 (ss715639408, ss715649359) were common to the traits of pod weight and yield/plant (as g seed/plant). Thus, Kamfwa et al. (2015) suggested that the same gene might have pleiotropic effects on these two traits, or that two distinct genes that reside on the same linkage disequilibrium (LD) block are associated with the same SNP.

For seeds traits, 100-seed weight has the main role for gigantism features, and together with seed length, seed height and seed width, characterises the domestication process in common bean, mainly in terms of the changes in seed shape and size. One-hundred seed weight is one of the main traits that affect the yield, and there is a wide variability for this trait among the domesticated individuals. Many efforts have been made to understand the genetic control of this quantitative trait. Four significant QTLs were mapped on chromosomes Pv01, Pv07 and Pv11 on the Midas × G12873 molecular linkage map (Koinange et al. 1996). The QTLs with the highest significant effects were those on chromosomes Pv01 and Pv07 ($P = 0.001$), in proximity to the *Phs* locus for the phaseolin gene, explaining 18% and 27% of the total observed phenotypic variance, respectively. Using similar QTL mapping approaches, other studies identified several QTLs for 100-seed weight on many of the common bean chromosomes (i.e. Pv02, Pv03, Pv04, Pv06, Pv07, Pv08, Pv09, Pv10, Pv11), both when wild and domesticated (Blair et al. 2006) or only domesticated (Tar'an et al. 2002; Pérez-Vega et al. 2010) accessions were compared. One of the QTLs on chromosome Pv07 (*sw7.1*) was linked to the phaseolin locus (Blair et al. 2006), in agreement to previous observations (Koinange et al. 1996). A QTL (*SW6*) identified on chromosome Pv06 (Pérez-Vega et al. 2010) mapped near QTLs for seed length (*SL6*) and seed height (*SH6*), in the same location as QTL *SW6.1* (Blair et al. 2006). Moreover, Pérez-Vega et al. (2010) identified additional QTLs for seed weight (*SW8.1* and *SW8.2*) on chromosome Pv08, where also QTL for seed height (*SH8*) and seed length (*SL8*) were detected. Cumulatively, these three QTLs for seed weight explained 54% of the total phenotypic variance, with high correlation both between 100-seed weight and seed length ($r = 0.83$; $P < 0.05$), and between seed weight and seed height ($r = 0.74$; $P < 0.05$). Further, QTLs for seed size-related traits (i.e. length, width) were also identified on chromosomes Pv02, Pv03, Pv06, Pv07 and Pv10.

More recently, Schmutz et al. (2014) performed diversity and population differentiation analyses on a set of wild and landrace individuals, and they identified different genomic regions as putatively under selection (i.e. 1835 Mesoamerican, 748 Andean physically mapped candidate genes for the domestication process). Among these genes, they highlighted *Phvul.008G168000* (Mesoamerican candidate gene) that encodes nitrate reductase, an enzyme that has an important role in plant nitrogen assimilation and seed growth. Interestingly, this gene was physically mapped to chromosome Pv08, near *SW8.2*, the QTL that was previously associated with 100-seed weight (Pérez-Vega et al. 2010).

To validate the Mesoamerican candidate genes that were under selection and to discover the genetic architecture of the seed weight trait, Schmutz et al. (2014) further examined their candidate genes for seed weight using a GWAS approach on a set of 271 modern common bean varieties from the Mesoamerican gene pool. Three genes were confirmed by GWAS among the 15 candidate genes previously shown putatively under selection in relation to seed weight. The GWAS approach also placed several domestication candidates related to seed weight, which showed extensive linkage disequilibrium, on one sweep window on chromosome Pv07 (Schmutz et al. 2014).

These data show that searching for candidate genes in genomic regions under selection can be a reliable approach, and GWAS analysis represents a powerful tool for validation of candidate genes, especially when the possibility to compare genetic and physical maps can be exploited.

Bellucci et al. (2014a) used RNA-seq techniques and identified 2364 transcripts (representing $\sim$9% of the total transcriptome) that showed signatures of selection between wild and domesticated Mesoamerican accessions. They focussed on the transcripts with higher selection index and analysed the functions of these genes while searching for homologies with genes related to the domestication process in different species. An interesting example here was related to fruit size (Bellucci et al. 2014a), a contig homolog of YABBY5 (*YAB5*) that was putatively under selection. YABBY5 is a transcription factor implicated in the regulation of seed shattering in cereal species, including sorghum (*Sorghum bicolor*), rice and maize (Lin et al. 2012). A YAB-like transcription factor (*FASCIATED*), which increases the number of locules, was also associated with the control of carpel number and fruit development in tomato (Cong et al. 2008). Indeed, the process that leads to larger fruit in the domesticated races consists of two main aspects: increased cell division and a greater number of organs in the fruit (Fig. 2.2).

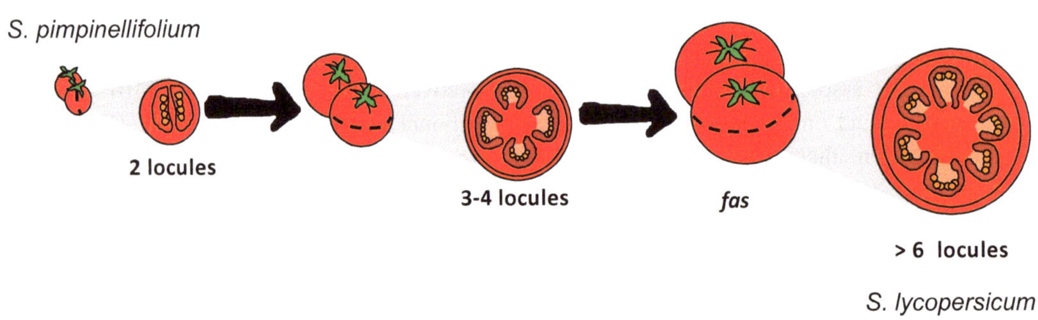

Fig. 2.2 *FASCIATED (fas)*: an example of a domestication gene related to gigantism, a homologue of a contig that is putatively under selection that was identified by Bellucci et al. (2014a). Locule number evolution during tomato domestication, from *S. pimpinellifolium* (two locules), which is considered as the wild ancestor of tomato, to the cultivated *S. lycopersicum* (as described in Muños et al. 2011)

2.5.5 Earliness

A common feature among crops is that they flower and mature earlier than in the wild, which represent a key factor in domestication, also considering that simultaneous flowering can guarantee simultaneous maturation and harvesting. The number of days to flowering is a trait that describes the number of days between planting and flowering, while the days to maturity corresponds to the days needed for the maturation of the pods. Therefore, different studies have been dedicated to these phenological traits. Ten QTLs associated with days to flowering were detected across chromosomes Pv01 (Koinange et al. 1996; Blair et al. 2006), Pv02, Pv06, Pv09, Pv11 (Blair et al. 2006) and Pv08 (Koinange et al. 1996), using a molecular linkage mapping approach with populations derived from a cross between wild and domesticated genotypes. Interestingly, one QTL on chromosome Pv01 (Koinange et al. 1996) explained 38% of the total phenotypic variance for flowering time. At the same time, it was close to *fin*, a gene for determinacy, and co-localised with QTLs for the number of nodes on the main stem, the number of pods and the days to maturity. Likewise, other QTLs were detected on the same chromosomes in populations derived from a cross between domesticated individuals; Pérez-Vega et al. (2010) mapped indeed three QTLs on chromosomes Pv01, Pv02 and Pv08. Interestingly, in this population, the QTL on chromosome Pv01 was the closest to the *fin* locus, confirming the observation of Koinange et al. (1996), both regarding the QTL position and the co-localisation of genes for determinacy and phenology traits.

Further interesting examples of co-localisation between genes for earliness and other domestication traits were reported by Koinange et al. (1996), who mapped the number of days to flowering and the sensitivity to photoperiod (*Ppd locus*) in the same genomic region on chromosome Pv01. Moreover, Blair et al. (2006) mapped two QTLs for days to flowering to chromosome Pv06, close to *V*, a flower colour locus (Nodari et al. 1993; McClean et al. 2002), and *PvSHP1*, a homologous gene to *Shatterproof*, which controls seed shattering in *A. thaliana* (Nanni et al. 2011). However, the most significant QTLs reported by Blair et al. (2006) were those on chromosome Pv09 (*df9.1*, *df9.2*), which suggested a parallelism with the observation of Tar'an et al. (2002). Indeed, using a population that was derived from a cross between two domesticated lines, Tar'an et al. (2002) identified a significant QTL for days to flowering on chromosome Pv09, close to the growth habit locus (*GH*) that encodes for the determinate/indeterminate phenotype in this population. They thus provided evidence of the co-localisation of genes for determinacy and phenology traits, although on a different chromosome to Koinange et al. (1996) and Pérez-Vega et al. (2010) who co-mapped days to flowering and determinacy on chromosome Pv01. Through the GWAS approach and by analysing domesticated genotypes, significant QTLs for days to flowering were detected on chromosome Pv01 by Kamfwa et al. (2015) and Moghaddam et al. (2016), which confirmed the observations of Koinange et al. (1996), Blair et al. (2006) and Pérez-Vega et al. (2010). Other QTLs have also been found on chromosome Pv08 (Kamfwa et al. 2015), as previously reported (Koinange et al. 1996; Pérez-Vega et al. 2010). Considering the days to maturity, two QTLs were reported on chromosome Pv01 (Koinange et al. 1996), one on each of chromosomes Pv05 and Pv07 (Blair et al. 2006) and one on chromosome Pv08 (Koinange et al. 1996).

The co-localisation observed between QTLs for days to flowering and days to maturity on chromosomes Pv01 and Pv08 by Koinange et al. (1996) is of particular interest. One QTL on chromosome Pv01 (Koinange et al. 1996) is mapped close to the *fin* locus for determinacy, and it explained 30% of the total phenotypic variance. In contrast, Blair et al. (2006) did not find QTLs on chromosome Pv01 and identified QTLs for days to maturity on different chromosomes than those reported for days to flowering.

Considering studies conducted only on domesticated genotypes, further QTLs were found on chromosomes Pv01, Pv02 and Pv06

(Pérez-Vega et al. 2010), and on chromosomes Pv09 and Pv10 (Tar'an et al. 2002). Likewise, in the study by Pérez-Vega et al. (2010), days to flowering and days to maturity co-mapped on chromosome Pv01, close to the *fin* locus, and on chromosome Pv02. This was also observed by Tar'an et al. (2002) on chromosome Pv09, which suggested that neighbouring genes control these two traits, or that pleiotropic effects might be involved in the genetic control of days to flowering and days to maturity.

Recent results provided by GWAS analysis confirmed a significant QTL for days to maturity on chromosome Pv01 (Kamfwa et al. 2015). In this case, the significant SNP reported for days to maturity (ss715646578; SNP position 48340819) was also significant for days to flowering, which confirmed that these two traits co-map in populations with different genetic backgrounds. In addition to the genetic association between these two traits, significant phenotypic correlation was detected between days to flowering and days to maturity in different populations ($r = 0.64$, Tar'an et al. 2002; $r = 0.67$, Pérez-Vega et al. 2010; $r = 0.70$, Kamfwa et al. 2015). Bellucci et al. (2014a) used RNA-seq on a set of wild and domesticated accessions, and among the transcripts that showed selection signatures, they identified a homologue of the *Vernalisation* genes of *Arabidosis*, which are involved in the vernalisation pathway to promote flowering in *A. thaliana*. These genes have a crucial function; indeed, a long cold period can be needed to promote flowering in many species, and some plants bloom only after winter. The *Arabidopsis Vernalisation* genes act by repression of *FLC*, which is a floral repressor, and after a long period of cold, *FLC* mRNA decreases due to the increased expression of genes such as *VRN1*, which thus indirectly controls the flowering time. Moreover, Schmutz et al. (2014) reported several genomic regions that were associated with the domestication process, and among these, they identified homologous genes for *VRN1* (*Phvul.003G033400*) on chromosome Pv03, and *VRN2* (*Phvul.002G000500*) on chromosome Pv02, as a Mesoamerican candidate gene.

Also, in this case, these studies on earliness demonstrate that searching for genes associated with domestication and looking for selection signals across the genome between wild and domesticated individuals appears to be a very promising approach.

2.5.6 Photoperiod Sensitivity

The response to photoperiod is a trait that is strictly dependent on the environment in which the species originated, evolved and adapted, with the regulation of flowering time with respect to day length. Long-day plants bloom when the length of the day tends to increase, with >12 h of daylight, while short-day species flower when the length of the day is <12 h. Based on this sensitivity to the photoperiod, a species cannot be cultivated at all latitudes, unless it is day-neutral or indifferent to the photoperiod. While domesticated individuals have become insensitive to the photoperiod, individuals introduced into areas to which they are not adapted bloom later or do not bloom at all. In common bean, this trait was measured as the delay in flowering under a day length of 16 h, compared to 12 h. This descriptor is appropriate to detect the delay in flowering in wild individuals that flower only under short days, compared to domesticated individuals in which no delay in flowering has been highlighted under long days, as observed by Koinange et al. (1996). A molecular linkage mapping approach to the RIL population of Midas × G12873 detected two QTLs for photoperiod sensitivity, on chromosomes Pv01 and Pv11 (Koinange et al. 1996). The QTL on chromosome Pv01 mapped to the same region as number of nodes on the main stem, 100-seed weight, days to flowering and days to maturity (Koinange et al. 1996). These corresponded to the *Ppd* locus for photoperiod sensitivity (Wallace et al. 1993) and were linked to the *fin* locus for determinacy (Koinange et al. 1996), and they explained 44% of the total phenotypic variance. The QTL mapped to chromosome Pv11 was close to the marker D1479 and explained 17% of the total phenotypic variance.

More recently, Bellucci et al. (2014a) used RNA-seq technology for a genome-wide analysis and reported several candidate genes related to the photoperiod response. These genes appeared to be related to domestication because they specifically investigated the function of the genes putatively under selection during domestication. Among these, a homologous sequence to *GIGANTEA* (*GI*) was found to be under selection. In *A. thaliana*, this gene has an important role in flowering and in the regulation of other genes, such as *CONSTANS* (*CO*) and *FLOWERING TIME* (*FT*), to induce flowering under long days. Interestingly, *CO* and *FT* (the target genes of *GI*; Fig. 2.3) were reported as targets of selection during domestication in rice and sunflower (Blackman et al. 2011; Takahashi and Shimamoto 2011; Wu et al. 2013), and *FT* is the target gene of the *Floral repressor FLC*, which in turn is repressed by the *Arabidopsis Vernalisation* genes, which were also found as under selection in Bellucci et al. (2014a).

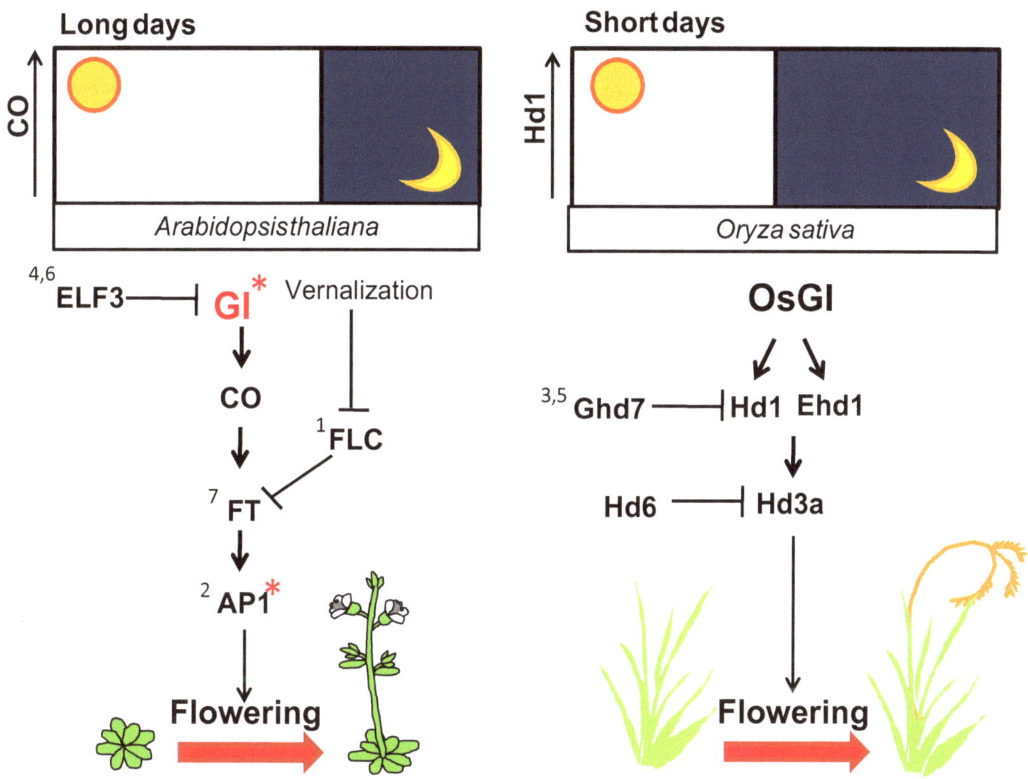

Fig. 2.3 *GIGANTEA*: an example of a domestication gene related to photoperiod sensitivity, reported by Bellucci et al. (2014a) as among the genes with high selection coefficient. *GIGANTEA* (*GI*; red) acts upstream of the *CO* (*CONSTANS*) and *FT* (*FLOWERING LOCUS T*) genes, as its rice ortolog *Hd1*, in the flowering pathway, to induce flowering in under long-day (*Arabidopsis*) and short-day (*Oryza sativa*) conditions. In addition, *CO* controls the responses of the downstream genes, *AP1* (*APETALA 1*) in *Arabidopsis* and *Hd3a* in rice (an *Arabidopsis FT* homologue). Also, a homologue of *AP1*, *Vrn1* (red asterisk) was found among the genes under selection in Bellucci et al. (2014a). The numbers indicate homologous genes that have been reported as targets of selection in other crop species: 1. rapeseed *BnFLC.A10*; 2. wheat *Vrn1*; 3. wheat *Vrn2*; 4. lentil *SN*; 5. maize *ZmCCT*; 6. pea *HR*; 7. sunflower *HaFT1*

2.5.7 Productivity

Artificial selection led towards a reduction in the number of pods and seeds in common bean, although without reducing yield, which increased in the cultivated plants. Among the domestication syndrome traits, the harvest index is a widely used descriptor to quantify plant productivity, as the measure of the ratio between seed yield and plant biomass. However, yield depends on several factors, like the size and number of seeds per plant, and it can be measured as yield/plant and yield/surface area. Using the molecular linkage mapping approach, two significant QTLs for harvest index were found on chromosomes Pv01 and Pv08, in the same region where QTLs for 100-seed weight (Pv01) and number of pods (Pv08) were detected (Koinange et al. 1996). Working on domesticated individuals, Tar'an et al. (2002) used a linkage mapping approach to map one QTL on chromosome Pv06, in contrast to Kamfwa et al. (2015), who identified two significant SNPs on chromosome Pv03 using a GWAS approach. The different genetic structures of the populations used in these studies might explain these divergent data, which confirms that yield has wide variability between wild and domesticated individuals and among domesticated varieties.

When dissecting yield into its component parts, other QTLs were detected in different studies: three QTLs were found for seeds per plant, one on chromosome Pv06 and two on chromosome Pv07 (Blair et al. 2006), while Kamfwa et al. (2015) used a GWAS approach to identify two QTLs, on chromosomes Pv03 and Pv05. For the trait of seed yield (kg/ha), nine QTLs were detected, one on chromosome Pv02, two on chromosome Pv03, four on chromosome Pv04 and two on chromosome Pv09 (Blair et al. 2006). Similarly, Kamfwa et al. (2015) used a GWAS approach to identify two QTLs for seed yield, on chromosomes Pv03 and Pv09, and Tar'an et al. (2002) mapped three QTLs for the same trait, one of which was on chromosome Pv09. Moreover, Kamfwa et al. (2015) detected a significant SNP for the trait of yield per plant on chromosome Pv09, and two QTLs on chromosome Pv08, one of which (*ss715639408*; position Ch8:5150618) was also significant for pod weight and plant biomass. Using different approaches and populations with different genetic backgrounds, the major components of yield have therefore been identified on chromosomes Pv01 (Koinange et al. 1996), Pv03 (Blair et al. 2006; Kamfwa et al. 2015), Pv08 (Koinange et al. 1996; Kamfwa et al. 2015) and Pv09 (Tar'an et al. 2002; Blair et al. 2006; Kamfwa et al. 2015). The pod harvest index is a further component of yield, which is measured as the ratio between seed weight and weight of the fertile pods. This was mapped by Kamfwa et al. (2015) to chromosome Pv04 using a GWAS approach on a set of domesticated accessions.

From these studies, it has emerged that the increased productivity observed in the domesticated individuals compared with their wild progenitors is due to several traits. These traits might in turn be under the control of multiple genes, which suggest both a relationship between the yield components and complex genetic control for the harvest index.

2.5.8 Seed and Pod Pigmentation

Although the selection process is generally correlated with a reduction in genetic diversity at target loci going from the wild to the domesticated individuals, domestication has led to an increase in the phenotypic variability for some domestication-related traits (Bellucci et al. 2014a). In the case of the domestication of beans, human selection has led to seeds and pods with a wide range of colours and colour patterns. This process has increased the diversification between the wild and domesticated individuals, and it has also contributed to the diversification and increased variability within the domesticated forms. As an example, in the commercial varieties, the two domesticated lines from which Pérez-Vega et al. (2010) developed their mapping population, Xana and Cornell 49242, showed marked differences for seed traits. Xana is described as a white and large-seeded line, while Cornell 49242 has small and black seeds.

Also Midas and G12873, the domesticated and wild parental lines of the MG RIL population used by Koinange et al. (1996), are very different in terms of seed colours. Midas is a domesticated snap bean that has white seeds, while G12873 is a wild Mesoamerican accession that has pigmented seeds, with an agouti colour.

A locus for seed colour, *P*, was identified on chromosome Pv07 in proximity to the *Phs* locus (phaseolin) and in a region near to the QTL for 100-seed weight (Koinange et al. 1996). The *y* locus for pod colour (i.e. green vs yellow) was also mapped to chromosome Pv02. These traits were treated as qualitative, with the hypothesis that the seed and fruit colours are both controlled by a single gene. Later, McClean et al. (2002) also investigated the genetic basis of the colour patterns for the seeds of the common bean, and they phenotypically mapped several loci that had been previously identified as associated with seed colour. They also developed different molecular markers (i.e. RAPD, STS) that were associated with these genes and mapped them in the core linkage map (Freyre et al. 1998). The loci *G*, *V*, *C* and *Gy* for seed colour were located to chromosomes Pv04, Pv06 and Pv08, while for the seed coat pattern they mapped *T*, *Bip*, *Ana*, *J* and *Z* to chromosomes Pv03, Pv09 and Pv10. All of these genes interacted with each other for the determination of the wide range of colouration and colour pattern of the seeds, while the gene *P* controlled the absence or presence of the pigmentation. In more detail, a dominant allele at the *P* locus (*PP*, *P-*) determined the presence of colour in the flowers and seeds (Emerson 1909), while the recessive genotype, *pp*, results in white flowers and seeds, as for the domesticated accession 'Midas' in Koinange et al. (1996).

2.6 Conclusions

With the release of the reference genomes of both the Mesoamerican and Andean genotypes, a new era of genetic and genomics studies has begun for the common bean and the other *Phaseolus* spp. More insight into the common bean genomics can be achieved by exploiting the reference genome sequences and the derived new tools to focus on the major phenotypic changes that occurred during domestication and the successive episodes of improvement, including modern plant breeding. Identification of the molecular basis of the domestication syndrome would also be a major step towards our understanding of the evolutionary processes and provide a useful lesson to improve the breeding of novel varieties. Along with the other *Phaseolus* spp., the common bean remains an ideal model to study the molecular implications of the convergent phenotypic evolution that occurred under domestication due to the multiple independent domestication events between and within species that occurred for *Phaseolus* spp. Similarly, analysis of the evolution after domestication, in terms of the introduction of the common bean into Europe (BEAN_ADAPT project, www.beanadapt.org), continues to offer novel opportunities to dissect out the genetic architecture of environmental adaptation in crop species.

Acknowledgements This work was conducted within the BEAN ADAPT project, founded through the ERA-CAPS Programme, 2014 Call, Expanding the European Research Area in Molecular Plant Sciences II.

References

Acosta-Gallegos JA, Kelly JD, Gepts P (2007) Pre-breeding in common bean and use of genetic diversity from wild germplasm. Crop Sci 47:44–59

Angioi SA, Rau D, Attene G, Nanni L, Bellucci E, Logozzo G, Negri V, Spagnoletti Zeuli PL, Papa R (2010) Beans in Europe: origin and structure of the European landraces of *Phaseolus vulgaris* L. Theor Appl Genet 121:829–843

Angioi SA, Rau D, Nanni L, Bellucci E, Papa R, Attene G (2011) The genetic make-up of the European landraces of the common bean. Plant Genet Resour 9:197–201

Asfaw A, Blair MW, Almekinders C (2009) Genetic diversity and population structure of common bean (*Phaseolus vulgaris* L.) landraces from the East African highlands. Theor Appl Genet 120:1–12

Becerra-Velásquez VL, Gepts P (1994) RFLP diversity in common bean (*Phaseolus vulgaris* L.). Genome 37:256–263

Beebe S, Rengifo J, Gaitan E, Duque MC, Tohme J (2001) Diversity and origin of Andean landraces of common bean. Crop Sci 41:854–862

Bellucci E, Bitocchi E, Ferrarini A, Benazzo A, Biagetti E, Klie S, Minio A, Rau D, Rodriguez M, Panziera A, Venturini L, Attene G, Albertini E, Jackson SA, Nanni L, Fernie AR, Nikoloski Z, Bertorelle G, Delledonne M, Papa R (2014a) Decreased nucleotide and expression diversity and modified coexpression patterns characterize domestication in the common bean. Plant Cell 26:1901–1912. doi:10.1105/tpc.114.124040

Bellucci E, Bitocchi E, Rau D, Rodriguez M, Biagetti E, Giardini A, Attene G, Nanni L, Papa R (2014b) Genomics of origin, domestication and evolution of *Phaseolus vulgaris*. In: Tuberosa R, Graner A, Frison E (eds) Genomics of plant genetic resources. Springer, Netherlands, pp 483–507

Bitocchi E, Bellucci E, Giardini A, Rau D, Rodriguez M, Biagetti E, Santilocchi R, Spagnoletti Zeuli P, Gioia T, Logozzo G, Attene G, Nanni L, Papa R (2013) Molecular analysis of the parallel domestication of the common bean (*Phaseolus vulgaris*) in Mesoamerica and the Andes. New Phytol 197:300–313

Bitocchi E, Nanni L, Bellucci E, Rossi M, Giardini A, Spagnoletti Zeuli PS, Logozzo G, Stougaard J, McClean P, Attene G, Papa R (2012) Mesoamerican origin of the common bean (*Phaseolus vulgaris* L.) is revealed by sequence data. Proc Natl Acad Sci USA 109:E788–E796

Bitocchi E, Rau D, Bellucci E, Rodriguez M, Murgia ML, Gioia T, Santo D, Nanni L, Attene G, Papa R (2017) Beans (*Phaseolus* ssp.) as a model for understanding crop evolution. Front Plant Sci 8:722. doi:10.3389/fpls.2017.00722

Bitocchi E, Rau D, Benazzo A, Bellucci E, Goretti D, Biagetti E, Panziera A, Laidò G, Rodriguez M, Gioia T, Attene G, McClean P, Lee RK, Jackson SA, Bertorelle G, Papa R (2016) High level of nonsynonymous changes in common bean suggests that selection under domestication increased functional diversity at target traits. Front Plant Sci 7:2005

Blackman BK, Rasmussen DA, Strasburg JL, Raduski AR, Burke JM, Knapp SJ, Michaels SD, Rieseberg LH (2011) Contributions of flowering time genes to sunflower domestication and improvement. Genetics 187:271–287

Blair MW, González LF, Kimani M, Butare L (2010) Genetic diversity, inter-gene pool introgression and nutritional quality of common beans (*Phaseolus vulgaris* L.) from central Africa. Theor Appl Genet 121:237–248

Blair MW, Iriarte G, Beebe S (2006) QTL analysis of yield traits in an advanced backcross population derived from a cultivated Andean × wild common bean (*Phaseolus vulgaris* L.) cross. Theor Appl Genet 112:1149–1163

Brown CH, Clement CR, Epps P, Luedeling E, Wichmann S (2014) The paleobiolinguistics of the common bean (*Phaseolus vulgaris* L.). Ethnobiol Lett 5:104

Burle ML, Fonseca JR, Kami JA, Gepts P (2010) Microsatellite diversity and genetic structure among common bean (*Phaseolus vulgaris* L.) landraces in Brazil, a secondary center of diversity. Theor Appl Genet 121:801–813

Chacón SMI, Pickersgill B, Debouck DG, Arias JS (2007) Phylogeographic analysis of the chloroplast DNA variation in wild common bean (*Phaseolus vulgaris* L.) in the Americas. Plant Syst Evol 266:175–195

Checa OE, Blair MW (2008) Mapping QTL climbing ability and component traits in common bean (*Phaseolus vulgaris* L.). Mol Breed 22:201–215

Choi JY, Platts AE, Fuller DQ, Hsing YI, Wing RA, Purugganan MD (2017) The rice paradox: multiple origins but single domestication in Asian rice. Mol Biol Evol 34:969–979

Cong B, Barrero LS, Tanksley SD (2008) Regulatory change in YABBY-like transcription factor led to evolution of extreme fruit size during tomato domestication. Nat Genet 40:800–804

Debouck DG, Hidalgo R, Fernandez FO, Correa E, Smithson JB (1986) Morphology of the common bean plant *Phaseolus vulgaris*. Centro Internacional de Agricultura Tropical, CIAT, p 56 (Series 04 EB-09.01)

Debouck DG, Toro O, Paredes OM, Johnson WC, Gepts P (1993) Genetic diversity and ecological distribution of *Phaseolus vulgaris* in northwestern South America. Econ Bot 47:408–423

Desiderio F, Bitocchi E, Bellucci E, Rau D, Rodriguez M, Attene G, Papa R, Nanni L (2013) Chloroplast microsatellite diversity in *Phaseolus vulgaris*. Front Plant Sci 3:312

Doust AN, Devos KM, Gadberry MD, Gale MD, Kellogg EA (2004) Genetic control of branching in foxtail millet. Proc Natl Acad Sci USA 101:9045–9050

Emerson RA (1909) Factors for mottling in beans. J Hered 5:368–376

Evans A (1973) Exploitation of the variability in plant architecture in *Phaseolus vulgaris*. In Potentials of field bean and other food legumes in Latin America. Centro Internacional de Agricultura Tropical, CIAT, p 279

Fernández F, Gepts P, López M (1986) Stages of development of the common bean plant. Centro Internacional de Agricultura Tropical, CIAT, p 32 (Series04 EB-09.03)

Foucher F, Morin J, Courtiade J, Cadioux S, Ellis N, Banfield MJ, Rameau C (2003) Determinate and late flowering are two terminal flowerl/centroradialis homologs that control two distinct phases of flowering initiation and development in pea. Plant Cell 15:2742–2754

Freyre R, Ríos R, Guzmán L, Debouck DG, Gepts P (1996) Ecogeographic distribution of *Phaseolus* spp. (Fabaceae) in Bolivia. Econ Bot 50:195–215

Freyre R, Skroch PW, Geffroy V, Adam-Blondon AF, Shirmohamadali A, Johnson WC, Llaca V, Nodari RO, Pereira PA, Tsai S-M, Tohme J, Dron M, Nienhuis J, Vallejos CE, Gepts P (1998) Towards an integrated linkage map of common bean. 4. Development of a core linkage map and alignment of RFLP maps. Theor Appl Genet 97:847–856

Gepts P, Bliss FA (1985) F1 hybrid weakness in the common bean: differential geographic origin suggests two gene pools in cultivated bean germplasm. J Hered 76:447–450

Gepts P, Debouck DG (1991) Origin, domestication, and evolution of the common bean, *Phaseolus vulgaris*. In: Voysest O, Van Schoonhoven A (eds) Common beans: research for crop improvement. CAB, Oxon, UK, pp 7–53

Gepts P, Osborn TC, RashkaK Bliss FA (1986) Phaseolin–protein variability in wild forms and landraces of the common bean (*Phaseolus vulgaris*): evidence for multiple centers of domestication. Econ Bot 40: 451–468

Gepts P, Papa R (2002) Evolution during domestication. In: Encyclopedia of life sciences. London: Nature Publishing Group. Macmillan Publishers Ltd., UK, pp 1–7

Gioia T, Logozzo G, Kami J, Spagnoletti Zeuli P, Gepts P (2013b) Identification and characterization of a homologue to the *Arabidopsis INDEHISCENT* gene in common bean. J Hered 104:273–286

Gioia T, Logozzo G, Attene G, Bellucci E, Benedettelli S, Negri V, Papa R, Spagnoletti Zeuli P (2013a) Evidence for introduction bottleneck and extensive inter-gene pool (Mesoamerica x Andes) hybridization in the European common bean (*Phaseolus vulgaris* L.) germplasm. PLoS ONE 8:e75974

Glémin S, Bataillon T (2009) A comparative view of the evolution of grasses under domestication. New Phytol 183:273–290

Johnson WC, Gepts P (1988) Segregation for performance in recombinant inbred populations resulting from inter-gene pool crosses of common bean (*Phaseolus vulgaris* L.). Euphytica 106:45–56

Kamfwa K, Cichy KA, Kelly JD (2015) Genome-wide association study of agronomic traits in common bean. Plant Genome 8(2):1–12. doi:10.3835/plantgenome2014.09.0059

Kaplan L, Lynch TF (1999) *Phaseolus* (Fabaceae) in archaeology: AMS radiocarbon dates and their significance for pre-Columbian agriculture. Econ Bot 53 (3):261–272

Kami J, Velásquez VB, Debouck DG, Gepts P (1995) Identification of presumed ancestral DNA sequences of phaseolin in *Phaseolus vulgaris*. Proc Natl Acad Sci USA 92:1101–1104

Koenig R, Gepts P (1989) Allozyme diversity in wild *Phaseolus vulgaris*: further evidence for two major centers of diversity. Theor Appl Genet 78:809–817

Koenig RL, Singh SP, Gepts P (1990) Novel phaseolin types in wild and cultivated common bean (*Phaseolus vulgaris*, Fabaceae). Econ Bot 44:50–60

Koinange EMK, Gepts P (1992) Hybrid weakness in wild *Phaseolus vulgaris* L. J Hered 83:135–139

Koinange EMS, Singh SP, Gepts P (1996) Genetic control of the domestication syndrome in common bean. Crop Sci 36:1037–1045

Kwak M, Gepts P (2009) Structure of genetic diversity in the two major gene pools of common bean (*Phaseolus vulgaris* L, Fabaceae). Theor Appl Genet 118:979–992

Kwak M, Velasco DM, Gepts P (2008) Mapping homologous sequences for determinacy and photoperiod sensitivity in common bean (*Phaseolus vulgaris*). J Hered 99:283–291

Lamprecht H (1932) Beitrage zur genetic von *Phaseolus vulgaris*. Hereditas 16:169–211

Leakey CLA (1988) Genotypic and phenotypic markers in common bean. In: Gepts P (ed) Gentic resources of *Phaseolus* beans. Kluwer Academic Publisher, Dordrecht, Holland, pp 245–327

Liljegren S, Ditta G, Eshed Y, Savidge B, Bowmanand J, Yanofsky M (2000) Control of fruit dehiscence in *Arabidopsis* by the *SHATTERPROOF* MADS-box genes. Nature 404:766–769

Liljegren SJ, Roeder AH, Kempin SA, Gremski K, Østergaard L, Guimil S, Reyes DK, Yanofsky MF (2004) Control of fruit patterning in *Arabidopsis* by *INDEHISCENT*. Cell 116:843–853

Lin Z, Li X, Shannon LM, Yeh CT, Wang ML, Bai G, Peng Z, Li J, Trick HN, Clemente TE, Doebley J (2012) Parallel domestication of the *Shattering1* genes in cereals. Nat Genet 44:720–724

Londo JP, Chiang YC, Hung KH, Chiang TY, Schaal BA (2006) Phylogeography of Asian wild rice, *Oryza rufipogon*, reveals multiple independent domestications of cultivated rice, *Oryza sativa*. Proc Natl Acad Sci USA 103:9578–9583

Lu J, Tang T, Tang H, Huang J, Shi S, Wu CI (2006) The accumulation of deleterious mutations in rice genomes: a hypothesis on the cost of domestication. Trends Genet 22:126–131

Martin GB, Adams MW (1987a) Landraces of *Phaseolus vulgaris* (Fabaceae) in northern Malawi I. Regional variation. Econ Bot 41:190–203

Martin GB, Adams MW (1987b) Landraces of *Phaseolus vulgaris* (Fabaceae) in northern Malawi II. Generation and maintenance of variability. Econ Bot 41:204–215

McClean PE, Lee R, Otto C, Gepts P, Bassett M (2002) Molecular and phenotypic mapping of genes controlling seed coat pattern and color in common bean (*Phaseolus vulgaris* L.). J Hered 93:148–152

McCouch S (2004) Diversifying selection in plant breeding. PLoS Biol 2:e347

Meyer RS, DuVal AE, Jensen HR (2012) Patterns and processes in crop domestication: an historical review and quantitative analysis of 203 global food crops. New Phytol 196:29–48

Moghaddam SM, Mamidi S, Osorno JM, Lee R, Brick M, Kelly J, Miklas P, Urrea C, Song Q, Cregan P, Grimwood J (2016) Genome-wide association study identifies candidate loci underlying agronomic traits in a Middle American diversity panel of common bean. Plant Genome 9(3). doi:10.3835/plantgenome2016.02.0012

Molina J, Sikora M, Garud N, Flowers JM, Rubinstein S, Reynolds A, Huang P, Jackson S, Schaal BA, Bustamante CD, Boyko AR (2011) Molecular evidence for a single evolutionary origin of domesticated rice. Proc Natl Acad Sci USA 108:8351–8356

Muños S, Ranc N, Botton E, Bérard A, Rolland S, Duffé P, Carretero Y, Le Paslier MC, Delalande C, Bouzayen M, Brunel D (2011) Increase in tomato locule number is controlled by two single-nucleotide polymorphisms located near *WUSCHEL*. Plant Physiol 156:2244–2254

Murgia ML, Attene G, Rodriguez M, Bitocchi E, Bellucci E, Fois D, Nanni L, Gioia T, Albani DM, Papa R, Rau D (2017) A comprehensive phenotypic investigation of the "pod-shattering syndrome" in common bean. Front Plant Sci 8:251

Nanni L, Bitocchi E, Bellucci E, Rossi M, Rau D, Attene G, Gepts P, Papa R (2011) Nucleotide diversity of a genomic sequence similar to *SHATTERPROOF* (*PvSHP1*) in domesticated and wild common bean (*Phaseolus vulgaris* L.). Theor Appl Genet 123:1341–1357

Nodari RO, Tsai SM, Gilbertson RL, Gepts P (1993) Towards an integrated linkage map of common bean. II. Development of an RFLP-based linkage map. Theor Appl Genet 85:513–520

Olson MV (1999) When less is more: gene loss as an engine of evolutionary change. Am J Human Genet 64:18–23

Osakabe Y, Arinaga N, Umezawa T, Katsura S, Nagamachi K, Tanaka H, Ohiraki H, Yamada K, Seo SU, Abo M, Yoshimura E (2013) Osmotic stress responses and plant growth controlled by potassium transporters in *Arabidopsis*. Plant Cell 25:609–624

Papa R, Acosta J, Delgado-Salinas A, Gepts P (2005) A genome-wide analysis of differentiation between wild and domesticated *Phaseolus vulgaris* from Mesoamerica. Theor Appl Genet 111:1147–1158

Papa R, Bellucci E, Rossi M, Leonardi S, Rau D, Gepts P, Nanni L, Attene G (2007) Tagging the signatures of domestication in common bean (*Phaseolus vulgaris*) by means of pooled DNA samples. Ann Bot 100:1039–1051

Papa R, Gepts P (2003) Asymmetry of gene flow and differential geographical structure of molecular diversity in wild and domesticated common bean (*Phaseolus vulgaris* L.) from Mesoamerica. Theor Appl Genet 106:239–250

Papa R, Nanni L, Sicard D, Rau D, Attene G (2006) The evolution of genetic diversity in *Phaseolus vulgaris* L. In: Motley TJ, Zerega N, Cross H (eds) New approaches to the origins, evolution and conservation of crops. Darwin's Harvest. Columbia University Press, New York

Pérez-Vega E, Pañeda A, Rodríguez-Suárez C, Campa A, Giraldez R, Ferreira JJ (2010) Mapping of QTLs for morpho-agronomic and seed quality traits in a RIL population of common bean (*Phaseolus vulgaris* L.). Theor Appl Genet 120:1367–1380

Prakken R (1934) Inheritance of colours and pod characters in *Phaseolus vulgaris* L. Genetica 16:177–296

Rendón-Anaya M, Montero-Vargas JM, Saburido-Álvarez S, Vlasova A, Capella-Gutierrez S, Ordaz-Ortiz JJ, Aguilar OM, Vianello-Brondani RP, Santalla M, Delaye L, Gabaldón T, Gepts P, Winkler R, Guigó R, Delgado-Salinas A, Herrera-Estrella A (2017) Genomic history of the origin and domestication of common bean unveils its closest sister species. Genome Biol 18:60

Repinski SL, Kwak M, Gepts P (2012) The common bean growth habit gene *PvTFL1y* is a functional homolog of *Arabidopsis TFL1*. Theor Appl Genet 124:1539–1547

Rodriguez M, Rau D, Bitocchi E, Bellucci E, Biagetti E, Carboni A, Gepts P, Nanni L, Papa R, Attene G (2016) Landscape genetics, adaptive diversity, and population structure in *Phaseolus vulgaris*. New Phytol 209:1781–1794

Rossi M, Bitocchi E, Bellucci E, Nanni L, Rau D, Attene G, Papa R (2009) Linkage disequilibrium and population structure in wild and domesticated populations of *Phaseolus vulgaris* L. Evol Appl 2: 504–522

Santalla M, Rodiño AP, De Ron AM (2002) Allozyme evidence supporting southwester Europe as a secondary center of genetic diversity for common bean. Theor Appl Genet 104:934–944

Schmutz J, McClean PE, Mamidi S, Wu GA, Cannon SB, Grimwood J, Jenkins J, Shu S, Song Q, Chavarro C, Torres-Torres M, Geffroy V, Moghaddam SM, Gao D, Abernathy B, Barry K, Blair MW, Brick MA, Chovatia M, Gepts P, Goodstein DM, Gonzales M, Hellsten U, Hyten DL, Jia G, Kelly JD, Kudrna D, Lee R, Richard MMS, Miklas PN, Osorno JM, Rodrigues J, Thareau V, Urrea CA, Wang M, Yu Y, Zhang M, Wing RA, Cregan PB, Rokhsar DS, Jackson SA (2014) A reference genome for common bean and genome-wide analysis of dual domestications. Nat Genet 46:707–713

Shannon S, Meeks-Wagner DR (1991) A mutation in the *Arabidopsis TFL1* gene affects inflorescence meristem development. Plant Cell 3:877–892

Singh SP (1982) A key for identification of different growth habits of *Phaseolus vulgaris* L. Ann Rep Bean Improv Coop 25:92–95

Singh SP, Gutiérrez JA, Molina A, Urrea C, Gepts P (1991) Genetic diversity in cultivated common bean. II. Marker-based analysis of morphological and agronomic traits. Crop Sci 31:23–29

Singh SP, Molina A (1996) Inheritance of crippled trifoliolate leaves occurring in interracial crosses of common bean and its relationship with hybrid dwarfism. J Hered 87:464–469

Takahashi Y, Shimamoto K (2011) *Heading date 1* (*Hd1*), an ortholog of *Arabidopsis CONSTANS*, is a possible target of human selection during domestication to diversify flowering times of cultivated rice. Genes Genet Syst 86:175–182

Tanksley SD, McCouch SR (1997) Seed banks and molecular maps: unlocking genetic potential from the wild. Science 277:1063–1066

Tar'an B, Michaels TE, Pauls KP (2002) Genetic mapping of agronomic traits in common bean. Crop Sci 42:544–556

Tarrago MN (1980) El proceso de agriculturización en el Noroeste Argentino, zona Valliserrana. In: Actas del V Congreso Nacional Arqueología Argentina, vol. 1. Instituto de Investigaciones Arqueológicas y Museo, Universidad de San Juan, San Juan, Argentina, pp 181–217

Toro O, Tohme J, Debouck DG (1990) Wild bean (*Phaseolus vulgaris* L.): description and distribution. Centro Internacional de Agricultura Tropical, CIAT, Cali, Colombia

Vitte C, Ishii T, Lamy F, Brar D, Panaud O (2004) Genomic paleontology provides evidence for two distinct origins of Asian rice (*Oryza sativa* L.). Mol Genet Genom 272:504–511

Vlasova A, Capella-Gutiérrez S, Rendón-Anaya M, Hernández-Oñate M, Minoche AE, Erb I, Câmara F, Prieto-Barja P, Corvelo A, Sanseverino W, Westergaard G (2016) Genome and transcriptome analysis of the Mesoamerican common bean and the role of gene duplications in establishing tissue and temporal specialization of genes. Genome Biol 17:32

Wallace DH, Yourstone KS, Masaya PN, Zobel RW (1993) Photoperiod gene control over partitioning between reproductive and vegetative growth. Theor Appl Genet 86:6–16

Wu W, Zheng XM, Lu G, Zhong Z, Gao H, Chen L, Wu C, Wang HJ, Wang Q, Zhou K, Wang JL (2013) Association of functional nucleotide polymorphisms at *DTH2* with the northward expansion of rice cultivation in Asia. Proc Natl Acad Sci USA 110:2775–2780

Zhang X, Blair MW, Wang S (2008) Genetic diversity of Chinese common bean (*Phaseolus vulgaris* L.) landraces assessed with simple sequence repeats markers. Theor Appl Genet 117:629–640

3 Cytogenetics and Comparative Analysis of *Phaseolus* Species

Artur Fonsêca and Andrea Pedrosa-Harand

Abstract

The genus *Phaseolus* includes approximately 75 species, most of which have $2n = 22$ small meta- or submetacentric chromosomes. This considerable karyotypic stability has been further reinforced by comparative cytogenetic mapping of single-copy sequences using Bacterial Artificial Chromosomes (BACs) and the Fluorescent in situ Hybridization (FISH) technique. These physical maps have revealed complete macrosynteny among the investigated species and only a few breaks in collinearity due to chromosomal inversions, except for *Phaseolus leptostachyus*, with $2n = 20$ and several rearrangements. The variation in the repetitive fraction of the genome is much greater, however, as expected. Considerable variation has been found in the number of 35S rDNA sites in the representatives of the Vulgaris group, and repetitive pericentromeric and subtelomeric sequences vary among *Phaseolus* species. This repetitive genome fraction has also been investigated in *Phaseolus vulgaris* through an epigenetic approach, revealing a higher degree of complexity in the heterochromatin than previously thought. The available molecular tools for the common bean have permitted a systematic analysis of the synteny, not only between the common bean and other cultivated and wild species of the genus, but also at the intergeneric level, contributing to a more detailed understanding of the cytogenetics of the group.

Keywords

BAC-FISH · Chromatin · Comparative cytogenetics
Cytogenetic map · Epigenetics · Molecular cytogenetics
Repetitive DNA

A. Fonsêca · A. Pedrosa-Harand (✉)
Laboratory of Plant Cytogenetics and Evolution,
Department of Botany, Federal University of
Pernambuco, Recife, PE, Brazil
e-mail: andrea.pedrosaharand@pesquisador.cnpq.br

M. Pérez de la Vega et al. (eds.), *The Common Bean Genome*,
Compendium of Plant Genomes, https://doi.org/10.1007/978-3-319-63526-2_3

3.1 Introduction

The genus *Phaseolus* L. includes around 75 species distributed in the New World between southwestern Canada and northern Argentina, with a center of diversity in Mexico, where more than 90% of the species are found (Mercado-Ruaro et al. 2009). The most prominent species of this genus are the economically important *Phaseolus vulgaris* L. (common bean), *Phaseolus lunatus* L. (lima bean or butter bean), *Phaseolus coccineus* L. (runner bean, scarlet runner bean or multiflora bean), *Phaseolus acutifolius* A. Gray (tepary bean), and *Phaseolus polyanthus* Grenm. (syn. *Phaseolus dumosus* Macfad.) (Gepts 1996).

3.2 Phylogenetic Relationships Among *Phaseolus* Species

The most recent phylogeny for *Phaseolus*, based on nuclear (ITS/5.8S) and plastidial (*trn*K) markers, was proposed by Delgado-Salinas et al. (2006) and indicated that the genus is monophyletic, with two principal clades (A and B). The clade A is composed of the Pauciflorus, Pedicellatus, and Tuerckheimii groups, as well as four other species that did not have a well-supported phylogenetic position (*Phaseolus glabellus* Piper, *Phaseolus macrolepis* Piper, *Phaseolus microcarpus* Mart, and *Phaseolus oaxacanus* Rose). The clade B, by contrast, encompasses all the cultivated species, including the Filiformis, Vulgaris, Lunatus, Leptostachyus, and Polystachios groups. While most of the groups have their own unique morphological characteristics, at least two—Pedicellatus and Tuerckheimii—can only be distinguished through molecular analyzes.

The phylogenetic relationships found within the genus are largely supported by biogeographic and morphological patterns (Delgado-Salinas et al. 2006). The clade A species are distributed primarily in Mexico and Panama between altitudes of 545 and 690 m above sea level, and do not occur on oceanic islands. These species flower only during the rainy season (except for *P. microcarpus*, which flowers in dry or humid conditions), are sensitive to habitat disturbance, and do not normally tolerate long periods of cold weather. The species of the B clade are more widely distributed, ranging between southwestern Canada and South America, at altitudes of between 654 and 737 m. These species may flower in either the dry or rainy seasons, are mostly not sensitive to environmental disturbances, and some are able to tolerate long periods at low temperatures (Delgado-Salinas et al. 2006).

3.3 Classic Cytogenetics

The first chromosomal studies of *Phaseolus* involved *P. acutifolius* A. Gray, *P. coccineus* L., *P. lunatus* L., and *P. vulgaris* L., all of which have $2n = 22$ (Karpetschenko 1925). Subsequent studies confirmed that all the species of the genus are diploid, with a basic chromosome number of $x = 11$, and predominantly metacentric and submetacentric chromosomes. One exception is the three species of the Leptostachyus group, *P. leptostachyus* Benth., *P. micranthus* Hook. & Arn., and *P. macvaughii* A. Delgado (Mercado-Ruaro and Delgado-Salinas 1996, 1998), which present $2n = 20$ (Table 3.1), forming a monophyletic clade.

Measurements of the DNA content of *Phaseolus* species have recorded relatively low values, ranging from 0.45 to 0.95 pg/1C (Table 3.1; Bennett and Leitch 2012). Considering the 30 species analyzed so far and that have been included in the phylogeny of the genus (Table 3.1), the DNA content found in different *Phaseolus* genotypes did not vary systematically in accordance with their phylogenetic relationships, although some groups may have relatively lower or higher mean DNA content, such as the Leptostachyus and the Pauciflorus groups,

Table 3.1 Chromosome numbers and DNA C-value of *Phaseolus* species grouped according to the main clades of the genus phylogeny (Delgado-Salinas et al. 2006)

Species	n^a	1C (pg)[b]
Clade A		
Pauciflorus group		
P. nelsonii Maré., Masch. & Stain	11	–
P. parvulus Greene	11	–
P. pauciflorus Sessé & Moc. ex G. Don.	11	0.73
P. pluriflorus Maréchal, Mascherpa & Stainer	11	0.95
Pedicellatus group		
P. esperanzae Seaton	11	0.60
P. grayanus Wooton & Standl.	11	0.95
P. neglectus Herm.	11	0.70
P. pedicellatus Benth.	11	0.58
Tuerkheimii group		
P. chiapasanus Piper	11	–
P. hintonii A. Delgado	11	0.68
P. oligospermus Piper	11	0.53
P. xanthotrichus Piper	11	0.60
Unresolved clade A species		
P. glabellus Piper	11	0.90
P. macrolepis Piper	–	0.65
P. microcarpus Mart.	11	0.52
Clade B		
Filiformis group		
P. angustisssimus A. Gray	11	0.66
P. filiformis Benth.	11	0.58
Leptostachyus group		
P. leptostachyus Benth.	10	0.53
P. macvaughii A. Delgado	10	0.45
P. micranthus Hook. & Arn.	10	0.60
Lunatus group		
P. augusti Harms	11	0.58
P. lunatus L.	11	0.70
P. pachyrrhizoides Harms	11	0.63
Polystachios group		
P. jaliscanus Piper	11	0.68
P. maculatus Scheele	11	0.65
P. marechalii A. Delgado	11	0.73
P. polystachios Britton, (L.) Sterns & Poggenb.	11	–
P. ritensis Jones	11	–
P. sonorensis Standl.	11	0.60
P. xolocotzii A. Delgado	11	0.68

Table 3.1 (continued)

Species	n^a	1C (pg)^b
Vulgaris group		
P. acutifolius A. Gray	11	0.75
P. coccineus L.	11	0.68
P. parviflorus G. Freytag	11	0.65
P. polyanthus Greenm.	11	0.73
P. vulgaris L.	11	0.60

[a]Haploid chromosome numbers from Mercado-Ruaro and Delgado-Salinas (1996, 1998)
[b]C-values from Bennett and Leitch (2012)

respectively. There has been no significant loss or gain in DNA content following domestication. Furthermore, Andean and Mesoamerican genomes are markedly similar, and intraspecific variation is minimal, except in the case of *Phaseolus xanthotrichus*, in which variation was reported to be 22% (Nagl and Treviranus 1995).

All *Phaseolus* species have small and morphologically similar chromosomes, which for a long time restricted cytogenetic analyzes to chromosome counts (Mercado-Ruaro and Delgado-Salinas 1996, 1998). In most cases, the differential staining of euchromatin and heterochromatin in small mitotic plant chromosomes is insufficient to distinguish chromosome pairs reliably (Ohmido et al. 2007). In *P. vulgaris*, for example, which has a genome of approximately 600 Mpb or 0.6–0.66 pg/1C (Table 3.1, Arumuganathan and Earle 1991; Bennett and Leitch 2012) and chromosomes varying between 1.5 and 3.0 μm (Sarbhoy 1978), the constitutive heterochromatin observed by C banding is ineffective for the differentiation of the chromosomes, given that blocks of similar intensity were found in the proximal and the terminal regions of most of the chromosomes (Mok and Mok 1976; Zheng 1991; Moscone et al. 1999).

Interspecific karyotypic comparisons using CMA (chromomycyn A3) and DAPI (4′-6-diamidino-2-phenylindole) fluorochromes not only confirmed the presence of pericentromeric blocks of heterochromatin in most chromosomes, but also characterized them as $CMA^+/DAPI^-$, or rich in GC (guanine and cytosine). Up until the present time, the available studies have focused predominantly on the species of the clade B, i.e., *P. vulgaris* (Fonsêca et al. 2010), *P. lunatus* (Bonifácio et al. 2012), *P. leptostachyus* (Fonsêca et al. 2015), and *P. macvaughii* (Frerraz et al. unpublished data), but included a single clade A species, *P. microcarpus* (Fonsêca and Pedrosa-Harand 2013). In addition to these pericentromeric blocks, all *Phaseolus* species have at least one pair of chromosomes with terminal CMA^+ bands, always colocalized to the 35S ribosomal DNA (rDNA) sites. Despite the overall homogeneity of the banding pattern observed in the genus, the weak CMA^+ pericentromeric banding found in only six pairs of chromosomes in *P. microcarpus* (clade A) supports the separation of this species from those of the clade B, all of which present intense pericentromeric banding on most of their chromosomes (Fig. 3.1a).

3.4 Molecular Cytogenetics

When a given DNA sequence is labeled and used as a probe, it will hybridize to the corresponding sequence found on a chromosome and can be detected by microscopy, indicating its location in the karyotype. This technique, fluorescent in situ hybridization (FISH), is one of the most important diagnostic techniques in plant cytogenetics. The fluorescent detection is widely used in present-day cytogenetic studies due to its sensitivity, and also permits the use of multiple probes of different colors, without the need for radioactive markers, as used in the early studies (Guerra 2004).

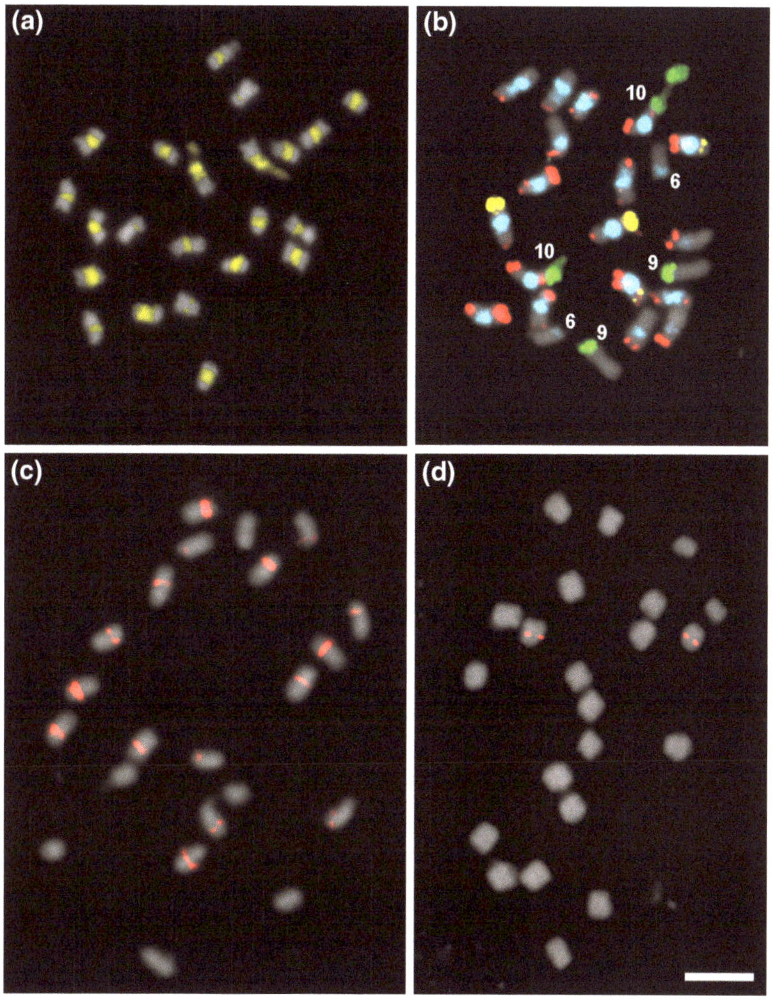

Fig. 3.1 Heterochromatin characterization and distinct in situ hybridization patterns in *Phaseolus* species. **a** CMA/DAPI banding pattern of *P. lunatus* Vermelhinha cultivar (GL0135) chromosomes. **b** In situ locatization of 35S rDNA (green), BACs 12M3 (blue, pericentromeric), 63H6 (red, subtelomeric), and 255F18 (yellow) on *P. vulgaris* BAT93 cultivar, evidencing the major heterochromatic regions of the species. NOR-bearing chromosomes 6, 9, and 10 are indicated. The minor 35S rDNA site on chromosome 6 is not seen in this image, but this chromosome pair can be recognized by its small size and acrocentric morphology. **c** Nazca satellite DNA sequence predominant in centromeric regions of eight chromosome pairs of *P. vulgaris* BAT93 cultivar. **d** In situ locatization of the single-copy BAC 163I7 on chromosome pair 7 of *P. vulgaris* BAT93 cultivar. Chromosomes are counterstained with DAPI and visualized in gray. Scale bar = 5 μm

The initial identification and characterization of the karyotypes of four *Phaseolus* species, including two *P. vulgaris* cultivars, were based on banding techniques associated with FISH (Moscone et al. 1999). The locations of the 5S and 18S-5.8S-25S rRNA genes (35S rDNA) differed among all samples, including the two *P. vulgaris* cultivars analyzed (‘Wax’ and ‘Saxa’). While a certain degree of similarity was found in their distribution, five 35S rDNA sites were observed in ‘Saxa,’ and seven in ‘Wax.’ At the interspecific level, by contrast, *P. vulgaris* and *P. coccineus* were very similar in their heterochromatic banding. In both species, there were close similarities in the location of the rDNA on two chromosomes pairs, indicating homoeology between them. A probable homoeology was also found in a pair of chromosomes in *P. lunatus* and *P. acutifolius* (Moscone et al. 1999).

In addition to the ribosomal DNA, other sequences found in the *Phaseolus* genome are repeated hundreds or even thousands of times. Repeat sequences evolve more rapidly in comparison with single-copy or low-copy coding sequences, primarily because they are not subject to strong selective pressures, and repeat sequence number is the principal factor responsible for the variation in the size of plant genomes (Schmidt and Heslop-Harrison 1998; Meyers et al. 2001). These sequences are typically found in two types—in tandem repeat sequences, such as microsatellites, minisatellites, and satellite DNA (e.g., rDNA) or dispersed repetitive sequences, usually transposable elements (Schweizer et al. 1990). Satellite DNA sequences are generally associated with rDNA sites and the heterochromatic centromeric and subtelomeric regions, while the dispersed repetitive sequences tend to be found throughout the genome, or enriched in specific chromosome regions, such as the pericentromeres (Jelinek and Schmid 1982; Guerra 2004).

In a study of 37 Andean and Mesoamerican accessions of *P. vulgaris*, Pedrosa et al. (2006) found that the number and positions of the 5S rDNA sites were highly conserved among accessions, whereas the 35S rDNA varied significantly within the species. This variation was not associated with domestication. While the Andean accessions, wild or domesticated, presented six to nine sites, the Mesoamerican ones had only three or four per haploid genome. This variation is probably the result of ectopic amplifications, deletions, and recombinations that occurred independently over the course of their evolutionary history. The variation found within the same accession indicates that these structural alterations continue occurring in the species.

As in *P. vulgaris*, a pair of 35S rDNA was observed on chromosome 6 of *P. lunatus* and *P. microcarpus* (Almeida and Pedrosa-Harand 2010; Fonsêca and Pedrosa-Harand 2013). While major sites were also found on chromosome 9 and 10 of the BAT93 cultivar (Fig. 3.1b), the available data indicate that the 35S rDNA site on chromosome 6 was probably retained from the ancestral karyotype of the genus. The origin of additional sites on other chromosomes appears to have been the result of events restricted to a clade of the Vulgaris group, which includes *P. coccineus* that also has three 35S rDNA sites (Moscone et al. 1999). By contrast, *P. acutifolius*, which is found in the other clade of this group, presents only one site (Moscone et al. 1999), presumably the conserved one on chromosome 6.

Repetitive pericentromeric sequences correspond to approximately 34% of the *P. vulgaris* genome and have been mapped, in different amounts, on all the chromosome pairs of the species (Fig. 3.1b; Fonsêca et al. 2010). Despite the relatively conserved nature of these sequences in *P. lunatus* and *P. microcarpus*, as judged by the cross-hybridization of the common bean probes, some of the pericentromeric probes did not reveal evidence of hybridization, indicating that the composition of the repetitive elements that make up the pericentromeres may vary among species (Fonsêca and Pedrosa-Harand 2013). The composition of the centromeric sequences of *P. vulgaris* has been investigated and centromeres are known to be composed of the CentPv1 (also known as Nazca) and CentPv2 satellite sequences, which have evolved independently, and are located on two distinct sets of chromosomes (Fig. 3.1c; Iwata et al. 2013). Nevertheless, there are still no data on their conservation in the different species or their evolution within the genus.

The CC4 satellite DNA of *P. vulgaris*, which is similar to the intergenic spacer sequence (IGS-like) of the 35S rDNA, is the only pericentromeric sequence that has been studied in detail in *Phaseolus*. Analyzes have revealed reduced similarity between the CC4 and the IGS sequences of *P. vulgaris* and *P. coccineus* and indicated the presence of pericentromeric sites in two to four pairs of chromosomes. It is interesting to note that these sequences hybridized only with the IGS sequences of the 35S rDNA in *Phaseolus* species that are not part of the Vulgaris group, and even with species of other genera. This has led to the conclusion that the 35S rDNA diverged and became homogenized by concerted evolution as a distinct CC4 sequence in the Vulgaris group, in which CC4 has become an independent satellite (Almeida et al. 2012).

The occurrence of repetitive sequences in the subtelomeric regions of almost all the *P. vulgaris* chromosomes confirmed the presence of a subterminal heterochromatin, which is nevertheless not as abundant as the pericentromeric heterochromatin (Pedrosa-Harand et al. 2009; Fonsêca et al. 2010). In addition to the 35S rDNA sites, this pattern is related to the presence of a satellite DNA sequence denominated *khipu* (with a 528 bp repeat unit, Fig. 3.1b), which was identified following the sequencing of the principal cluster of disease resistance genes in the species, located on the extremity of the short arm of chromosome 4 (David et al. 2009). The plant telomeric sequence (TTTAGGG) was found at the extremity of all the chromosomes, and in *P. microcarpus*, it was also observed in the interstitial region, representing the first evidence of a paracentric inversion of chromosome 3 in this species (Fonsêca and Pedrosa-Harand 2013).

One other repetitive sequence, denominated *Pv*Meso, has been isolated and characterized in *Phaseolus*, where it is restricted to the subterminal region of chromosome 7 in the Mesoamerican accessions of *P. vulgaris* (Fig. 3.1b). The data indicated that a repeat of chromosome 11 was amplified in chromosome 7 after the isolation of the Andean and Mesoamerican gene pools (Ribeiro et al. 2011).

3.5 Chromatin and Epigenetics

The cytogenetic studies of the genus *Phaseolus* have shown a marked tendency for the distribution of the repetitive sequences in two distinct locations, the pericentromeric and subtelomeric regions. These regions are typically characterized by constitutive heterochromatin, which is condensed over most of the cell cycle. The C-banding is the main technique used in the identification of that heterochromatin, based on differential staining of dark bands after fragmentation and removal of chromosomal DNA. In *P. vulgaris*, the pericentromeric and subtelomeric regions are associated with the post-FISH $DAPI^+$ bands (Fonsêca et al. 2010), which are equivalent to the C-bands, and are intimately associated with an increased degree of chromosome compaction in mitotic prophase and meiotic pachytene chromosomes. Chromosomal DNA is associated with histone and non-histone proteins, in the highly organized structure known as chromatin. The euchromatin fraction is typically associated with uncondensed regions of higher gene expression, with single-copy or moderately repetitive sequences. The heterochromatin, in turn, is associated with the more condensed chromosomal regions, rich in repetitive sequences, and detected as C-bands. These different properties of the chromatin are associated primarily with distinct levels of the covalent modifications of the histones and methylation of the DNA (Pfluger and Wagner 2007; Zemach and Grafi 2007).

The acetylation of histones affects the interactions between the DNA and the rest of the nucleosome components (Luger et al. 1997), influencing transcription, the cell cycle, and other functions (Spencer and Davie 1999). In contrast with mammals, the acetylated isoforms of H3 and H4 do not always coincide in plants, although these acetylated forms, such as the acetylation of lysine 5 of the H4 (H4K5ac), are rare in the heterochromatin (Fuchs and Schubert 2012). However, the mono-, di-, or trimethylation of lysines 9 and 27 in the H3 histones (H3K9me1/2, H3K27me1/2/3) and of lysine 20 in the H4 histones (H4K20me1/2/3), as well as the hypermethylation of the cytosines of the CG and CHG DNA sites (5mC), have generally been associated with the heterochromatin and gene silencing (Sumner 2003; Dhar et al. 2009; Marques et al. 2011).

Few data are available on the modified histones or methylated DNA of plants, and are rare for the Fabaceae, although the pioneering study of Frediani et al. (1986) investigated the constitution of the heterochromatin in *P. coccineus*, combining data on the methylation of the DNA, C-banding, and in situ hybridization. The results indicated that there is no direct relationship between the presence of constitutive heterochromatin or repetitive sequences and the methylation pattern of the DNA. In at least three chromosomes, methylation did not occur in all the regions of the heterochromatin. Some of the unmethylated bands of heterochromatin are made up of repetitive sequences, as observed in the interstitial band of the long arm of chromosome 6 (Frediani et al. 1986). A similar situation has been observed in the mitotic chromosomes of *Vicia faba*, in which many of the regions of methylated cytosine are not related to the heterochromatin, in particular in the telomeric and subtelomeric regions (Frediani et al. 1996). The pattern of methylation also varied between homologous chromosomes. In pairs 3, 4, and 6, for example, bands present on one chromosome were often absent from its homologue. This differentiation may be related to either the accessibility of the antibody or to differential gene silencing (Frediani et al. 1996).

Fonsêca and Pedrosa-Harand (2014) recently found that the condensation patterns of *P. vulgaris* chromatin were related to repetitive sequences and epigenetic modifications. The epigenetic markers for the modification of the histones and the DNA revealed that H3K4me3 and H4K5ac, in general, were associated with the euchromatic regions, while H3K27me1, H3K9me2, and 5mC were associated with the heterochromatic regions. However, the 35S rDNA sites, centromeric regions, and most of the terminal blocks of heterochromatin were hypomethylated, including the block associated with a cluster of genes responsible for resistance to anthracnose. No association was found between the regulation of the activity of these clusters of resistance genes and the modifications of the chromatin within the chromosomes. Overall, then, the different heterochromatic domains varied in their epigenetic patterns, emphasizing the complexity, and heterogeneity of the heterochromatin in this species (Fonsêca and Pedrosa-Harand 2014).

3.6 Integrated Genetic and Cytogenetic Maps

The first cytogenetic map established for the common bean used as FISH probes RFLP markers, known as *Bngs* (Vallejos et al. 1992), which were amplified by PCR using the cloning vector primers. Markers located close to one another in the genetic map were labeled together and permitted the first identification of the 11 chromosome pairs of *P. vulgaris*, contributing to the production of the first chromosome map of the species (Pedrosa et al. 2003). However, as each chromosome-specific probe was made up of a set of unique sequences distributed in a given chromosome segment, it was not possible to compare physical or genetic distances along multiple points on the chromosome.

Bacterial artificial chromosomes (BACs) are widely used as chromosome-specific probes in FISH studies. This is because, in comparison with other specific probes (such as plasmid inserts or PCR products), they are easily visualized because of the strength of their signal. This is due to their capacity to store inserts of hundreds of kilobases. In addition to having low levels of chimerism and high cloning efficiency, they are relatively easy to isolate and manipulate (Song et al. 2000). Furthermore, the use of BACs for cytogenetic mapping permits the association to contig physical maps and genomic data, as well as comparative analyzes within and between genera.

A number of BAC libraries have become available for the common bean in recent years, making it possible to build a more complete map, using BAC clones selected with genetically mapped markers as probes (Fig. 3.1d; Pedrosa-Harand et al. 2009; Fonsêca et al. 2010). Forty-three anchor points were established between the genetic map and the cytogenetic map, relating the genetic distances in cM (centimorgans) to the physical distances in base pairs (Pedrosa-Harand et al. 2009; Fonsêca et al. 2010). These analyzes revealed low levels of recombination over large pericentromeric regions, in which single-copy and repetitive sequences were intermingled. In addition, all 11 chromosome pairs of *P. vulgaris* BAT93 cultivar could be identified using only four hybridized BACs, three of which containing repetitive sequences (Fonsêca et al. 2010).

3.7 Comparative Cytogenetics

Once the cytogenetic map for the common bean was developed using BAC-FISH, these same BACs could be used in other *Phaseolus* species and closely related genera in order to infer the chromosomal synteny and collinearity in species for which no genomic data existed. The cross-hybridization of single-copy BACs has provided important evidence on events of chromosome evolution occurring at different taxonomic levels (Lysak et al. 2006; Szinay et al. 2012).

The BAC markers established for *P. vulgaris* BAT93 cultivar were used as probes for comparative studies in *P. lunatus* (clade B) and *P. microcarpus* (clade A), as well as G19833, another accession of the common bean which had its genome sequenced (Altrock et al. 2011; Bonifácio et al. 2012; Fonsêca and Pedrosa-Harand 2013). These studies found a complete synteny among the three karyotypes, albeit with a slight loss of colinearity. At least five chromosomal inversions were identified among species, on chromosomes 2, 3, 6, 9, and 10 (Bonifácio et al. 2012; Fonsêca and Pedrosa-Harand 2013). It was possible to interpret the types of inversion through a combination of chromosome morphology and the location of the chromosome-specific markers. While the inversion of chromosome 6 was paracentric, in all the other chromosomes (2, 3, 9, and 10) it was pericentric (Fig. 3.2; Bonifácio et al. 2012; Fonsêca and Pedrosa-Harand 2013).

In addition to these inversion events, changes in the location of the rDNA sites and the distribution of the pericentromeric and subtelomeric sequences provided evidence for the principal alterations that have occurred during the chromosomal evolution of the genus. When considered in a phylogenetic context (Delgado-Salinas et al. 2006), it is possible to recognize chromosomal alterations that are exclusive to certain lineages (karyotypic autapomorphies), with a majority of events being observed in the *P. vulgaris* lineage (Fig. 3.2; Fonsêca and Pedrosa-Harand 2013). Despite the predominance of chromosome inversions observed so far, Fonsêca et al. (2015) have now identified a nested chromosome fusion (NCF) in *P. leptostachyus*, which is associated with a process of descending dysploidy (decrease in haploid chromosome number between related species) in the Leptostachyus group, characterized by its unique chromosome number $2n = 20$. Furthermore, several translocations, so far not detected in the genus, have also been identified in this species.

3.8 Conclusions and Final Considerations

Advances in molecular cytogenetics, combined with the available tools for the common bean, have contributed to the status of *Phaseolus* as one of the best-studied plant genera in cytogenetic terms. The sequencing of its genome (Andean and Mesoamerican) will permit even more detailed cytogenomic analyzes, based on the identification of new sequences, mapped chromosomally, and associated with modifications of the chromatin, which may contribute to a better understanding of the functional features of

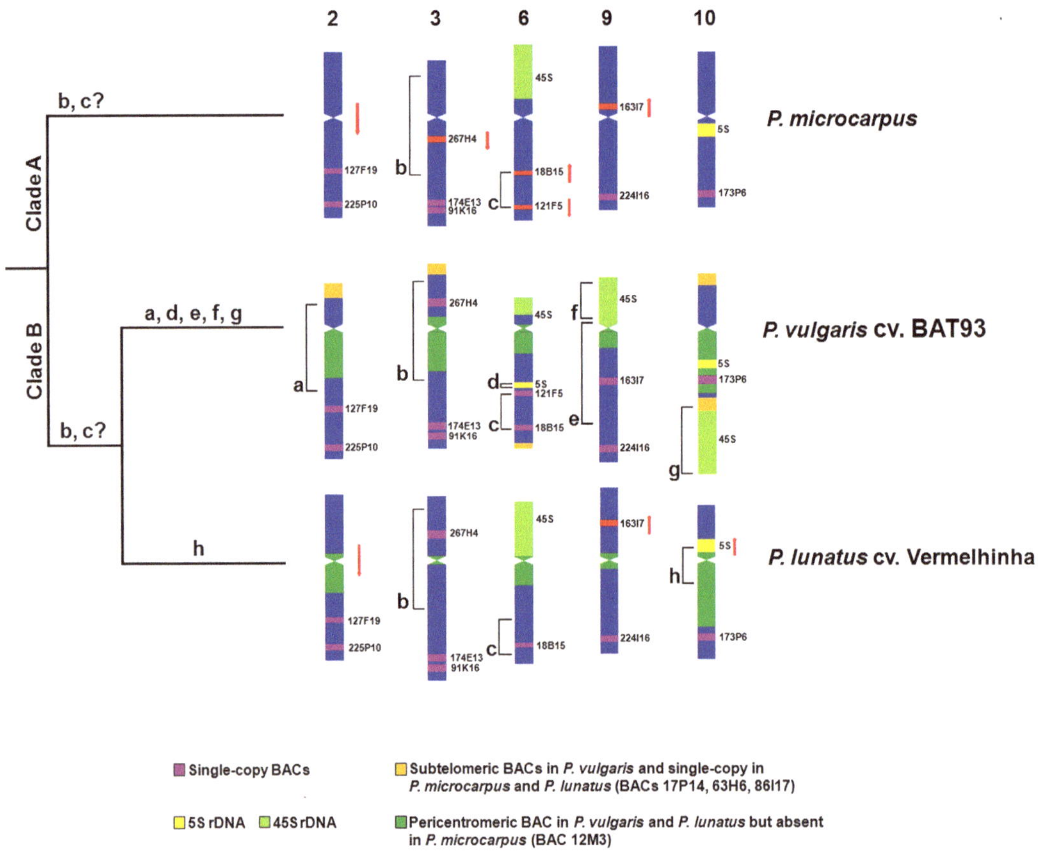

Fig. 3.2 Schematic comparative analysis of five chromosomes of *Phaseolus microcarpus* (clade A) and *Phaseolus lunatus* (clade B) involved in rearrangements in relation to *Phaseolus vulgaris* (clade B) considering the proposed phylogeny for the genus (modified from Delgado-Salinas et al. 2006). Blocks in red indicate the BACs involved in rearrangements and the arrows indicate the changes in their positions or of centromeres in relation to *P. vulgaris*. Letters a–j suggest putative synapomorphic events (indicated by brackets) for each clade represented, except for b and c, which depend on an outside group to its positioning between clades A and B. Reproduced with permission from Fonsêca and Pedrosa-Harand (2013) Karyotype stability in the genus *Phaseolus* evidenced by the comparative mapping of the wild species *Phaseolus microcarpus*. Genome 56: 335–343

the chromosomes. These cytogenetic analyzes may easily be extended to species of lesser commercial value, wild forms, and even related genera (Vasconcelos et al. 2015). This will permit a more systematic understanding of the chromosomal evolution of the genus *Phaseolus* and may eventually contribute to the more effective exploitation of many species.

Acknowledgements We thank the CIAT and the Embrapa Recursos Genéticos e Biotecnologia seed banks, particularly A. Torres, M. Wetzel, and L. Gonçalves Pereira Neto, for providing *Phaseolus* seeds. A.P.-H. thanks the Conselho Nacional de Desenvolvimento Científico e Tecnológico (CNPq) and A.F. the Fundação de Amparo à Ciência e Tecnologia de Pernambuco (FACEPE), Brazil, for financial support.

References

Almeida C, Pedrosa-Harand A (2010) Contrasting rDNA evolution in lima bean (*Phaseolus lunatus* L.) and common bean (*P. vulgaris* L., Fabaceae). Cytogenet Genome Res 132:212–217

Almeida C, Fonsêca A, dos Santos KGB et al (2012) Contrasting evolution of a satellite DNA and its ancestral IGS rDNA in *Phaseolus* (Fabaceae). Genome 55:683–689

Almeida C, Pedrosa-Harand A (2013) High macro-collinearity between lima bean (*Phaseolus lunatus* L.) and the common bean (*P. vulgaris* L.) as revealed by comparative cytogenetic mapping. Theor Appl Genet 126:1909–1916

Altrock S, Fonsêca A, Pedrosa-Harand A (2011) Chromosome identification in the Andean common bean accession G19833 (*Phaseolus vulgaris* L., Fabaceae). Genet Mol Biol 34:459–463

Arumuganathan K, Earle ED (1991) Nuclear DNA content of some important plant species. Plant Mol Biol Rep 9:208–218

Bennett MD, Leitch IJ (2012) Plant DNA C-values database (release 6.0, December 2012). http://data.kew.org/cvalues/. Accessed 14 Feb 2015

Bonifácio EM, Fonsêca A, Almeida C et al (2012) Comparative cytogenetic mapping between the lima bean (*Phaseolus lunatus* L.) and the common bean (*P. vulgaris* L.). Theor Appl Genet 124:1513–1520

David P, Chen NW, Pedrosa-Harand A et al (2009) A nomadic subtelomeric disease resistance gene cluster in common bean. Plant Physiol 151:1048–1065

Delgado-Salinas A, Bibler R, Lavin M (2006) Phylogeny of the genus *Phaseolus* (Leguminosae): a recent diversification in an ancient landscape. Syst Botany 31:779–791

Dhar MK, Fuchs J, Houben A (2009) Distribution of Eu-and heterochromatin in *Plantago ovata*. Cytogenet Genome Res 125:235–240

Fonseca A, Ferraz ME, Pedrosa-Harand A (2015) Speeding up chromosome evolution in *Phaseolus*: multiple rearrangements associated with a one-step descending dysploidy. Chromosoma 125(3):413–421. doi:10.1007/s00412-015-0548-3

Fonsêca A, Ferreira J, dos Santos TRB et al (2010) Cytogenetic map of common bean (*Phaseolus vulgaris* L.). Chromosome Res 18:487–502

Fonsêca A, Pedrosa-Harand A (2013) Karyotype stability in the genus *Phaseolus* evidenced by the comparative mapping of the wild species *Phaseolus microcarpus*. Genome 56:335–343

Fonsêca A, Richard MM, Geffroy V, Pedrosa-Harand A (2014) Epigenetic analyses and the distribution of repetitive DNA and resistance genes reveal the complexity of common bean (*Phaseolus vulgaris* L., Fabaceae) heterochromatin. Cytogenet Genome Res 143:168–178

Frediani M, Tagliasacchi AM, Durante M, Avanzi S (1986) Distribution of 5-methylcytosine-rich regions in the polytene chromosomes of *Phaseolus coccineus* embryo suspensor as shown by the immunoperoxidase technique. Exp Cell Res 167:337–342

Frediani M, Giraldi E, Castiglione MR (1996) Distribution of 5-methylcytosine-rich regions in the metaphase chromosomes of *Vicia faba*. Chromosome Res 4:141–146

Fuchs J, Schubert I (2012) Chromosomal distribution and functional interpretation of epigenetic histone marks in plants. Plant Cytogenet. Springer, New York, pp 231–253

Gepts P (1996) Origin and evolution of cultivated *Phaseolus* species. In: Pickersgill B, Lock JM (eds) Advances in legume systematics 8: Legumes of economic importance. Royal Gardens, Kew, pp 65–74

Guerra M (2004) Hibridização in situ: princípios básicos. In: Conceitos e aplicações na citogenética. Sociedade Brasileira de Genética. 1[a] edição, Ribeirão Preto, pp 1–32

Iwata A, Tek AL, Richard M et al (2013) Identification and characterization of functional centromeres of the common bean. Plant J 76:47–60

Jelinek WR, Schmid CW (1982) Repetitive sequences in eukaryotic DNA and their expression. Annu Rev Biochem 51:813–844

Karpechenko GD (1925) On the chromosomes of the Phaseolineae. Bull Appl Bot 14:143–148

Luger K, Mäder AW, Richmond RK et al (1997) Crystal structure of the nucleosome core particle at 2.8 Å resolution. Nature 389:251–260

Lysak MA, Berr A, Pecinka A et al (2006) Mechanisms of chromosome number reduction in *Arabidopsis thaliana* and related Brassicaceae species. Proc Natl Acad Sci USA 103:5224–5229

Marques A, Fuchs J, Ma L et al (2011) Characterization of eu- and heterochromatin of *Citrus* with a focus on the condensation behavior of 45S rDNA chromatin. Cytogenet Genome Res 134:72–82

Mercado-Ruaro P, Delgado-Salinas A (1996) Karyological studies in several Mexican species of *Phaseolus* L. and *Vigna* Savi (Phaseolinae, Fabaceae). Adv Legume Syst 8:83–87

Mercado-Ruaro P, Delgado-Salinas A (1998) Karyotypic studies on species of *Phaseolus* (Fabaceae: Phaseolinae). Am J Bot 85:1–9

Mercado-Ruaro P, Delgado-Salinas A, Chiang F (2009) Taxonomic re-assessment of *Phaseolus dasycarpus* (Leguminosae): Systematic position, chromosome studies and re-description. Brittonia 61:8–13

Meyers BC, Tingey SV, Morgante M (2001) Abundance, distribution, and transcriptional activity of repetitive elements in the maize genome. Genome Res 11:1660–1676

Mok DWS, Mok MC (1976) A modified Giemsa technique for identifying bean chromosomes. J Hered 67:187–188

Moscone EA, Klein F, Lambrou et al (1999) Quantitative karyotyping and dual-color FISH mapping of 5S and 18S-25S rDNA probes in the cultivated *Phaseolus* species (Leguminosae). Genome 42:1224–1233

Nagl W, Treviranus A (1995) A flow cytometric analysis of the nuclear 2C DNA content in 17 *Phaseolus* species (53 genotypes). Bot Acta 108:403–406

Ohmido N, Sato S, Tabata S, Fukui K (2007) Chromosome maps of legumes. Chromosome Res 15:97–103

Pedrosa A, Vallejos C, Bachmair A, Schweizer D (2003) Integration of common bean (*Phaseolus vulgaris* L.) linkage and chromosomal maps. Theor Appl Genet 106:205–212

Pedrosa-Harand A, de Almeida CCS, Mosiolek M et al (2006) Extensive ribosomal DNA amplification during

Andean common bean (*Phaseolus vulgaris* L.) evolution. Theor Appl Genet 112:924–933

Pedrosa-Harand A, Kami J, Gepts P et al (2009) Cytogenetic mapping of common bean chromosomes reveals a less compartmentalized small-genome plant species. Chromosome Res 17:405–417

Pfluger J, Wagner D (2007) Histone modifications and dynamic regulation of genome accessibility in plants. Curr Opin Plant Biol 10:645–652

Ribeiro T, dos Santos KG, Fonsêca A, Pedrosa-Harand A (2011) Isolation and characterization of a new repetitive DNA family recently amplified in the Mesoamerican gene pool of the common bean (*Phaseolus vulgaris* L., Fabaceae). Genetica 139:1135–1142

Sarbhoy RK (1978) Cytogenetical studies in the genus *Phaseolus* Linn.: 1 and 2. somatic and meiotic studies in fifteen species of *Phaseolus*: part 1. Cytologia 43:161–170

Schmidt T, Heslop-Harrison JS (1998) Genomes, genes and junk: the large-scale organization of plant chromosomes. Trends Plant Sci 3:195–199

Schweizer D, Strehl S, Hagemann S (1990) Plant repetitive DNA elements and chromosome structure. Chromosomes Today 10:33–113

Song J, Dong F, Jiang J (2000) Construction of a bacterial artificial chromosome (BAC) library for potato molecular cytogenetics research. Genome 43:199–204

Spencer VA, Davie JR (1999) Role of covalent modifications of histones in regulating gene expression. Gene 240:1–12

Sumner AT (2003) Chromosomes: organization and function. Blackwell Publishing Company, Berlin, p 287

Szinay D, Wijnker E, van den Berg et al (2012) Chromosome evolution in *Solanum* traced by cross-species BAC-FISH. New Phytol 195:688–698

Vallejos CE, Sakiyama NS, Chase CD (1992) A molecular marker-based linkage map of *Phaseolus vulgaris* L. Genetics 131:733–740

Vasconcelos EV, de Andrade Fonsêca AF, Pedrosa-Harand, A et al (2015) Intra-and interchromosomal rearrangements between cowpea [*Vigna unguiculata* (L.) Walp.] and common bean (*Phaseolus vulgaris* L.) revealed by BAC-FISH. Chromosome Res, 1–14

Zemach A, Grafi G (2007) Methyl-CpG-binding domain proteins in plants: interpreters of DNA methylation. Trends Plant Sci 12:80–85

Zheng JY (1991) Giemsa C-banding patterns in several species of *Phaseolus* L. and *Vigna* Savi, Fabaceae. Cytologia 56:459–466

Genetic Mapping and QTL Analysis in Common Bean

4

Ana M. González, Fernando J. Yuste-Lisbona, Antonia Fernández-Lozano, Rafael Lozano and Marta Santalla

Abstract

Common bean (*Phaseolus vulgaris* L.) is the most important legume for direct human consumption and a well-studied crop species in terms of genetics, genomics and breeding. Genome maps are important tools that are an integral part of genetic resource conservation and breeding programmes. Several maps have been developed or are being developed in common bean. Different types of molecular markers such as RFLP, AFLP, SSR, CAPS, RGA and EST have been developed and mapped onto the 11 common bean chromosomes. Markers have been used extensively for identification and mapping of genes and QTL for many biologically and agriculturally important traits, including disease resistance genes, photoperiod sensitivity, growth habit, pod size, seed weight, pigmentation, phenology and abiotic stress tolerance, and occasionally for germplasm screening, fingerprinting and marker-assisted breeding. MAS has been employed mainly for improving simply inherited traits and not much for improving complex traits. The utility of MAS in common bean breeding has been restricted largely due to inaccurate estimation of main QTL, epistatic and QTL × environmental interaction effects. GWAS has also proved to be a powerful tool for investigating complex traits and developing new markers for breeding. The huge amount of sequence information available for common bean via whole-genome sequencing projects facilitates in the next years the development of a rapid and

A. M. González · M. Santalla (✉)
Grupo de Biología de Agrosistemas, Misión Biológica de Galicia-CSIC, PO. Box 28, 36080 Pontevedra, Spain
e-mail: msantalla@mbg.csic.es

F. J. Yuste-Lisbona · A. Fernández-Lozano · R. Lozano
Centro de Investigación En Biotecnología Agroalimentaria (BITAL), Universidad de Almería, 04120 Almería, Spain

M. Pérez de la Vega et al. (eds.), *The Common Bean Genome*,
Compendium of Plant Genomes, https://doi.org/10.1007/978-3-319-63526-2_4

cost-effective generation of high-density functional maps, which could also lead to the direct gene tagging for QTL mapping of important agronomic traits, improving the efficiency of common bean breeding programmes via MAS.

Keywords
Phaseolus vulgaris · Quantitative trait locus · Epistasis · Diseases
Abiotic stress · Agronomic traits · Genome-wide association

4.1 Introduction

Continuous progress made in the last two decades on phenotypic and DNA marker analyses has provided a set of useful tools both for genetic research and for plant breeding. They have also led to the construction of genetic linkage maps for most of the crop species, particularly for legume species such as common bean (*Phaseolus vulgaris* L.), where genes involved in valuable traits have been located. Currently, selectable DNA marker development entails the main goal for most public research institutions and private companies working on plant breeding. Mapping DNA markers does allow not only for an efficient genotype selection but also for the detection of quantitative trait loci (QTL) for interesting traits. QTL analysis combines linkage analysis and molecular and statistical genetics, providing consistent information about the chromosome regions contributing to the variance and the inheritance pattern of such traits. This basic genetic information can be used by plant breeders to accelerate introgression of desirable traits and to manage environmental interactions.

The genome sequence of the common bean has already been published from the results of two sequencing consortia, which have focused their works on 'G19833' and 'BAT93' genotypes representing the Andean (Schmutz et al. 2014) and Mesoamerican (Vlasova et al. 2016) gene pools, respectively. In addition, a Canadian consortium has also sequenced a third genotype ('OAC-Rex9') and the genome information is also available (http://www.beangenomics.ca/research/projects/view/draft-genome-sequence-for-common-bean-i-p-vulgaris-i/). Hopefully, the integration of linkage maps, QTL and genomic tools will be essential not only for the development of more accurate tools useful for genomics-assisted breeding, but also for the map-based cloning approaches devoted to the isolation of genes controlling important traits.

In this review, genetic mapping approaches performed in common bean are summarized, from the more classical ones to the more recent maps based on molecular and genomic data. Similarly, the more relevant contributions of QTL analysis are also reviewed despite the elevated number of quality reports recently published on this topic. Particular attention has been paid to epistatic and environmental interactions as the single-locus QTL only reveal part of the genetic determinants underlying phenotypic variance. Finally, a section is devoted to mapping results from genome-wide association study (GWAS) as it provides an alternative to linkage mapping for the dissection of complex traits.

4.2 The Beginnings of Genetic Mapping in Common Bean

The first genetic analysis of common bean was conducted by Gregor Mendel in the mid-nineteenth century (Mendel 1866). It was performed in a progeny from *P. vulgaris* and *Pseudomys nanus* (= *P. vulgaris*, bush type) and was aimed to corroborate results on the inheritance of growth habit, pod colour and shape that Mendel had obtained in pea (*Pisum sativum* L.). Later, Shaw and Norton (1918) used intraspecific crosses and determined that pigmentation and pigmentation patterns of the seed coat were controlled by multiple independent factors. The first report of a linkage in common bean was performed by Tjebbes and Kooiman (1921), who reported the so-called non-constant mottling, which is due to the tight repulsion linkage between the *B* gene (now *C*, a colour factor) and the *S* gene (now *M*, a cis-acting factor of the *C* locus). A few years later, Sax (1923) began to identify the multiple components that determine the inheritance of pattern and colour of the seed. Indeed, metabolic control of seed colour in common bean was one of the first QTL to be identified. Differences for seed weight were associated with one or both factors that determine the pattern and colour of the seed. Later, an association between seed weight and phaseolin protein type was observed on the linkage group (LG) 07 (Johnson et al. 1996). Vallejos and Chase (1991) reported a linkage between isozyme loci *Adh-1* and *Got2* and seed size; Weeden and Liang (1985) also observed an association between isozyme loci *EST-2* and white flower. The second half of the twentieth century provided much more evidence for genetic linkage affecting a wide variety of traits. Among others, the *I* allele, which confers resistance to all known strains of the bean common mosaic virus (BCMV) and four related potyviruses, was found to be linked to seed coat (Temple and Morales 1986; Kyle and Dickson 1988) and hilum-region-darkening allele *B* (Park and Tu 1986). A genetic linkage of coloured seed coat to resistance to *Pythium* and/or *Rhizoctonia* root rots was also reported (Dickson and Petzoldt 1986). Likewise, genetic linkage was described between maturity and indeterminate growth habit by Valladares-Sánchez et al. (1979), among genes for rust resistance by Stavely (1984) and between arcelin and lectin genes by Osborn et al. (1986). Since then, tagging of many other traits with molecular markers has been reported.

Lamprecht (1961) published the first genetic linkage map for common bean, which consisted mainly of morphological markers distributed over eight LGs. This linkage map was rudimentary, with many loci that could only be reliably evaluated in a limited number of populations or which were subject to epistasis. Lamprecht's map was extended with additional isozymes, seed proteins and induced mutations (Bassett 1988; Gepts 1988; Koenig et al. 1990; Vallejos and Chase 1991). These classical maps showed a reduced genomic coverage and scarce usefulness for marker-assisted selection (MAS), but they provided a point of reference for subsequently developed DNA-based linkage maps.

A problem encountered in establishing the classical maps was the use of different gene symbols for the same gene by different researchers (Bassett 1991). A subcommittee of the *Phaseolus* Genetics Committee addressed this lack of coordination among geneticists and formulated guidelines for gene designation and nomenclature (Myers and Bassett 1993; Bassett and Myers 1999). As a result, an updated list of genes for *P. vulgaris* was published (Bassett 2004). Another problem was that many previously described mutants could not be tested due to the lack of a seed source. To solve this drawback, in 1987, the *Phaseolus* Genetics Committee (Gepts 1988) advocated for a repository of genetic stocks that M Bassett established and which is currently maintained by the USDA-ARS National Plant Germplasm System (NPGS, Pullman, WA, USA).

4.3 Development of DNA-Based Linkage Maps in Common Bean

The progress of genetic linkage mapping in common bean is closely related to the development of different generations of molecular markers. Random DNA markers such as Restriction Fragments Length Polymorphism (RFLP), Random Amplification Polymorphic DNA (RAPD), Amplified Fragments Length Polymorphism (AFLP) or Simple Sequence

Repeat (SSR) have been the basis for most of the common bean genetic maps currently published. However, the increasing availability of high-throughput molecular marker technology and the reduction of costs of marker development and genotyping technologies have provided a wealth of sequence information. Thus, the incessant evolution in genomic research is driving a trend away from random DNA markers towards those called functional markers; whereas the former are derived at random from polymorphic sites in the genome, functional markers are specifically developed from the transcribed genomic regions (Andersen and Lubberstedt 2003). Thereby, throughout this section, common bean genetic maps are described as first- or second-generation genetic maps according to the type of molecular marker used for linkage mapping (i.e. random DNA or functional markers, respectively).

4.3.1 First-Generation Genetic Maps

The first DNA-based genetic maps of common bean were mainly based on RFLP markers (Vallejos et al. 1992; Nodari et al. 1993a). Divergent parents were chosen for both maps in order to maximize polymorphism at the nucleotide level, as well as the phenotypic variation. The mapping population used by Vallejos et al. (1992) consisted of a backcross progeny from the 'XR-235-1-1' (Mesoamerican) × 'Calima' (Andean) cross, whereas Nodari et al. (1993a) used a F_2 population derived from the 'BAT 93' (Mesoamerican) × 'Jalo EEP558' (Andean) cross. The map developed by Vallejos et al. (1992) included the pigmentation gene *P*, 224 RFLP, nine seed proteins and nine isozyme markers, which were sorted into 11 LGs covering 960 cM of the bean genome. This map was later expanded to 980 cM by adding seven additional markers (Vallejos 1994). Subsequently, Vallejos et al. (2001) increased the number of markers up to 294; however, the map coverage was reduced up to 900 cM as a bigger stringency was used for placement of markers on the map. The map developed by Nodari et al. (1993a) was constructed using 108 RFLPs (from *Pst*I and *Eco*RI-*Bam*HI genomic libraries), seven isozymes, seven RAPDs and 18 marker loci corresponding to known genes that were selected after hybridization, as well as three phenotypic traits. These markers were distributed into 15 LGs spanning 827 cM of the genome. Gepts et al. (1993) rapidly improved this map, which finally included 204 markers grouped into 13 LGs covering 1060 cM. The following genetic map published was developed by Adam-Blondon et al. (1994) from a BC_1 population derived from the 'Ms8EO2' × 'Corel' cross, which in addition to 51 RFLPs included 100 RAPDs, two sequence-characterized amplified regions (SCARs) and four morphological markers, covering 567.5 cM of the common bean genome. Furthermore, Adam-Blondon et al. (1994) carried out the first effort to align LGs with the map published by Vallejos et al. (1992), as 19 of the 51 RFLP markers were shared, which established a preliminary correspondence between both maps.

In successive years, the initial F_2 'BAT 93' × 'Jalo EEP558' mapping population was advanced to a Recombinant Inbred Line (RIL) one in order to perform the core linkage map of common bean (Freyre et al. 1998; Hanai et al. 2010); additionally, many RIL populations were developed and used for genetic mapping studies. Koinange et al. (1996) used a RIL population from the 'Midas' (Andean cultivar) × 'G12873' (Mesoamerican wild bean) cross to create a map composed of 77 RFLP and 5 isozyme markers in order to identify the major alleles and QTL that differentiate the wild from cultivated beans. Furthermore, RIL populations from intra-gene pool crosses were also used for linkage mapping. Jung et al. (1996) used a RIL population obtained from the cross between two Mesoamerican genotypes, 'BAC 6' × 'HT 7719', and mapped 75 RAPD markers distributed into 9 LGs covering 545 cM. Likewise, a RIL population obtained between two Andean genotypes, 'PC-50' × 'XAN 159', was used by Jung et al. (1997) to map 168 RAPD markers distributed into 10 LGs covering 426 cM, to study the common bacterial blight disease resistance. Subsequently, more than twenty-five RIL mapping populations have been developed to map individual or multiple traits, most of them created from inter-gene pool crosses, which include divergent parents showing high genetic polymorphism (Broughton et al. 2003; Kelly et al. 2003). A complete description of the main mapping populations used for linkage map construction purpose is provided in Table 4.1.

Whereas RAPD and AFLP markers were used for saturating previous RFLP maps and to create new genetic maps from additional populations (Miklas et al. 1996, 1998, 2000; Ariyarathne et al. 1999; Tar'an et al. 2001, 2002; Vallejos et al. 2001; Johnson and Gepts 2002), RFLP markers were also useful to anchor different genetic maps. The 'BAT93' × 'Jalo EEP558' RIL population as well as RFLP and RAPD markers of three previous maps (Vallejos et al. 1992; Nodari et al. 1993a; Adam-Blondon et al. 1994; Vallejos 1994) were used to create the core linkage map (Freyre et al. 1998). This map comprised a total of 563 markers, including 120 RFLPs and 430 RAPDs, in addition to a few isozyme and phenotypic marker loci, which were grouped into 11 LGs spanning 1226 cM (Freyre et al. 1998). Later, Vallejos et al. (2001) integrated three linkage maps based on three RIL mapping populations obtained from the 'XR-235-1-1' (Mesoamerican) × 'Calima' (Andean), 'Jamapa' (Mesoamerican) × 'Calima' (Andean) and 'Eagle' (Andean) × 'Puebla 152' (Mesoamerican) crosses that allowed for the placement of 230 RFLPs and 464 RAPDs on the map.

Nevertheless, as with other species, the development of single-locus PCR-based markers such as SSRs or SCARs brought about incomparable progress for common bean genetic mapping research; these quickly replaced RFLPs as the markers of choice for comparing and integrating genetic maps. Among their advantages, SSR markers are multiallelic, codominant, highly polymorphic and have an abundant distribution in plant genomes (Kalia et al. 2011). Yu et al. (2000) published the first successful assignment of 15 SSRs to a framework map based on RAPD and RFLP markers. Subsequently, Blair et al. (2003) developed a total of 150 SSRs: 81 were anonymous genomic or non-coding SSRs and 69 were developed from expressed sequence tag (EST) databases. In this study, 100 SSRs were integrated in a base map developed from the 'DOR364' (Mesoamerican) × 'G19833' (Andean) population in order to anchor two existing linkage maps. The base map comprised a total of 246 loci (78 SSR, 48 RFLP, 102 RAPD and 18 AFLP markers) spanning 1720 cM, with an average distance between SSR loci of 19.5 cM. Thereby, the linkage map developed by Blair et al. (2003) could be classified as the first second-generation genetic map of common bean.

4.3.2 Second-Generation Genetic Maps

In the past few years, common bean genome and EST sequencing programmes have generated large amounts of sequence data, which led to the acceleration in the identification of functional markers. Nowadays, approximately 168,500 common bean sequences have been deposited in the GenBank nucleotide database (http://www.ncbi.nlm.nih.gov/nuccore/; July 2016), and the vast majority of them are EST sequences (~129,000). These have resulted in new research opportunities, such as data mining approaches for searching repetitive motifs (e.g. SSR), single nucleotide polymorphism (SNP) and insertion/deletion (InDel). A large-scale sequence analysis was carried out by Ramírez et al. (2005), who examined over 21,000 EST sequences derived from different cDNA libraries of the Mesoamerican ('Negro Jamapa') and the Andean ('G19833') gene pools. This analysis allowed for the identification of 529 SNPs in 214 kb of contigs, giving one SNP every 387 bp. More recently, data mining approaches have led to the detection of a huge number of polymorphisms from both coding and non-coding regions. Thus, for example, Zou et al. (2014) used 36 common bean genotypes to construct DNA libraries for next-generation sequencing (NGS). By analysing 76 million

Table 4.1 Main common bean populations used for common bean linkage mapping studies

Mapping population	Gene pool	Markers mapped	Reference
XC[a] (68 BC_1)[b]	MA[c]	294 loci (224 RFLPs, 9 seed proteins, 9 isozymes, the gene *P*), 11 LGs, 960 cM	Vallejos et al. (1992, 2001), Vallejos (1994),
MsCo (128 BC_1)	MA	157 loci (51 RFLPs, 100 RAPDs, 2 SCARs, 4 morphological markers), 11 LGs, 567.5 cM	Adam-Blondon et al. (1994)
MiG12 (65 RIL)	AM	82 loci (77 RFLPs, 5 isozymes), 15 LGs, 1,111 cM	Koinange et al. (1996)
DX (79 RIL)	MA	155 loci (147 RAPDs, 2 SCARs, 1 ISSR, and the *R, V, Asp* and two rust resistance genes), 11 LGs, 930 cM	Miklas et al. (1996, 1998, 2000)
BH (128 RIL)	MM	75 RAPDs, 8 LGs, 545 cM	Jung et al. (1996)
PXA (70 RIL)	AA	168 RAPDs, 10 LGs, 426 cM	Jung et al. (1997), Park et al. (2001)
BA (78 RIL)	MM	174 loci (172 RAPDs, 2 SCARs), 11 LGs, 755 cM	Ariyarathne et al. (1999)
EP (75 RIL)	AM	361 RAPDs, 11 LGs, 825 cM	Vallejos et al. (2001)
JaCa (76 RIL)	MA	243 loci (155 RAPDs, 88 RFLPs), 11 LGs, 950 cM	Vallejos et al. (2001)
S95 (142 RIL)	MA	115 loci (49 AFLPs, 43 RFLPs, 11 SSRs, 9 RAPDs, 1 SCAR, 2 morphological markers), 12 LGs, 1,717 cM	Tar'an et al. (2001, 2002)
CDRKY (150 RIL)	AM	192 AFLPs, 15 LGs, 862 cM	Johnson and Gepts (2002)
WOSp (110 F_5)	AM	105 loci (99 RAPDs, 3 SSRs, 3 SCARs), 8 LGs, 641 cM	Beattie et al. (2003)
BG21 (94 RIL)	MM	115 loci (26 SSRs, 89 RAPDs), 8 LGs, 611.2 cM	Frei et al. (2005)
G23G19 (84 RIL)	MA	149 loci (79 SSRs, 57 RAPDs, 11 SCARs, and 1 biochemical and 1 morphological markers), 11 LGs, 1,175 cM.	Ochoa et al. (2006)
IG24 (157 $BC_2F_{3:5}$)	AM	84 loci (80 SSRs, 1 SCAR, 3 morphological markers), 11 LGs, 869,5 cM	Blair et al. (2006)
JulCa (103 F_2)	MA	103 loci (21 RAPDs, 82 AFLPs), 12 LGs, 1,983.6 cM	Yaish et al. (2006)
G19AND (75 RIL)	AA	167 loci (64 SSRs, 11 RAPDs, 91 AFLPs, 1 phenotypic trait), 11 LGs, 1,105 cM	Cichy et al. (2009a)
G14G48 (110 RIL)	MM	114 loci (68 SSRs, 46 RAPDs), 11 LGs, 915.4 cM	Blair et al. (2010)
DB (113 RIL)	MM	291 loci (22 AFLPs, 98 RAPDs, 153 SSRs, 18 ESTs), 11 LGs, 1,788 cM	Blair et al. (2012), Galeano et al. (2011, 2012)
XCo (104 RIL)	AM	349 loci (175 AFLPs, 115 SSRs, 30 SCARs, 12 RAPDs, 13 proteins, 4 genes), 11 LGs, 1,042 cM	Pérez-Vega et al. (2010), Casañas et al. (2013), Trabanco et al. (2014)

(continued)

Table 4.1 (continued)

Mapping population	Gene pool	Markers mapped	Reference
IACAL (380 RIL)	MA	292 SSRs, 11 LGs, 2,058 cM	Campos et al. (2011), Oblessuc et al. (2012, 2013, 2014)
P1037 (185 RIL)	AA	229 loci (86 AFLPs, 98 SSRs, 42 SNPs, 2 SCARs and *P* locus), 11 LGs, 858.4 cM	Yuste-Lisbona et al. (2012, 2014a, b), González et al. (2015)
BJ (70 F_2, 70 RIL)	MA	428 loci (300 gene-based, 103 core and 24 other markers), 11 LGs, 1,545.5 cM	Nodari et al. (1993a), Gepts et al. (1993), Freyre et al. (1998), Gepts (1999), Yu et al. (2000), Hougaard et al. (2008), Hanai et al. (2010), McConnell et al. (2010)
DG (87 RIL)	MA	534 gene-based markers, 11 LGs, 2,400 cM.	Blair et al. (2003), Córdoba et al. (2010a, b); Galeano et al. (2011, 2012)
StRe (267 F_2, 85 RIL)	MA	7,276 SSRs and SNPs, 11 LGs	Schmutz et al. (2014)
SEA5CAL (125 RIL)	MA	2,122 SNPs, 11 LGs, 1,351 cM	Mukeshimana et al. (2014)

[a]Mapping population acronyms: BA = Belneb-RR-1 × A55; BH = BAC6 × HT7719; BJ = BAT93 × JaloEEP558; BG21 = BAT881 × G21212; BA = Belneb-RR-1 × A55; CDRKY = CDRK × Yolano; DB = DOR364 × BAT477; DG19 = DOR364 × G19833; DX = DOR364 × XAN176; EP = Eagle × Puebla152; G14G48 = G14519 × G4825; G19AND = G19833 × AND696; G23G19 = G2333 × G19839; IACAL = IAC-UNA × CAL 143; IG24 = ICACerinza × G24404; JaCa = Jamapa × Calima; JuCa = Jules × Canela; MiG12 = Midas × G12873; MsCo = Ms8EO2 × Corel; P1037 = PMB0225 × PHA1037; PX = PC50 × XAN159; SEACAL = SEA5 × CAL96; StRe = Stampede × Red Hawk; S95 = OACSeaforth × OAC95; Xco = Xana × Cornell49242; XC = XR-235-1-1 × Calima; WOSp = WO3391 × OAC Speedvale
[b]RIL = Recombinant Inbred Line; BC_1 = backcross first generation
[c]M = Mesoamerican; A = Andean

sequence reads generated by the Illumina's HiSeq 2000 Sequencing System, they identified a total of 43,698 putative SNPs and 1267 putative InDels and located 24,907 SNPs and 692 InDels in 8835 and 637 genes, respectively. Likewise, Müller et al. (2014) analysed the ends of 52,270 Bacterial Artificial Chromosome (BAC) libraries from the Mesoamerican breeding line 'BAT93' and identified a total of 3789 SSR loci with a distribution of one SSR per 8.36 kbp. Meanwhile, Wu et al. (2014), using a RNA-seq approach for the discovery of drought-responsive genes, identified a total of 10,482 SSR and 4099 SNP loci in transcripts of the 'Long 22-0579' (Mesoamerican) and 'Naihua' (Andean) common bean cultivars. However, despite the gigantic number of SSR and SNP polymorphisms identified to date, most of them have remained untapped as a source of functional markers and need future validation for practical use in common bean breeding and research.

Common bean linkage maps have been progressively incorporating functional markers. For instance, 108 markers based on genes known to be involved in the nodulation process in model legumes (Galeano et al. 2012) were evaluated by Ramaekers et al. (2013) in the RIL population generated from the cross between the Mesoamerican 'G2333' and the Andean 'G19839' genotypes. This mapping population has been previously used in several genetic studies (Ochoa et al. 2006; Checa and Blair 2008; Caldas and Blair 2009); thus, the existing genetic map was improved through the mapping of 42 out of 108 nodulation gene-based markers. The final linkage map consisted of a total of 207 markers (57 RAPDs, 106 SSRs, 42 SNPs, one SCAR and one isozyme) grouped into 11 LGs with a total map length of 1601 cM. Using this improved genetic map, Ramaekers et al. (2013) performed QTL analysis for the symbiotic nitrogen fixation capacity, and candidate genes were tentatively identified among the nodulation markers.

Nowadays, the RIL populations derived from the 'BAT93' × 'Jalo EEP558' and 'DOR364' × 'G19833' inter-gene pool crosses are considered as core mapping populations since both populations have been widely used for genetic mapping studies and QTL identification (Freyre et al. 1998; McClean et al. 2002, 2010; Blair et al. 2003, 2009a; Liao et al. 2004; Beebe et al. 2006; Hougaard et al. 2008; Caldas and Blair 2009; López-Marín et al. 2009; Hanai et al. 2010; McConnell et al. 2010; Galeano et al. 2011, 2012). Markers with putative gene functions have also been included in the extension of both core linkage maps. For the RIL population 'BAT93' × 'Jalo EEP558', EST libraries from anthracnose-infected common bean leaves (Melotto et al. 2005) were screened for microsatellites by Hanai et al. (2010), yielding a set of 140 EST-SSR markers. In addition, Resistance Gene Analogs (RGAs)-based markers were also developed. The merging of the data of the 285 new loci (50 EST-SSR, 32 RGA and 203 AFLP markers) mapped by Hanai et al. (2010) with the data of 143 markers previously mapped by Freyre et al. (1998) resulted in a map which comprised 413 loci. These loci were placed across 11 LGs and spanned a genetic distance of 1259 cM with an average distance between neighbouring loci of 3.0 cM. Likewise, the previous genetic map of the RIL population 'DOR364' × 'G19833' was updated by Galeano et al. (2012), who developed a total of 313 intron-based EST-SNP markers. Thus, the final genetic map consisted of 534 marker loci distributed into 11 LGs with a full map length of 2400 cM.

In addition, in order to map individual or multiple traits, new mapping populations have been developed in the last few years. Thus, a RIL population derived from an inter-gene pool cross between 'Xana' (Andean) and 'Cornell 49242' (Mesoamerican) was used to develop a genetic map including 349 markers (175 AFLPs, 115 SSRs, 30 SCARs, 12 RAPDs, 13 loci codifying for seed proteins and four genes) distributed into 11 LGs, with a total length of 1042 cM (Pérez-Vega et al. 2010; Casañas et al. 2013; Trabanco et al. 2014). Likewise, the 'PMB0225' × 'PHA1037' Andean intra-gene pool RIL population has been used to study the inheritance of different agronomic and resistance traits (Yuste-Lisbona et al. 2012, 2014a, b; González et al. 2015). The last version of this genetic map consisted of 229 loci (86 AFLPs, 98 SSRs, 42 SNPs, 2 SCARs and the *P* locus), which were distributed into 11 LGs and spanned 858.4 cM (González et al. 2015). Moreover, the 'IAC-UNA' (Mesoamerican) × 'CAL 143' (Andean) RIL population has been used to detect loci controlling growth habit and disease resistance (Campos et al. 2011; Oblessuc et al. 2012, 2013, 2014). The updated version of this map had 292 SSR markers distributed into 11 LGs spanning a total map length of 2058 cM (Oblessuc et al. 2014). Furthermore, in order to assign markers to chromosomes and construct the LGs, SSR markers were located in the *P. vulgaris* chromosomes using the native Phytozome's BLAST and default algorithm parameters (http://www.phytozome.net/). As a result, the Oblessuc et al. (2014) map was more consistent with the genome sequence, and some markers mapped to different chromosomes in relation to the previous analysis (Campos et al. 2011; Oblessuc et al. 2012, 2013).

The International Center for Tropical Agriculture (CIAT), as part of the Harvest Plus challenge programme on Biofortification, has developed different mapping populations in order to improve the iron and zinc concentration in both gene pools (Blair et al. 2009a, 2010, 2011; Cichy et al. 2009a, b). In addition to the 'DOR364' × 'G19833' inter-gene pool RIL population, a genetic map of the 'G19833' × ' AND696' Andean intra-gene pool RIL population was developed by Cichy et al. (2009a). This linkage map consisted of a total of 167 markers (64 SSRs, 11 RAPDs, 91 AFLPs and 1 phenotypic trait) with 11 LGs and a total length of 1105 cM. Likewise, another Andean genetic map was created with the RIL population derived from the 'G21242' × 'G21078' cross. The genetic map was created using a total of 74 SSRs so as to anchor the map to previously published reference maps and 42 RAPDs, which were distributed into 11 LGs and spanned 726 cM (Blair et al. 2011). Furthermore, the 'G14519' 'G4825' Mesoamerican RIL population was used to examine the inheritance of seed iron and zinc concentrations (Blair et al. 2010). The genetic map for the 'G14519' × 'G4825' population was constructed with a total of 68 SSRs and 46 RAPDs grouped into 11 LGs spanning 915.4 cM of the common bean genome.

Additionally, a new linkage map was developed by Galeano et al. (2011) using the 'DOR364' × 'BAT477' Mesoamerican intra-gene pool population. This map was constructed by evaluating a total of 2706 molecular markers (including SSR, SNP and gene-based markers) and consisted of 291 loci distributed into 11 LGs with a total map length of 1788 cM. In order to create a consensus map for fine mapping and synteny analysis in common bean, the 'DOR364' × 'BAT477' map was merged with the previously existing linkage maps of both 'BAT93' × 'JALO EEP558' and 'DOR364' 'BAT477' core populations. Thereby, the consensus map consisted of a total of 1,060 markers distributed into 11 LGs and a total map length of 2041 cM with an average distance between adjacent loci of 1.9 cM (Galeano et al. 2012). The common bean consensus map includes a higher number of loci than most single cross maps, thus increasing the number of potentially useful markers across divergent genetic backgrounds and providing broader genome coverage.

Functional genetic maps based on genes involved in physiological processes potentially underlying important agronomic traits allow for the identification of candidate genes through translational genomics. Thereby, Kwak et al. (2008) identified common bean homologues of 12 *Arabidopsis thaliana* genes related to floral transition and flowering pathways. Seven out of 12 genes could be mapped using the 'BAT93' 'JaloEEP558' and 'Midas' × 'G12873' RIL populations. Thus, three *Terminal Flower 1* homologues (*PvTFL1x*, *PvTFL1y* and *PvTFL1z*) were mapped. *PvTFL1y* co-segregated with the phenotypic locus for determinacy growth habit (*fin*) on LG01, whereas *PvTFL1z* mapped near or at a second determinacy locus on LG07 (Kolkman and Kelly 2003). In addition, a *Zeitlupe* homologue mapped close to a QTL for flowering time on LG09 (Kwak et al. 2008). These results support the role of functional maps including genes of known function as an important component of the candidate gene approach. However, further studies are needed to confirm the role of these homologues as potential candidate genes.

Moreover, functional maps are useful for synteny studies among different species. Sequence data from legumes are available in the Legume Information System (LIS: http://phavu.comparative-legumes.org/gb2/gbrowse/Pv1.0/; Dash et al. 2015) which is focussed on legume comparative analysis. Thus, in order to investigate the syntenic relationship between *P. vulgaris*, *A. thaliana*, *Medicago truncatula* and *Lotus japonicus*, a gene-based map was developed by McConnell et al. (2010) using the 'BAT93' × 'JaloEEP558' RIL population. The map included a total of 420 loci (304 gene-based markers, 103 core markers and 13 colour gene markers), which were sorted into 11 LGs and spanned 1545.5 cM. The genetic map information and the marker sequences were used as a query in a 'tblastx' analysis with the genome sequence of each of the species. The results showed that while only short blocks of synteny were observed with *A. thaliana*, large-scale macrosyntenic blocks were observed with *M. truncatula* and *L. japonicus*. These syntenic relationships are in accordance with the results previously obtained by Hougaard et al. (2008), who carried out the first attempt at estimating the extent of synteny and collinearity among these species based on 104 legume anchor-marker loci representing single-copy genes. Similarly, this gene-based map was used by McClean et al. (2010) to understand syntenic relationship between common bean and soybean. Genetically positioned transcript loci of common bean were mapped in relation to the soybean 1.01 genome assembly (http://soybase.org/gbrowse/cgi-bin/gbrowse/gmax1.01/). In nearly every case, each common bean locus mapped into two positions in soybean, a result consistent with the duplicate polyploidy history of soybean. Furthermore, by this genetic/physical synteny approach, McClean et al. (2010) were also able to electronically position ~15,000 common bean sequences (primarily EST contigs and EST singletons) onto the common bean map using the shared syntenic blocks as reference points. Therefore, this extensive gene-based map significantly expands the genomic resources available for common bean and provides a framework for comparative genetics and genomics of legumes.

Currently, the genomes of the Andean 'G19833' (http://www.phytozome.net/commonbean.php/; Schmutz et al. 2014) and the Mesoamerican 'BAT93' are available (http://denovo.cnag.cat/genomes/bean; Vlasova et al. 2016), while the genome of the Andean 'OAC-Rex' is underway (Canadian team, http://www.beangenomics.ca/research/projects/view/draft-genome-sequence-for-common-bean-i-p-vulgaris-i/). In this way, a large number of specific disease resistance genes have been identified and located in the genome, constituting a valuable material to design new functional molecular markers in common bean (Meziadi et al. 2015). Hence, the huge amount of sequence information available for common bean via whole-genome sequencing projects facilitates the development of an almost unlimited number of genetic markers suitable for high-throughput genotyping and easily transferable across different mapping populations. The PhaseolusGenes database was developed as part of the BeanCAP project (http://www.beancap.org/; http://phaseolusgenes.bioinformatics.ucdavis.edu/), including phenotypic, genotypic and molecular marker data collected from publications and projects throughout the world and a genome browser in order to place markers on assembled common bean and soybean genomes. The BeanCAP project has also carried out the design of BeadChips that are being used to genotype bean populations (Song et al. 2015). Two of these BeadChips (BARCBEAN6K_1 with 5,232 SNP markers and BARCBEAN6K_2 with 5,514 SNP markers) have been used by Schmutz et al. (2014) in the 'G19833' genome sequencing. The resulting assembled sequence was organized into 11 chromosomes by integration with a map of 7015 SNP markers, typed on 267 F_2 lines from the 'Stampede' (Mesoamerican) × 'Red Hawk' (Andean) cross, and a similar set of SNP and 261 SSR markers, typed on 88 F_5-RIL population derived from the same cross. Thus, the final genetic map contained 7276 SSR and SNP markers arranged in 11 LGs. Another BeadChip (BARCBEAN6K_3 with 5389 SNP markers) has been recently used by Mukeshimana et al. (2014). A total of 2122 SNP markers were mapped in the 'SEA5' (Mesoamerican) × 'CAL96' (Andean) RIL population. The genetic map spanned 1351 cM and covered all 11 LGs with an average distance of 0.64 cM between markers. The Mukeshimana et al. (2014) results showed that SNP marker order and location in the 'SEA5' × 'CAL96' map generally agreed with order and chromosome assignment in the 'Stampede' × 'Red Hawk' common bean map. Therefore, such high-throughput genotyping approaches allow for the rapid and cost-effective generation of high-density functional maps, which could also lead to the direct gene tagging for QTL mapping of important agronomic traits, improving the efficiency of common bean breeding programmes via MAS.

4.4 Molecular Mapping of Simple and Complex Traits

QTL mapping has become very popular in bean genetics and breeding research, where QTL have been identified for numerous agronomical and biological important complex traits. This section will summarize the most important genes and QTL, which have been described and mapped during the past decades in common bean.

4.4.1 Genes and QTL Involved in Biotic Stress Resistance

Identification of genetic markers associated with disease resistance in common bean started in 1970s with the pioneering work of Coyne and his co-workers who identified an association between the common bacterial blight (CBB) resistance and late flowering (Coyne et al. 1973). Since then, numerous genetic markers have been used to map major genes and QTL conferring resistance to common bean diseases. An overview of the main resistance genes and QTL as well as their location on the common bean linkage map is summarized in Table 4.2.

4.4.1.1 Virus Diseases

Resistance to different pathogroups of the potyviruses such as bean common mosaic virus (BCMV) and bean common mosaic necrosis virus (BCMNV) is conferred by four different recessive loci: *bc-1*, *bc-2*, *bc-3* and *bc-*

Table 4.2 Details of QTL mapping studies performed for the mapping of major genes and QTL for different biotic stress resistance in common bean

Disease[a]	Gene/QTL name[b]	LG[c]	Mapping population[d]	Reference
BCMV/BCMNV	*I*	2	BA	Ariyarathne et al. (1999)
	bc-1^{2}, *bc-u*	3	DG, OS	Strausbaugh et al. (1999)
	bc-3	6	CIAT breeding lines	Johnson et al. (1997)
BGYMV	*bgm-1*	3	DS	Blair et al. (2007)
	BGMV	4, 7	DX	Miklas et al. (2000)
ANT	*Co-1, Co-w, Co-x*, $Co\text{-}1^{65-X}$, $Co1^{73-X}$, $SDC^{23}\text{-}1$, $PDC^{1545}\text{-}1$, $PAUDPC^{1545}\text{-}1$	1	BJ, XCo, P1037	Geffroy et al. (2008), Campa et al. (2014), González et al. (2015)
	Co-u, $CoPv02c^{3-X}$, $CoPv02c^{7-X}$, $CoPv02c^{19-X}$, $CoPv02C^{449-X}$	2	BJ, XCo	Geffroy et al. (2008), Campa et al. (2014)
	Co-13, Co-17, $PAUDPC^{1545}\text{-}3$, $LDC^{23}\text{-}3$	3	JMex, JCo, SM, P1037	Gonçalves-Vidigal et al. (2009), Trabanco et al. (2015), González et al. (2015)
	Co-3, Co-10, Co-9, Co-y, Co-z, Co-15, $Co\text{-}3c^{3-X}$, $Co\text{-}3c^{7-X}$, $Co\text{-}3c^{19-X}$, $Co\text{-}3c^{449-X}$, $Co3c^{453-X}$, $SDC^{23-}4$, $SAUDPC^{23-}4$, $LDC^{23}\text{-}4.1$, $LAUDPC^{23}\text{-}4$	4	RuOu, Aou, BJ, CoCo, XCo, P1037	Geffroy et al. (2000), Gonçalves-Vidigal et al. (2013), Campa et al. (2014), González et al. (2015), Sousa et al. (2015)
	$SDC^{1545}\text{-}5$, $SAUDPC^{1545}\text{-}5$, $PDC^{1545}\text{-}5$, $PAUDPC^{1545}\text{-}5$, $LDC^{1545}\text{-}5$, $LAUDPC^{1545-}5$	5	P1037	González et al. (2015)
	Co-5, Co-6, $LDC^{1545}\text{-}7$, $LAUDPC^{1545}\text{-}7$	7	TM, ABMi, P1037	Campa et al. (2007, 2009), González et al. (2015)
	Co-4, $SDC^{1545}\text{-}8$, $SAUDPC^{1545}\text{-}8$, $LDC^{1545}\text{-}8$, $LAUDPC^{1545}\text{-}8$	8	SM, P1037	Trabanco et al. (2015), González et al. (2015)
	$CoPv09c^{453-C}$, $LDC^{23}\text{-}9$, $LAUDPC^{23}\text{-}9$	9	XCo, P1037	Campa et al. (2014), González et al. (2015)
	Co-2, $Co\text{-}2^{6-C}$, $Co\text{-}2^{39-C}$, $Co\text{-}2^{38-C}$, *and* $Co\text{-}2^{357-C}$	11	MsCo, XCo	Adam-Blondon et al. (1994), Campa et al. (2014)
FRR	$P7_{1550}$, $P7_{700}$, $P10_{1600}$, $G6_{1100}$, $D3_{600}$, $I18_{1800}$, $I18_{1700}$, $AG2_{800}$, $G17_{900}$, $G3_{800}$, $G3_{2000}$, $P9_{1550}$, $Y11_{600}$, $O12_{800}$, $S8_{500}$, $V12_{1100}$	1, 2, 3, 4, 5, 6, 7	MF	Schneider et al. (2001)
	$UBC218_{1200}/UBC503_{640}$, $UBC503_{640}/UBC211_{1000}$	6	ACNY	Chowdhury et al. (2002)
	$AL20_{850}/G8_{1400}$, $O12_{800}/AL20_{850}$, $S19_{1000}/S19_{1100}$, $G17_{900}/AL20_{350}$, $AL20_{700}/G6_{2000}$, $AJ4_{350}/X3_{3054}$, $AN19_{1300}/H4_{1200}$	1, 5, 7, 9, 8	CNSL	Román-Avilés and Kelly (2005)
	$FRR3.1^{km}$	3	K32MLB, K20MLB	Kamfwa et al. (2013)
	ARR2.1, ARR4.1, ARR6.1, FRR3.1,FRR7.1	1, 3, 4, 6, 7		Hagerty et al. (2015)

(continued)

Table 4.2 (continued)

Disease[a]	Gene/QTL name[b]	LG[c]	Mapping population[d]	Reference
WM	*WM1.1, WM7.1*	1, 7	AG	Miklas et al. (2001)
	WM2.1, WM4.1, WM5.1, WM8.1	2, 4, 5, 8	PX	Park et al. (2001)
	WM2.2, WM7.2	2, 7	BuN	Kolkman and Kelly (2003)
	WM2.3, WM5.2, WM7.2, WM8.4	2, 5, 7, 8	IBR	Ender and Kelly (2005)
	WM1.2, WM2.4, WM8.2, WM8.3, WM9.1	1, 2b, 8, 9	GCO	Maxwell et al. (2007)
	WM2.2, WM4.2, WM5.3, WM5.4, WM6.1, WM7.3, WM8.4	2, 4, 5, 6, 7	R31	Soule et al. (2011)
	WM2.2, WM8.3	2, 8	BV	Soule et al. (2011)
	WM3.3, WM7.5, WM9.2, WM11.1	3, 7, 9, 11	TPI95, TPI50	Mkwaila et al. (2011)
	WM1.3, WM3.2, WM6.2, WM7.4, WM8.5, WM11.12	1, 3, 6, 7, 8, 11	XCo	Pérez-Vega et al. (2012)
Rust	*Ur-9*	1	PC	Miklas et al. (2002)
	Pu-a	3	PX	Jung et al. (1998)
	Ur-5, Ur-Dorado108, Ur-ON (Ur-14)	4	DX, OUB, OUM	Miklas et al. (2000), Souza et al. (2011)
	Ur-4	6	BJ	Miklas et al. (2002)
	Ur-12	7	PX	Jung et al. (1998)
	Ur-13, Crg	8	KB, Sierra mutagenized seed	Kalavacharla et al. (2000), Mienie et al. (2005)
	Ur-3, Ur-11, Ur-Dorado53, Ur-6, Ur-7, Ur-BAC6	11	P07Be, P32Be, DX, BH, BA	Stavely (1998), Miklas et al. (2000, 2002)
CBB	*D1, D2, D5, D9*	1,5,7,9	BJ	Nodari et al. (1993b)
	CBB-1LL, CBB-2LL, CBB-2S, CBB-2P, CBB-2FL	1, 2, 3, 4, 5, 6	BH	Jung et al. (1996)
	CBLEAF, CBPOD	1, 2, 9, 10	BA	Ariyarathne et al. (1999)
	FT-1, FT-2, LDT-2, Pod-1, Pod-2, Seed-1, Seed-2	1, 4, 5, 9	PX	Jung et al. (1997)
	Bng40, Bng139	7, 8	XC	Yu et al. (1998)
	CBB-GH-leaf, CBB-GH-pod, CBB-field	7, 10	DX	Miklas et al. (2000)
	SU91, SAP6, $Xa11.4^{OV1,OV3}$	8, 10, 11	OV1, OV3	Viteri et al. (2015)
HB	*Pse1, Pse2, Pse3, Pse4, Pse6*	2, 4, 10	CWU, ZCW, BA	Miklas et al. (2009, 2011, 2014)
	Stem, 96LFA, 98LFA, 96BBS, 98BBS	1, 2, 3, 4, 5, 6, 7, 8, 9, 11	BA	Jung et al. (2003)
	Rpsar-1, Rpsar-2	8, 11	BJ	Fourie et al. (2004)
	HB83, HB16	2, 3, 4, 5, 9, 10	BA	Ariyarathne et al. (1999)
	$Psp4^{812XC}$, $Psp6.1^{812XC}$, $Psp6.1^{684XC}$, $Psp6.2^{684XC}$	4, 6	XCo	Trabanco et al. (2014)

(continued)

Table 4.2 (continued)

Disease[a]	Gene/QTL name[b]	LG[c]	Mapping population[d]	Reference
	$SAUDPC^{3}$-2, $PLAUDPC^{3}$-2, PDC-^{3}2, PDC^{4}-2, PDC^{5}-2, $PAUDPC^{3}$-2, $PAUDPC^{4}$-2, SDC^{7}-6	2, 6	P1037	González et al. (2016)
ALS	*Phg-1*	1	AOu	Gonçalves-Vidigal et al. (2011)
	Phg-2	8	BJ	Miklas et al. (2006)
	Phg_{G5686A}, Phg_{G5686B}, Phg_{G5686C}, $Phg_{G10909A}$, $Phg_{G10909B}$	4, 8, 9	G56Sp, G10Sp	Mahuku et al. (2009, 2011)
	Phg-ON (Phg-3)	4	RuOu, Aou	Gonçalves-Vidigal et al. (2013)
	ALS	4, 10	DG19	López et al. (2003)
	$ALS2.1^{UC}$, $ALS3.1^{UC}$, $ALS4.1^{GS,UC}$ $ALS4.2^{GS,UC}$, $ALS5.1^{UC}$, $ALS5.2^{UC}$, $ALS10.1^{DG,UC}$	2, 3, 4, 5, 10	UC, GS	Oblessuc et al. (2012)

[a]*ALS* = angular leaf spot; *ANT* = anthracnose; *BGYMV* = bean golden yellow mosaic virus; *BCMV* = bean common mosaic virus; *BCMNV* = bean common mosaic necrosis virus; *CBB* = common bacterial blight; *FRR* = *Fusarium* root rot; *HB* = halo blight; *WM* = white mould

[b]*Co* = anthracnose; *Ur* = rust; *Pse* and *Rpsar* = halo blight; *Phg* and *ALS* = angular leaf spot; *I* and *bc* = BCMV/BCMNV resistance loci

[c]LG = linkage group

[d]Mapping population acronyms: ABMi = AB136 × Michelite; ACNY = A.C. Compass × NY2114-12; AG = A55 × G122; AOu = AND277 × Ouro Negro; BA = Belneb-RR-1 × A55; BH = BAC6 × HT7719; BJ = BAT93 × JaloEEP558; BV = Benton × VA19; BuN = Bunsi × Newport; CoCo = Corinthiano × Cornell 49-242; CNSL = C97407 × Negro San Luís; CWU = Canadian Wonder × UI-3; DG19 = DOR364 × G19833; DS = DOR476 × SEL1309; DX = DOR 364 × XAN 176; EP = Eagle × Puebla 152; EEP = Eagle*2 × Puebla 152; EPH = Eagle*2 × Hystyle; GCO = G122 × CO72548; G10Sp = G10909 × Sprite; G56Sp = G5686 × Sprite; H95 = HR67 × OAC95; IBR = ICA Bunsi × Raven; JCo = JLP × Cornell 49242; JMex = JLP × Mexico 222; JuCa = Jules × Canela; KB = Kranskop × Bonus; K32MLB = K132 × MLB-49-89A; K20MLB = K20 × MLB-49-89A; MF = Montcalm × FR266; MsCo = Ms8EO2 × Corel; OS = Olathe × Sierra; OUB = Ouro Negro × Belmidak RR-3; OUM = Ouro Negro × Mexico309; OUS = Ouro Negro × US Pinto 111; OV1 = Othello × VAX 1; OV3 = Othello × VAX 3; P07Be = P94207 × Beltsville; P32Be = P94232 × Beltsville; P1037 = PMB0225 × PHA1037; PC = PC50 × Chichara-83-109; PX = PC50 × XAN159; R31 = Raven × I9365-31; RNSL = Red Hawk × Negro San Luís; RuOu = Ruda × Ouro Negro; S95 = OACSeaforth × OAC95; SM = SEL1308 × MDRK; TPI95 = Tacana × PI318695; TPI50 = Tacana × PI313850; UC = IAC-UNA × CAL143; XC = XR-235-1-1 × Calima; XCo = Xana × Cornell49242; ZCW = ZAA12 × Canadian Wonder

u (Drijfhout 1978). In addition, the dominant *I* gene confers immune resistance to all strains of BCMV through a hypersensitive response (Ariyarathne et al. 1999). As regards the begomovirus such as bean golden yellow mosaic virus (BGYMV), a Mesoamerican source of partial resistance, is conditioned by the recessive *bgm-1* gene (Blair et al. 2007). The Andean-derived recessive *bgm-2* gene (Velez et al. 1998) and dominant *Bgp-1* gene, which confers resistance to pod deformation and requires the presence of *bgm-1* for complete expression (Molina Castañeda and Beaver 1998), have also been reported. Furthermore, two independent QTL have been identified for BCMV resistance, explaining 60% of the phenotypic variation (Miklas et al. 2000). One of these QTL was located on LG07 close to the *Asp* and *Phs* loci, together with other QTL or major genes conditioning resistance to CBB, white mould, anthracnose and stem blight (Nodari et al. 1993a; Miklas et al. 2000, 2001).

4.4.1.2 Fungal Pathogens

With respect to fungal diseases, over 40 genes have been described as conferring resistance to anthracnose (labelled as *Co-*), caused by the fungus *Colletotrichum lindemuthianum*. Anthracnose resistance is related to the presence of closely linked race-specific loci, which comprise different single, duplicate or complementary dominant genes, except for the recessive *co-8* (Ferreira et al. 2013; Campa et al. 2014). Moreover, a major QTL located on LG04 explained 70% of resistance to race 45 co-localized with the genes *Co-9*, *Co-y* and *Co-z* at the end of LG04 (Geffroy et al. 2000). Clusters of nucleotide-binding site–leucine-rich repeat (NBS-LRR) genes have been identified on this region of the LG04, as well as on LG11, which co-localize with previously mapped *Co-3* and *Co-2* genes, respectively (Schmutz et al. 2014). Likewise, 17 out of 26 main-effect QTL and 20 epistatic interactions were detected by González et al. (2015) harbouring NBS-LRR genes.

Several studies demonstrated that resistance to Fusarium root rot (FRR, fungus: *Fusarium* species) is controlled by several genes located at different loci. Over 30 QTL for FRR resistance (many minor in effect) have been reported in RIL populations derived from several resistance sources (Table 4.2). Most of the QTL detected by Schneider et al. (2001) were located on LGs 02 and 03, close to a region where defence response genes, *Pgip* and *ChS*, and pathogenesis-related protein genes, *PvPR-1* and *PvPR-2*, have been positioned. Additionally, Román-Avilés and Kelly (2005) identified two QTL on LGs 02 and 05, the former located near the QTL previously detected by Schneider et al. (2001) on LG02. Likewise, Kamfwa et al. (2013) mapped the *FRR3.1*^{KM} close to the *PvPR-1* gene on LG03.

The *Sclerotinia sclerotiorum* fungus is the causal agent of white mould (WM). Genetic resistance to WM is quantitatively inherited with low–to-moderate heritability (Park et al. 2001). Single major and numerous weak QTL for WM resistance have been identified. Among them, Miklas et al. (2001) reported a single major-effect QTL located on LG07 that accounted for 38% of the total phenotypic variation and was closely linked to the *Phs* locus. Association among WM physiological resistance and disease avoidance traits was also investigated by Miklas et al. (2013), whose results showed 13 WM resistant QTL associated with disease avoidance traits.

Resistance to bean rust, caused by the *Uromyces appendiculatus* fungus, is mainly controlled by major single dominant genes (named as *Ur-1* to *Ur-14*). Clustering is observed for rust resistance and other disease resistance genes. Thus, *Co-1* and *Ur-9* genes co-localize on LG01 and *Co-3/Co-9*, *Co-10*, *Ur-5* and *Ur-Dorado-108* co-localize on LG04 (Miklas et al. 2006), while *Ur-3*, *Ur-11* and *Ur-Dorado53* map close to *Co-2* on LG11 (Miklas et al. 2002). Similarly, *Ur-13* is located on LG08 near the *Phg-2* gene for resistance to angular leaf spot (Garzon et al. 2014).

4.4.1.3 Bacterial Diseases

Studies on CBB genetics, disease caused by *Xanthomonas axonopodis* pv. *phaseoli*, reported quantitative inheritance with largely additive effects. Thus, more than twenty minor and major QTL for CBB resistance have been identified across all 11 LGs (Nodari et al. 1993b; Jung et al. 1996, 1997; Yu et al. 1998; Ariyarathne et al. 1999; Miklas et al. 2000; Viteri et al. 2015). Although specific genes associated with resistance to CBB have not been identified, genomic regions which are likely to contain genes for resistance to this disease have been found. Thus, for example, Miklas et al. (2003) detected one genomic region on LG10 associated with the RAPD marker *AP6*^{820} that explained up 60% of the phenotypic variance for CBB resistance.

Both qualitative and quantitative responses to halo blight (HB), caused by *Pseudomonas syringae* pv. *phaseolicola* (*Psp*), have been described (Ariyarathne et al. 1999; Fourie et al. 2004; Miklas et al. 2009, 2011, 2014; Trabanco et al. 2014; González et al. 2016). Five dominant (*Pse-1*, *Pse-2*, *Pse-3*, *Pse-4* and *Pse-6*) and one recessive (*pse-5*) genes were identified among the set of differential cultivars by means of complementary tests (Ferreira et al. 2013; Miklas et al. 2014). Furthermore, two independent genes that confer *AvrRpm1*-specific resistance (*Rpsar-1* and *Rpsar-2*) were located near genes that confer resistance to the *C. lindemuthianum* fungus (Fourie et al. 2004). Quantitative response has also been observed; thus, González et al. (2016) detected 76 main-effect QTL that explained up to 41% of the phenotypic variation for HB resistance, although they also identified 101 epistatic QTL, which suggest that epistasis plays an important role in the genetic control of this trait. Additionally, Trabanco et al. (2014) searched for candidate genes associated with HB resistance and identified 16 candidate genes in the physical positions in which the QTL *Psp6.1*^{812XC}, *Psp6.1*^{684XC} and *Psp6.2*^{684XC} were mapped.

These candidate genes carried sequences homologous to the resistance genes *RPM1*, *FLS2*, *RPG1*/*RPG1*-B and *Pto*, all of which confer resistance to *P. syringae* in different species.

Resistance to the angular leaf spot (ALS) (caused by the *Pseudocercospora griseola* Sacc. fungus) is mediated by several independent genes, which possess one or more alleles conferring resistance to several races of the fungus (Miklas et al. 2006; Mahuku et al. 2009, 2011; Gonçalves-Vidigal et al. 2011, 2013). Quantitative resistance has also been reported (López et al. 2003; Oblessuc et al. 2012). López et al. (2003) found a cluster of RGA on LG10 associated with one major QTL for resistance to different ALS isolates, explaining from 47 to 64% of the phenotypic variance depending on the isolate used. Moreover, seven QTL on five LGs were detected by Oblessuc et al. (2012). Among these, $ALS10.1^{DG,UC}$ on LG10 presented major effects, explaining between 16 and 22% of the phenotypic variance for ALS resistance. The QTL $ALS4.1^{GS,UC}$ was fine-mapped with two closely linked SNP markers (Marker50 and 4M437) to a region on LG04 containing 36 candidate genes (Keller et al. 2015).

4.4.2 Genes and QTL Involved in Abiotic Stress Resistance

To date, several studies have reported QTL that may play a role in mitigating the negative effects of abiotic stresses in common bean (Table 4.3). The importance of abiotic stress is unquestionable, especially in low-input agricultural systems of underdeveloped countries, where conventional breeding may be insufficient because of the fact that global climate change increases the frequency and severity of abiotic constraints. It is important to unravel molecular mechanisms in response to abiotic stress in common bean, which would help accelerate genetic improvement through MAS.

4.4.2.1 Drought Tolerance

Drought is the most important abiotic stress that limits crop productivity worldwide (Lauer et al. 2012). Although some drought-responsive genes have been reported in common bean (Blair et al. 2016), breeding for drought is complex due to the number of traits involved, quantitative inheritance and environmental influence (Mir et al. 2012).

In the absence of an effective linkage map, Schneider et al. (1997) studied the genetics of the response to drought across a broad range of environments in the 'Sierra' × 'AC1028' and 'Sierra' × 'Lef-2RB' populations using RAPD markers and multiple regression analyses. Nine markers were reported for drought resistance, although they were located on non-anchored LGs. A RIL population from the 'SEA 5' × 'MD 23-24' cross was evaluated under drought and irrigated conditions in two seasons, and common and specific QTL for drought were identified (Beebe et al. 2007). The most significant result was that in no case were one locus' alleles specifically adapted to the contrary environments (i.e., one allele to drought conditions and the other allele to favourable conditions). This implies that yield under drought and yield under well-watered conditions are not mutually exclusive and can be combined. Five QTL were detected by composite interval mapping for seed yield under drought irrigated conditions over 3 years in an intra-gene pool RIL population derived from 'BAT477' × 'DOR364' cross (Blair et al. 2012). Positive alleles for the QTL came from each parent, indicating that both contributed to yield in the drought treatment. The same mapping population was analysed with a mixed model methodology to dissect QTL of root traits associated with contrasting water availability (Asfaw and Blair 2012), and nine QTL were mapped for drought stress tolerance on six of the 11 LGs. Mukeshimana et al. (2014) used an inter-gene pool RIL population derived from the 'SEA5' × 'CAL96' cross for the identification of QTL for performance under drought stress. A mapping population from the

Table 4.3 Details of QTL mapping studies performed for the mapping of major genes and QTL for different abiotic stress tolerance in common bean

Stress[a]	Gene/QTL name[b]	LG[c]	Population[d]	Reference
Drought	$OA08_{780}$, $OA04_{560}$, $OX11_{680}$, $OZO8_{750}$, $OXI8_{980}$	unknown	SL, SAC	Schneider et al. (1997)
	Yld4.1, Yld6.1, Yld8.1, Yld8.2, Yld10.1	4, 6, 8, 10	DB	Blair et al. (2012)
	Cbm3.1, Ppi3.1, Hri3.1, Stc5.1, Stc6.1, Scr6.1, Yld8.1, Sbr9.1	3, 5, 6, 8, 9,	DB	Asfaw et al. (2012)
	$PHI1.1^{SC}$, $NP3.1^{SC}$, $SW3.1^{SC}$, $SW7.2^{SC}$, $SY9.2^{SC}$,	1, 3, 7, 9	SC	Mukeshimana et al. (2014)
	$SY1.1^{BR}$, $SY2.1^{BR}$	1, 2	BR	Trapp et al. (2015)
Zn Deficiency	*Znd*	unknown	MT	Singh and Westermann (2002)
	QTL1-Zn	4	BG	Guzmán-Maldonado et al. (2003)
	BM154/BM184	9	VA	Gelin et al. (2007)
	Zn-ICPa3, Zn-ICPa7, Zn-ICPa11	3, 7, 11	DG	Blair et al. (2009a)
	SeedZn	1, 5, 6, 11	AG	Cichy et al. (2009a)
	QZnPoAA2.1, QZnPoAA3.1, QZnPoAA6.1, QZnDaAA8.1, QZnPaAA6.1, QZnPaAA8.2,	2, 3, 6, 8,	G14G48	Blair et al. (2010)
	Zn-AAS2c, Zn-AAS7c, Zn-AAS8c	2, 7, 8	G42G78	Blair et al. (2011)
	Zn_cont3.1, Zn_cont5.1, Zn_cont5.2, Zn_cont7.1	3, 5, 7	CCCG	Blair and Izquierdo (2012)
Fe Deficiency	*QTL1-Fe, QTL2-Fe*	2, 3	BG	Guzmán-Maldonado et al. (2003)
	Fe-ICPa4, Fe-ICPa6, Fe-ICPa7, Fe-ICPa8.1, Fe-ICPa8.2, Fe-ICPa11.1	4, 6, 7, 8, 11	DG	Blair et al. (2009a)
	SeedFe	1, 5, 6, 8, 9, 11	AG	Cichy et al. (2009a)
	QFeDaAA4.1, QFePaAA6.1, QFePoAA6.1, QFePaAA7.1,	4, 6, 7	G14G48	Blair et al. (2010)
	Fe-AAS2a, Fe-AAS6b, Fe-AAS6c	2, 6	G42G78	Blair et al. (2011)
	Fe7.1, Fe_cont8.1	7, 8	CCCG	Blair and Izquierdo (2012)
P Efficiency Al Toxicity	*Pup4.1, Pup10.1*	4, 10	DG	Yan et al. (2004)
	Pup3.1, Pup4.1, Pup7.1, Pup9.1, Pup10.1, Pup11.1	3, 4, 7, 9, 10, 11	DG	Liao et al. (2004)
	Pup4.1, Pup10.1	4, 10	DG	Beebe et al. (2006)
	LPAdvNoF.1, LPAdvNoF.2	2, 9	G23G19	Ochoa et al. (2006)
	PupLP, PueLP, PupLP	7, 8, 11	AG	Cichy et al. (2009b)
	Tsp2.1, Npc6.1, Npc7.1, Npc10.1, Tsp2.1, Tsp2.1 Tsp11.1	2, 6, 7, 10, 11	G23G19	Blair et al. (2009b)
	Srl2.1, Nrt3.1, Nrt5.1, Ard6.1, Srl7.1, Ard7.1, Trl9.1, Nrt9.3, Nrt11.1, Rdw11.1, Trl11.1, Trl11.2	2, 3, 5, 6, 7, 9, 11	DG	López-Marín et al. (2009)

(continued)

Table 4.3 (continued)

Stress[a]	Gene/QTL name[b]	LG[c]	Population[d]	Reference
SNF	*D1, D3.1, D3.2, D7*	1, 3, 7	BJ	Nodari et al. (1993b)
	D1 N-, D1 N +, D3.1 N-, D3.2 N-, D3.3 N +, D4 N +, D7 N-, D7 N+	1, 3 4, 7	BJ	Tsai et al. (1998)
	NN_A	2, 3, 5, 6, 7, 9, 11	BJ	Souza et al. (2000)
	NN_B	3, 5, 7, 10	BJ	Souza et al. (2000)
	%NROOT, %NPLANT_1, %NPLANT_2, NROOT_1, NROOT_2, NPLANT_1, NPLANT_2	1, 3, 4, 10	G23G19	Ramaekers et al. (2013)

[a]Zn = zinc; Fe = iron; P = phosphorus; Al = aluminium; SNF = symbiotic nitrogen fixation
[b]*Yld, SY*: yield; *Cbm*: canopy biomass; *Ppi*: pod portioning index; *Sbr*: stem biomass reduction; *Hri*: harvest index; *Stc*: stem TNC; *Scr*: SPAD chlorophyll metre reading; *NP*: number of pods per plant; *SW*: seed weight; *PHI*: pod harvest index; *ICP*: mineral concentration with inductively coupled plasma-optical emission spectrometry; *AA*: mineral concentration with atomic absorption method (trials Popayán, Darién and Palmira); *AAS*: mineral concentration with absorption spectroscopy method; *Pup*: phosphorus uptake; *LPAdvNoF:* adventitious root traits under low-phosphorus availability in the field; Pue: P use efficiency under low-phosphorus availability; PupLP: phosphorus uptake under low-phosphorus availability; *Npc*: Net P content; *Tsp*: total seed phosphorus; *Ard*: average root diameter; *Nrt*: average number of root tips; *Srl*: specific root length; *Rdw:* root dry weight; *Trl*: total root length; *D*: number of *Rhizobium* nodules QTL; NN_A and NN_B: number of *Rhizobium* nodules trials A and B; *%NROOT, %NPLANT, NROOT, NPLANT*: %N and total N content of root and total plant
[c]LG = linkage group
[d]Mapping population acronyms: AG = AND696 × G19833; BJ = BAT93 × Jalo EEP558; BG = Bayo Baranda × G-22837; BR = Buster × Roza; CCCG = Cerinza × (Cerinza × (Cerinza × G10022; DB = DOR364 × BAT477; DG = DOR364 × G19833; G14G48 = G14519 × G4825; G23G19 = G2333 × G19839; G42G78 = G21242 × G21078; MT = Matterhorn × T-39; SL = Sierra × Lef-2RB; SAC = Sierra × AC1028, SC = SEA5 × CAL96; SMD = SEA 5 × MD 23-24; VA = Voyager × Albion

'Buster' × 'Roza' cross was tested for yield under multiple stresses (intermittent drought, compaction and low fertility) across several location-years, resulting in the detection of two major QTL (located on LGs 01 and 02), which explained up to 37% of the phenotypic variance for seed yield (Trapp et al. 2015).

4.4.2.2 Tolerance to Zinc and Iron Deficiency

Zinc (Zn) and iron (Fe) deficiency is one of the most widespread crop micronutrient deficiencies and is capable of causing severe yield reductions. The inheritance of Fe and Zn concentration in common bean seeds has been suggested to be quantitative in most studies (Guzmán-Maldonado et al. 2003; Blair et al. 2009a, 2010; Cichy et al. 2009a), even while a few initial reports suggested that a single dominant gene (named with the symbol *Znd*) was involved in the tolerance to Zn deficiency (Singh and Westermann 2002). An interesting feature of several studies (Cichy et al. 2009a; Blair et al. 2009a, 2010, 2011) was that a number of QTL for Fe and Zn co-localized or overlapped, suggesting a possibly pleiotropic locus effect for mineral uptake. This provides further support for the suggestion that the same genetic and molecular mechanisms are controlling both Zn and Fe mobilization, uptake, distribution and accumulation in the plant (Clemens et al. 2002).

Using AFLP markers in the 'Bayo Baranda' 'G-22837' cross, Guzmán-Maldonado et al. (2003) found a locus that accounted for 15% of the phenotypic variation associated with Zn content. Gelin et al. (2007) identified a locus on LG09 that accounted for 18% of the seed Zn accumulation. An inter-gene pool RIL population of the cross 'DOR364' × 'G19833' was used to scan for Fe and Zn accumulation loci (Blair et al. 2009a). QTL clustered on the upper half of LG11, explaining up to 48% of phenotypic variance. QTL for Fe and Zn content also co-localized on LGs 01, 06 and 11 in a RIL population developed from a 'AND696' × ' G19833' cross (Cichy et al. 2009a). A new QTL for Fe and Zn concentrations was mapped on LG06 in the inter-gene pool RIL population from 'G14519'× 'G4825' cross (Blair et al. 2010). Other QTL for both mineral concentrations were found on LGs 02, 03, 04, 07 and 08, which were also mostly novel compared to loci found in previous studies. In addition, the evaluation of a $BC_2F_{3:5}$ introgression line population derived from genotype 'G10022' backcrossed into 'Cerinza' allowed for the identification of four QTL associated with Fe and Zn content on LGs 03, 05 and 07 (Blair and Izquierdo 2012).

4.4.2.3 Phosphorus Use Efficiency

Among the edaphic stresses, phosphorus (P) deficiency is the primary constraint to common bean production in the tropics and subtropics, limiting seed yield to 60% of the bean-producing areas of Latin America and Africa (Wortmann et al. 1998). Root hair length, adventitious rooting and basal root growth angle in low-P soils were shown to be under the control of QTL (Miguel 2004).

The RIL population derived from the 'DOR364' (P inefficient) × 'G19833' (P efficient) cross has been widely used to study the morphological, physiological and genetic mechanisms underlying P efficiency. Yan et al. (2004) detected an association between root hair growth, acid exudation and P uptake, as well as two QTL for P uptake on LGs 04 and 10, which were closely linked to three QTL for root-exudation. The same mapping population was used to detect QTL associated with root gravitropism and their influence in the acquisition of P (Liao et al. 2004). QTL for P uptake were closely linked to QTL for shallow basal root length on LGs 04, 07 and 11. Beebe et al. (2006) confirmed in the same population that P acquisition was associated with basal root development and specific root length. A RIL population derived from the 'G2333' × 'G19839' cross was used to identify a total of 19 QTL for adventitious root traits (Ochoa et al. 2006). Two QTL for the number of adventitious roots under low P were mapped on LGs 02 and 09 and explained 61% of total phenotypic variation. In low-P conditions, two P-uptake QTL on LGs 07 and 11, and one P use efficiency QTL on LG08 were identified in the 'AND696' × 'G19833' Andean mapping population (Cichy et al. 2009b). A total of six QTL, three under each high and medium P soil, were mapped on LGs 06, 07 and 11 and on LGs 02, 07 and 10, respectively, in the mapping population from 'G2333' × 'G19839' cross (Blair et al. 2009b). The QTL on LG11 co-localized with a QTL for the number of adventitious roots (Ochoa et al. 2006), suggesting that this trait may have led to increased P uptake.

Several candidate genes induced by low P have been isolated in various plant species including legumes. In common bean, a total of 3,165 ESTs belonging to P-starved root cDNA library were reported by Ramírez et al. (2005) and a limited number (575) was registered in the NCBI database (http://www.ncbi.nlm.nih.gov/dbEST). An in silico approach for the identification of genes involved in the adaptation of common bean and other legumes to P-deficiency has also been reported (Graham et al. 2006). Over 240 putative P starvation-responsive genes were identified in a cDNA library (Tian et al. 2007). Full-length cDNAs for three genes, representing *PvIDS4*-like, *PvPS2* and *PvPT1,* were cloned and characterized. The open reading frames contained a SPX domain, a putative phosphatase and a P transporter, respectively. It is also worth mentioning that Hernández et al. (2007) have completed a transcript profiling of bean plants grown under P-deficient and P-sufficient conditions and showed 126 genes with a significant differential expression, of which 62% were induced in P-deficient roots. Finally, variations in the microRNA 399-mediated *PvHO2* regulation within the *PvPHR1* transcription factor were found in two contrasting genotypes (P-tolerant 'BAT477' and

P-sensitive 'DOR364') (Valdés-López et al. 2008; Ramírez et al. 2013).

4.4.2.4 Tolerance to Aluminium Toxicity

Other edaphic constraints include toxicities of aluminium (Al) associated with acid soil together with low calcium (Ca) availability (Rao 2001). Genetic variation exists for Al tolerance among common bean genotypes (Rangel et al. 2005). The most frequently measured effect of Al excess is inhibition of root elongation (Rangel et al. 2005). While root traits in the presence of Al are controlled by many genes in common bean, QTL for root morphological traits identified under the stress of Al were located on six genomic regions of the 'DOR364' × 'G19833' RIL population (López-Marín et al. 2009). A total of 12 QTL were involved in specific mechanisms of Al resistance, two of them in the same genomic regions, where QTL for the length of shallow basal roots and P acquisition efficiency were identified by Liao et al. (2004).

Rangel et al. (2010) hypothesized that the expression of a citrate transporter and the enhanced synthesis of citrate are crucial for sustained Al resistance in common bean. Eticha et al. (2010) corroborated these results, showing that the Al-induced expression of a citrate transporter gene family *MATE* (multidrug and toxin extrusion family protein) in root apices is a prerequisite for citrate exudation and Al resistance in common bean. In addition, Al-induced inhibition of root elongation was positively correlated with the expression of an *ACCO* (1-aminocyclopropane-1-carboxylic acid oxidase) gene in the root apex (Eticha et al. 2010). The expression of *MATE* and *ACCO* genes has been used as a sensitive indicator of Al impact on the root apex in common bean (Yang et al. 2011).

4.4.2.5 Symbiotic Nitrogen Fixation (SNF)

Enhancing the natural capacity for biological nitrogen (N) fixation has proved to help overcome the loss of soil fertility (Hungria and Vargas 2000). Studies on symbiotic nitrogen fixation (SNF), nitrogenase activity and nodulation-related traits (Nodari et al. 1993b; Tsai et al. 1998; Souza et al. 2000; Ramaekers et al. 2013; Kamfwa et al. 2015b) suggest that these traits have a complex inheritance with the involvement of multiple genes. Most of these studies focused on QTL analyses of nodulation traits and shoot dry weight under N-fixing conditions.

The first QTL study on nodule number (NN) trait in common bean was performed by Nodari et al. (1993b), who reported four genomic regions for NN in a RIL population derived from the 'BAT93' × 'Jalo EEP558' cross; all QTL together accounted for 50% of the phenotypic variation. One of these four genomic regions influenced both NN resistance and CBB resistance. Tsai et al. (1998) screened the same RIL population, but at two different soil N levels. They confirmed previous findings by Nodari et al. (1993b) and reported three QTL for NN in high N levels, one QTL associated with *CH18* (chitinase) and the other two QTL with *CHS* (chalcone synthase) and *PAL-1* (phenylalanine ammonia lyase). Souza et al. (2000) identified also in the same RIL population seven QTL under low N conditions and five under high N conditions, accounting for 34 and 28% of the total phenotypic variation, respectively. In accordance with Nodari et al. (1993b), Souza et al. (2000) indicated that NN resistance and CBB resistance in common bean have overlapping QTL on LGs 02, 03, 07 and 11. Ramaekers et al. (2013) conducted a QTL analysis of SNF and related traits under greenhouse and field conditions in a RIL population derived from the 'G2333' × 'G19839' cross, which resulted in two QTL for per cent N fixed in greenhouse located on LGs 01 and 04.

DNA sequence comparison of markers closely linked to these QTL allowed for the detection of some potential candidate genes. One of these genes encodes an auxin-responsive transcription factor and explained differences in N accumulation between climbing and bush beans. Alternatively, an AP2/ERF-domain-containing transcription factor underlies the QTL for the total amount of symbiotic N fixed in the field

(Ramaekers et al. 2013). Recently, Kamfwa et al. (2015b) detected 11 significant SNPs (five on LG03 and six on LG09) for nitrogen derived from atmosphere (Ndfa) in the shoot at flowering and for Ndfa in the seed in an Andean diversity panel of 259 common bean genotypes. Two genes *Phvul.007G050500* and *Phvul.009G136200* that code for leucine-rich repeat receptor-like protein kinases were identified as candidate genes for Ndfa.

4.4.3 Genes and QTL for Agronomic and Quality Traits

Agronomic and quality-related traits are almost always quantitative traits in plant species. Table 4.4 includes many of the research studies that have been carried out to identify QTL for agronomic and quality traits in common bean.

4.4.3.1 Plant Height and Traits Related to Vegetative Growth

Determinacy is controlled by the *FIN* gene that is located on LG01, where the dominant allele causes an indeterminate growth habit (Koinange et al. 1996). *PvTFL1y* is homologous to the *Arabidopsis TFL1* (Kwak et al. 2008) and is responsible for determinacy in common bean (Repinski et al. 2012; González et al. 2016). The *TOR* gene controls twining and correlates with *FIN*, suggesting that either *FIN* had a pleiotropic effect on twining or *TOR* was tightly linked to *FIN* (Koinange et al. 1996). Likewise, Koinange et al. (1996) reported pleiotropic effects or tight linkage of *FIN* on plant height, number of days to flowering and maturity, number of pods and harvest index. Another gene for growth habit was located at the end of LG11 and was significantly associated with QTL for days to flowering and maturity (Tar'an et al. 2002). Growth habit

Table 4.4 Details of QTL mapping studies performed for the mapping of major genes and QTL for agronomical and quality traits in common bean

Trait	Gene/QTL name[a]	LG[b]	Population[c]	Reference
Plant height	*FIN, TOR*	1	MG	Koinange et al. (1996)
	GH	11	S95	Tar'an et al. (2002)
	HABIT	4	A55G	Kolkman and Kelly (2003)
	PH, TB, TN, Ag	1, 2, 9, 11	S95	Tar'an et al. (2002)
	BP	7	A55G	Kolkman and Kelly (2003)
	Angle, Height	3, 4, 5	WO	Beattie et al. (2003)
	ph1.1, ph6.1, ph6.2, pw6.1, pw6.2, pw7.1, ph7.1	1, 6, 7	CCCG	Blair et al. (2006)
	Plh1-2, Plh1-1, Plh1-3, Plh1-4, Plh2-1, Plh2-3, Plh2-2, Int2, Int3, Int4, Int2, Int3, Int1, Cab1-1, Cab1-4, Cab1-2, Cab1-1, Cab1-3, Cab1-5, Cab1-5, Cab2-1, Cab2-1, Brn1	3, 4, 5, 7, 8, 10, 11	G23G19	Checa and Blair (2008)
	Ph1.1 [AG], *Ph1.2* [AG], *Ph1.3* [AG], *Ph7.1* [AG]	1, 7	A55G	Chavarro and Blair (2010)
Days to flowering	*PPD, HR*	1	MG, RM, RR	Koinange et al. (1996), Gu et al. (1998), Kwak et al. (20080
	Tip	unmapped	GC, FR	White et al. (1996)
	PvTFl1y, PvTFL1z, PvGI, PvZTL, PvFLD	1, 7, 9, 11	BJ, MG	Kwak et al. (2008)
	DF1.1, DF1.2, PD	1	MG	Koinange et al. (1996)
	DF11	11	S95	Tar'an et al. (2002)
	df1.1, df2.1, df6.1, df6.2, df9.1, df9.2, df11.1	1, 2, 6, 9, 11	CCCG	Blair et al. (2006)
	Df1.1, Df1.2, Df1.3, Df2.1, Df2.2, Df2.3, Df3.1, Df3.2, Df7.1	1, 2, 3, 7	A55G	Chavarro and Blair (2010)
	DF1, DF2, DF8, DE1, DE2, DE6.1, DE6.2	1, 2, 6, 8	Xco	Pérez-Vega et al. (2010)
	Df4.1, Df4.2, Df5.1, Df5.2, Df5.3, Df6.1, Df7.1, Df11.1	4, 4, 5, 5, 5, 6, 7, 11	DB	Blair et al. (2012)

(continued)

Table 4.4 (continued)

Trait	Gene/QTL name[a]	LG[b]	Population[c]	Reference
Seed size	*SW*	1, 3, 4, 7	BJ	Nodari (1992)
	SW	1, 7, 7, 11	MG	Koinange et al. (1996)
	SW, SL, SH	2, 3, 4, 5, 6, 7, 8, 11	PX	Park et al. (2000)
	QTL1-SM, QTL2-SM, QTL3-SM, QTL4-SM, QTL5-SM	1, 2, 3, 4, 5	BG	Guzmán-Maldonado et al. (2003)
	sw2.1, sw2.2, sw3.1, sw6.1, sw7.1, sw8.1, sw8.2, sw9.1, sw10.1, sw11.1	2, 3, 6, 7, 8, 9, 10, 11	CCCG	Blair et al. (2006)
	Swf3.1, Swf4.1, Swf11.1	3, 4, 11	DG	Beebe et al. (2006)
	SW6, SW8.1, SW8.2, SL2, SL3, SL6, SL8, SL10, SH6, SH8, WI3, WI6, WI7	2, 3, 6, 7, 8, 10	Xco	Pérez-Vega et al. (2010)
	SW1^{PP}, *SW6*^{PP}, *SW9.1*^{PP}, *SW9.2*^{PP}, *SL1.1*^{PP}, *SL1.2*^{PP}, *SL2.1*^{PP}, *SL6*^{PP}, *SL7*^{PP}, *SL10*^{PP}, *SWI2*^{PP}, *SWI7*^{PP}, *SWI9*^{PP}, *ST2*^{PP}, *ST9*^{PP}	1, 2, 6, 7, 9, 10	P1037	Yuste-Lisbona et al. (2014a)
Pod size	*PL*	1, 2, 7	MG	Koinange et al. (1996)
	Podlength, podheight, podwidth	2, 6, 8, 10	MO	Davis et al. (2006)
	PH, PL, PWT	1, 3, 4, 5, 6, 7, 9, 10	RO	Hagerty (2013)
	PL1^{PP}, *PL4*^{PP}, *PL11*^{PP}, *PWI1*^{PP}, *PWI4*^{PP}, *PT4.1*^{PP}, *PSI1.1*^{PP}, *PSI1.2*^{PP}, *PSI4*^{PP}, *PBL1.1*^{PP}, *PBL1.2*^{PP}, *PBL1.3*^{PP}, *PBL4*^{PP}	1, 4, 11	P1037	Yuste-Lisbona et al. (2014b)
Yield	*NP*	1, 8, 4	MG	Koinange et al. (1996)
	PPP, SPP, SY	2, 5, 9, 11	S95	Tar'an et al. (2002)
	SYD, HI	5, 6, 7, 11,	CDRKY	Johnson and Gepts (2002)
	PP, Y	2, 3, 5	WO	Beattie et al. (2003)
	pp7.2, pp9.2, pp11.3, sp6.1, sp7.1, sp7.2, yld2.1, yld3.1, yld3.2, yld4.1, yld4.2, yld4.3, yld4.4, yld9.1, yld9.2	2, 3, 4, 6, 7, 9, 11	CCCG	Blair et al. (2006)
	yield2004, yield2005, yield2006	3, 5, 10, 11	J115	Wright and Kelly (2011)
	nppr4.1, nppr5.1, npp4.1, npp10.1, yld3.1, yld3.2, yld4.1, yld10.1	3, 4, 5, 10	G23G19	Checa and Blair (2012)
	PHI1.1^{SC}, *SY3.3*^{SC}, *SY9.1*^{SC}, *SY9.2*^{SC}, *NP3.1*^{SC}, *NP8.1*^{SC}	1, 3, 8, 9	SC	Mukeshimana et al. (2014)
Colour	*P, Asp*	7	MG	Koinange et al. (1996)
	Gy, C, R, J, G, B, Rk	2, 4, 8, 10	W593	Bassett et al. (2002)
	Ana, Bip, C, G, V, Gy, Z, T	3, 4, 6, 8, 9, 10	BJ	McClean et al. (2002)
	Prp	8	BJ	Kelly and Vallejo (2004)
	QTL1-TA, QTL2-TA, QTL3-TA, QTL4-TA	2, 3, 4	BG	Guzmán-Maldonado et al. (2003)
	Cst1a, Cit1a, Ctt1a, Cst1b, Cst2b, Cit1b, Ctt1b, Ctt2b, Ctt3b, Cst1c, Cit1c, Ctt1c	3, 6, 7, 8, 9, 10	BJ	Caldas and Blair (2009)
	Color2005, Color2006, Color2007	1, 3, 5, 8, 11	J115	Wright and Kelly (2011)
	PSC3^{PP}, *PSC4*^{PP}, *PSC7.1*^{PP}, *PSC7.2*^{PP}, *PSC9*^{PP}, *SSC4*^{PP}, *SSC7*^{PP}, *SSC8.2*^{PP}, *SSC9*^{PP}, *PC2*^{PP}, *PC6*^{PP}, *PC7.1*^{PP}, *PC8*^{PP}	2, 3, 4, 6, 7, 8, 9	P1037	Yuste-Lisbona et al. (2014a, 2014b)

(continued)

Table 4.4 (continued)

Trait	Gene/QTL name[a]	LG[b]	Population[c]	Reference
Quality	*St*	2	MG	Koinange et al. (1996)
	PvIND	2	BJ, MG	Gioia et al. (2012)
	PvSHP1	6	BJ, MG	Nanni et al. (2011)
	APP, SPLT	8, not assigned	MCDRK, MCELRK	Posa-Macalincang et al. (2002)
	podstring	6	MO	Davis et al. (2006)
	WA3, WA4, CP3, CP7	3, 4, 7	Xco	Pérez-Vega et al. (2010)
	Texture2005, Texture2006, Visual appearance2005, Visual appearance2006, Washed-drainedweight2006	3, 5, 6, 8, 10, 11	J115	Wright and Kelly (2011)
	SpA, SpB, SpE, SpI, SpJ, Phs, SpF, SpG, SpK, SpL, SpM, SpC, SpD	3, 4, 5, 7	Xco	Campa et al. (2011)
	ct1.3, ct1.1, ct9.2	1, 9	CL	Vasconcelos et al. (2012)
	PST, PFB	1, 4	RO	Hagerty (2013)
	A5xc, AA9xc, As1xc, As7xc, Ca1xc, Ca7xc, Ca9xc, DF6xc, DF7xc, Mg7xc, Pt5xc, Pt7xc, S1xc, S2xc, S4xc, S9.1xc, S9.2xc, UA5xc, UA7xc	5, 9, 1, 7, 1, 7, 9, 6, 7, 7, 5, 7, 1, 2, 4, 9, 9, 5, 7	Xco	Casañas et al. (2013)
	Anthavg, L10, b11, L11, color11, W_uptake, Soak_Anth, text11, hc11, text10, wdc10, hc10, wdc11, a11, b11, L10	4, 5, 7, 11	BS	Cichy et al. (2014)

[a]*PH, H, ph, Plh* = plant height; *TN* = total nodes; *TB, Brn* = total branches; *Ag, A* = branch angle; *BP* = branching pattern; *pw* = plant width; *Int* = internode length; *Cab* = climbing ability; *DF, df*: days to flowering; *PD*: photoperiod-induced delay influencing; *DE*: days to end of flowering; *SW, sw, Swf*: seed weight; *SL*: seed length; *SH*: seed height; *SM*: seed mass; *WI, SWI*: seed width; *ST* = seed thickness; *PL*: pod length; *PH*: pod height; *PWT*: pod wall thickness; *PT*: pod thickness; *PSI*: pod size; *PBL*: pod beak length; *NP, PPP, PP, pp, npp*: number of pods per plant; *SPP, sp*: seeds per plant; *Y, SYD, yld, SY*: seed yield; *HI*: harvest index; *npr*: number of pods per raceme; *PHI*: pod harvest index; *TA*: tannins; *Cst*: condensed soluble tannins; *Cit*: condensed insoluble tannins; *Ctt*: condensed total tannins; *PSC*: primary seed colour; *SSC*: secondary seed colour; *PC*: pod colour; *WA* = water absorption; *CP*: coat proportion; *ct*: cooking time; *PBF*: pod fibre; *PST*: strings; *A*: amylose; *AA*: apparent amylose; *As*: Ashes; *Ca*: calcium; *DF*: dietary fibre; *Mg*: magnesium; *Pt*: protein; *S*: starch; *UA*: uronic acids; *W uptake*: water uptake; *Soak-Anth*: anthocyanin concentration of the soak water after 12 h; *Anthavg*: anthocyanins; *wdc*: washed drained weight coefficient; *text*: texture; *HC*: hydration coefficient; *L*: lightness; *b*: blue/yellow; *a*: red/green

[b]LG = linkage group

[c]Mapping population acronyms: A55G = A 55 × G 122; BJ, BG = Bayo Baranda × G-22837; BAT93 x JaloEEP558; BS = Black Magic × Shiny Crow; CCCG = Cerinza × Cerinza × (Cerinza × G24404); DB = DOR364 × BAT477; CDRKY = California Dark Red Kidney × Yolano; CL = CNFM7875 × Laranja; DG = DOR364 × G19833; FR = Flor de Mayo × Rojo 70; GC = Gordo × de Celaya; G23G19 = G2333 × G19839; J115 = Jaguar × 115 M; MG = Midas × G12873; MO = Minuette × OSU5630; MCDRK = Montcalm × California Dark Red Kidney 82; MCELRK = Montcalm × California Early Light Red Kidney; S95 = OACSeaforth × OAC95; P1037 = PMB0225 × PHA1037; PX, PC50 × XAN159; RM = Redkloud × MAM; RO = RR6950 × OSU5446; RR = Redkloud Rojo; SC = SEA5 × CAL96; W593 = Wagenar × BC3 5-593; Xco = Xana × Cornell49242; WO = WO3391 × OAC Speedvale

(locus *habit*) mapped on LG04 (Kolkman and Kelly 2003). Tar'an et al. (2002) mapped one QTL for total branches on LG02 and one QTL for branch angle on LG11. QTL were detected on LG07 for branching pattern (Kolkman and Kelly 2003). Beattie et al. (2003) mapped two QTL for plant height on LGs 03 and 04 and three QTL for branch angle on LGs 03 and 05. QTL for plant height were found at the *ATA5* locus on LG01, at the *V* locus on LG06 and at the *Phs* locus on LG07 (Blair et al. 2006). QTL were found for plant height, climbing ability, internode length and branch number on LGs 03, 04, 08 and 11 (Checa and Blair 2008). Chavarro and Blair (2010) discovered a cluster of QTL for different plant height traits on LG01.

4.4.3.2 Flowering Date and Photoperiod Response

Flowering time is a key issue contributing to the adaptation and range of expansion of the short-day common bean (Vadez et al. 2012; Weller and Ortega 2015). Two loci, *PPD* and *HR*, are known to affect photoperiod response in common bean. *PPD* has been mapped within

5 cM of the *FIN* locus on LG01 (Koinange et al. 1996; Kwak et al. 2008), where recessive alleles confer reduced photoperiod response and early flowering under long days. This region is syntenic with the region in soybean containing the *E3/PHYA3* gene (McClean et al. 2010) and, as expected, contains the bean *E3* orthologue, suggesting this as an attractive candidate for the *PPD* locus. The second locus *HR* is less well-defined but is positioned towards the other end of the same linkage group (Gu et al. 1998), a region containing homologues of *ELF3* and the *FTa/c* cluster. Mapping *HR* was difficult because homozygous *ppd* is epistatic over *HR*, and thus, genotypes, *ppdHR* and *ppdhr*, give the same insensitive response to photoperiod. In addition, the expression of *HR* is influenced by the environment, producing an overlap of intermediate and highly sensitive genotypes. A third locus identified as a QTL on LG9 is located near the bean orthologue of *ZEITLUPE*, an important gene for circadian clock regulation in *Arabidopsis* (Tar'an et al. 2002; Kwak et al. 2008). QTL controlling flowering time and other flowering-related traits have now been identified in common bean (Koinange et al. 1996; Tar'an et al. 2002, Chavarro and Blair 2010; Pérez-Vega et al. 2010). Blair et al. (2006) detected two QTL on LG09, explaining 13 and 22% of the phenotypic variation, while Chavarro and Blair (2010) found a cluster of QTL for flowering time on LG01 close to *FIN* genomic region. Clusters of QTL for days to flowering were also found by Pérez-Vega et al. (2010) and González et al. (2016) on LG01 (close to the *FIN* gene) and LG02 (close to the *I* gene), as well as Blair et al. (2012) on LGs 04, 05, 06, 07 and 11.

4.4.3.3 Seed and Pod Size

The Danish plant scientist Wilhelm Johannsen (1911) concluded that a genetic effect could influence seed size in self-fertilizing beans, detecting segregation for seed size in a progeny from a large × small seed size population. A few years later, seed size was described as a polygenic trait in common bean (Sax 1923) with at least 10 genes involved in its genetic control (Motto et al. 1978). Quantitative inheritance of seed size has been reported by Vallejos and Chase (1991). QTL for seed size were mapped on LGs 01, 03, 04, 07 and 11 (Nodari 1992; Koinange et al. 1996). Park et al. (2000) found QTL for seed size traits on LGs 02, 03, 04, 05, 06, 07, 08 and 11. QTL on LG07 span the *PHS* locus that codes for phaseolin seed protein (Nodari 1992; Koinange et al. 1996; Park et al. 2000). Guzmán-Maldonado et al. (2003) reported five QTL for seed weight, explaining 42% of the phenotypic variation, while Blair et al. (2006) identified 10 QTL across eight LGs, explaining from 4 to 17% of the phenotypic variation, which agrees with the previous studies (Koinange et al. 1996; Park et al. 2000; Tar'an et al. 2002). Seed size QTL mapped near the upper end on LGs 02 and 06; the lower end on LGs 03, 07, 08 and 10; and near the centre on LGs 06 and 08 (Park et al. 2000; Blair et al. 2006; Pérez-Vega et al. 2010). Yuste-Lisbona et al. (2014a) detected QTL for seed weight on LGs 01, 02, 06, 07, 09 and 10 that were consistent with QTL mapped by Park et al. (2000) and Pérez-Vega et al. (2010).

Four genes (*Ea*, *Eb*, *Ia* and *Ib*) control pod cross-sectional shape although the exact genetics is uncertain (Leakey 1988). QTL for pod size have been reported by Koinange et al. (1996) on LGs 01, 02 and 07. QTL for pod length and height clustered together on LGs 01 and 03 (Hagerty 2013). Yuste-Lisbona et al. (2014b) detected 17 QTL for pod size traits, which were distributed throughout most of the LGs except for LG02. Most of the QTL affecting pod size clustered on LGs 01 and 04, which indicates that these genomic regions may contain linked genes or a gene with pleiotropic effects governing these traits.

4.4.3.4 Pod and Seed Yield

Yield is a quantitative trait influenced by many genes and is primarily conditioned by three components: number of pods per plant, number of seeds per pod and 100 seed weight (Adams 1967). Koinange et al. (1996) identified QTL for the number of pods per plant on LGs 01 (associated with *FIN*), 04, 08 and 11. Tar'an et al. (2002) mapped three QTL for seed yield on LGs 05, 09 and 11, explaining 28% of the total phenotypic variation, one QTL for the number of pods per plant on LG02 and one QTL for seeds per pod on LG05. Six QTL for seed yield and four QTL for harvest index were detected on LGs 05, 06, 07 and 11 (Johnson and Gepts 2002). Beattie et al. (2003) reported three yield QTL on LGs 03 and 05 and three QTL for the number of pods per plant on LGs 02, 03 and 05. Blair et al. (2006) reported nine QTL for seed yield on LGs 02, 03, 04 and 09, accounting from 9 to 21% of the phenotypic variation, and three QTL for the number of pods per plant, explaining up to 64% of the phenotypic variation. Likewise, they reported three QTL for the number of seeds per plant located on LGs 06, 07, 09 and 11, which explain from 15 to 29% of phenotypic variation. Wright and Kelly (2011) reported seven QTL for seed yield on LGs 03, 05, 10 and 11. Checa and Blair (2012) identified four QTL for yield on LGs 03, 04 and 10; two QTL for the number of pods per raceme on LGs 04 and 05; and two QTL for the number of pods per plant on LGs 04 and 10. Mukeshimana et al. (2014) reported QTL for seed yield on LGs 03 and 09. Three QTL for the number of pods per plant and pod harvest index mapped on LGs 01, 03 and 08. In spite of the different procedures used, several studies found QTL associated with the number of pods per plant and seed yield on LG03 (Beattie et al. 2003; Blair et al. 2006; Wright and Kelly 2011; Checa and Blair 2012). Recently, Qi (2015) characterized the *A. thaliana* homologue of *BnMicEmUp/AT1G74730* gene in common bean (*Phvul.009G190100*), which encodes a cbZIP transcription factor that could affect seed yield.

4.4.3.5 Seed and Pod Colour

The genetic control of the different patterns and colours of bean seeds has been studied by Beninger et al. (2000). The *P* gene determines the presence or absence of flavonoids in the seed coat, and the specific colour depends on the epistatic interactions of the alleles at the other genes (Erdmann et al. 2002). The *Asp* gene controls the shine of the seed coat. Both genes are located on LG07 (Koinange et al. 1996). In *P_* individuals, a multiallelic serie at *V* gene controls flower colour, with the genotypes *V_* (purple) > v^{lae}*_* (pink) > *vv* (white) (Beninger et al. 1999, 2000). Alleles at other genes (*Gy*, *C*, *R*, *J*, *G*, *B* and *Rk*) interact with *V* and with each other to determine the many colours found in the seed coat (Bassett et al. 2002). *Ana*, *Ane*, *Bip*, *L*, *T* and *Z* are genes of the pattern and colour of the seed (McClean et al. 2002). The *J* locus for seed coat shininess is located on LG10 (Freyre et al. 1998; Galeano et al. 2011). The colour modifying *B* gene is linked to the *I* gene for BCMV resistance on LG02 (Nodari et al. 1993b). The seed coat colour genes *C*, *G*, *V* and *Gy* have been mapped on LGs 08, 04, 06 and 08, respectively (McClean et al. 2002). The seed pattern of *Z* and *T* genes was located on LGs 03 and 09, and the *Bip* and *L* loci on LG10. Wax bean pod colour is controlled by a single recessive gene (*y*), but may be affected by a second gene (*arg*) and perhaps other modifiers (Currence 1931). *P* and *V* genes control solid purple colouring or purple stripes depending on the allele at the [*C Prp*] complex locus (Bassett 1996; Bassett et al. 2005). The *Prp* (purple pod) locus was located on LG 08 (Kelly and Vallejo 2004).

The pigments responsible for variations in seed colour are flavonoids, principally flavonol glycosides, anthocyanins and condensed tannins (Beninger et al. 1999). Four QTL for tannin content were detected (Guzmán-Maldonado et al. 2003), explaining 42% of the phenotypic variation. Caldas and Blair (2009) found twelve QTL for tannin content, explaining from 10 to 64% of the phenotypic variation. Yuste-Lisbona et al. (2014a) showed that seed colour is controlled by a QTL located near the *P* locus, explaining from 27 to 42% of the phenotypic variation. In addition, QTL for pod colour were found on LGs 06 and 08, which may correspond to the *Prp* and *V* genes, and a major QTL on LG07, where the locus *P* was previously identified (Erdmann et al. 2002; Koinange et al. 1996; Vallejos et al. 1992; Yuste-Lisbona et al. 2014a). A QTL analysis revealed that the region near the *Asp* gene (seed coat shininess) on LG07 contained 141 genes, the best gene candidate for *Asp* being a FAE1/Type III polyketide synthase-like protein that acts as a fatty acid elongase (Cichy et al. 2014).

4.4.3.6 Other Quality Traits

Despite the genetic complexity of most common bean quality traits, many studies have recently reported interesting molecular and functional results about this topic. Koinange et al. (1996) found that the lack of pod suture fibres was controlled by a major gene (*St* locus) on LG02. Gioia et al. (2012) amplified a gene homologous to *INDESHICENT* (*IND*, a factor required for silique shattering in *Arabidopsis*) that mapped next to the *St* locus. The homologue *PvSHP1* (*SHATTERPROOF-1* in *Arabidopsis*) was mapped on LG06, linked to seed colour gene *V* (Nanni et al. 2011) and close to a QTL for pod string, explaining 26% of the phenotypic variation (Davis et al. 2006). Hagerty (2013) detected a pod suture string QTL on LG01 and a pod fibre QTL on LG04 clustered to QTL for pod length, height and thickness. QTL for seed weight and length have been mapped on LG08 (Park et al. 2000; McClean et al. 2002).

In the last two decades, several markers and QTL have been associated with organoleptic quality traits of different nature, from water absorption and coat proportion (Pérez-Vega et al. 2010) to cooking time (Vasconcelos et al. 2012). It is interesting to note that major QTL for colour retention mapped in overlapping positions to four flavonoid biosynthesis genes (two of which code for chalcone synthase proteins), while other minor effect QTL co-localized with anthocyanin-related genes (Wright and Kelly 2011). In addition, five QTL associated with content of ash, calcium, dietary fibre, magnesium and uronic acid were mapped on LG07, close to *P* locus and QTL for content of tannins and seed coat proportion (Caldas and Blair 2009; Pérez-Vega et al. 2010). Two QTL for content of ash and calcium were detected close to the *FIN* gene (LG01), and one QTL for protein content was located on LG07 close to the *Phs* cluster (Campa et al. 2011).

4.5 Epistatic and Environmental Interactions Among QTL

The goal of QTL mapping is to identify the genes/regions responsible for generating differences between individuals within a polymorphic population. Such phenotypic variation in the population can be divided into three components: (i) the contribution of QTL main effects, (ii) the role of QTL × QTL interactions or epistatic effects and (iii) the influence of QTL × environmental interaction effects. Inaccurate estimation of these effects further reduces the power and precision of QTL detection. Nonetheless, most of the QTL reports on common bean have not taken into account the identification of both epistatic and environmental effects.

Epistasis is considered an integral part of the genetic architecture of quantitative traits (Parvez et al. 2007), and in autogamous plants, it is expected to have significant effects on traits controlled by several genes/QTL, as pointed out by Holland (2001). Therefore, not only can epistasis be considered the major barrier to inferring the genetic basis of a given trait, but it also hampers the efficiency of breeding programmes. A direct implication of epistasis is that the fitness of individual alleles could be affected (increase or decrease) when they are found together in a given genotype (Holland 2007). If alleles involved in positive epistatic interactions are not transferred together to the cultivar that is being developed, improvement will be unsuccessful due to the presence of epistatic effects (Lark et al. 1995). Thus, any attempt to use QTL for improved plant performance and adaptation to different environmental conditions should take into account such epistatic effects, involving selection methods which tend to accumulate favourable allele combinations in the same genotype. Hence, the identification of QTL and the elucidation of their genetic control (main and epistatic effects) are essential for the development of efficient MAS programmes aimed at improving breeding efficiency.

The presence of epistasis can greatly obscure the mapping between genotype and phenotype. The effects of QTL may be masked by interactions with other loci, which can make mapping difficult (Phillips 2008). According to Asins (2002), the lack of information about QTL × QTL interaction could be explained by the plant material employed in the experiments. Epistatic interactions can hardly be detected in F_2 or BC populations. The reason is that in F_2 generations, even if large populations are used, there are insufficient individuals with two-locus double homozygotes, whereas in BC, every generation reduces the number of gene combinations while increasing genes from the recurrent genotype. Thus, the appropriate segregant populations would be RIL or doubled haploid (DH) populations, as additive and epistatic interactions effects can be detected but not dominant or over-dominant effects.

Johnson and Gepts (2002) found a reduced average fitness in the progeny of an inter-gene pool RIL population 'California Dark Red Kidney' × 'Yolano' that could be attributed to a break-up of co-adapted gene complexes or low viability of preferred epistatic relationships. They found that digenic epistatic interactions clearly played an important role for the number of days to maturity, average daily biomass, seed yield accumulation and harvest index. Both independently acting and digenic epistatic QTL of similar magnitude were identified. A total of 22 epistatic interactions were detected for the four traits evaluated. Each of the interactions accounted on average for 10% of the variation in the traits. In addition, eight interactions included a locus that also had a significant effect as independently acting QTL. Hence, the results obtained by Johnson and Gepts (2002) showed that, in addition to independent QTL action, epistatic QTL interactions play an important role in the cross-analysis.

The importance of epistatic QTL in the genetic control of pod size and colour traits has been recently revealed by Yuste-Lisbona et al. (2014b), who used an Andean intra-gene pool RIL population from a cross between a cultivated common bean ('PMB0225') and an exotic 'nuña' bean ('PHA1037'). A common feature of the epistatic interactions detected for pod-related traits is that most of them occur between QTL with main additive effects, but QTL that showed only epistatic effects were also detected. Thus, 12 out of 18 epistatic QTL identified were previously detected as main-effect QTL. Interestingly, 6 out of the 12 epistatic interactions detected were identified for pod colour, whose interactions explained 13.3% of the phenotypic variance observed, indicating the significant role of epistasis in the genetic control of this trait. This complex genetic inheritance is in accordance with the results obtained by McClean et al. (2002), who reported the existence of many genes that exhibit epistatic interactions that define the many colours observed within the species. Likewise, the role of epistatic effects in the genetic control of popping ability and others seed quality traits has also been studied using the same RIL population (Yuste-Lisbona et al. 2012, 2014a). Overall, the results showed that digenic epistatic interactions clearly play a significant role in the genetic control of these traits in the Andean common bean intra-gene pool.

A qualitative digenic model of inheritance, discerning an interaction between two QTL conditioning disease resistance in plants, was reported by Vandemark et al. (2008). Two QTL based on the closest markers such as *BC420* and *SU91* are of particular interest to breeding programmes focused on enhancing resistance to CBB in common bean, which is caused by *Xanthomonas axonopodis* pv. *phaseoli* (Xap). Results mainly showed that the expression of *BC420* was epistatically suppressed by a homozygous recessive *su91/su91* genotype and the highest level of disease resistance was conferred by genotypes with at least a single resistance allele at both QTL (*BC420/-*; *SU91/-)*. The observed recessive epistatic interaction between the two QTL suggests that *SU91* is essential for the expression of an effective resistance mechanism. Moreover, this finding emphasized the need for breeders to correctly identify plants that are homozygous for both *SU91* and *BC420* loci, since breeding materials that are not fixed for *BC420* may produce moderately resistant progeny in subsequent generations, while plants that are not fixed for *SU91* may produce susceptible progeny.

Recently, new insights into the role of epistasis in anthracnose resistance were provided by González et al. (2015). A total of 39 epistatic QTL (21 for resistance to race 23 and 18 for resistance to race 1545) involved in 20 epistatic interactions (eleven and nine interactions for resistance to races 23 and 1545, respectively) were identified in an Andean intra-gene pool RIL population. Depending on the race and organ tested, the total phenotypic variation explained by epistatic interactions ranged from 3 to 15%. Most of the epistatic interactions detected were due to loci without detectable QTL additive main effects, which showed the importance of the epistatic effects in genetic resistance to anthracnose.

What is more, in addition to epistatic effects, QTL × environmental interaction effects similarly complicate the use of MAS as genetic variance at one QTL may be sufficiently large in one environment but not in another. For instance, Jung et al. (2003) clearly showed the discrepancies among different environments regarding the locations and effects of QTL for bacterial brown spot resistance in the same mapping population. Thus, mapping QTL under natural infection in the field and artificial inoculation in growth chamber in two years revealed the existence of four QTL on LGs 02, 03, 04 and 09 in 1996, whereas two QTL on LGs 02 and 08 were detected in 1998. Only the genomic region on LG02 was significantly associated with bacterial brown spot resistance over both years (see Table 4.2). Similarly, depending on the experimental conditions, different QTL were identified by Beattie et al. (2003). Of the 21 QTL identified for plant architecture and yield traits, only 10 QTL (48%) were detected across all environments and, in most cases, these were the QTL with the largest influence on a given trait.

Asfaw et al. (2012) identified QTL for traits related to photosynthate mobilization across different drought stress and non-stress environments. The results showed that when using composite interval mapping for each individual environment, many QTL were detected, but these tend to be site-specific. However, when using a multienvironmental approach, only a small number of stable QTL and a high QTL × environmental interaction effects were identified. In addition, Asfaw and Blair (2012) detected root length QTL with significant QTL × environmental interaction effects under drought stress *versus* non-stress conditions. Interestingly, the QTL × environmental interaction effects were not attributed to the contrasting effects of the parental alleles between non-stress and stress environment, rather they were attributed to the differential expression of paternal alleles in different environments.

Long-day and short-day natural photoperiod conditions have been used by Yuste-Lisbona et al. (2012, 2014a, b) in order to extend the knowledge of the QTL × environmental interaction effects involved in common bean seed and pod quality traits. Among main-effect QTL detected by Yuste-Lisbona et al. (2014b), 11 QTL only exhibited significant genetic main effects, while 6 showed both significant genetic main effects and environmental interaction effects. As regards the epistatic interaction effects, only 2 out of 12 digenic interactions had environmental interaction effects. For seed shape and weight, as well as seed coat colour, 12 out of 32 main-effect QTL detected showed environmental interaction effects. Furthermore, only 6 out of 26 epistatic interactions identified had environmental interaction effects (Yuste-Lisbona et al. 2014a). Likewise, Yuste-Lisbona et al. (2012) showed that popping ability of 'nuña' bean is controlled by several QTL, which only have individual additive effects or may also be involved in epistatic or environmental interactions. Overall, even though most of the QTL detected were consistent over environment, some of them were subject to environmental modification. Despite this, QTL with differential effect on long-day and short-day environments were not found for seed and pod quality traits.

Finally, QTL × environmental interaction effects have also been reported for common bean resistance to angular leaf spot by Oblessuc et al. (2012). They revealed the existence of seven

QTL with variable magnitudes of phenotypic effects depending on the environments. One major QTL (*ALS10.1*) is highlighted with stable effect across environments. In addition, two QTL with minor effects (*ALS5.2* and *ALS4.2*) showed an interesting QTL × environmental interaction. *ALS5.2* revealed a greater resistance effect under greenhouse conditions, but only a small effect in the field experiments, whereas *ALS4.2* presented an opposite interaction with a greater resistance effect only under field conditions but not in the greenhouse. Hence, it is necessary to perform trials in different environmental conditions in order to draw conclusions about the genetic architecture of quantitative traits. However, a few multienvironmental QTL analyses have been carried out in common bean.

4.6 Genome-Wide Association Study (GWAS) Mapping

The association mapping (AM) exploits historical recombination events and has become a powerful alternative to linkage mapping for the dissection of complex trait variation at the sequence level (Zhu et al. 2008). There are two kinds of AM approaches: (i) candidate gene (CG) association mapping, which relates polymorphisms in selected candidate genes that have putative roles in controlling phenotypic variation for specific traits, and (ii) genome-wide association study (GWAS) mapping, also named genome scan, which surveys genetic variation in the whole genome to find associations with various complex traits (Risch and Merikangas 1996). The former implies good understanding of the biochemistry and genetics of the trait, while the latter requires a large number of well-distributed molecular markers and a broader reference population for the identification of numerous causative genetic polymorphisms with previously unappreciated biological function.

Advances in common bean genomics such as the sequenced genome (Schmutz et al. 2014) and the application of high-throughput and efficient genotyping platforms (Hyten et al. 2010; Goretti et al. 2014; Gujaria-Verma et al. 2016) have created the opportunity to conduct GWAS to dissect the genetic architecture of several complex traits in common bean. Moreover, common bean has been recognized as a valuable target for GWAS because of its extensive genetic diversity (Blair et al. 2009a). An extra advantage of the GWAS design for common bean is the homozygous nature of most varieties, which makes it possible to employ a 'genotype or sequence once and phenotype many times over' strategy, whereby once the lines are genomically characterized, the genetic data can be reused many times over across different phenotypes and environments. Additionally, in AM, unlike conventional QTL mapping, it is important to consider population structure and kinship among individuals, since false associations may be detected due to the confounding effects of population admixture (Oraguzie et al. 2007). Therefore, the divided population structure for common bean has made it necessary to consider the Andean and Mesoamerican gene pools as separate subgroups for AM. Several statistical methods have been proposed to account for population structure and familial relatedness, structured association (Falush et al. 2003), genomic control (Devlin and Roeder 1999), mixed model approach (Yu et al. 2006) and principal component approach (Price et al. 2006).

Several examples can be found in the literature in the 2010s to identify significant associations between agronomical and resistant traits and common polymorphisms in or near genes. GWAS not only identified previously reported QTL, but also resulted in narrower genomic regions than the regions reported as containing these QTL (see, e.g. Kamfwa et al. 2015b; Perseguini et al. 2015, 2016). Moreover, the results have also provided unprecedented views into the contribution of common variants to complex traits and new valuable markers for breeding that can now be used in common bean in future programmes. However, the time of GWAS is actually the beginning of a new age: one characterized by many new regions of the genome worthy of pursuit as candidate genes to explore. Advances in GWAS methodology and continued improvements in different genetic and genomic techniques would eventually make it possible to realize the potential offered by AM in identifying as many as possible of new genes underlying complex traits (Korte and Farlow 2013).

The application of GWAS for common bean was originally assessed in Shi et al. (2011). In this work, 395 dry bean lines of different market classes were genotyped with 132 SNPs and evaluated for association with CBB resistance. Twelve SNP markers co-localized with or close to previously identified CBB-QTL, and eight new resistance loci were identified. Later on, a panel including 93 genotypes, mainly of Andean origin, was genotyped with 110 SNPs and 24 SSRs, and several flowering and pod features were characterized (Galeano et al. 2012). From GWAS, four putative proteins (i.e. acyl-acp thioesterase, auxin response factor 2, transcription factor bhlh96-like and oxygen-evolving enhancer protein chloroplastic-like protein) were found to be associated with several traits. A whole-genome sequencing approach was conducted for a 280-member panel of modern Mesoamerican cultivars (34,799 SNPs) in order to understand the genetic architecture of days to flower, days to maturity, growth habit, canopy height, lodging and seed weight. About 30 candidate genes were detected by GWAS; among them, most of the components of cytokinin biosynthesis pathways, multiple-component phosphorelay regulatory systemand genes relative to the *Arabidopsis* flowering pathway were identified (Schmutz et al. 2014; Moghaddam et al. 2016). Similar agronomical traits were subjected to GWAS analysis by Nemli et al. (2014) in 66 common bean genotypes of different geographic regions. In addition, an Andean diversity panel of 237 genotypes of common bean was conducted to gain insight into the genetic architecture of phenology, biomass, yield components and seed yield traits (Kamfwa et al. 2015a). Interestingly, the *phyA* gene, which codes for phytochrome, was identified as a candidate gene involved in the genetic control of these traits. In addition, significant SNPs for seed yield were also identified on LGs 03 and 09, co-localizing with QTL for yield from the previous studies on LG09 (Tar'an et al. 2002; Blair et al. 2006; Wright and Kelly 2011; Checa and Blair 2012; Mukeshimana et al. 2014). These QTL were stable and expressed in diverse genetic backgrounds, which makes them useful tools for MAS breeding of yield in common bean. Taken together, GWAS results may provide markers and genes that are useful for common bean genetics, trait selection, breeding applications and genetic dissection of novel traits to widely characterize common bean germplasm diversity. Furthermore, the markers detected will be interesting for future association studies, wherein marker–trait associations are compared. However, further elucidation of gene function has not yet been achieved, but it must be acquired by further functional and experimental analysis.

4.7 Perspectives and Future Direction

Since the early 1990s, numerous studies have identified molecular markers linked to QTL involved in the inheritance of agronomically important traits in common bean (Kelly et al. 2003). QTL mapping approaches have proved to be enormously useful to identify loci of large effect and dissect the genetic basis of fairly simple traits. Following the discovery of promising loci and identification of molecular markers, MAS has been used to transfer single genes in adapted cultivars (Yu et al. 2000; Kelly et al. 2003; Faleiro et al. 2004) and to develop multiple-introgression lines with improved resistance (Mutlu et al. 2005a, b; Miklas et al. 2006). However, most QTL mapping studies used small population size and low marker density, which allows only for an approximate mapping of the chromosomal region. Therefore, identification of reliable QTL is a preliminary step in developing a MAS programme for genetic improvement. So as to transfer QTL in selective breeding or to identify functional genes, the identified major QTL should be fine-mapped to a higher level of resolution and verified or validated in additional genetic backgrounds and environments by developing advanced segregating populations with large number of recombinants in the region of interest. However, it is difficult to fine map several minor QTL associated with highly complex traits, such as drought tolerance and yield. Different factors may contribute to such failure or to an unexpected result in MAS: the magnitude of inconsistency of minor QTL, most QTL effects will have limited transferability across populations, epistatic and genotype-by-environment interactions, and population sizes of 500–1000 are needed for mapping QTL in order to eliminate the effects of sampling error (Bernardo 2008).

A new generation of genetic mapping populations must be designed with the aim to overcome many of the limitations of biparental QTL mapping and association mapping. A widely adopted strategy to estimate the position and effect of a mapped QTL more accurately is to create a new experimental population by crossing nearly isogenic lines (NILs) that differ only in the allelic constitution at the short chromosome segment harbouring the QTL (QTL-NILs) (Yamashita et al. 2014). In such populations, because of the absence of other segregating QTL, the target QTL becomes the only genetic source of variation, and the phenotypic means of the QTL genotypic classes can be statistically differentiated and genotypes recognized. Other populations combine the controlled crosses of QTL mapping with multiple parents and multiple generations of intermating. In this sense, MAGIC (multiparent advanced generation intercross) (Huang et al. 2015) and NAM (nested association mapping) populations (Buckler et al. 2009) would be an ideal resource to generate high-density maps using germplasm of direct relevance to the breeders. These designs require trade-offs among the amount of genetic variation sampled, the resolution of genetic mapping, the confounding effects of population substructure and the effort required to generate the mapping population. Another alternative genetic mapping strategy with higher resolution includes association analysis (Myles et al. 2009).

The advent of fast-evolving DNA sequencing technology has given a new direction in the field of common bean genomics by enabling sequencing of whole genome, extracting precious genomic information and resequencing in quick time and under manageable cost. Reduction of cost for sequencing leads to the development next-next- or third-generation sequencing technologies such as single-molecule real-time (SMRT) sequencing capable of generating longer sequence read (Thudi et al. 2012). A combination of GWAS and next-generation-mapping populations will improve the ability to connect phenotypes and genotypes, while genomic selection could take advantage of all these data for rapid selection and implementation in common breeding programmes. In addition to this, the next frontier in mapping and identification of candidate genes involved in complex traits is high-throughput phenotyping. Due to the development of NGS technologies, genomic resources are rapidly accumulating, but phenotypic data collected in a global context remain scarce. Automated platforms must be developed for phenotyping in growth chambers and controlled environments to provide new technologies for high-throughput phenotyping. The combination of these approaches and the promise of improved and cheaper genomic technologies will provide an opportunity to apply our understanding of the past to the future of common bean improvement.

Acknowledgements This work was financially supported by the Ministerio de Economía y Competitividad (AGL2014-51809-R project), Instituto Nacional de Investigación y Tecnología Agraria y Alimentaria (RF2012-C00026-C02-01 and RF2012-00026-C02-02 projects), Junta de Andalucía (Grant P12-AGR-01482 funded by Programa de Excelencia) and UE-FEDER Program. The authors would also like to thank Campus de Excelencia Internacional Agroalimentario-CeiA3 and Contrato Programa Xunta de Galicia-BAS for partially supporting this work.

References

Adam-Blondon A, Sévignac M, Dron M, Bannerot H (1994) A genetic map of common bean to localize specific resistance genes against anthracnose. Genome 37:915–924

Adams M (1967) Basis of yield component compensation in crop plants with special reference to the field bean, *Phaseolus vulgaris*. Crop Sci 7:505–510

Andersen J, Lubberstedt T (2003) Functional markers in plants. Trends Plant Sci 8:554–560

Ariyarathne H, Coyne D, Jung G, Skroch PV, Vidaver AK, Steadman JR, Miklas PN, Bassett MJ (1999) Molecular mapping of disease resistance genes for halo blight, common bacterial blight, and bean common mosaic virus in a segregating population of common bean. J Am Soc Hort Sci 124:654–662

Asfaw A, Blair M (2012) Quantitative trait loci for rooting pattern traits of common beans grown under drought stress versus non-stress conditions. Mol Breed 30:681–695

Asfaw A, Blair M, Struik P (2012) Multi environment quantitative Trait Loci analysis for photosynthate acquisition, accumulation, and remobilization traits in common bean under drought stress. G3(2):579–595

Asins MJ (2002) Present and future of QTL analysis in plant breeding. Plant Breed 121:281–291

Bassett M (1988) Linkage mapping of marker genes in common bean. In: Gepts P (ed) Genetic resources of *Phaseolus* beans. Kluwer, Dordrecht, The Netherlands, pp 329–353

Bassett M (1991) A revised linkage map of common bean. HortSci 26:834–836

Bassett M (1996) List of genes—*Phaseolus vulgaris* L. Annu Rep Bean Improv Coop 39:1–19

Bassett M (2004) List of genes: *Phaseolus vulgaris* L. Annu Rep Bean Improv Coop 47:1–24

Bassett M (2005) A new gene (*Prpi-2*) for intensified anthocyanin expression (IAE) syndrome in common bean and a reconciliation of gene symbols used by early investigators of gene symbols for purple pod and IAE syndrome. J Am Soc Hortic Sci 130:550–554

Bassett M, Myers J (1999) Report of BIC genetic committee. Annu Rep Bean Improv Coop 42 vi

Bassett M, Lee R, Otto C, McClean PE (2002) Classical and molecular genetic studies of the strong greenish yellow seed coat color in 'Wagenaar' and 'Enola' common bean. J Am Soc Hort Sci 127:50–55

Beattie A, Larsen J, Michaels T, Pauls K (2003) Mapping quantitative trait loci for a common bean (*Phaseolus vulgaris* L.) ideotype. Genome 46:411–422

Beebe S, Rojas-Pierce M, Yan X, Blair MW, Pedraza F, Muñoz F, Tohme J, Lynch JP (2006) Quantitative trait loci for root architecture traits correlated with phosphorus acquisition in common bean. Crop Sci 46:413–423

Beebe S, Rao I, Terán H, Cajiao C (2007) Breeding concepts and approaches in food legumes: The example of common bean. pp. 23–29. In: Food and forage legumes of Ethiopia: Progress and prospects. Proceedings of the workshop on food and forage legumes. Addis Ababa, Ethiopia

Beninger C, Hosfield G, Bassett M (1999) Flavonoid composition of three genotypes of dry bean (*Phaseolus vulgaris* L.) differing in seed coat color. J Am Soc Hortic Sci 124:514–518

Beninger C, Hosfield G, Bassett M, Owens S (2000) Chemical and morphological expression of the *B* and *Asp* seedcoat genes in *Phaseolus vulgaris* L. J Am Soc Hortic Sci 125:52–58

Bernardo R (2008) Molecular markers and selection for complex traits in plants: learning from the last 20 years. Crop Sci 48:1649–1664

Blair MW, Izquierdo P (2012) Use of the advanced backcross-QTL method to transfer seed mineral accumulation nutrition traits from wild to Andean cultivated common beans. Theor Appl Genet 125:1015–1031

Blair MW, Pedraza F, Buendia HF, Gaitán-Solís E, Beebe SE, Gepts P, Tohme J (2003) Development of a genome-wide anchored microsatellite map for common bean (*Phaseolus vulgaris* L.). Theor Appl Genet 107:1362–1374

Blair MW, Iriarte G, Beebe S (2006) QTL analysis of yield traits in an advanced backcross population derived from a cultivated Andean × wild common bean (*Phaseolus vulgaris* L.) cross. Theor Appl Genet 112:1149–1163

Blair MW, Rodríguez L, Pedraza F, Morales F, Beebe S (2007) Genetic mapping of the bean golden yellow mosaic geminivirus resistance gene *bgm-1* and linkage with potyvirus resistance in common bean (*Phaseolus vulgaris* L.). Theor Appl Genet 114:261–271

Blair MW, Astudillo G, Grusak M, Graham R, Beebe SE (2009a) Inheritance of seed iron and zinc content in common bean (*Phaseolus vulgaris* L.). Mol Breed 23:197–207

Blair MW, Sandoval TA, Caldas GV, Beebe SE, Páez MI (2009b) Quantitative trait locus analysis of seed phosphorus and seed phytate content in a recombinant inbred line population of common bean. Crop Sci 49:237–246

Blair MW, Medina J, Astudillo C, Rengifo J, Beebe SE, Machado G, Graham R (2010) QTL for seed iron and zinc concentrations in a recombinant inbred line population of Mesoamerican common beans (*Phaseolus vulgaris* L.). Theor Appl Genet 121:1059–1071

Blair MW, Astudillo C, Rengifo J, Beebe SE, Graham R (2011) QTL for seed iron and zinc concentrations in a recombinant inbred line population of Andean common beans (*Phaseolus vulgaris* L.). Theor Appl Genet 122:511–523

Blair MW, Galeano C, Tovar E, Muñoz Torres MC, Castrillón AV, Beebe SE, Rao IM (2012) Development of a Mesoamerican intra-genepool genetic map for quantitative trait loci detection in a drought tolerant. Mol Breed 29:71–88

Blair MW, Cortés AJ, This D (2016) Identification of an *ERECTA* gene and its drought adaptation associations with wild and cultivated common bean. Plant Sci 242:250–259

Broughton WJ, Hernandez G, Blair M, Beebe S, Gepts P, Vanderleyden J (2003) Beans (*Phaseolus* spp.) model food legumes. Plant Soil 252:55–128

Buckler ES, Holland JB, Bradbury PJ, Acharya CB, Brown PJ, Browne C, Ersoz E, Flint-Garcia S,

Garcia A, Glaubitz JC, Goodman MM, Harjes C, Guill K, Kroon DE, Larsson S, Lepak NK, Li H, Mitchell SE, Pressoir G, Peiffer JA, Rosas MO, Rocheford TR, Romay MC, Romero S, Salvo S, Sanchez Villeda H, da Silva HS, Sun Q, Tian F, Upadyayula N, Ware D, Yates H, Yu J, Zhang Z, Kresovich S, McMullen MD (2009) The genetic architecture of maize flowering time. Science 325:714–718

Caldas GV, Blair MW (2009) Inheritance of seed condensed tannins and their relationship with seed-coat color and pattern genes in common bean (*Phaseolus vulgaris* L.). Theor Appl Genet 119:131–142

Campa A, Pérez-Vega E, Giraldez R, Ferreira JJ (2007) Inheritance of race-specific resistance to anthracnose in the differential cultivar AB136. Annu Rep Bean Improv Coop 50:87–88

Campa A, Giraldez R, Ferreira J (2009) Genetic dissection of the resistance to nine anthracnose races in the common bean differential cultivars MDRK and TU. Theor Appl Genet 119:1–11

Campa A, Pañeda A, Pérez-Vega E, Giraldez R, Ferreira JJ (2011) Mapping and use of seed protein loci for marker-assisted selection of growth habit and photoperiod response in 'nuña' bean (*Phaseolus vulgaris* L.). Euphytica 179:383–391

Campa A, Rodríguez-Suárez C, Giraldez R, Ferreira J (2014) Genetic analysis of the response to eleven *Colletotrichum lindemuthianum* races in a RIL population of common bean (*Phaseolus vulgaris* L.). BMC Plant Biol 14:115

Campos T, Oblessuc PR, Sforca DA, Cardoso JMK, Baroni RM, Sousa ACB, Carbonell SAM, Chioratto AF, Garcia AAF, Rubiano LB, Souza AP (2011) Inheritance of growth habit detected by genetic linkage analysis using microsatellites in the common bean (*Phaseolus vulgaris* L.). Mol Breed 27:549–560

Casañas F, Pérez-Vega E, Almirall A et al (2013) Mapping of QTL associated with seed chemical content in a RIL population of common bean (*Phaseolus vulgaris* L.). Euphytica 192:279–288

Chavarro M, Blair M (2010) QTL Analysis and effect of the *fin* locus on tropical adaptation in an inter-genepool common bean population. Tropical Plant Biol 3:204–218

Checa O, Blair M (2008) Mapping QTL for climbing ability and component traits in common bean (*Phaseolus vulgaris* L.). Mol Breed 22:201–215

Checa O, Blair W (2012) Inheritance of yield-related traits in climbing beans (*Phaseolus vulgaris* L.). Crop Sci 52:1998–2013

Chowdhury MA, Yu K, Park SJ (2002) Molecular mapping of root rot resistance in common beans. Annu. Rep. Bean Improv. Coop. 45:96–97

Cichy KA, Blair MW, Galeno C, Snapp SS, Kelly JD (2009a) QTL analysis of root architecture traits and low phosphorus tolerance in an Andean bean population. Crop Sci 49:59–68

Cichy KA, Caldas G, Sieglinde S, Matthew W (2009b) QTL analysis of seed iron, zinc, and phosphorus levels in an Andean bean population. Crop Sci 49:1742–1750

Cichy KA, Fernandez A, Kilian A, Kelly JD, Galeano CH, Shaw S, Brick M, Hodkinson D, Troxtell E (2014) QTL analysis of canning quality and color retention in black beans (*Phaseolus vulgaris* L.). Mol Breed 33:139–154

Clemens S, Palmgren MG, Kramer U (2002) A long way ahead: understanding and engineering plant metal accumulation. Trends Plant Sci 7:309–315

Córdoba JM, Chavarro C, Rojas F, Muñoz C, Blair MW (2010a) Identification and mapping of simple sequence repeat markers from common bean (*Phaseolus vulgaris* L.) bacterial artificial chromosome end sequences for genome characterization and genetic-physical map integration. Plant Genome 3:154–165

Córdoba JM, Chavarro C, Schlueter JA, Jackson SA, Blair MW (2010b) Integration of physical and genetic maps of common bean through BAC-derived microsatellite markers. BMC Genom 11:436

Coyne DP, Shuster ML, Hill K (1973) Genetic control of reaction to common blight bacterium in bean (*Phaseolus vulgaris*) as influenced by plant age and bacterial multiplication. J Am Soc Hortic Sci 98:94–99

Currence T (1931) A new pod colour in snap beans. J Hered 22:21–23

Dash S, Campbell JD, Cannon EK, Cleary AM, Huang W, Kalberer SR, Karingula V, Rice AG, Singh J, Umale PE, Weeks NT, Wilkey AP, Farmer AD, Cannon SB (2015) Legume information system (LegumeInfo. org): a key component of a set of federated data resources for the legume family. Nucl Acids Res 44:1181–1188

Davis J, Kean D, Yorgey B, Fourie D, Miklas PN, Myers JR (2006) A molecular marker linkage map of snap bean (*Phaseolus vulgaris*). Annu Rep Bean Improv Coop 49:73–74

Devlin B, Roeder K (1999) Genomic control for association studies. Biometrics 55:997–1004

Dickson M, Petzoldt R (1986) *p* gene in beans (*Phaseolus vulgaris* L.): a gene for horizontal mediocrity. In: 22nd In Hort Congress, Davis, CA. HortSci 21 (Abstract 339), 338

Drijfhout E (1978) Genetic interaction between *Phaseolus vulgaris* L. and bean common mosaic virus with implications for strain identification and breeding for resistance. Centre for Agriculture Publishing and Documentation, (pp. 1–98). Wageningen, The Netherlands

Ender M, Kelly JD (2005) Identification of QTL associated with white mold resistance in common bean. Crop Sci 45:2482–2490

Erdmann P, Lee R, Basset M, McClean PE (2002) A molecular marker tightly linked to *P*, a gene required for flower and seed coat color in common bean (*Phaseolus vulgaris* L.), contains the *ty3-gypsy* retrotransposon *tpv3g*. Genome 45:728–736

Eticha D, Zahn M, Bremer M, Yang Z, Rangel AF, Rao IM, Horst WJ (2010) Transcriptomic analysis reveals differential gene expression in response to aluminium in common bean (*Phaseolus vulgaris*) genotypes. Ann Bot 105:1119–1128

Faleiro F, Ragagnin V, Moreira M, Barros E (2004) Use of molecular markers to accelerate the breeding of common bean lines resistant to rust and anthracnose. Euphytica 183:213–218

Falush D, Stephens M, Pritchard JK (2003) Inference of population structure using multilocus genotype data: linked loci and correlated allele frequencies. Genetics 164(4):1567–1587

Ferreira J, Campa A, Kelly J (2013) Organization of genes conferring resistance to anthracnose in common bean. In: Ferreira J, Campa A, Kelly J, Varshney R, Tuberosa R (eds) Translational genomics for crop breeding. Wiley, New York (pp 151–181)

Fourie D, Miklas PN, Ariyaranthe H (2004) Genes conditioning halo blight resistance to races 1, 7 and 9 occur in a tight cluster. Annu Rep Bean Improv Coop 47:103–104

Frei A, Blair MW, Cardona C, Beebe SE, Gu H, Dorn S (2005) QTL mapping of resistance to *Thrips palmi* Karny in common bean. Crop Sci 45:379–387

Freyre R, Skroch P, Geffroy V, Adam-Blondon AF, Shirmohamadali A, Johnson WC, Llaca V, Nodari RO, Pereira PA, Tsai SM, Tohme J, Dron M, Nienhuis J, Vallejos CE, Gepts P (1998) Towards an integrated linkage map of common bean. 4. Development of a core map and alignment of RFLP maps. Theor Appl Genet 97:847–856

Galeano CH, Fernández AC, Franco-Herrera N, Cichy KA, McClean PA, Vanderleyden J, Blair MW (2011) Saturation of an intra-gene pool linkage map: towards a unified consensus linkage map for fine mapping and synteny analysis in common bean. PLoS ONE 6:12

Galeano CH, Cortés A, Fernández A, Soler A, Franco-Herrera N, Makunde G, Vanderleyden J, Blair MW (2012) Gene-based single nucleotide polymorphism markers for genetic and association mapping in common bean. BMC Genet 13:48

Garzon LN, Blair MW (2014) Development and mapping of SSR markers linked to resistance-gene homologue clusters in common bean. The Crop J 2(4):183–194

Geffroy V, Sévignac M, De Oliveira JC, Fouilloux G, Skroch P, Thoquet P, Gepts P, Langin T, Dron M (2000) Inheritance of partial resistance against *Colletotrichum lindemuthianum* in *Phaseolus vulgaris* and co-localization of Quantitative Trait Loci with genes involved in specific resistance. Mol Plant-Microbe Interact 13:287–296

Geffroy V, Sévignac M, Billant P, Dron M, Langin T (2008) Resistance to *Colletotrichum lindemuthianum* in *Phaseolus vulgaris*: a case study for mapping two independent genes. Theor Appl Genet 116:407–415

Gelin JP, Forster S, Grafton KF et al (2007) Analysis of seed zinc and other minerals in a recombinant inbred population of navy bean (*Phaseolus vulgaris* L.). Crop Sci 47:1361–1366

Gepts P (1988) Provisional linkage map of common bean. Annu Rep Bean Improv Coop 31:20–25

Gepts P (1999) Development of an integrated linkage map. In: Singh S (ed) Developments in plant breeding common bean improvement in the twenty-first century. Kluwer,Dordrecht, The Netherlands, pp 53–91

Gepts P, Nodari R, Tsai SM, Koinange EMK, LLaca V, Gilbertson R, Guzman P (1993) Linkage mapping in common bean. Annu Rep Bean Improv Coop 36:24–38

Gioia T, Logozzo G, Kami J, Zeuli PS, Gepts P (2012) Identification and characterization of a homologue to the *Arabidopsis INDEHISCENT* gene in common bean. J Heredity 104:273–286

Gonçalves-Vidigal MC, Vidigal Filho PS, Medeiros AF, Pastor-Corrales MA (2009) Common bean landrace Jalo Listras Pretas is the source of a new Andean anthracnose resistance gene. Crop Sci 49:133–138

Gonçalves-Vidigal MC, Cruz A, Garcia A, Kami J, Vidigal Filho PS, Sousa LL, McClean P, Gepts P, Pastor-Corrales MA (2011) Linkage mapping of the *Phg-1* and *Co-1*^{4} genes for resistance to angular leaf spot and anthracnose in the common bean cultivar AND 277. Theor Appl Genet 122:893–903

Gonçalves-Vidigal MC, Cruz A, Lacanallo G, Vidigal Filho PS, Sousa LL, Pacheco CM, McClean P, Gepts P, Pastor-Corrales MA (2013) Co-segregation analysis and mapping of the anthracnose *Co-10* and angular leaf spot *Phg-ON* disease-resistance genes in the common bean cultivar Ouro Negro. Theor Appl Genet 126:2245–2255

González A, Yuste-Lisbona F, Rodiño A, De Ron AM, Capel C, García-Alcázar M, Lozano R, Santalla M (2015) Uncovering the genetic architecture of *Colletotrichum lindemuthianum* resistance through QTL mapping and epistatic interaction analysis in common bean. Front Plant Sci 6:141

González A, Yuste-Lisbona F, Godoy L, Fernández-Lozano A, Rodiño A, De Ron AM, Lozano R, Santalla M (2016) Exploring the quantitative resistance to *Pseudomonas syringae* pv. *phaseolicola* in common bean (*Phaseolus vulgaris* L.) Mol Breed 36:166

González A, Yuste-Lisbona F, Rodiño A, Saburido S, Bretones S, De Ron AM, Lozano R, Santalla M (2016b) Major contribution of flowering time and vegetative growth to plant production in common bean as deduced from a comparative genetic mapping. Front Plant Sci 7:1940

Goretti D, Bitocchi E, Bellucci E, Rodriguez M, Rau D, Gioia T, Attene G, McClean P, Nanni L, Papa R (2014) Development of single nucleotide polymorphisms in *Phaseolus vulgaris* and related *Phaseolus* spp. Mol Breed 33:531–544

Graham M, Ramírez M, Valdés-López O, Lara M, Tesfaye M, Vance CP, Hernandez G (2006) Identification of candidate phosphorus stress induced genes in *Phaseolus vulgaris* through clustering analysis across several plant species. Funct Plant Biol 33:789–797

Gu W, Zhu J, Wallace D, Singh SP, Weeden NF (1998) Analysis of genes controlling photoperiod sensitivity in common bean using DNA markers. Euphytica 102:125–132

Gujaria-Verma N, Ramsay L, Sharpe AG, Sanderson LA, Debouck DG, Tar'an B, Bett KE (2016) Gene-based SNP discovery in tepary bean (*Phaseolus acutifolius*) and common bean (*P. vulgaris*) for diversity analysis and comparative mapping. BMC Genomics 17:239

Guzmán-Maldonado S, Martínez O, Acosta-Gallegos J, Guevara-Lara F, Paredes-López O (2003) Putative quantitative trait loci for physical and chemical components of common bean. Crop Sci 43:1029–1035

Hagerty C (2013) Mapping QTL for root rot resistance, root traits, and morphological trait in a common bean recombinant inbred population. Master of Science in Horticulture. Oregon State University. http://ir.library.oregonstate.edu/xmlui/handle/1957/38263

Hagerty CH, Cuesta-Marcos A, Cregan PB, Song Q, McClean P, Noffsinger S et al (2015) Mapping and root rot resistance and root architecture quantitative trait loci in common bean. Crop Sci 55:1969–1977

Hanai LR, Santini L, Camargo LE, Fungaro MH, Gepts P, Tsai SM, Vieira ML (2010) Extension of the core map of common bean with EST-SSR, RGA, AFLP, and putative functional markers. Mol Breed 25:25–45

Hernández G, Ramírez M, Valdés-López O et al (2007) Phosphorus stress in common bean: root transcript and metabolic responses. Plant Physiol 144:752–767

Holland JB (2001) Epistasis and plant breeding. Plant Breed Rev 21:27–82

Holland JB (2007) Genetic architecture of complex traits in plants. Curr Opin Plant Biol 10:156–161

Hougaard BK, Madsen LH, Sandal N, Moretzsohn MC, Fredslund J, Schauser L, Nielsen AM, Rohde T, Sato S, Tabata S, Bertioli DJ, Stougaard J (2008) Legume anchor markers link syntenic regions between *Phaseolus vulgaris*, *Lotus japonicus*, *Medicago truncatula*, and *Arachis*. Genetics 179:2299–2312

Huang BE, Verbyla K, Verbyla A, Raghavan C, Singh VK, Gaur P, Leung H, Varshney RK, Cavanagh CR (2015) MAGIC populations in crops: current status and future prospects. Theor Appl Genet 128(6):999–1017

Hungria M, Vargas M (2000) Environmental factors affecting N_2 fixation in grain legumes in the tropics, with an emphasis on Brazil. Field Crops Res 65:151–164

Hyten DL, Song Q, Fickus EW, Quigley CV, Lim JS, Choi IY, Hwang EY, Pastor-Corrales M, Cregan PB (2010) High throughput SNP discovery and assay development in common bean. BMC Genom 11:475

Johannsen W (1911) The genotype conception of heredity. Am Nat 45:129–159

Johnson W, Gepts P (2002) The role of epistasis in controlling seed yield and other agronomic traits in an Andean × Mesoamerican cross of common bean (*Phaseolus vulgaris* L.). Euphytica 125:69–79

Johnson W, Menéndez C, Nodari R, Koinange EMK, Magnusson S, Singh SP, Gepts P (1996) Association of a seed weight factor with the phaseolin seed storage protein locus across genotypes, environments, and genomes in *Phaseolus-Vigna* spp.: Sax (1923) revisited. J Agric Genomics 2

Johnson W, Guzmán P, Mandala D, Mkandawire ABC, Temple S, Gilbertson RL, Gepts P (1997) Molecular tagging of the *bc-3* gene for introgression into Andean common bean. Crop Sci 37:248–254

Jung G, Coyne D, Scroch P, Nienhuis J, Bokosi J, Ariyarathne HM, Steadman JR, Beaver JS, Kaeppler SM (1996) Molecular markers associated with plant architecture and resistance to common blight, web blight, and rust in common beans. J Am Soc Hort Sci 121:794–803

Jung G, Skroch P, Coyne D, Nienhuis J, Arnaud-Santana E, Ariyarathne HM, Kaeppler SM, Bassett MJ (1997) Molecular-markers-based genetic analysis of tepary bean derived common bacterial blight resistance in different developmental stage of common bean. J Am Soc Hort Sci 122:329–337

Jung G, Coyne D, Bokosi J, Steadman JR, Nienhuis J (1998) Mapping genes for specific and adult plant resistance to rust and abaxial leaf pubescence and their genetic relationship using random amplified polymorphic DNA (RAPD) markers in common bean. J Am Soc Hortic Sci 123:859–863

Jung G, Ariyarathne H, Coyne DP, Nienhuis J (2003) Mapping QTL for bacterial brown spot resistance under natural infection in field and seedling stem inoculation in growth chamber in common bean. Crop Sci 43:350–357

Kalavacharla V, Stavely JR, Myers JR, McClean PE (2000) *Crg*, a gene required for *Ur-3*-mediated rust resistance in common bean, maps to a resistance gene analog cluster. Mol Plant-Microbe Interact 13:1237–1242

Kalia RK, Rai MK, Kalia S, Singh R, Dhawan AK (2011) Microsatellite markers: an overview of the recent progress in plants. Euphytica 177:309–334

Kamfwa K, Mwala M, Okori P, Gibson P, Mukankusi C (2013) Identification of QTL for *Fusarium* Root Rot resistance in common bean. J Crop Improv 27:406–418

Kamfwa K, Cichy K, Kelly J (2015a) Genome-Wide association study of agronomic traits in common bean. The Plant Genome 8:2

Kamfwa K, Cichy KA, Kelly JD (2015b) Genome-wide association analysis of symbiotic nitrogen fixation in common bean. Theor Appl Genet 128:1999–2017

Keller B, Manzanares C, Jara C, Lobaton JD, Studer B, Raatz B (2015) Fine-mapping of a major QTL controlling angular leaf spot resistance in common bean (*Phaseolus vulgaris* L.). Theor Appl Genet 128:813–826

Kelly J, Vallejo V (2004) A comprehensive review of the major genes conditioning resistance to anthracnose in common bean. HortSci 39:1196–1207

Kelly J, Gepts P, Miklas P, Coyne D (2003) Tagging and mapping of genes and QTL and molecular marker-assisted selection for traits of economic importance in bean and cowpea. Field Crops Res 82:135–154

Koenig R, Singh S, Gepts P (1990) Novel phaseolin types in wild and cultivated common bean (*Phaseolus vulgaris*, *Fabaceae*). Econ Bot 44:50–60

Koinange E, Singh S, Gepts P (1996) Genetic control of the domestication syndrome in common bean. Crop Sci 36:1037–1045

Kolkman JM, Kelly JD (2003) QTL conferring resistance and avoidance to white mold in common bean. Crop Sci 43:539–548

Korte A, Farlow A (2013) The advantages and limitations of trait analysis with GWAS: a review. Plant Methods 9:29

Kwak M, Velasco D, Gepts P (2008) Mapping homologous sequences for determinacy and photoperiod sensitivity in common bean (*Phaseolus vulgaris*). J Hered 99:283–291

Kyle M, Dickson M (1988) Linkage of hypersensitivity to five viruses with the *B* locus in *Phaseolus vulgaris* L. J Hered 79:308–311

Lamprecht H (1961) Weitere Kopplungsstudien an *Phaseolus vulgaris* mit einer Ubersicht ber die Koppelungsgruppen. Agr Hort Genet 19:319–332

Lark KG, Chase K, Adler F, Mansur LM, Orf JH (1995) Interactions between quantitative trait loci in soybean in which trait variation at one locus is conditional upon a specific allele at another. Proc Natl Acad Sci USA 92:4656–4660

Lauer J, Bijl C, Grusak M, Baenziger PS, Boote K, Lingle S, Carter T, Kaeppler S, Boerma R, Eizenga G, Carter P, Goodman M, Nafziger E, Kidwell K, Mitchell R, Edgerton MD, Quesenberry K, Willcox MC (2012) The scientific grand challenges of the 21st century for the Crop Science Society of America. Crop Sci 52:1003–1010

Leakey C (1988) Genotypic and phenotypic markers in common bean. In: Gepts P (ed) Genetic resources of *Phaseolus* Beans. Kluwer, Dordrecht, Netherlands, pp 245–347

Liao H, Yan X, Rubio G, Beebe SE, Blair MW, Lynch JP (2004) Genetic mapping of basal root gravitropism and phosphorus acquisition efficiency in common bean. Funct Plant Biol 31:959–970

López CE, Acosta IF, Jara C, Pedraza F, Gaitán-Solís E, Gallego G, Beebe S, Tohme J (2003) Identifying resistance gene analogs associated with resistance to different pathogens in common bean. Phytopathology 93:88–95

López-Marín H, Rao I, Blair M (2009) Quantitative trait loci for root morphology traits under aluminum stress in common bean (*Phaseolus vulgaris* L.). Theor Appl Genet 119:449–458

Mahuku G, Iglesias AM, Jara C (2009) Genetics of angular leaf spot resistance in the Andean common bean accession G5686 and identification of markers linked to the resistance genes. Euphytica 167:381–396

Mahuku G, Henriquez MA, Montoya C, Jara C, Teran H, Beebe S (2011) Inheritance and development of molecular markers linked to angular leaf spot resistance genes in the common bean accession G10909. Mol Breed 28:57–71

Maxwell JJ, Brick MA, Byrne PF, Schwartz H, Shan X, Ogg JB, Henson R (2007) Quantitative trait loci linked to white mold resistance in common bean. Crop Sci 47:2285–2294

McClean PE, Lee R, Otto C, Gepts P, Bassett MJ (2002) Molecular and phenotypic mapping of genes controlling seed coat pattern and color in common bean (*Phaseolus vulgaris* L.). J Hered 93:148–152

McClean PE, Mamidi S, McConnell M, Chikara S, Lee R (2010) Synteny mapping between common bean and soybean reveals extensive blocks of shared loci. BMC Genom 11:184

McConnell M, Mamidi S, Lee R, Chikara S, Rossi M, Papa R, McClean P (2010) Syntenic relationships among legumes revealed using a gene-based genetic linkage map of common bean (*Phaseolus vulgaris* L.). Theor Appl Genet 121:1103–1116

Melotto M, Monteiro-Vitorello CB, Bruschi AG, Camargo LE (2005) Comparative bioinformatic analysis of genes expressed in common bean (*Phaseolus vulgaris* L.) seedlings. Genome 48:562–570

Mendel G (1866) Versuche über Plflanzenhybriden. Verhandlungen des naturforschenden Vereines in Brünn. Bd. IV für das Jahr 1865, Abhandlungen, 3–47

Meziadi C, Richard MM, Derquennes A, Thareau V, Blanchet S, Gratias A, Pflieger S, Geffroy V (2015) Development of molecular markers linked to disease resistance genes in common bean based on whole genome sequence. Plant Sci 242:351–357

Mienie CMS, Liebenberg MM, Pretorius ZA, Miklas PN (2005) SCAR markers linked to the common bean rust resistance gene *Ur-13*. Theor Appl Genet 111:972–979

Miguel M (2004) Genotypic variation in root hairs and phosphorus efficiency in common bean (*Phaseolus vulgaris* L.). P. Pennsylvania State University: University Park (ed) Horticulture

Miklas PN, Johnson E, Stone V, Beaver JS, Montoya C, Zapata M (1996) Selective mapping of QTL conditioning disease resistance in common bean. Crop Sci 36:1344–1351

Miklas PN, Stone V, Urrea CA, Johnson E, Beaver JS (1998) Inheritance and QTL analysis of field resistance to ashy stem blight. Crop Sci 38:916–921

Miklas PN, Delorme R, Stone V, Stavely J, Steadman J, Bassett Beaver J (2000) Bacterial, fungal, virus disease loci mapped in a recombinant inbred common bean population ('Dorado/XAN176'). J Am Soc Hort Sci 125:476–481

Miklas PN, Johnson W, Delorme R, Gepts P (2001) QTL conditioning physiological resistance and avoidance to white mold in dry bean. Crop Sci 41:309–315

Miklas PN, Pastor-Corrales M, Jung G, Coyne D, Kelly J, Mcclean P, Gepts P (2002) Comprehensive linkage map of bean rust resistance genes. Annu Rep Bean Improv Coop 45:125–129

Miklas PN, Coyne D, Grafton K, Mutlu N, Reiser J, Lindgren D, Singh SP (2003) A major QTL for

common bacterial blight resistance derives from the from the common bean great northern landrace cultivar Montana No. 5. Euphytica 131:137–146
Miklas PN, Kelly J, Beebe S, Blair M (2006) Common bean breeding for resistance against biotic and abiotic stresses: from classical to MAS breeding. Euphytica 147:105–131
Miklas PN, Fourie D, Wagner J, Larsen RC, Mienie CMS (2009) Tagging and mapping *Pse-1* gene for resistance to halo blight in common bean host differential cultivar UI-3. Crop Sci 49:41–48
Miklas PN, Fourie D, Trapp J, Larsen RC, Chavarro C, Blair MW, Gepts P (2011) Genetic characterization and molecular mapping *Pse-2* gene for resistance to halo blight in common bean. Crop Sci 51:2439–2448
Miklas PN, Porter LD, Kelly JD, Myers JM (2013) Characterization of white mold disease avoidance in common bean. Eur J Plant Pathol 135:525–543
Miklas PN, Fourie D, Trapp J, Davis J, Myers JR (2014) New loci including conferring resistance to halo bacterial blight on chromosome Pv04 in common bean. Crop Sci 54:2099–2108
Mir RR, Zaman-Allah M, Sreenivasulu N et al (2012) Integrated genomics, physiology and breeding approaches for improving drought tolerance in crops. Theor Appl Genet 125:625–645
Mkwaila W, Terpstra KA, Ender M, Kelly JD (2011) Identification of QTL for resistance to white mold in wild and landrace germplasm of common bean. Plant Breeding 130:665–672
Moghaddam SM, Mamidi S, Osorno JM, Lee R, Brick M, Kelly J, Miklas P, Urrea C, Song Q, Cregan P, Grimwood J, Schmutz J, McClean PE (2016) Genome-wide association study identifies candidate loci underlying agronomic traits in a middle american diversity panel of common bean. Plant Genome 9(3):1–21
Molina A, Beaver J (1998) Inheritance of normal pod development in bean golden mosaic resistant common beans. Annu Rep Bean Improv Coop 41:3–4
Motto M, Soressi GP, Salamini F (1978) Seed size inheritance in a cross between wild and cultivated common beans (*Phaseolus vulgaris L.*). Genetica 49:31–36
Mukeshimana G, Butare G, Cregan P, Blair MW, Kelly JD (2014) Identification of Quantitative Trait Loci associated with drought tolerance in common bean using SNP markers. Crop Sci 54:923–938
Müller BS, Sakamoto T, de Menezes IP, Prado GS, Martins WS, Brondani C, de Barros EG, Vianello RP (2014) Analysis of BAC-end sequences in common bean (*Phaseolus vulgaris* L.) towards the development and characterization of long motifs SSRs. Plant Mol Biol 86:455–470
Mutlu N, Miklas P, Reiser J, Coyne D (2005a) Backcross breeding for improved resistance to common bacterial blight in pinto bean (*Phaseolus vulgaris* L). Plant Breed 124:282–287
Mutlu N, Miklas PN, Steadman JR, Vidaver A, Lindgren D, Reiser J, Pastor Corrales M (2005b) Registration of pinto bean germplasm line ABCP-8 with resistance to common bacterial blight. Crop Sci 45:806–807
Myers J, Bassett M (1993) Inheritance, allelism, and morphological characterization of unifoliate mutations in common bean. J Hered 84:17–20
Myles S, Peiffer J, Brown PJ, Ersoz ES, Zhang Z, Costich DE, Buckle ES (2009) Association mapping: critical considerations shift from genotyping to experimental design. Plant Cell 21:2194–2202
Nanni L, Bitocchi E, Bellucci E, Rossi M, Rau D, Attene G, Gepts P, Papa R (2011) Nucleotide diversity of a genomic sequence similar to *SHATTERPROOF* (*PvSHP1*) in domesticated and wild common bean (*Phaseolus vulgaris* L.). Theor Appl Genet 123:1341–1357
Nemli S, Asciogul T, Kaya H, Kahraman A, Eşiyok D, Tanyolac B (2014) Association mapping for five agronomic traits in the common bean (*Phaseolus vulgaris* L.). J Sci Food Agric 94:3141–3151
Nodari RO (1992). Towards an integrated linkage map of common bean (*Phaseolus vulgaris* L.). Ph.D. diss. University of California, Davis
Nodari RO, Tsai S, Gilbertson R, Gepts P (1993a) Towards an integrated linkage map of common bean. II. Development of an RFLP-based linkage map. Theor Appl Genet 84:186–192
Nodari RO, Tsai S, Guzmán P, Gilbertson RL, Gepts P (1993b) Toward an integrated linkage map of common bean. III. Mapping genetic factors controlling host-bacteria interactions. Genetics 134:341–350
Oblessuc PR, Baroni RM, Garcia AAF, Chioratto AF, Carbonell SAM, Camargo LEA, Benchimol LL (2012) Mapping of angular leaf spot resistance QTL in common bean (*Phaseolus vulgaris* L.) under different environments. BMC Genet 13:e50
Oblessuc PR, Cardoso Perseguini JMK, Baroni RM, Chiorato AF, Carbonell SA, Mondego JM, Vidal RO, Camargo LE, Benchimol-Reis LL (2013) Increasing the density of markers around a major QTL controlling resistance to angular leaf spot in common bean. Theor Appl Genet 126:2451–2465
Oblessuc PR, Baroni RM, Pereira GS, Chioratto AF, Carbonell SAM, Briñez B, Da Costa ESL, Garcia AAF, Camargo LEA, Kelly JD, Benchimol-Reis LL (2014) Quantitative analysis of race-specific resistance to *Colletotrichum lindemuthianum* in common bean. Mol Breed 34:1313–1329
Ochoa I, Blair M, Lynch J (2006) QTL Analysis of adventitious root formation in common bean under contrasting phosphorus availability. Crop Sci 46:1609–1621
Oraguzie NC, Rikkerink EHA, Gardiner SE, De Silva HN (2007) Association mapping in plants. Springer, New York, NY
Osborn T, Blake T, Gepts P, Bliss F (1986) Bean arcelin. 2. Genetic variation, inheritance and linkage relationships of a novel seed protein of *Phaseolus vulgaris* L. Theor Appl Genet 71:847–855
Park SJ, Tu JC (1986) Association between BCMV resistant *I* gene and eye color of cv. Steuben. Annu Rep Bean Improv Coop 29:4–5

Park SO, Coyne DP, Jung G, Skroch PW, Arnaud-Santana E, Steadman JR, Ariyarathne HM, Nienhuis J (2000) Mapping of QTL for seed size and shape traits in common bean. J Am Soc Hortic Sci 125:466–475

Park SO, Coyne DP, Steadman JR, Skroch PW (2001) Mapping of QTL for resistance to white mold diseases in common bean. Crop Sci 41:1253–1262

Parvez AS, Rather AG, Warsi MZK (2007) Implications of epistasis in maize breeding. Int J Plant Breed Genet 1:1–11

Pérez-Vega E, Pañeda A, Rodríguez-Suárez C, Campa A, Giraldez R, Ferreira JJ (2010) Mapping of QTLs for morpho-agronomic and seed quality traits in a RIL population of common bean (*Phaseolus vulgaris* L.). Theor Appl Genet 120:1367–1380

Pérez-Vega E, Pascual A, Campa A, Giraldez R, Miklas PN, Ferreira JJ (2012) Mapping quantitative trait loci conferring partial physiological resistance to white mold in the common bean RIL population Xana × Cornell 49242. Mol Breed 29:31–41

Perseguini JMKC, Silva GMB, Rosa JRBF, Gazaffi R, Marçal JF, Carbonell SA, Chiorato AF, Zucchi MI, Garcia AA, Benchimol-Reis LL (2015) Developing a common bean core collection suitable for association mapping studies. Genet Mol Biol. 38(1):67–78

Perseguini JMKC, Oblessuc PR, Rosa JRBF, Gomes KA, Chiorato AF, Carbonell SA, Garcia AA, Vianello RP, Benchimol-Reis LL (2016) Genome-wide association studies of anthracnose and angular leaf spot resistance in common bean (*Phaseolus vulgaris* L.). PLoS ONE 11(3):e0150506

Phillips PC (2008) Epistasis—the essential role of gene interactions in the structure and evolution of genetic systems. Nat Rev Genet 9:855–867

Posa-Macalincag M, Hosfield G, Grafton K (2002) Quantitative trait loci (QTL) analysis of canning quality traits in kidney bean (*Phaseolus vulgaris* L.). J Amer Soc Hort Sci 127:608–615

Price AL, Patterson NJ, Plenge RM et al (2006) Principal components analysis corrects for stratification in genome-wide association studies. Nat Genet 38:904–909

Qi Y (2015) Characterization of a putative yield-related gene in common bean *(Phaseolus vulgaris L.)*. Master of Science in Plant Agriculture. Guelph, Ontario, Canada: The University of Guelph, http://hdl.handle.net/10214/8727

Ramaekers L, Galeano C, Garzón N, Vanderleyden J, Blair MW (2013) Identifying quantitative trait loci for symbiotic nitrogen fixation capacity and related traits in common bean. Mol Breed 31:163–180

Ramírez M, Graham M, Blanco-López L, Silvente S, Medrano-Soto A, Blair MW, Hernández G, Vance CP, Lara M (2005) Sequencing and analysis of common bean ESTs. Building a foundation for functional genomics. Plant Physiol 137:1211–1227

Ramírez M, Flores-Pacheco G, Reyes J, Alvarez AL, Drevon JJ, Girard L, Hernández G (2013) Two common bean genotypes with contrasting response to phosphorus deficiency show variations in the microRNA 399-mediated *PvPHO2* regulation within the *PvPHR1* signaling pathway. Int J Mol Sci 14:8328–8344

Rangel AF, Mobin M, Rao IM, Horst WJ (2005) Proton toxicity interferes with the screening of common bean (*Phaseolus vulgaris* L.) genotypes for aluminium resistance in nutrient solution. J Plant Nutr Soil Sci 168:607–616

Rangel AF, Rao IM, Braun H-P, Horst WJ (2010) Aluminum resistance in common bean (*Phaseolus vulgaris*) involves induction and maintenance of citrate exudation from root apices. Physiol Plant 138:176–190

Rao I (2001) Role of physiology in improving crop adaptation to abiotic stresses in the tropics: the case of common bean and tropical forages. In: Pessarakl M (ed) Handbook of plant and crop physiology. NY, USA: Marcel Dekker (pp 583–613)

Repinski S, Kwak M, Gepts P (2012) The common bean growth habit gene *PvTFL1y* is a functional homolog of *Arabidopsis TFL1*. Theor Appl Genet 124:1539–1547

Risch N, Merikangas K (1996) The future of genetic studies of complex human diseases. Science 273:1516–1517

Román-Avilés B, Kelly JD (2005) Identification of quantitative trait loci conditioning resistance to *Fusarium* root rot in common bean. Crop Sci 45:1881–1890

Sax K (1923) The association of size differences with seed coat pattern and pigmentation in *Phaseolus vulgaris*. Genetics 8:552–560

Schmutz J, McClean P, Mamidi S, Wu GA, Cannon SB, Grimwood J, Jenkins J, Shu S, Song Q, Chavarro C, Torres-Torres M, Geffroy V, Moghaddam SM, Gao D, Abernathy B, Barry K, Blair M, Brick MA, Chovatia M, Gepts P, Goodstein DM, Gonzales M, Hellsten U, Hyten DL, Jia G, Kelly JD, Kudrna D, Lee R, Richard MMS, Miklas PN, Osorno JM, Rodrigues J, Thareau V, Urrea CA, Wang M, Yu Y, Zhang M, Wing RA, Cregan PB, Rokhsar DS, Jackson SA (2014) A reference genome for common bean and genome-wide analysis of dual domestications. Nat Genet 46:707–713

Schneider KA, Brothers M, Kelly J (1997) Marker assisted selection to improve drought resistance in common bean. Crop Sci 37:51–60

Schneider KA, Grafton KF, Kelly JD (2001) QTL analysis of resistance to *Fusarium* root rot in bean. Crop Sci 41:535–542

Shaw JK, Norton JB (1918) The inheritance of seed-coat color in garden beans. Massachusetts Agr Exp Sta Bul 185:59–104

Shi C, Navabi A, Yu K (2011) Association mapping of common bacterial blight resistance QTL in Ontario bean breeding populations. BMC Plant Biol 11:52

Singh SP, Westermann DT (2002) A single dominant gene controlling resistance to soil zinc deficiency in common bean. Crop Sci 42:1071–1074

Song Q, Jia G, Hyten DL, Jenkins J, Hwang EY, Schroeder SG, Osorno JM, Schmutz J, Jackson SA,

McClean PE, Cregan PB (2015) SNP assay development for linkage map construction, anchoring whole-genome sequence, and other genetic and genomic applications in common bean G3(5):2285–2290

Soule M, Porter L, Medina J, Santana GP, Blair MW, Miklas PN (2011) Comparative QTL map for white mold resistance in common bean, and characterization of partial resistance in dry bean lines VA19 and 19365-31. Crop Sci 51:123–139

Sousa LL, Gonçalves AO, Gonçalves-Vidigal MC, Lacanallo GF, Fernandez AC, Awale HE, Kelly JD (2015) Genetic characterization and mapping of anthracnose resistance of common bean landrace cultivar Corinthiano. Crop Sci 55:1900–1910

Souza A, Boscariol R, Moon D, Camargo LEA, Tsai SM (2000) Effects of *Phaseolus vulgaris* QTL in controlling host-bacteria interactions under two levels of nitrogen fertilization. Gen Mol Biol 23:155–161

Souza TLPO, Dessaune SN, Sanglard DA, Moreira MA, de Barros EG (2011) Characterization of the rust resistance gene present in the common bean cultivar Ouro Negro, the main rust resistance source used in Brazil. Plant Pathol 60:839–845

Stavely JR (1984) Genetics of resistance to *Uromyces phaseoli* in a *Phaseolus vulgaris* line resistant to most races of the pathogen. Phytopathology 74:339–344

Stavely JR (1998) Recombination of two major dominant rust resistance genes that are tightly linked in repulsion. Annu Rep Bean Improv Coop 41:17–18

Strausbaugh C, Myers J, Forster R, McClean P (1999) *bc-1* and *bc-u*—two loci controlling bean common mosaic virus resistance in common bean are linked. J Am Soc Hort Sci 124:644–648

Tar'an B, Michaels TE, Pauls KP (2001) Mapping genetic factors affecting the reaction to *Xanthomonas axonopodis* pv. *phaseoli* in *Phaseolus vulgaris* L. under field conditions. Genome 44: 1046–1056

Tar'an B, Michaels TE, Pauls KP (2002) Genetic mapping of agronomic traits in common bean. Crop Sci 42:544–556

Temple S, Morales F (1986) Linkage of dominant hypersensitive resistance to bean common mosaic virus to seed color in *Phaseolus vulgaris* L. Euphytica 35:331–333

Thudi M, Li Y, Jackson SA, May GD, Varshney RK (2012) Current state-of-art of sequencing technologies for plant genomics research. Brief Funct Genomics 11:3–11

Tian J, Venkatachalam P, Liao H, Yan X, Raghothama K (2007) Molecular cloning and characterization of phosphorus starvation responsive genes in common bean (*Phaseolus vulgaris* L.). Planta 227:151–165

Tjebbes K, Kooiman H (1921) Erfelijkheidsonderzoekingen bij boonen. IV. Over den strepingsfactor, een geval van volkomen afstooting tusschen de twee factoren. Genetica 3:28–49

Trabanco N, Asensio-Manzanera M, Pérez-Vega E, Ibeas A, Campa A, Ferreira JJ (2014) Identification of quantitative trait loci involved in the response of common bean to *Pseudomonas syringae* pv. *phaseolicola*. Mol Breed 33:577–588

Trabanco N, Campa A, Ferreira J (2015) Identification of a new chromosomal region involved in the genetic control of resistance to anthracnose in common bean. The Plant Genome 8(2):1–11

Trapp JJ, Urrea CA, Cregan PB, Miklas PN (2015) Detection of major QTL for yield under multiple abiotic and terminal drought stress in a recombinant inbred dry bean population. In: Genetics of drought tolerance in common bean (*Phaseolus vulgaris* L.). Doctoral Thesis. Department of Crop and Soil Sciences. Washington State University

Tsai S, Nodari R, Moon D, Camargo LEA, Vencovsky R, Gepts P (1998) QTL mapping for nodule number and common bacterial blight in *Phaseolus vulgaris* L. Plant Soil 204:135–145

Vadez V, Jens D, Warkentin T, Asseng S, Ratnakumar P, Rao KPC, Gaur PM, Munier-Jolain N, Larmure A, Voisin AS, Sharma HC, Pande S, Sharma M, Krishnamurthy L, Zaman MA (2012) Adaptation of grain legumes to climate change: a review. Agron Sustain Dev 32:31–44

Valdés-López O, Arenas-Huertero C, Ramírez M, Girard L, Sánchez F, Vance CP, Luis Reyes J, Hernández G (2008) Essential role of MYB transcription factor: *PvPHR1* and microRNA: *PvmiR399* in phosphorus-deficiency signalling in common bean roots. Plant, Cell Environ 31:1834–1843

Valladares-Sánchez NE, Coyne DP, Schuster ML (1979) Differential reaction of leaves and pods of *Phaseolus* germplasm to strain of *Xanthomonas phaseoli* and transgressive segregation for tolerance from crosses of susceptible germplasm. J Am Soc Hort Sci 104:648–654

Vallejos CE (1994) *Phaseolus vulgaris*—The common bean. In: Phillips R, Vasil I (eds) DNA-based markers in plants. Kluwer, Dordrecht, pp 261–270

Vallejos CE, Chase CD (1991) Linkage between isozyme markers and a locus affecting seed size in *Phaseolus vulgaris* L. Theor Appl Genet 81:413–419

Vallejos CE, Sakiyama NS, Chase CD (1992) A molecular marker-based linkage map of *Phaseolus vulgaris* L. Genetics 131:733–740

Vallejos CE, Skroch P, Nienhuis J (2001) *Phaseolus vulgaris*—The common bean integration of RFLP and RAPD- based linkage maps. In: Phillips R, Vasil I (eds) DNA—Based markers in plants. Kluwer, Dordrecht, pp 301–317

Vandemark G, Fourie D, Miklas P (2008) Genotyping with real-time PCR reveals recessive epistasis between independent QTL conferring resistance to CBB in dry bean. Theor Appl Genet 117:513–522

Vasconcelos RA, Nascimento P, Zaczuk P et al (2012) QTL mapping for the cooking time of common beans. Euphytica 186(3):779–792

Velez J, Bassett M, Beaver J, Molina A (1998) Inheritance of resistance to bean golden mosaic virus in common bean. J Am Soc Hort Sci 123: 628–631

Viteri D, Cregan P, Trapp J, Miklas P, Shre S (2015) A new common bacterial blight resistance QTL in VAX 1 common bean and interaction of the new QTL, SAP6, and SU91 with bacterial strains. Crop Sci 54:1598–1608

Vlasova A, Capella-Gutiérrez S, Rendón-Anaya M, Hernández-Oñate M, Minoche AE, Erb I, Câmara F, Prieto-Barja P, Corvelo A, Sanseverino W7, Westergaard G, Dohm JC, Pappas GJ Jr, Saburido-Alvarez S, Kedra D, Gonzalez I, Cozzuto L, Gómez-Garrido J, Aguilar-Morón MA, Andreu N, Aguilar OM, Garcia-Mas J, Zehnsdorf M, Vázquez MP, Delgado-Salinas A, Delaye L, Lowy E, Mentaberry A, Vianello-Brondani RP, García JL, Alioto T, Sánchez F, Himmelbauer H, Santalla M, Notredame C, Gabaldón T, Herrera-Estrella A, Guigó R (2016) Genome and transcriptome analysis of the Mesoamerican common bean and the role of gene duplications in establishing tissue and temporal specialization of genes. Genome Biol 17:32

Weeden NF, Liang CY (1985) Detection of a linkage between white flower color and *EST-2* in common bean. Annu Rep Bean Improv Coop 28:87–88

Weller J, Ortega R (2015) Genetic control of flowering time in legumes. Front Plant Sci 6:207

White J, Kornegay J, Cajiao C (1996) Inheritance of temperature sensitivity of the photoperiod response in common bean (*Phaseolus vulgaris* L.). Euphytica 9:5–8

Wortmann C, Kirkby R, Eledu C, Allen D (1998) Atlas of common bean *(Phaseolus vulgaris L.)* production in Africa. Cali, Colombia: (CIAT), Centro Internacional de Agricultura Tropical

Wright E, Kelly J (2011) Mapping QTL for seed yield and canning quality following processing of black bean (*Phaseolus vulgaris* L.). Euphytica 179:471–484

Wu J, Wang L, Li L, Wang S (2014) De novo assembly of the common bean transcriptome using short reads for the discovery of drought-responsive genes. PLoS ONE 9:e109262

Yaish M, Sosa D, Vences F, Vaquero F (2006) Genetic mapping of quantitative resistance to race 5 of *Pseudomonas syringae* pv. *phaseolicola* in common bean. Euphytica 179:417–425

Yamashita Y, Takeuchi T, Okuyama M et al (2014) Development and validation of DNA markers linked to *Sdvy-1*, a common bean gene conferring resistance to the yellowing strain of Soybean dwarf virus. Breed Sci 64:404–408

Yan X, Liao H, Beebe S, Blair MW, Lynch JP (2004) QTL mapping of root hair and acid exudation traits and their relationship to phosphorus uptake in common bean. Plant Soil 265:17–29

Yang Z, Eticha D, Rotter B, Rao IM, Horst WJ (2011) Physiological and molecular analysis of polyethylene glycol-induced reduction of aluminium accumulation in the root tips of common bean (*Phaseolus vulgaris*). New Phytol 192:99–113

Yu Z, Stall R, Vallejos C (1998) Detection of genes for resistance to common bacterial blight of beans. Crop Sci 38:1290–1296

Yu K, Park S, Poysa V, Gepts P (2000) Integration of simple sequence repeat (SSR) markers into a molecular linkage map of common bean (*Phaseolus vulgaris* L.). J Hered 91:429–434

Yu J, Pressoir G, Briggs WH et al (2006) A unified mixed-model method for association mapping that accounts for multiple levels of relatedness. Nat Genet 38:203–208

Yuste-Lisbona F, Santalla M, Capel C, García-Alcázar M, De la Fuente M, Capel J, De Ron AM, Lozano R (2012) Marker-based linkage map of Andean common bean (*Phaseolus vulgaris* L.) and mapping of QTLs underlying popping ability traits. BMC Plant Biol 12:136

Yuste-Lisbona F, González A, Capel C, García-Alcázar M, Capel J, De Ron AM, Lozano R, Santalla M (2014a) Genetic analysis of single–locus and epistatic QTLs for seed traits in an Adapted × Nuña RIL population of common bean (*Phaseolus vulgaris* L.). Theor Appl Genet 127:897–912

Yuste-Lisbona F, González A, Capel C, García-Alcázar M, Capel J, De Ron AM, Santalla M, Lozano R (2014b) Genetic variation underlying pod size and color traits of common bean depends on quantitative trait loci with epistatic effects. Mol Breed 33:939–952

Zhu C, Gore M, Buckler ES, Jiangming YU (2008) Status and prospects of association mapping in plants. The Plant Genome 1:5–20

Zou X, Shi C, Austin RS, Merico D, Munholland S, Marsolais F, Navabi A, Crosby WL, Pauls KP, Yu K, Cui Y (2014) Genome-wide single nucleotide polymorphism and Insertion-Deletion discovery through next-generation sequencing of reduced representation libraries in common bean. Mol Breed 33:769–778

Requirement of Whole-Genome Sequencing

5

Martha Rendón-Anaya and Alfredo Herrera-Estrella

Abstract

Crop plants that sustain modern civilizations, including common bean (*Phaseolus vulgaris*), were domesticated and improved by thousands of years of human selection, which transformed wild ancestors into high-yielding domesticated descendants. Understanding how the genome of crop species has been shaped through time, with and without human intervention, is a fascinating field of research. In addition, defining the loci and associated polymorphisms behind the emergence of domestication and improvement traits in *P. vulgaris* is of major importance. Uncovering intra- and inter-species introgression events that could indicate transferred genes, which together with domestication protein-coding and non-coding genes that have given rise to domestication and adaptive traits are required for future improvement strategies. Such strategies, in our view, will depend to a significant extent on crop re-wilding, given the local adaptations undergone by their wild relatives and climate change. Essential tools for reaching these goals have recently been developed, such as the complete genome sequences ($\sim$600 Mb) of a Mesoamerican and an Andean accession, as well as a large gene expression atlas. Further, there are significant re-sequencing efforts for both wild and domesticated genotypes, which will play a major role in the future of this crop. Altogether, this information will allow the genetic dissection of the characters involved in the domestication and adaptation of the crop and their further implementation in breeding strategies faced with an ever-expanding human population and unpredictable environmental challenges.

Keywords

Genome · Domestication · Introgression · Evolution
Diversification · Speciation

M. Rendón-Anaya · A. Herrera-Estrella (✉)
Laboratorio Nacional de Genómica para la Biodiversidad, CINVESTAV-Irapuato, 36821 Irapuato, Mexico
e-mail: alfredo.herrera@cinvestav.mx

M. Pérez de la Vega et al. (eds.), *The Common Bean Genome*,
Compendium of Plant Genomes, https://doi.org/10.1007/978-3-319-63526-2_5

5.1 *Phaseolus vulgaris* Reference Genome

The common dry bean, *Phaseolus vulgaris*, is the most important food legume for direct consumption in the world; it is a major source of calories and protein in many developing countries throughout the world (FAO: http://faostat.fao.org/) providing as much as 15% of the total daily calories and more than 30% of the protein intake per day. Among major food crops, it has one of the highest levels of variation in growth habit, seed characteristics (size, shape and colour), maturity and adaptation.

The distribution of populations currently, classified as wild-growing *P. vulgaris*, is very extensive as it stretches from northern Mexico and southern USA to northwestern Argentina (from approximately 28°N. Lat. to 27°S. Lat., or ~10,000 km long). Genetic diversity of wild beans has evidenced a complex population structure, defining up to five gene pools (Blair et al. 2012a) corresponding to segments of the geographical range of *P. vulgaris* distribution in America, including Mesoamerican (Mexican), Guatemalan, Colombian, Central Andean (Ecuador, Northern Peru) and Southern Andean (northern Argentina, Bolivia and Southern Peru). Such observations raise the question of the differential origins, if any, of this broad distribution. Even though the New World origin of the genus was established by phylogenetic studies using nuclear and chloroplast markers (Delgado-Salinas et al. 2006), and the origin of *P. vulgaris* has been strongly debated. For many years, the most accepted hypothesis regarding the origins of common bean indicated that, from a core area on the western Andes in northern Peru and Ecuador, wild beans were dispersed north (to Colombia, Central America and Mexico) and south (to southern Peru, Bolivia and Argentina), and indigenous people independently domesticated this crop during pre-Colombian times (Kwak and Gepts 2009). In this regard, radiocarbon dating and the evidence of starch grains in human teeth found at archaeological sites have placed common bean cultivation and consumption in South America, Northern Peru and Mexico between 4300 and 8000 B.P. (Kaplan and Lynch 1999; Piperno and Dillehay 2008; Mensack et al. 2010). The hypothesis of an Andean origin of the species relied on phylogenetic inferences using phaseolin, the major seed storage protein, that in the wild populations from northern Peru and Ecuador (Debouck et al. 1993) shows an ancestral form (type I) because of the absence of tandem direct repeats in its genes; similar to those loci coding for phaseolin in the two most closely related species to *P. vulgaris*: *Phaseolus dumosus* and *Phaseolus coccineus* (Kami et al. 1995). However, recent studies using other molecular markers contradict this theory and place Mesoamerica as a more probable centre of origin of the species. Using five different loci from 49 Mesoamerican, 47 Andean and 6 Peruvian wild *P. vulgaris*, four different genetic clusters were observed in Mesoamerica, whereas only one was clustered with the Andean accessions and one from Peru (Bitocchi et al. 2012). The phylogenetic results indicate in the first place that the Peruvian accessions are closer to one of the Mesoamerican clusters and that there was a strong bottleneck in South America before the domestication process took place. These results are not surprising for several reasons. First, using Amplified Fragment Length Polymorphisms (AFLPs), several genomic loci and chloroplast markers, it was observed that there is a very low genetic diversity in wild and domesticated *P. vulgaris* of Andean origin, (Chacón et al. 2005; Mensack et al. 2010; Mamidi et al. 2011); whereas in the Mesoamerican varieties, the genetic diversity is higher and it is accompanied by a lower linkage disequilibrium (LD) estimation compared to the Andean values (Rossi et al. 2009). Additionally, although the genus *Phaseolus* extends from México to Argentina, a large majority of species is found in Mexico, which suggests that *P. vulgaris* originated in Mesoamerica by sympatric or allopatric speciation and latter, migrated to the south of the continent. The fact that the ancestral phaseolin type has not been found in Mesoamerican accessions might be due to a sampling limitation or maybe it is really extinct from these populations. Nevertheless, without whole-genome analyses, it has been extremely complex to clearly establish the history and evolution of *P. vulgaris*.

The *Phaseoli* are diploid plants ($2n = 22$) with estimated genome sizes of about 600 Mbp, and 30% GC content. Interestingly, *Phaseolus leptostachyus* is an exception to the rule, having only $2n = 20$ chromosomes (Mercado-Ruaro and Delgado-Salinas 1998). For a long time, common bean genetic resources were limited to linkage maps using reference populations that combined Mesoamerican and Andean genotypes, such as DOR364 × G19833 (Blair et al. 2002; Córdoba et al. 2010) or BAT93 × Jalo EEP558 (Grisi et al. 2007), that were continuously enriched with new microsatellites and SNPs. These genetic maps were useful for the identification of several QTLs associated with resistance traits (Kelly et al. 2003; Garzon and Blair 2014) or even the popping ability of nuña beans (Yuste-Lisbona et al. 2012). Recently, the genomes of two varieties of *P. vulgaris* of Mesoamerican (Vlasova et al. 2016) and Andean origin (Schmutz et al. 2014) have been sequenced and provide an excellent reference for further comparative studies. The Mesoamerican variety BAT 93, which was developed at the "Centro Internacional de Agricultura Tropical" (CIAT), is highly homozygous and possesses resistance genes for the Bean common mosaic virus, *Xanthomonas campestris*, *Uromyces appendiculatus* and *Colletotrichum lindemuthianum*. The final BAT93 genome sequence assembly encompassed 549.6 Mb, which is close to the previously estimated genome size (Arumuganthan and Earle 1991; Bennett and Leitch 1995). The assembly included 97% of the conserved core eukaryotic genes, reflecting its completeness. Ab initio predictions in combination with RNA-Seq data, as well as public EST and cDNA sequences allowed the prediction of 30,491 protein-coding genes (PCGs) whose 66,634 transcripts encode 53,904 unique proteins in its genome. Non-coding RNAs were identified using a combination of RNA-Seq and in silico homology modelling methods, leading to the identification of 2529 small RNAs belonging to plant known families and 1033 long non-coding RNAs (lncRNAs). Similarly, Schmutz et al. (2014) sequenced an inbred landrace line of *P. vulgaris* (G19833) derived from the Andean pool (Race Peru) and assembled it onto 11 chromosome-scale pseudomolecules that ultimately represent ~80% of the 587-Mb genome. The genome was found to encode 27,000 genes and contains a high proportion of recent transposon insertions. Approximately, 91% of *P. vulgaris* genes were found located within soybean (*Glycine max*) synteny blocks.

The comparison of the protein-coding sequences of equivalent genes derived from both genomic projects showed that 1186 pairs had sequence identity lower than 95%, and these were enriched in defence response and terpene synthase activity. Even if one could expect to find differences in terms of the gene clustering or copy-number variations of resistance genes given the BAT93 selection for less disease susceptibility, it was noteworthy that no resistance-gene cluster was specific to any of the two sequenced accessions. This observation indicates that genomic clustering of resistance genes predates the split of the Mesoamerican and Andean gene pools, and that differences in pathogen susceptibility might be due to gene polymorphisms, rather than a gene presence/absence effect.

Furthermore, the availability of *P. vulgaris* genome sequences, complemented with genomic, phylogenomic and metabolomic signals from 29 re-sequenced genomes from 12 different *Phaseolus* species that represent most of the phylogenetic clade diversity in the genus, reinforced the hypothesis of a Mesoamerican origin of *P. vulgaris*, but also revealed a particular speciation event in the Peruvian-Ecuadorian region of tropical Andes that predates the split of Mesoamerican and Andean *P. vulgaris* gene pools (Rendón-Anaya et al. 2017). The wild populations located in the Amotape-Huancabamba Depression (northern Peru and south of Ecuador), one of the most biodiverse regions in the Neotropics (Richter et al. 2009; Luebert and Weigend 2014), were shown to have diverged enough to be tagged as a "sister species" to *P. vulgaris*, a fact that would have remained hidden without the generation of genomic resources.

As described along this section, the elucidated common bean genome sequences may prove useful to breeders of this species but also to evolutionary biologists and crop geneticists.

5.2 Understanding Common Bean Domestication Through Genome Sequencing

The transition from hunting–gathering to agriculture is one of the major milestones in human evolution. An important consequence of this transition has been the domestication of crop plants and farm animals (Purugganan and Fuller 2009; Larson et al. 2014). Agriculturalists, from prehistoric times until present, have improved their crops and livestock by choosing the best individuals as parents for the next generations. Different geographic areas can be distinguished as centres of domestication, including the Fertile Crescent, China, Mesoamerica, Andes/Amazonia, eastern USA, Sahel, tropical West Africa, Ethiopia and New Guinea. Expansions of crops, livestock, people and technologies tended to occur more rapidly along east–west axes than along north–south axes since locations at the same latitude share similar climates, habitats and hence require less evolutionary change or adaptation of domesticates, technologies and cultures than do locations at different latitudes (Gepts 2004). Some New World crops are represented by distinct but related species in North/South America and Mesoamerica, suggesting that related species were domesticated independently in these areas; this is the case of common bean *P. vulgaris*, lima bean *Phaseolus lunatus*, chilli peppers *Capsicum annuum*, and squashes *Cucurbita pepo*, among other crops (Diamond 2002).

Domestication provides an experimental model to study evolution in general, with several advantages, including the existence of ancestral populations, an established time frame (~10,000 years), identifiable traits under selection and often the availability of advanced experimental tools for domesticated plants or animals (e.g. Gepts 2014). In this perspective, *Phaseolus* species are of interest because of the multiple domestications that have taken place in this genus. Indeed of the 70–80, wild species that have been described, no less than five species have been domesticated in contrasting ecogeographic settings: common bean (*P. vulgaris*), lima bean (*P. lunatus*), runner bean (*P. coccineus*), tepary bean (*P. acutifolius*) and year bean (*P. dumosus*). In addition, the first two species were domesticated at least twice independently, in Mesoamerica and in the Andes mountains. In addition, some domestication traits may have been selected multiple times, as shown by the determinacy trait in common bean (Kwak et al. 2012). This is in contrast with other crops, such as maize (single domestication) or rice and wheat (three domestications), which have been subjected to less domestication events. Thus, the multiple domestication phenomenon in *Phaseolus* provides an opportunity to examine to what extent similar selection pressures have led to convergent evolution at the molecular level (Lenser and Theißen 2013). Conversely, comparative genomics can illustrate the differential genetic control of adaptation to contrasting environments in which the different *Phaseolus* species were domesticated.

Many morphological and physiological changes (determinate growth habit, the lack of seed dispersal, dormancy or lack of toxicity) are repeated traits in different domesticated crops and thus have been used to define the concept of the "domestication syndrome" (Gepts 2004). The conservation and inheritance of such traits were originally based on a Mendelian strategy and more recently, on the identification of quantitative trait loci (QTL) that represent blocks of genes that have dramatic effects on adaptation (Hancock 2005). Therefore, domestication, considered as the outcome of a selection process that leads to increased adaptation of plants and animals to cultivation and utilization by humans, can be evaluated under a population genetics perspective (reviewed by Morell et al. 2011). One of the main consequences of domestication is the loss of genetic variability, compared to that observed in the wild ancestors (Gepts and Papa 2002). This is partially explained by the reduced size of founding populations and successive bottlenecks, after which, only few allelic combinations are passed on to future generations; there is an important loss of heterozygosis and effective recombination, and thus, substantial LD can be generated. Generally, LD decays more rapidly in outcrossing species as compared to selfing ones because recombination is less effective in auto-pollinated individuals, which are more likely to be homozygous (Morell et al. 2011).

A second effect of plant domestication is the modification of breeding systems: outcrossing plants are often forced to follow a self-pollinating system (Hancock 2005). This change in the mating system produces a decrease in population sizes since lethal alleles are expressed as homozygous. Once these lethal alleles are eliminated from the population, the individual fitness increases and thus, the size of the population is balanced. At the same time, homozygosis becomes more frequent in the population and the genetic diversity is greatly affected.

Several lines of evidence from traditional (allozymes or seed proteins) to more recent molecular markers [Restriction Fragment Length Polymorphisms (RFLPs), Random Amplified Polymorphic DNAs (RAPDs), AFLPs (reviewed by McClean et al. 2004) and Single Nucleotide Polymorphisms (SNPs), Cortés et al. 2011; Bitocchi et al. 2013] converge in the establishment of two geographically and genetically isolated gene pools, one in Mesoamerica and one in the Andes, from which, two independent domestication events took place starting ~8000 years ago, followed by local adaptations and further expansions. This scenario is not atypical in crops, as other plants have been domesticated more than once, offering the possibility of studying parallel evolution of independent lineages. Such an example is given by rice, *Oryza sativa*, with its two cultivated subspecies, *indica* and *japonica*, whose genomes clearly display independent origins from their wild relatives but share genomic segments bearing important agronomic traits that arose only once in one population and spread across cultivars through introgression and artificial selection (He et al. 2011).

The recent publication of a *P. vulgaris* genome of Andean origin (Schmutz et al. 2014) allowed for the first time to have a large-scale screening of the effects of artificial selection on both gene pools. The estimation of genetic diversity losses and differentiation index (F_{st}) on four re-sequenced pooled populations representing Mesoamerican and Andean landraces, suggested that different sets of genes, 1835 in Mesoamerica and 748 in the Andean region, were selected during both independent domestication events, with only 57 of them shared by both processes. Even within gene pools, domestication candidates were not shared by subpopulations, suggesting that similar phenotypes in cultivated accessions were achieved following independent evolutionary trajectories. At the genomic level, 74 and 60 Mb, respectively, were shown to be affected by artificial selection. Although relevant, certain aspects of this approach have to be carefully considered: the fact that pooled populations were sequenced means that some biases could have been introduced in terms of over/under-representation of polymorphisms particular to certain individuals within each subpopulation. In addition, the estimators that authors propose to identify signals of domestication (Tajima's D, F_{st} and π) are sensitive to population structures and are not necessarily direct indicators of the effects of artificial selection if gene flow is not considered as part of the genomic dynamics of the landraces.

An independent screening using the genome of the Mesoamerican variety BAT93 as a reference and individual genome sequencing of 12 domesticated and wild accessions from both gene pools, produced contrasting results. By defining haplotypes strongly associated with the domesticated phenotype, 599 PCG and 52 lncRNAs with haplotypes shared between domesticated genotypes from Mesoamerica and the Andes, and 628 PCGs and 45 lncRNAs with haplotypes specific to Mesoamerican domesticated accessions were identified (Rendón-Anaya et al. 2017). Screening of protein definitions associated with the domestication gene candidates identified enriched GO categories that could be easily linked to the emergence of domestication traits (seed size, photoperiod sensitivity, regulation of reproductive processes, plant architecture and hormone signalling, among others).

A strikingly small overlap was, however, observed between the sets of domestication PCGs produced by the above-mentioned studies. Even though several methodological differences could be the source of such observation (sequencing pooled vs. individual genomes; coverage; sample size; population genetics parameters), the absence of gene flow estimations in the model reported by Schmutz et al. (2014) could be determinant as it was considered by Rendón-Anaya et al. (2017), as some of the reported loci that differentiate landraces from wild genotypes may not be the outcome of artificial selection but rather represent admixture with other gene pools. Nevertheless, the addition of more genotypes to this analysis should clarify if more genes were differentially selected for each gene pool and how population expansions and adaptations to different habitats have altered genetic diversity in other loci in the genome.

So far, the description of genome-wide screenings of selection highlights the importance of generating reference genomes of common bean from each gene pool, as different conclusions, some even unexpected, have been drawn regarding the effects of domestication. These analyses converge to one important observation: domestication of common bean has affected, intentionally or by hitchhiking, protein-coding genes and many different kinds of regulatory elements contained in intergenic segments with selection signatures that, all together, have produced the phenotypes that we observe in cultivated lines. Given the availability of non-coding RNA predictions for *P. vulgaris* reference genomes, their association to the domestication process could be easily evaluated through transcriptional correlations between domestication candidates and long/small non-coding RNAs encoded within selective sweeps, combined with estimations of genetic diversity losses in such regulators and their targets. A more detailed description of the biological processes and transcriptional patterns involved in the emergence of domestication traits will be accelerated as genomic data from more accessions is generated.

Ultimately, domestication gene candidates (Fig. 5.1) should be experimentally validated. Even though a few examples of successful transformation (Kwapata et al. 2012) and virus-based gene silencing (Díaz-Camino et al. 2011) have been reported, the recalcitrance of common bean for genetic transformation has made functional studies really challenging. The generation of mutant populations of *P. vulgaris* trough fast neutron radiation (O'Rourke et al. 2013) and TILLING [targeted induced local lesions in genomes (Porch et al. 2009)] protocols has been used as an alternative for the identification of genes behind visual phenotypic differences. Thus, the combination of such strategies and genome re-sequencing of mutant plants could eventually facilitate the direct association of coding and non-coding loci to the emergence of domestication and adaptation traits.

5.3 Going Back to the Wild

Domesticated crops have experienced strong human-mediated selection during improvement, aimed at developing high-yielding varieties. Traditional breeding programs tend to concentrate on specific genotypes, which combine traits of interest and may be used as progenitors in several crosses. However, high-throughput SNP genotyping in crops, such as maize or wheat (Cavanagh et al. 2013), has evidenced small differences in terms of the amount of genetic diversity between modern cultivars and landraces, since minor bottlenecks occurring during improvement do not dramatically alter allele frequency as stringent domestication bottlenecks do. Undoubtedly, genetic diversity represents a necessary condition for further evolution in response to selection pressures and thus, it also represents the raw material to develop improved breeds or cultivars (Gepts and Papa 2002). However, the combination of the loss of adaptive

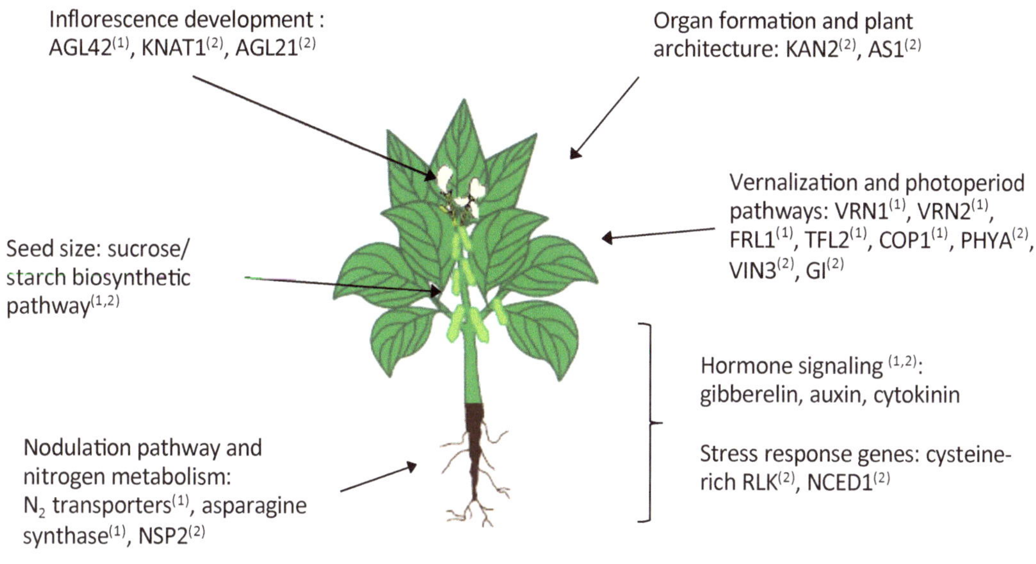

Fig. 5.1 Examples of domestication gene candidates proposed by two independent genomic studies: (1) Schmutz et al. (2014), (2) Rendón-Anaya et al. (2017)

alleles through drift and fixation of deleterious alleles due to artificial selection necessarily constrains our ability to expand the cultivation of domesticated species into environments beyond those in which domestication occurred, e.g. into more extreme climates, marginal soils, degraded agricultural landscapes or into sustainable systems with reduced agricultural inputs. Thus, systematic efforts to bring genetic diversity from wild relatives into crop plants to incorporate a wider range of useful adaptations for disease resistance, abiotic stress tolerance and other agronomic challenges are required in order to increase the resiliency and productivity of agriculture. Furthermore, given the fact that no biotechnological tools (efficient transformation systems) are available to easily manipulate common bean plants and that seed improvement relies on traditional breeding, it becomes essential to identify molecular processes behind wild *P. vulgaris* adaptation to variable environments and stress conditions to facilitate targeted breeding and successfully introduce new allelic variation into domesticated lines.

The first step towards a more conscious use of the genetic reservoir contained in wild populations of *P. vulgaris*, consists in understanding parallels and contrasts between natural and artificial selections, how they have shaped genetic diversity and altered expression profiles in wild versus domesticated populations. A useful strategy so far employed in other crops is transcriptome sequencing of wild and domesticated relatives to describe how selection on quantitative traits has affected gene expression networks. In the case of maize, the expression profiling of 18,242 genes (using an expression array) for 38 diverse maize genotypes and 24 teosinte (the wild relative of maize) genotypes, revealed more than 600 genes having significantly different expression levels in maize compared with teosinte. Moreover, more than 1100 genes showed significantly altered co-expression profiles, reflective of substantial rewiring of the transcriptome since domestication (Swanson-Wagner et al. 2012). Although limited information on the functional consequences of the expression changes can be drawn, differentially expressed genes show a significant enrichment for genes previously identified through population genetic analyses as likely targets of selection during maize domestication and improvement. Another example is the comparison of transcriptomes from wild and domesticated cotton accessions during fibber formation, which revealed that wild cottons allocate greater resources to stress response pathways, while domestication led to reprogrammed resource allocation towards increased fibber growth, possibly through modulating stress response networks (Yoo et al. 2014). Transcriptomic tools have been also used to answer intriguing points regarding the emergence of domestication traits in common bean. RNA-Seq data obtained from 10 domesticated and eight wild Mesoamerican *P. vulgaris* accessions at the first true-leaf stage revealed that domestication not only affected the level of nucleotide diversity in about 9% of the genes but also changed expression patterns of certain loci (Bellucci et al. 2014). Differentially expressed transcripts in wild accessions compared to the domesticated ones were enriched in putatively selected genes and the loss of expression diversity appeared significantly higher in selected genes compared to neutral loci. These observations could be linked to domestication but could be also explained by hitchhiking of regulatory elements.

The unanswered question remains: how to introduce such allelic variation from wild relatives into cultivated lines of common bean? More subtle, but no less important, are the evolutionary effects that arise from spontaneous and/or intentional mating of domesticated plants with their wild relatives. We would be mistaken if we regarded domesticated plant taxa as evolutionary discrete from their wild relatives. Such hybridization may lead to gene flow: the incorporation of genes into the gene pool of one population from one or more populations, also known as introgression (Dowling and Secor 1997). Even though the hybrid descendants can be less viable, it is possible that farmer selection of introgressants may have played a very important role in early development of crops, as agriculture spread out of the centres of origin, when the alterations resulted in useful genetic combination.

The mating system is an important determinant of the genetic variation that is maintained in plants: outcrossing species usually show higher genetic diversity, compared to selfing species, in which heterozygosis is rapidly lost. Hybridization has been defined as the interbreeding of individuals from two populations, or groups of populations, which are distinguishable on the basis of one or more heritable characters (Dowling and Secor 1997), and it occurs when there are incomplete reproductive barriers between two taxa (Counterman and Noor 2006). The fraction of species that hybridize is variable, but on average around 10% of animal and 25% of plant species are known to hybridize with at least one other species (Mallet 2007), even if they are distantly related (Weissmann et al. 2005). Hybridization can operate in different directions: reducing taxon diversity by eliminating the boundaries between species, particularly if gene flow occurs into one or both parental taxa (which might facilitate adaptive evolution; Fig. 5.2a); generating new taxa by homoploid or allopolyploid hybrid speciation (Fig. 5.2b) as in the case of *P. dumosus* that derived from the hybridization of *P. vulgaris* and *P. coccineus* (Mina-Vargas et al. 2016); and merging the two hybridizing taxa (Pastorini et al. 2009; Schneider et al. 2011). The geographic pattern and spatial scale of introgression will depend on many factors, including the environmental context in which hybridization occurs, how far individuals disperse, and the nature of selection. Therefore, studies of natural hybrids and their genetic composition can give important insights into evolutionary processes and the adaptation of species.

In many cases, domesticated plants and their wild progenitor can hybridize and their progeny is viable and fertile. These hybridizations are the first step in the formation of weedy populations that combine traits of domesticated and wild types (Fig. 5.2a). Such hybridizations can result in adaptive introgressions, as it has been documented between maize and wild teosinte (*Zea mays* ssp. *Mexicana*), where the incorporation of adaptive *Mexicana* alleles into maize during its expansion allowed this crop to grow in the highlands of central Mexico. More recently, a genome-wide scan of introgression signal was documented for cassava (*Manihot esculenta*) cultivars, whose domestication started around

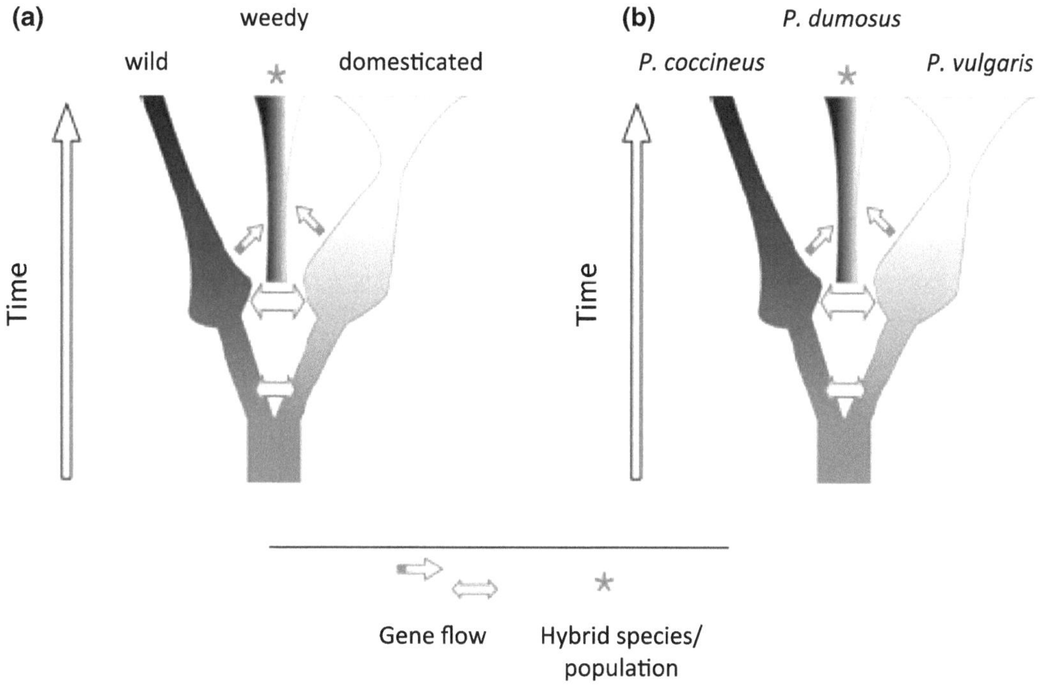

Fig. 5.2 Hybridization and gene flow reintroduce genetic diversity. **a** Hybridization inside gene pool one originates weedy populations with high levels of genetic diversity. **b** Hybridization between two *Phaseolus* species produced a stable hybrid species, *P. dumosus* (modified from Abbott et al. 2013)

6000 years ago in the Amazonian basin. Sequencing wild (*M. esculenta* ssp. *flabellifolia*) and domesticated cassava genomes and comparing them to related species (*M. glaziovii*) did not only evidence a strong maternal bottleneck but interspecific introgressions were shown to introduce variation into the nuclear genome, particularly in farmer varieties in Africa, where it was introduced only 500 years ago and spread by undocumented crosses (Bredeson et al. 2016).

In the case of common bean, the richness of wild and domesticated populations growing in sympatry, particularly in Mesoamerica, has facilitated intra-species hybridizations (gene pool 1) leading to adaptive introgression events. This has been an ongoing phenomenon that occurs naturally and under the human influence all along the domestication process. In different geographic zones, farmers that still maintain traditional cultivating systems usually interchange seeds and plant several different landraces in the same complexes in order to ensure some harvest, regardless of the annual growth and environmental conditions. Therefore, it has been possible to maintain a high diversity that increases through spontaneous crossing among landraces. Indeed, a higher molecular diversity has been observed within domesticated seeds planted under traditional cultivating systems than in the local wild populations or the original breeding lines in several regions in Mexico, like Oaxaca (Worthington et al. 2012), Yucatán (Martínez-Castillo et al. 2006), Guanajuato and Michoacán (de la Cruz et al. 2005; Zizumbo-Villarreal et al. 2005), and Peru and Colombia (Beebe et al. 1997). Furthermore, the protection of wild populations in the plots by traditional farmers can lead to hybridization of wild and domesticated populations, thereby generating weedy plants. In the same way, this protection favours backcrossing of weedy with domesticated plants and subsequently the establishment of segregants with high morphological similarity to the domesticated individuals.

Measuring AFLP diversity, it has been proposed that differentiation of sympatric wild and domesticated populations is higher around domestication genes than in other loci in the genome; these observations suggest that selection in the presence of introgression is a major evolutionary factor maintaining the identity of wild and domesticated populations (Papa et al. 2005).

Even though gene flow can occur in both directions, from domesticated to wild populations and vice versa. A genome-wide screening of introgression signatures between *P. vulgaris* genotypes in Mesoamerica (Rendón-Anaya et al. 2017) confirmed a remarkable asymmetry of gene flow between wild and domesticated common bean subpopulations, as previously measured using microsatellite diversity (Papa and Gepts 2003). Such asymmetry could be due to the fact that introgression in domesticated genotypes from wild neighbours is usually limited through selection against hybrids where wild traits, which are dominant or semi-dominant, are easily recognized by farmers (Koinange and Gepts 1992). These observations imply that genetic admixture and a possible mosaic genomic structure might be more frequent than expected following the preferential autogamy of the species.

Normally, the introgression of traits from wild or weedy germplasm is difficult in modern breeding programs due to the prevalence of non-domesticated traits governed by dominant genes (Beebe et al. 1997). However, knowing that traditional farming systems have made of domestication a dynamic process resulting from selection, hybridization and reselection over many years, we suggest that the variability so generated could be useful beyond the site where it occurs by continuous screenings to recover promising recombinants and introgressants that would complement modern breeding programs. Unfortunately, the use of wild relatives as a genetic resource has been taken into account from an old fashion optic, just by looking for particular phenotypes of agromorphological interest. Once a population with a desirable characteristic is identified, breeders cross them with modern varieties or cultivars, in order to introduce such traits from the wild donor. This strategy can potentially work with efficiency if the selected trait is monogenic, that is, one or only a few genes in proximity, such as pathogen resistance. A survey of the use of wild germplasm in crop improvement over the last decades (Hajjar and Hodgkin 2007), including rice, wheat, maize, barley, sorghum, millet, cassava, potato, chickpea, cowpea, lentil, soybean, bean, pigeon pea, banana and groundnut, revealed that over 80% of the reported beneficial traits conferred by genes derived from wild relatives, are involved in pest and disease resistance. Similarly, the stabilized hybrid *Helianthus annuus* ssp. *texanus* captured alleles that provide herbivore resistance from wild *Helianthus debilis* (Whitney et al. 2006), and tomato cultivars introgressed several chromosomal segments from wild *Solanum pimpinellifolium*, enhancing fruit colour (Tomato Genome Consortium 2012). The identification of PCGs encoded within introgressed genomic regions between *P. vulgaris* subpopulations (Rendón-Anaya et al. 2017) also revealed an enrichment of functional terms associated with hormone-mediated signalling pathways, reproductive processes, post-embryonic development and the formation of reproductive organs. Genes involved in biotic and abiotic stress responses (WRKYs, leucine-rich repeat receptor kinases, and pathogenesis-related proteins) were also found as transferred in most of the *P. vulgaris* comparisons in both directions (from and towards domesticated accessions), implying that the continuous movement of such loci favoured the adaptation of common bean to different habitats.

Traditionally, crosses of *P. vulgaris* cultivars with wild accessions have been used to develop varieties possessing different resistance alleles. This is the case of arcelin, which confers moderate levels of resistance to bruchids (*Acanthoscelides obtectus* and *Zabrotes subfasciatus)*; cultivars resulting from crosses of elite lines (BAT93) and wild beans collected in Mexico (PI 417662) are web blight and common bacterial blight resistant, caused by *Thanatephorus cucumeris* (anamorph *Rhizoctonia solani*) and *Xanthomonas axonopodis* pv. *phaseoli*, respectively.

Other inbred backcross populations show higher nitrogen, iron, and calcium seed content, or display higher yields than the recurrent elite parent (reviewed in Acosta-Gallegos et al. 2007). Abiotic stress tolerance is another important example of desirable improvement in common bean that has been difficult to introduce. Efforts to increasing drought tolerance in common bean commercial varieties have been a priority for breeders since we face important and quick climate changes (Beebe et al. 2013). It has been reported that wild *P. vulgaris* accessions are distributed in a wide range of altitudes, different precipitation regimes and soil types. Thus, combining ecogeographical information, population structure, genomic and transcriptomic data could be useful for genome-wide genetic-environmental associations that could accelerate the selection of wild individuals to be included in breeding programs (Cortés et al. 2013).

Several traits, however, rely on the additive action of several loci, epistatic interactions and by tuning gene expression by other types of regulatory elements on the genome. Finding such genes and regulatory elements is a great challenge for plant breeders. A remarkable example is the improvement in fruit colour of tomato, attributed to lycopene synthesis. Wild tomatoes remain green even when ripe because they lack the enzyme of the very last step in the lycopene synthesis pathway. However, wild alleles have been shown to enhance earlier steps from this metabolic pathway that, when combined with an active form of the enzyme for lycopene synthesis from cultivated tomatoes, produces highly pigmented offspring. Recent genomic studies have shown that whole-genome triplication of *Solanum* species, added new gene family members that mediate important fruit-specific functions and that, within cultivated germplasm, particularly among the small-fruited cherry tomatoes, several chromosomal segments are more closely related to wild *S. pimpinellifolium* than to cultivated forms, supporting the hypothesis of recent admixture of these gene pools due to breeding (Tomato Genome Consortium 2012).

Thus, although wild germplasm is perceived to be a poor bet for the improvement of most traits based on phenotypic examination, it is quite possible that some favourable alleles are hidden in unexplored accessions. Massive genomic screenings, including SNP detection through individual genome sequencing and comparison of transcriptomic profiles and co-expression networks of wild and domesticated populations, are indispensable tools for finding those loci and construct more accurate genetic maps reflecting recombination hotspots and barrier loci for introgression. Implementing such strategies requires a major shift in the paradigm for using our genetic resources but should accelerate targeted breeding programs in the short term.

5.4 Genomic Introgression Outside *P. vulgaris*

Around 75 *Phaseolus* species have been identified. Interestingly, they have important morphological differences and grow in variable environmental conditions (altitude, temperature, humidity). A phylogenetic study performed by Delgado-Salinas and collaborators (Delgado-Salinas et al. 2006) revealed that *Phaseolus* species can be separated in eight different groups, and two independent clades can be distinguished. The first one, which includes *Phaseolus glabellus*, *Phaseolus oaxacanus*, *Phaseolus macrolepis*, *Phaseolus microcarpus* and species from Pauciflorus, Pedicellatus and Tuerckhemii groups, is geographically and ecologically limited and is distributed in Mexico and southern USA (Arizona, New Mexico and Texas). There is no evidence of domestication events of these species, probably due to their low abundance. In contrast, the second clade that includes Filiformis, Vulgaris, Lunatus, Leptostachyus and Polistachyos groups is widely distributed in America, from southern Canada to the Andean region, and it is speculated that its broad distribution and abundance made domestication of several of the species in this clade, such as *Phaseolus vulgaris* (common bean), *P. coccineus* (runner bean), *P. acutifolius* (tepary bean), *P. lunatus* (Lima bean) and *P. dumosus* (year-long bean) possible. The latter being an interesting case of an evolutionarily stable hybrid of *P. vulgaris* and *P. coccineus*, with its morphology, life cycle, adaptation, and reproductive mode intermediate between the two parental species (Llaca et al. 1994), with important agronomic and nutritional functions in Mexico, Central America and Colombia.

Domestication processes affecting the above-mentioned *Phaseolus* species have attracted the attention of some research groups. It has been suggested that a single domestication event of *P. acutifolius* occurred in the Sonoran desert region of Sinaloa since wild tepary accessions from this area were grouped with cultivated lines in distance-based trees using microsatellite sequences (Blair et al. 2012b). Two major gene pools have been defined for lima bean, Andean and Mesoamerican, the latter subdivided in at least two groups (MI and MII) (Andueza-Noh et al. 2013; Martínez-Castillo et al. 2014). Wild populations of the large-seeded Andean gene pool have a narrow distribution on the western slope of the Andes in Ecuador and northern Peru, while wild populations of the small-seeded Mesoamerican pool have a much broader distribution that included not only Mexico and Central America but also the eastern slope of the Andes from Colombia to Argentina. Given the high outcrossing rate of lima beans, introgression has played a very important role in determining the level of genetic diversity of wild and domesticated populations. Just as in common bean, gene flow is bidirectional and higher from domesticated to wild populations but highly variable when different regions are considered for sampling. This results in different levels of genetic diversity, maintaining higher values in those regions where introgressions are more frequent (Martínez-Castillo et al. 2007; Félix et al. 2014). Using chloroplast markers from 262 wild and domesticated accessions (Andueza-Noh et al. 2013), it was recently proposed that the MI group was domesticated in western central Mexico, (Nayarit, Jalisco, Colima, Michoacan and Guerrero), while MII in Guatemala, Honduras, Costa Rica and the Mesoamerican Mayan region. On the other hand, population structure analyses suggest that domestication of runner bean could have occurred independently in two areas, Mexico and Guatemala–Honduras, followed by extensive hybridizations (Spataro et al. 2011). Just as in common bean until the generation of reference genomes, many questions remain open in terms of the effects of artificial selection on different genomic features and rewiring of transcriptional networks of other domesticated *Phaseolus* species. Given that domestication syndrome traits are common to most cultivated *Phaseolus* species, it is possible to imagine some degree of convergence of domestication processes into similar loci, metabolic pathways, regulatory elements and expression tuning. However, important differences in ecological niches, different degrees of availability/proximity to wild populations and reproduction habits, open the possibility of identifying alternative outcomes of domestication compared to common bean.

Most *Phaseolus* species reproduce by self-pollination; however, there are examples of intermediate outcrossing in the genus. This is the case of *P. coccineus*, a species that is usually pollinated by hummingbirds, and *P. lunatus* that uses bees as natural pollinators. Not surprisingly, opposite to tepary, lima and common beans, different populations of Mesoamerican *P. coccineus* sampled in central Mexico and Chiapas display high and similar levels of genetic variation (determined with seven electrophoretic markers) without differences among wild and cultivated populations (Escalante et al. 1994). The same was concluded while comparing several SSRs from European and American populations of *P. coccineus* (Spataro et al. 2011). In spite of its preferential autogamy, *P. vulgaris* cannot be considered to have a closed reproductive system, as it maintains outcrossing rates between 1 and 70%, depending upon the experimental conditions (Wells et al. 1988; Ferreira et al. 2007). It is also possible to hypothesize that wild and domesticated *P. vulgaris* could have differential outcrossing patterns, such as soybean, in which cultivated soybean, *Glycine max*, has an outcrossing rate of approximately 1%, whereas the wild ancestor, *Glycine soja*, outcrosses at an average rate of 13% (reviewed by Flint-Garcia et al. 2003).

Overall, intra-species outcrosses correspond to a primary gene pool (GP-1); however, inter-species hybridizations have also been reported within the Vulgaris clade (Fig. 5.3). In the secondary gene pool (GP-2), in which hybridization is possible but hybrids are weak with low fertility; it has been observed in *P. coccineus*, *P. vulgaris*, *Phaseolus costaricensis*, *P. dumosus* (Blair et al. 2006). Within the tertiary gene pool (GP-3) in which embryo rescue is needed since hybrids are lethal or sterile, hybridization is possible in *Phaseolus parvifolius* and *P. acutifolius*. Even though it has been shown that no outcrossing events occur between the Lunatus and Vulgaris groups, it is possible to obtain viable descendants by crossing *P. lunatus* and *Phaseolus polystachios* plants.

The evaluation of morphoagronomic traits of the species belonging to each *Phaseolus* gene pool highlights the need to integrate them as genetic resources for breeding programs in the short term. Two cultivated species from GP-2, *P. coccineus* and *P. dumosus*, as well as wild *P. costaricensis*, arc vigorous vines with perennial or semi-perennial tendencies. Even though three incompatibility barriers in crosses between common beans and runner beans have been identified (blocked cotyledon lethal, crinkle leaf dwarf and dwarf lethal), runner beans and year-long beans are often found in cloud forests of Central America and Mexico, where climatic conditions are favourable for the development of fungal diseases such as rust (caused by *U. appendiculatus*), anthracnose (caused by

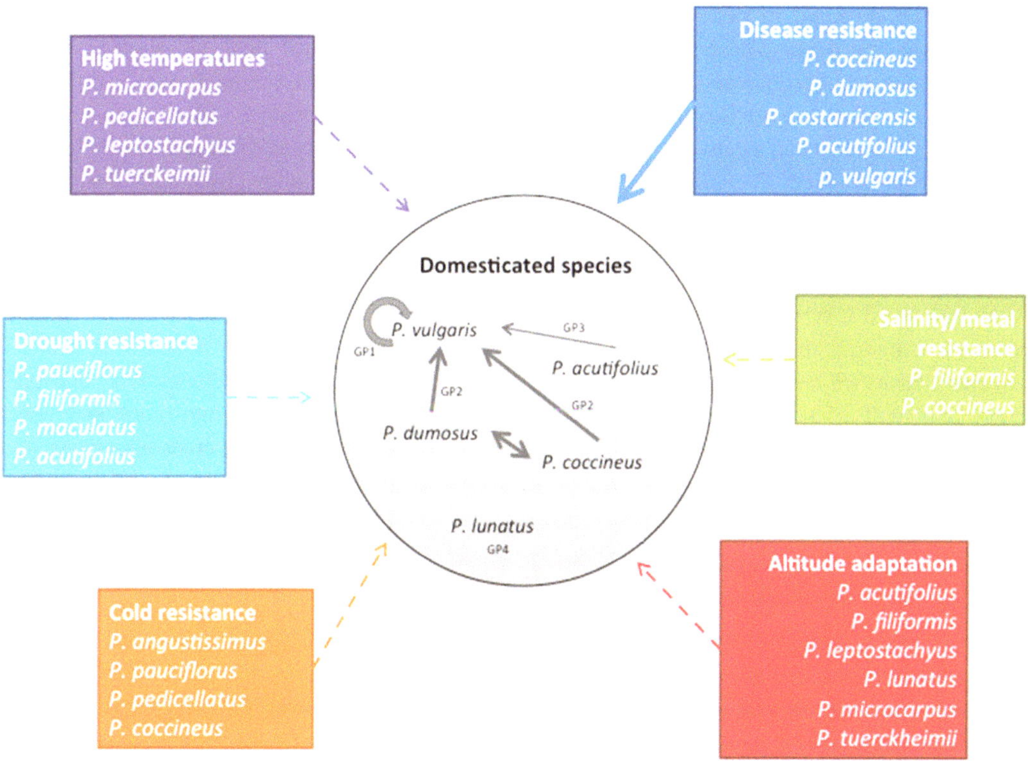

Fig. 5.3 Potential use of *Phaseolus* species in breeding programs of domesticated lines. Domesticated species belonging to different gene pools (GP) are contained in the inner circle. Arrows indicate the direction of genetic flux; their thickness represents the feasibility of hybridizations; pointed arrows indicate hybridizations for several species in each box which have not been explored

C. lindemuthianum) and web blight. Therefore, they have been employed as sources of resistance to a wide array of bean pathogens, although their use for other traits has been very limited (reviewed by Porch et al. 2013). There is also evidence of interspecific genomic introgression between *P. vulgaris* and these phylogenetically close species (Rendón-Anaya et al. 2017), that showed that PCGs related to cell wall biogenesis and organization, and pectin and cell wall polysaccharide metabolic processes, have been transferred from *P. coccineus* and *P. dumosus/P. costaricensis* into *P. vulgaris*, which could contribute to the acquisition of pathogen resistance in common bean (Miedes et al. 2014).

Using hydroponic systems, some accessions of *P. coccineus* were also shown to be very tolerant to aluminium-toxic acid soils (Butare et al. 2011). Field observations and subsequent greenhouse studies of root systems have revealed that runner beans have thick roots that might have a better potential to penetrate compacted soil than common beans. These traits could well contribute to drought resistance and merit further investigation. Moreover, wild populations of common bean and runner beans are often found growing together. The *P. vulgaris* × *P. coccineus* hybrid occurs naturally and can be easily made by controlled pollinations, whereas reciprocal crosses have met with limited success due to unidirectional compatibility, postzygotic barriers and F_1 hybrid sterility.

Tepary beans (*P. acutifolius)* are native to the desert highlands of northwest Mexico and the southwest of the USA. As such, they are extremely resistant to drought, heat and cold and have been viewed as a potential source of drought resistance for common beans. Their roots are very long and thin, giving them the ability to penetrate soil rapidly to access limited soil water reserves (Butare et al. 2011). Additionally, comparative transcript profiling under water deficit of common and tepary beans revealed a very high number of responsive genes in *P. acutifolius*, some of them with functional annotations directly associated with drought tolerance (Micheletto et al. 2007). In spite of crossing difficulties given that selection for common bean phenotype imposed by breeders eliminates much of the tepary bean introgressions during simple backcrosses, tepary beans have been used as a source of resistance for biotic constraints, especially common bacterial blight. The introduction of a novel congruity crossing method, however, enhances recombination to reduce the elimination of the tepary bean large introgressions (Haghighi and Ascher 1988), and thus, the observation of higher introgression rates estimated by AFLP sharing suggests that the use of *P. acutifolius* as a source of drought resistance alleles might be attainable (Muñoz et al. 2004). Tepary bean accessions have been identified with several other traits of potential value to common bean breeders including ashy stem blight and *Fusarium* wilt (*Fusarium oxysporum*) resistance, bean golden yellow mosaic virus (BGYMV) and bean rust resistance. Recently, a set of gene-based SNPs in tepary bean was developed and the derived genetic map was compared to the common bean genome assembly (Gujaria-Verma et al. 2016). This analysis showed a high collinearity of both genomes, which differ by a few intra-chromosomal rearrangements. The degree of similarity of both species at the genomic level should allow hybridizations and the eventual fixation of adaptive loci in cultivated genotypes.

Lima beans (*P. lunatus)* grow over an even wider range of environments than common beans, since they are very tolerant to heat and edaphic problems. It is tempting to introgress traits from lima beans into common beans. However, efforts to date to cross lima beans with common beans have resulted in no more than totally sterile F_1 plants.

Still, the wide diversity and geographic overlap of *Phaseolus* species in Mexico offer the possibility of viable outcrosses outside the Vulgaris clade that has not been extensively studied and represents an important reservoir of novel alleles (Fig. 6.3). *Phaseolus filiformis* no longer crosses successfully with the common bean while it has tolerance to salinity and extreme temperatures. Furthermore, the ecological description of 25 *Phaseolus* species distributed in Mexico (López Soto et al. 2005) highlights other candidate species to be evaluated as potential genomic resources. For example, it was observed that *P. leptostachyus* (together with *P. coccineus*) grows in the widest range of climatic conditions; *P. leptostachyus*, *P. microcarpus*, *P. tuerckheimii* and *P. pedicellatus* were observed in dry regions with high temperature; *P. tuerckheimii*, *P. lunatus* and *P. leptostachyus* were abundant in tropical climates with high degree of humidity (prone to fungal diseases). In terms of altitude adaptation, *P. acutifolius*, *P. filiformis*, *P. leptostachyus*, *P. lunatus*, *P. microcarpus* or *P. tuerckheimii* were collected from very low areas (sea-level), which contrast with *P. vulgaris* preference for altitude. In addition to *P. filiformis*, *P. maculatus* could be considered an alternative candidate to evaluate drought resistance, while *P. pauciflorus and P. pedicellatus* for cold adaptation.

The use of closely related species without full reproductive barriers to introduce interesting agronomic traits into domesticated plants has been explored for different species of tomato. By sequencing the transcriptomes of six tomato species (Koenig et al. 2013), including one domesticated accession (*Solanum lycopersicum*, M82), two related red-fruited wild species (*S. pimpinellifolium* and *S. galapagense*) and three green-fruited wild accessions from vastly differing habitats (*Solanum habrochaites*, a high altitude-adapted, chilling-tolerant accession; a high altitude drought-tolerant accession, *Solanum chmielewskii*; and *Solanum pennellii*, a desert-adapted accession), it was concluded that adaptation to extreme environments among tomato relatives caused alterations of transcriptional networks in parallel with positive selection at the sequence level for a number of genes related to environmental adaptation. Gene expression changes in *S. pennellii* in particular were highly accelerated relative to nucleotide divergence, suggesting that regulatory changes in morphological evolution are likely a genome-scale phenomenon. In contrast to adaptation to environmental conditions, artificial selection and domestication were found to be associated with a relatively small number of changes both at the sequence and transcriptional level, with only 51 genes having significant evidence of evolution under positive selection. Finally, the identification of introgression events between domesticated lines and *S. pimpinellifolium* was proposed to contribute to reduced genome-wide divergence in nucleotide sequence and divergence in gene expression between cultivated and wild accessions. Similar to the above-described example, a first attempt to understand species-specific transcriptional response to cold stress was conducted comparing SSH (subtraction suppression hybridization) generated cDNA libraries from *Phaseolus angustissimus*, a cold-tolerant species, and *P. vulgaris*. Transcriptome data generated under low temperature stress suggests that it is highly unlikely that the greater cold tolerance observed in *P. angustissimus* is mediated by the signalling mechanisms responsible for inducing cold acclimation and freezing tolerance in model systems such as *Arabidopsis thaliana* (lack of any significant representation of *COR* genes and predominant representation of myo-inositol-1L-phosphate synthase—MIPS; Vijayan et al. 2011).

A limiting step breeders and geneticists should overcome in the short term is the elucidation of the morphological and genetic mechanisms behind the establishment of reproductive barriers in the genus *Phaseolus* in order to improve hybridization strategies. Previous studies have tried to identify the genetic source of incompatibility between the Andean and Mesoamerican common bean gene pools, as certain crosses result in temperature-dependent hybrid weakness associated with a severe root phenotype. It has been found that such phenotype is controlled by the interaction of the root- and shoot-expressed semi-dominant alleles dosage-dependent lethal 1 (*DL1*) and *DL2*, which communicate via long-distance signalling (Hannah et al. 2007). Biochemical data showed that root death likely occurs by defence-related programmed cell death, as indicated by salicylic acid accumulation. *DL2*-expressing cotyledons supply a potent inhibitory signal that is sufficient to cause such death in *DL1*-expressing roots. However, deeper screenings of introgression events and the definition of coding and non-coding genomic elements displaying different degrees of mobility between *Phaseolus* populations might contribute to a complete understanding of the genetic basis of reproductive isolation in the genus. Furthermore, given that inter-species introgressions have been shown to decrease as phylogenetic distance increases (Rendón-Anaya et al. 2017), the use of intermediate populations should be evaluated to transfer adaptation loci from distant *Phaseolus* species into common bean populations.

5.5 Concluding Remarks

Thanks to genome sequencing strategies we now have more certainty of the origin of *P. vulgaris*, and of the genomic features affected during its domestication. However, we are only beginning to approach the use of the enormous versatility of these plants, which has allowed them to cope with challenging environmental conditions.

Systematic exploration of the biodiversity of plants promises to facilitate traditional breeding and biotechnology-based improvement of vegetable crops in key characteristics. In this regard, even though marker-assisted breeding programs have been successful in generating several common bean cultivars, the lack of biotechnological tools to manipulate *Phaseolus* species requires the design of more efficient strategies to incorporate a wider range of adaptations for disease resistance, abiotic stress tolerance and other agronomic challenges that are required in order to increase their resiliency and productivity. It is not enough to identify protein-coding genes directly affected by selection; a better understanding of how transcriptional networks are rewired following adaptation processes is needed. Furthermore, it is necessary to explore the genomes of wild relatives that represent immediate sources of genetic innovations. Consequently, more elaborated and complementary sequencing protocols at the genomic and transcriptomic levels are required to distinguish key regulatory elements in the genomes of agronomic advantageous species that could be targeted by introgression strategies.

The use of new sequencing technologies will, in the near future, allow us to obtain true reference genome sequences not only for *P. vulgaris* but also for many other species of the genus, which accompanied by more ambitious comparative genomics strategies will clarify how natural and artificial evolutions have shaped the genomes and transcriptional networks of *Phaseoli*. These new insights should accelerate our understanding of the molecular processes involved in common bean domestication and improvement and set the bases for the establishment of a new generation of marker-assisted breeding programs. The use of these tools, however, urgently requires training of breeders in regions where *Phaseolus* species represent a key crop for human nourishment, such as Africa and Latin America.

References

Abbott R, Albach D, Ansell S, Arntzen JW, Baird SJ, Bierne N et al (2013) Hybridization and speciation. J Evol Biol 26:229–246

Acosta-Gallegos JA, Kelly JD, Gepts P (2007) Prebreeding in common bean and use of genetic diversity from wild germplasm. Crop Sci 47:S44–S59

Andueza-Noh RH, Serrano Serrano ML, Chacón Sánchez MI, Sánchez-del Pino I, Camacho-Pérez L, Coello-Coello J, Mijangos Cortes J, Debouck D, Martínez-Castillo J (2013) Multiple domestications of the Mesoamerican gene pool of lima bean (*Phaseolus lunatus* L.): evidence from chloroplast DNA sequences. Genet Resour Crop Evol 60:1069–1086

Arumuganthan K, Earle E (1991) Nuclear DNA content of some important plant species. Plant Mol Biol Rep 9:208–218

Beebe S, Toro O, González AV, Chacón MI, Debouck DG (1997) Wild-weed-crop complexes of common bean (*Phaseolus vulgaris* L., Fabaceae) in the Andes of Peru and Colombia, and their implications for conservation and breeding. Genet Resour Crop Evol 44:73–91

Beebe S, Rao IM, Blair M, Acosta-Gallegos JA (2013) Phenotyping common beans for adaptation to drought. Front Physiol 4:35

Bellucci E, Bitocchi E, Ferrarini A, Benazzo A, Biagetti E, Klie S, Minio A, Rau D, Rodriguez M, Panziera A, Venturini L, Attene G, Albertini E, Jackson SA, Nanni L, Fernie AR, Nikoloski Z,

Bertorelle G, Delledonne M, Papa R (2014) Decreased nucleotide and expression diversity and modified coexpression patterns characterize domestication in the common bean. Plant Cell 26:1901–1912

Bennett MD, Leitch IJ (1995) Nuclear DNA amounts in angiosperms. Ann Bot (London) 76:113–116

Bitocchi E, Nanni L, Bellucci E, Rossi M, Giardini A, Zeuli PS, Logozzo G, Stougaard J, McClean P, Attene G, Papa R (2012) Mesoamerican origin of the common bean (*Phaseolus vulgaris* L.) is revealed by sequence data. Proc Natl Acad Sci USA 109:E788–E796

Bitocchi E, Bellucci E, Giardini A, Rau D, Rodriguez M, Biagetti E, Santilocchi R, Spagnoletti Zeuli P, Gioia T, Logozzo G, Attene G, Nanni L, Papa R (2013) Molecular analysis of the parallel domestication of the common bean (*Phaseolus vulgaris*) in Mesoamerica and the Andes. New Phytol 197:300–313

Blair MW, Pedraza F, Buendia HF, Gaitán-Solís E, Beebe SE, Gepts P, Tohme J (2002) Development of a genome-wide anchored microsatellite map for common bean (*Phaseolus vulgaris* L.). Theor Appl Genet 107:1362–1374

Blair MW, Iriarte G, Beebe S (2006) QTL analysis of yield traits in an advanced backcross population derived from a cultivated Andean x wild common bean (*Phaseolus vulgaris* L.) cross. Theor Appl Genet 112:1149–1163

Blair MW, Soler A, Cortés AJ (2012a) Diversification and population structure in common beans (*Phaseolus vulgaris* L.). PLoS ONE 7:e49488

Blair MW, Pantoja W, Carmenza Muñoz L (2012b) First use of microsatellite markers in a large collection of cultivated and wild accessions of tepary bean (*Phaseolus acutifolius* A. Gray). Theor Appl Genet 125:1137–1147

Bredeson JV, Lyons JB, Prochnik SE, Wu GA, Ha CM, Edsinger-Gonzales E, Grimwood J, Schmutz J, Rabbi IY, Egesi C, Nauluvula P, Lebot V, Ndunguru J, Mkamilo G, Bart RS, Setter TL, Gleadow RM, Kulakow P, Ferguson ME, Rounsley S, Rokhsar DS (2016) Sequencing wild and cultivated cassava and related species reveals extensive interspecific hybridization and genetic diversity. Nat Biotechnol 34:562–570

Butare L, Rao I, Lepoivre P, Polania J, Cajiao C, Cuasquer J, Beebe S (2011) New genetic sources of resistance in the genus *Phaseolus* to individual and combined aluminium toxicity and progressive soil drying stresses. Euphytica 181:385–404

Cavanagh CR, Chao S, Wang S, Huang BE, Stephen S, Kiani S, Forrest K, Saintenac C, Brown-Guedira GL, Akhunova A, See D, Bai G, Pumphrey M, Tomar L, Wong D, Kong S, Reynolds M, da Silva ML, Bockelman H, Talbert L, Anderson JA, Dreisigacker S, Baenziger S, Carter A, Korzun V, Morrell PL, Dubcovsky J, Morell MK, Sorrells ME, Hayden MJ, Akhunov E (2013) Genome-wide comparative diversity uncovers multiple targets of selection for improvement in hexaploid wheat landraces and cultivars. Proc Natl Acad Sci USA 110:8057–8062

Chacón MI, Pickersgill B, Debouck DG (2005) Domestication patterns in common bean (*Phaseolus vulgaris* L.) and the origin of the Mesoamerican and Andean cultivated races. Theor Appl Genet 110:432–444

Córdoba JM, Chavarro C, Schlueter JA, Jackson SA, Blair MW (2010) Integration of physical and genetic maps of common bean through BAC-derived microsatellite markers. BMC Genomics 11(1):436

Cortés AJ, Chavarro MC, Blair MW (2011) SNP marker diversity in common bean (*Phaseolus vulgaris* L.). Theor Appl Genet 123:827–845

Cortés AJ, Monserrate FA, Ramírez-Villegas J, Madriñán S, Blair MW (2013) Drought tolerance in wild plant populations: the case of common beans (*Phaseolus vulgaris* L.). PLoS ONE 8:e62898

Counterman BA, Noor MA (2006) Multilocus test for introgression between the cactophilic species *Drosophila mojavensis* and *Drosophila arizonae*. Am Nat 168:682–696

Debouck DG, Toro O, Paredes OM, Johnson WC, Gepts P (1993) Genetic diversity and ecological distribution of *Phaseolus vulgaris* in northwestern South America. Econ Bot 47:408–423

de la Cruz EP, Gepts P, Colunga García-Marín P, Zizumbo Villareal D (2005) Spatial distribution of genetic diversity in wild populations of *Phaseolus vulgaris* L. from Guanajuato and Michoacán, México. Genet Res Crop Evol 52:589–599

Delgado-Salinas A, Bibler R, Lavin M (2006) Phylogeny of the genus *Phaseolus* (Leguminosae): a recent diversification in an ancient landscape. Syst Bot 31:779–791

Diamond J (2002) Evolution, consequences and future of plant and animal domestication. Nature 418:700–707

Díaz-Camino C, Annamalai P, Sanchez F, Kachroo A, Ghabrial SA (2011) An effective virus- based gene silencing method for functional genomics studies in common bean. Plant Methods 7:16

Dowling T, Secor CL (1997) The role of hybridization and introgression in the diversification of animals. Annu Rev Ecol Syst 28:593–619

Escalante AM, Coello G, Eguiarte LE, Piñero D (1994) Genetic structure and mating systems in wild and cultivated populations of *Phaseolus coccineus* and *P. vulgaris*. Am J Bot 81:1096–1103

Félix DT, Coello-Coello J, Martínez-Castillo J (2014) Wild to crop introgression and genetic diversity in Lima bean (*Phaseolus lunatus* L.) in traditional Mayan milpas from Mexico. Conserv Genet 15:1315–1328

Ferreira J, Souza Carneiro J, Teixeira A, Lanes F, Cecon P, Borém A (2007) Gene flow in common bean (*Phaseolus vulgaris* L.). Euphytica 153:165–170

Flint-Garcia SA, Thornsberry JM, Buckler IV, ES (2003) Structure of linkage disequilibrium in plants. Annu Rev Plant Biol. 54:357–374

Garzon LN, Blair M (2014) Development and mapping of SSR markers linked to resistance-gene homologue clusters in common bean. Crop J 2:183–194

Gepts P (2004) Crop domestication as a long-term selection experiment. Plant Breed Rev 24:1–44

Gepts P (2014) The contribution of genetic and genomic approaches to plant domestication studies. Curr Opin Plant Biol 18:51–59

Gepts P, Papa, R (2002) Evolution during domestication. In: Encyclopedia of life sciences. Nature Publishing Group, London. doi:10.1038/npg.els.0003071

Grisi MCM, Blair MW, Gepts P, Brondani C, Pereira PA, Brondani RP (2007) Genetic mapping of a new set of microsatellite markers in a reference common bean (*Phaseolus vulgaris*) population BAT93 x Jalo EEP558. Genet Mol Res 6:691–706

Gujaria-Verma N, Ramsay L, Sharpe AG, Sanderson LA, Debouck DG, Tar'an B, Bett KE (2016) Gene-based SNP discovery in tepary bean (*Phaseolus acutifolius*) and common bean (*P. vulgaris*) for diversity analysis and comparative mapping. BMC Genomics 17:239

Haghighi KR, Ascher PD (1988) Fertile, intermediate hybrids between *Phaseolus vulgaris* and *P. acutifolius* from congruity backcrossing. Sex Plant Reprod 1:51–58

Hajjar R, Hodgkin T (2007) The use of wild relatives in crop improvement: a survey of developments over the last 20 years. Euphytica 156:1–13

Hannah MA, Krämer KM, Geffroy V, Kopka J, Blair MW, Erban A, Vallejos CE, Heyer AG, Sanders FE, Millner PA, Pilbeam DJ (2007) Hybrid weakness controlled by the dosage-dependent lethal (*DL*) gene system in common bean (*Phaseolus vulgaris*) is caused by a shoot-derived inhibitory signal leading to salicylic acid-associated root death. New Phytol 176:537–549

Hancock JF (2005) Contributions of domesticated plant studies to our understanding of plant evolution. Ann Bot 96:953–963

He Z, Zhai W, Wen H, Tang T, Wang Y, Lu X, Greenberg AJ, Hudson RR, Wu CI, Shi S (2011) Two evolutionary histories in the genome of rice: the roles of domestication genes. PLoS Genet 7:e1002100

Kami J, Velásquez VB, Debouck DG, Gepts P (1995) Identification of presumed ancestral DNA sequences of phaseolin in *Phaseolus vulgaris*. Proc Natl Acad Sci USA 92:1101–1104

Kaplan L, Lynch TF (1999) *Phaseolus* (Fabaceae) in archaeology: AMS radiocarbon dates and their significance for pre-Colombian agriculture. Econ Bot 53:261–272

Kelly JD, Gepts P, Miklas PN, Coyne DP (2003) Tagging and mapping of genes and QTL and molecular marker-assisted selection for traits of economic importance in bean and cowpea. Field Crops Res 82:135–154

Koenig D, Jiménez-Gómez JM, Kimura S, Fulop D, Chitwood DH, Headland LR, Kumar R, Covington MF, Devisetty UK, Tat AV, Tohge T, Bolger A, Schneeberger K, Ossowski S, Lanz C, Xiong G, Taylor-Teeples M, Brady SM, Pauly M, Weigel D, Usadel B, Fernie AR, Peng J, Sinha NR, Maloof JN (2013) Comparative transcriptomics reveals patterns of selection in domesticated and wild tomato. Proc Natl Acad Sci USA 110:E2655–E2662

Koinange EMK, Gepts P (1992) Hybrid weakness in wild *Phaseolus vulgaris* L. J Hered 83:135–139

Kwak M, Gepts P (2009) Structure of genetic diversity in the two major gene pools of common bean (*Phaseolus vulgaris* L., Fabaceae). Theor Appl Genet 118:979–992

Kwak M, Toro O, Debouck D, Gepts P (2012) Multiple origins of the determinate growth habit in domesticated common bean (*Phaseolus vulgaris* L.). Ann Bot 110:1573–1580

Kwapata K, Nguyen T, Sticklen M (2012) Genetic transformation of common bean (*Phaseolus vulgaris* L.) with the GUS color marker, the BAR herbicide resistance, and the barley (*Hordeum vulgare*) *HVA1* drought tolerance genes. Int J Agron 198960. doi:10.1155/2012/198960

Larson G, Piperno DR, Allaby RG, Purugganan MD, Andersson L, Arroyo-Kalin M, Barton L, Climer Vigueira C, Denham T, Dobney K, Doust AN, Gepts P, Gilbert MT, Gremillion KJ, Lucas L, Lukens L, Marshall FB, Olsen KM, Pires JC, Richerson PJ, Rubio de Casas R, Sanjur OI, Thomas MG, Fuller DQ (2014) Current perspectives and the future of domestication studies. Proc Natl Acad Sci USA 111:6139–6146

Lenser T, Theißen G (2013) Molecular mechanisms involved in convergent crop domestication. Trends Plant Sci 18:704–714

Llaca V, Delgado Salinas A, Gepts P (1994) Chloroplast DNA as an evolutionary marker in the *Phaseolus vulgaris* complex. Theor Appl Genet 88:646–652

López Soto JL, Ruiz Corral JA, Sánchez González J, Lépiz Ildefonso R (2005) Climatic adaptation of 25 wild bean species (*Phaseolus* spp.) in México. Rev Fitotecnia Mexicana 28:221–230

Luebert F, Weigend M (2014) Phylogenetic insights into Andean plant diversification. Front Ecol Evol 2:27

Mallet J (2007) Hybrid speciation. Nature 446:279–283

Mamidi S, Rossi M, Annam D, Moghaddam S, Lee R, Papa R, McClean P (2011) Investigation of the domestication of common bean (*Phaseolus vulgaris*) using multilocus sequence data. Funct Plant Biol 38:953–967

Martínez-Castillo J, Zizumbo-Villarreal J, Gepts P, Delgado-Valerio P, Colunga-García Marín P (2006) Structure and genetic diversity of wild populations of Lima bean (*Phaseolus lunatus* L.) from the Yucatan Peninsula, Mexico. Crop Sci 46:1071–1080

Martínez-Castillo J, Zizumbo-Villarreal J, Gepts P, Colunga-García Marín P (2007) Gene flow and genetic structure in the wild–weedy–domesticated complex of *Phaseolus lunatus* L. in its Mesoamerican center of domestication and diversity. Crop Sci 47:58–66

Martínez-Castillo J, Camacho-Pérez L, Villanueva-Viramontes S, Andueza-Noh RH, Chacón-Sánchez MI (2014) Genetic structure within the Mesoamerican gene pool of wild *Phaseolus lunatus*

(Fabaceae) from Mexico as revealed by microsatellite markers: implications for conservation and the domestication of the species. Am J Bot 101:851–864

McClean P, Gepts P, Kami J (2004) Genomics and genetic diversity in common bean. In: Wilson RF, Stalker HT, Brummer EC (eds) Legume crop genomics. AOCS Press, Champaign, IL, pp 60–82

Mensack MM, Fitzgerald VK, Ryan EP, Lewis MR, Thompson HJ, Brick MA (2010) Evaluation of diversity among common beans (*Phaseolus vulgaris* L.) from two centers of domestication using 'omics' technologies. BMC Genom 11:686

Mercado-Ruaro P, Delgado-Salinas A (1998) Karyotypic studies on species of *Phaseolus* (Fabaceae: Phaseolinae). Am J Bot 85:1–9

Micheletto S, Rodriguez-Uribe L, Hernandez R, Richins RD, Curry J, O'Connell MA (2007) Comparative transcript profiling in roots of *Phaseolus acutifolius* and *P. vulgaris* under water deficit stress. Plant Sci 173:510–520

Miedes E, Vanholme R, Boerjan W, Molina A (2014) The role of the secondary cell wall in plant resistance to pathogens. Front Plant Sci 5:358

Mina-Vargas AM, McKeown PC, Flanagan NS, Debouck DG, Kilian A, Hodkinson TR, Spillane C (2016) Origin of year-long bean (*Phaseolus dumosus* Macfady, Fabaceae) from reticulated hybridization events between multiple *Phaseolus* species. Ann Bot 118:957–969

Morell PL, Buckler ES, Ross-Ibarra J (2011) Crop genomics: advances and applications. Nat Rev Genet 13:85–96

Muñoz LC, Blair MW, Duque MC, Tohme J, Roca W (2004) Introgression in common bean × tepary bean interspecific congruity-backcross lines as measured by AFLP markers. Crop Sci 44:637–645

O'Rourke JA, Iniguez LP, Bucciarelli B, Roessler J, Schmutz J, McClean PE, Jackson SA, Hernandez G, Graham MA, Stupar RM, Vance CP (2013) A re-sequencing based assessment of genomic heterogeneity and fast neutron-induced deletions in a common bean cultivar. Front Plant Sci 4:210

Papa R, Gepts P (2003) Asymmetry of gene flow and differential geographical structure of molecular diversity in wild and domesticated common bean (*Phaseolus vulgaris* L.) from Mesoamerica. Theor Appl Genet 106:239–250

Papa R, Acosta-Gallegos JA, Delgado-Salinas A, Gepts P (2005) A genome-wide analysis of differentiation between wild and domesticated *Phaseolus vulgaris* from Mesoamerica. Theor Appl Genet 111:1147–1158

Pastorini J, Zaramody A, Curtis DJ, Nievergelt CM, Mundy NI (2009) Genetic analysis of hybridization and introgression between wild mongoose and brown lemurs. BMC Evol Biol 9:32

Piperno DR, Dillehay TD (2008) Starch grains on human teeth reveal early broad crop diet in northern Peru. Proc Natl Acad Sci USA 105:19622–19627

Porch TG, Blair MW, Lariguet P, Galeano C, Pankhurst CE, Broughton WJ (2009) Generation of a mutant population for TILLING common bean genotype BAT93. J Am HortSoc 134:348–355

Porch TG, Beaver JS, Debouck DG, Jackson SA, Kelly JD, Dempewolf H (2013) Use of wild relatives and closely related species to adapt common bean to climate change. Agronomy 3:433–461

Purugganan MD, Fuller DQ (2009) The nature of selection during plant domestication. Nature 457:843–848

Rendón-Anaya M, Montero-Vargas JM, Saburido-Álvarez S, Vlasova A, Capella-Gutierrez S, Ordaz-Ortiz JJ, Aguilar OM, Vianello-Brondani RP, Santalla M, Delaye-Arredondo L, Gabaldón T, Gepts P, Winkler R, Guigó R, Delgado-Salinas A, Herrera-Estrella A (2017) Genomic history of the origin and domestication of common bean unveils its closest sister species. Genome Biol 18:60

Richter M, Diertl KH, Emck P, Peters T, Beck E (2009) Reasons for an outstanding plant diversity in the tropical Andes of Southern Ecuador. Landscape Online 12:1–35

Rossi M, Bitocchi E, Bellucci E, Nanni L, Rau D, Attene G, Papa R (2009) Linkage disequilibrium and population structure in wild and domesticated populations of *Phaseolus vulgaris* L. Evol Appl 2:504–522

Schmutz J, McClean PE, Mamidi S, Wu GA, Cannon SB, Grimwood J, Jenkins J, Shu S, Song Q, Chavarro C, Torres-Torres M, Geffroy V, Moghaddam SM, Gao D, Abernathy B, Barry K, Blair M, Brick MA, Chovatia M, Gepts P, Goodstein DM, Gonzales M, Hellsten U, Hyten DL, Jia G, Kelly JD, Kudrna D, Lee R, Richard MM, Miklas PN, Osorno JM, Rodrigues J, Thareau V, Urrea CA, Wang M, Yu Y, Zhang M, Wing RA, Cregan PB, Rokhsar DS, Jackson SA (2014) A reference genome for common bean and genome-wide analysis of dual domestications. Nat Genet 46:707–713

Schneider JV, Schulte K, Aguilar JF, Huertas ML (2011) Molecular evidence for hybridization and introgression in the neotropical coastal desert-endemic *Palaua* (Malveae, Malvaceae). Mol Phylogenet Evol 60:373–384

Spataro G, Tiranti B, Arcaleni P, Bellucci E, Attene G, Papa R, Spagnoletti Zeuli P, Negri V (2011) Genetic diversity and structure of a worldwide collection of *Phaseolus coccineus* L. Theor Appl Genet 122:1281–1291

Swanson-Wagner R, Briskine R, Schaefer R, Hufford MB, Ross-Ibarra J, Myers CL, Tiffin P, Springer NM (2012) Reshaping of the maize transcriptome by domestication. Proc Natl Acad Sci USA 109:11878–11883

Tomato Genome Consortium (2012) The tomato genome sequence provides insights into fleshy fruit evolution. Nature 485:635–641

Vlasova A, Capella-Gutiérrez S, Rendón-Anaya M, Hernández-Oñate M, Minoche AE, Erb I, Câmara F, Prieto-Barja P, Corvelo A, Sanseverino W, Westergaard G, Dohm JC, Pappas GJ Jr, Saburido-Alvarez S, Kedra D, Gonzalez I, Cozzuto L, Gómez-Garrido J,

Aguilar-Morón MA, Andreu N, Aguilar OM, Garcia-Mas J, Zehnsdorf M, Vázquez MP, Delgado-Salinas A, Delaye L, Lowy E, Mentaberry A, Vianello-Brondani RP, García JL, Alioto T, Sánchez F, Himmelbauer H, Santalla M, Notredame C, Gabaldón T, Herrera-Estrella A, Guigó R (2016) Genome and transcriptome analysis of the Mesoamerican common bean and the role of gene duplications in establishing tissue and temporal specialization of genes. Genome Biol 17:32

Vijayan P, Parkin IA, Karcz SR, McGowan K, Vijayan K, Vandenberg A, Bett KE (2011) Capturing cold-stress-related sequence diversity from a wild relative of common bean (*Phaseolus angustissimus*). Genome 54:620–628

Weissmann S, Feldman M, Gressel J (2005) Sequence evidence for sporadic intergeneric DNA introgression from wheat into a wild *Aegilops* species. Mol Biol Evol 22:2055–2062

Wells WC, Isom WH, Waines JG (1988) Outcrossing rates of six common bean lines. Crop Sci 28:177–178

Whitney KD, Randell RA, Rieseberg LH (2006) Adaptive introgression of herbivore resistance traits in the weedy sunflower *Helianthus annuus*. Am Nat 167:794–807

Worthington M, Soleri D, Gepts P (2012) Genetic composition and spatial distribution of farmer-managed *Phaseolus* bean plantings: an example from a village in Oaxaca, Mexico. Crop Sci 52:1721–1735

Yoo MJ, Wendel JF (2014) Comparative evolutionary and developmental dynamics of the cotton (*Gossypium hirsutum*) fiber transcriptome. PLoS Genet 10: e1004073

Yuste-Lisbona FJ, Santalla M, Capel C, García-Alcázar M, La De, Fuente M, Capel J, De Ron AM, Lozano R (2012) Marker-based linkage map of Andean common bean (*Phaseolus vulgaris* L.) and mapping of QTLs underlying popping ability traits. BMC Plant Biol 12:136

Zizumbo-Villarreal D, Colunga-García Marín P, Payro de la Cruz E, Delgado-Valerio P, Gepts P (2005) Population structure and evolutionary dynamics of wild-weedy-domesticated. Crop Sci 45:1073–1083

Common Bean Genomes: Mining New Knowledge of a Major Societal Crop

6

Phillip E. McClean and Bodo Raatz

Abstract

Common bean, *Phaseolus vulgaris* L., is the most consumed grain legume in the world. The continued improvement of this crop is necessary as it has historically been pushed to more marginal lands as other crops displace it. Genomic technologies are providing valuable information aid this process. This includes a reference and draft genome of representatives of the Andean and Mesoamerican gene pools, respectively. Early applications such as new marker development, marker development, transcriptomics, and comparative genomics. Future applications such as the need for a denser gene chip for fine-mapping are highlighted.

Keywords

Common bean · *Phaseolus vulgaris* L. · Genomics · Sequencing Marker-assisted selection

6.1 Introduction

The development of reference genome sequences for important crop species impacts all efforts to improve the performance of that crop. Because of their importance as a food throughout the world, early sequencing efforts targeted major crops such as rice, sorghum and soybean while the discovery of candidate dicot and monocot genes were supported by the genome sequences of *Arabidopsis* and rice. A landmark was reached with the publication of the 50th plant genome in 2013 (Michael and Jackson 2013), and additional draft and reference genomes of crop species continue to be described in publications.

The draft sequence of the common bean (*Phaseolus vulgaris* L.) Andean landrace Chaucha Chuga (CIAT germplasm bank identifier, G19833) was released in 2014 (Schmutz et al. 2014). This release is indicative of what we are now seeing in the plant genomic community;

P. E. McClean (✉)
Department of Plant Sciences, North Dakota State University, Fargo, ND 58108, USA
e-mail: phillip.mcclean@ndsu.edu

B. Raatz
CIAT—International Center for Tropical Agriculture, 6713 Cali, Colombia

M. Pérez de la Vega et al. (eds.), *The Common Bean Genome*,
Compendium of Plant Genomes, https://doi.org/10.1007/978-3-319-63526-2_6

research focused on crops important to smallholder farmers. A plan to sequence the bean genome was developed by genomics, genetics, and bioinformatics experts in the USA, and that plan was circulated as a white paper to major funding agencies in the USA to solicit funding support that became available in 2009 (McClean et al. 2008; http://bic.css.msu.edu/_pdf/Bean_Genomics_Status_2008.pdf). The plan described specific applications including marker development, fine mapping of candidate genes, legume gene discovery, and the discovery of genes associated with adaptation. Funding for the USA project was provided by the United States Department of Agriculture, the USA Department of Energy Joint Genome Institute (DOE-JGI), and the American Recovery and Reinvestment Act of 2009, and the project was led by researchers at the University of Georgia, DOE-JGI, HudsonAlpha Biotechnology Institute, and North Dakota State University. More recently, in 2016, the sequence of the Middle American breeding line BAT 93 was released (Vlasova et al. 2016). This sequence was the output from the Genome-CYTED PhasIbeAm project that also began in 2009. This project was a collaboration of bean researchers from Argentina, Brazil, Mexico, and Spain that specialize in genetics, genomics, and bioinformatics. Funding for the project was provided by the Iberoamerican Programme for Science, Technology and Development the Ministry of Science, Technology and Productive Innovation of Argentina; the National Scientific and Technology Development Council of Brazil; Brazilian Agricultural Research Corporation; the National Council for Science and Technology of Mexico; and the Ministry of Economy and Competitiveness of Spain. The project was to support the development of varieties with greater production potential and higher quality. A third ongoing project sequenced the Canadian variety OAC Rex (http://www.beangenomics.ca/). This variety is unique for its common bacterial blight (CBB) resistance that was introgressed from the tepary bean, *Phaseolus acutifolius*. The sequence effort was a component of the "Applied Bean Genomics and Bioproducts" lead by researchers at the University of Guelph, University of Windsor, the University of Western Ontario, and Agriculture and Agri-Food Canada. The goal of that project was not only to sequence the genome, but also to provide tools to enhance CBB resistance, improve healthy components of bean seed, modified seed storage composition, and increase the appeal of beans to the consumer.

Here, we discuss the relevance of common bean, outline the sequencing projects results, provide a summary of the principle features of the two genome sequences, feature several applications of the sequence information, and provide a future perspective on how genome sequence data can impact future efforts to improve common bean.

6.2 The Why and How: Motivation and Tools for Sequencing Common Bean

6.2.1 Species Organization, Diversity, and Domestication

Recent research determined an ancestral wild bean population in south-central Mexico (Bitocchi et al. 2012) split about $\sim$110,000 years ago into the wild Mesoamerican and wild Andean gene pools (Mamidi et al. 2013; Schmutz et al. 2014). Subpopulations then spread to the north and south, and two waves of migration most likely led to the current Andean population. From these wild Middle American and Andean gene pools, the original landraces were derived. Multiple lines of evidence show that domestication occurred separately within each wild gene pool at $\sim$7000 years ago. Domesticated germplasm was the basis from which modern, cultivated Middle American and Andean varieties were developed (Mamidi et al. 2011). The Andean races are classified as Nueva Granada, Peru, and Chile. The Middle American races are Jalisco, Durango, Guatemala, and Mesoamerica (Singh et al. 1991; Beebe et al. 2000).

Common bean is characterized by many market classes that are distinguished by seed size, color, and pattern traits that are controlled by many genes (McClean et al. 2002). The principle bean market classes vary worldwide. In the USA, pinto, navy, black, great northern, and kidney beans are the major market classes but represent a small subset of the many classes grown throughout the world. Other classes such as red mottled, red kidney, sugar, large white, and purple speckled are grown in Africa. These market classes are derived from the Andean gene pool. In Central America and Mexico, the Middle American small red, navy, pinto, black, yellow-tan market classes predominate, and in Brazil, the carioca type is the preferred bean market class. Crosses are typically made between genotypes within a market class because it is time-consuming to recover a market class phenotype from intermarket-class crosses. This has led to limited sequence variability within each market class, and in general, variability is lowest for Andean-derived beans (McClean et al. 2004; McClean and Lee 2007; Schmutz et al. 2014). At the same time, a large amount of variability is observed between the two gene pools. This variability has been exploited extensively during the development of populations used to map important agronomic traits (Miklas et al. 2006).

6.2.2 Common Bean, a Major Societal Crop

Food legumes are a critical component of the nutritional and financial livelihood of smallholder farmers in eastern and southern Africa as well as Central and South America. In these regions, common bean is the principal food legume crop that provides essential food components to combat malnutrition (Katungi et al. 2009). When used as a source of proteins and micronutrients, it is beneficial to the health of children (Katungi et al. 2011; Ugen et al. 2012). In Central America and some countries in eastern Africa, the old milpa intercropping system of beans and maize (Cook 1919) is the predominant production practice for many smallholder bean farmers (<2 ha). In Tanzania and Guatemala, two countries in which the milpa system is utilized, beans provided about 10% of the daily intake of proteins between 2010 and 2013 (FAO STAT; http://faostat.fao.org/), while maize was the primary source of calories. In that same time frame in Rwanda, 31% of the daily protein intake came from beans; while in other major bean-producing African countries, the percentage ranged from 10 to 15%.

While common bean is universally appreciated as a major nutritional crop and critical to food security, its economic importance to smallholder farmers is underappreciated especially as a cash crop. In Africa and Central and South America, a large part of the smallholder farmer bean harvest is sold at the local market to raise cash for other family needs. This exacerbates the malnutrition problem because the beans are lost to the family as a nutritious food source leaving maize as the primary source of both calories and protein. For example, Guatemala, one of the poorest countries in the world with the fourth highest level of malnutrition-related stunting, grows large quantities of beans in the milpa system. Yet, only 7% of the diet is from beans in the milpa growing region. Current estimates suggest that a diet of 25% would alleviate much of the malnutrition and stunting problems. Increasing productivity here may positively affect this imbalance.

Common bean is the principle food legume for 400 million people in eastern Africa and 250 million in Central and South America. Despite recent efforts, yields here are typically 1/3 of full potential, and poor production performance is associated with multiple abiotic and biotic stresses. Drought is a significant factor that reduces yields in major bean-growing regions in Africa (Beebe et al. 2011; Munishi et al. 2015) and Central America (Gourdji et al. 2015), while most diseases are prevalent in both regions with different severity levels (Beebe et al. 2011).

These constraints, tied to biotic and abiotic stresses, affect both small and large holder production. But these constraints pose a much greater food security risk for smallholder farmers. Using the accepted definition of smallholder farms (<2 ha), 69% and 76% of the farms in

Colombia and Guatemala are held by smallholder farmers (Berdegué and Fuentealba 2011). In Eastern Africa, 75% of agriculture outputs are produced on smallholder farms (Salami et al. 2010). Often these farms support entire families. Since only <1 and <6% of production land is irrigated in the African and Central American/Caribbean countries, respectively, (FAO STAT; http://faostat.fao.org/), farm operations rely on natural rainfall.

The most important African bean producers are in the Eastern region of the continent (Burundi, Cameroon, Democratic Republic of the Congo, Ethiopia, Kenya, Malawi, Uganda, Tanzania, Rwanda, and Zimbabwe), a region that accounted for 5.3 million of the 14.5 million harvested hectares of beans from 2011 to 2013 (FAO STAT; http://faostat.fao.org/). Here, production is almost entirely in the hands of smallholder farmers who often plant less than a hectare of beans and who have limited access to capital to purchase production inputs that would improve yield. With a total population in 2010 estimated at ~400 million, ~200 million earned less than $2USD per day, and child stunting averaged more than 35%.

A recent Living Standards Measurement Study discovered that the poor in Uganda can spend only 50% or less on beans than the well-to-do suggesting that the poor would benefit from increased production and lower market prices. In Central America, where beans continue to be an important part of the diet, pockets of poverty still persist. About 2.4 million ha are cultivated in El Salvador, Guatemala, Mexico, and Nicaragua. Among a population of 161 million in these countries, 21 million earn less than $2 US per day. Child stunting here is in the range of 19%, and childhood anemia ranges from 19 to 38% (Graham et al. 2007).

Average yields of beans in these regions are almost uniformly below 800 kg/ha (FAOSTAT; http://faostat.fao.org/), although yield potential is above 3000 kg/ha. The trend line for yield since 1970 is essentially flat, whereas in North America (primarily the USA) yields have increased 15 kg/ha/year in the same time period. Poor yields are attributed to biotic and abiotic stresses, many of which are amenable to genetic improvement. Fungal, bacterial, and viral diseases continue to reduce yields, and drought and soil constraints are serious limitations. These regions can greatly benefit from the promise of new genetic materials that will be developed using the genomic tools available for common bean improvement.

6.2.3 Early Genomic Resources of Common Bean

Prior to the genome sequencing projects, researchers had developed a suite of resources that would later prove vital for sequencing the common bean genome. A set of 83,000 expressed sequence tags (EST) were available (Melotto et al. 2005; Ramírez et al. 2005; Thibivilliers et al. 2009), and these were collapsed into 11,000 contigs and 9000 singletons (McClean et al. 2010). These EST sequences were one important component of the gene modeling step of the sequencing project. EST data was collected from multiple tissues such as seedling shoots [with or without *Colletotrichum lindemuthianum* (anthracnose) infection], seedling leaves, nodules elicited by *Rhizobium tropici*, roots, leaves (three genotypes), and pods.

BAC libraries are important resources for a sequencing project because they assist with the assembly of contigs into large scaffolds. The most critical library was constructed from G19833, an Andean landrace from Peru (Schlueter et al. 2008). The availability of the library and early physical map developed from BAC-end sequencing was the reason this genotype was chosen to serve as the reference genotype. Eleven other BAC libraries were also available in the *Phaseolus* genus, ten from *P. vulgaris* and one from *Phaseolus lunatus* (Kami et al. 2006). The *Phaseolus* BAC libraries were a phylogenetically ordered set useful for evolutionary studies. DGD1962, a wild bean from northern Peru, represented the presumed ancestral gene pool of the species (Debouck et al. 1993; Kami et al. 1995). The remainder of the libraries represented the two evolutionary gene pools. Several libraries were constructed from wild Mexican beans of Mesoamerican origin that contain the three subfamilies of the APA seed proteins, which confer resistance to seed weevils. Single BAC clones had been fully sequenced, one around the *Co-4* locus for resistance to anthracnose (Melotto et al. 2004), and the other around the *APA* locus (Kami et al. 2006). Full sequences of BAC clones were important to test the accuracy of the genome assembly.

Over 25 common bean linkage maps were available in 2008 (Kelly et al. 2003; Miklas et al. 2006), and new maps are still being described. To maximize molecular polymorphism, the majority of mapping populations were derived from intergene pool crosses between domesticated Andean and Mesoamerican parents. For specialized purposes, some maps were developed by crossing parents within a gene pool, but polymorphism was low. A highly polymorphic core map utilizing a recombinant inbred population from the cross BAT 93 × Jalo EEP 558 (Nodari et al. 1993) was developed to coalesce the mapping data (Freyre et al. 1998). Some 600 markers had been mapped directly in this population (Freyre et al. 1998; Papa and Gepts 2003; Blair et al. 2003; Grisi et al. 2007), and shared markers, principally RFLPs and sequence-tagged markers, were used to collate linkage groups among the different maps. A major addition to the map was 300 gene-based markers (McConnell et al. 2010) that proved useful to correlate the genetic and physical maps. Utilizing the G19833 BAC library, a common bean physical map was constructed using High-Information-Content Fingerprinting and BAC-end sequencing (41,717 BACs, ~9× clone coverage; Schlueter et al. 2008). This physical map assembled into 1183 contigs and 6385 singletons and was anchored with more than 540 markers derived from RFLPs, genes, ESTs, and other sequences. Microsatellites discovered from the end-sequencing effort were also used to define the relationship between the genetic and physical maps (Córdoba et al. 2010).

6.3 Andean and Middle American Common Bean Genome Sequencing Projects: Results and Outputs

6.3.1 The Genome Sequence Papers

Two concurrent projects were initiated to sequence a representative genotype from each of the two gene pools. G19833 was selected for the USA project (Schmutz et al. 2014) because a BAC library and physical map (Schlueter et al. 2008) were initially available for that genotype at the time the project began. A single G19833 plant was grown, and leaf tissue was collected from that plant. Plants grown from the seed of that initial plant were used for sequencing. The primary sequence data was collected from 454 reads (18.6× genome coverage) from a linear library with an average insert size of 362 bp. That data set was augmented with 454 paired-end library read data with insert sizes ranging from ~3 to 12 kb. Sanger paired-end sequence data was collected from two 35 kb fosmid and three ~125 kb BAC libraries. The assembly was based on 21× coverage. During the early stages of the project, an Illumina Golden Gate assay with 827 SNPs was developed (Hyten et al. 2010). The SNP depth was increased during the genome project through resequencing of a family of divergent genotypes and mapping to early scaffold assemblies. The product of that effort was an Illumina Infinium chip (Song et al. 2015) that was used to score segregating populations from a Stampede x Red Hawk cross. The resultant genetic map (1784 genetic loci) was an integral part of the assembly process. The map contained 7015 SNP and 261 SSR, along with 25 indel markers that were used to anchor the map to the common bean linkage groups. The map development was partially funded the USDA Common Bean Coordinated Agricultural Project (BeanCAP; http://www.beancap.org/). The assembled contig data spanned 472 Mb (of the estimated 587 Mb genome; http://www.kew.org/cvalues/), while the assembled scaffold distance was 521 Mb. The L50 of the contig assembly was

~40 kb, while the L50 for the scaffolds was ~50 Mb. The chromosomal scale assembly represented 89% of the scaffold distance. About 45.4% of the total assembly were repeat elements. The largest class of repeats, the LTR retrotransposons, makes up 36.7% of the genome size. Gene modeling was based on RNA-seq data from 11 libraries representing multiple anatomical tissues sampled at different development stages. The RNA was collected, again, from tissues obtained by growing progeny of that initial plant used for sequencing. In addition, ~50 k transcript assemblies based on available EST sequences in NCBI GenBank were developed. Standard homology-based prediction software packages were employed to develop gene models. These approaches defined 31,638 transcripts derived from 27,197 gene models.

BAT 93, a breeding line developed by CIAT, was chosen by the Iberoamerican research group for sequencing (Vlasova et al. 2016). BAT 93 is a small seeded bean race representing the Middle American gene pool. This line has multiple resistances to major bean diseases and is a parent of a historical mapping population (Nodari et al. 1993). SOLiD, 454, and Sanger libraries provided the bulk of the sequencing reads, and those reads were augmented with a Sanger BAC-end paired sequence library. Approximately 51% of the reads were derived from short-read SOLiD pair libraries, and 34% of the reads were obtained from 454 paired-end libraries. The primary reads represented 133× coverage of the bean genome. Illumina reads (45×) were included to adjust for homopolymer errors. The final assembly was comprised of 68% 454 single-end reads, and 25% SOLiD paired reads. The total contig length was 428 kb (73% of estimated size; L50 = 18.1 kb), and the scaffold length was 495 kb (L50 = 0.43 Mb), while the physical distance of the chromosomes was 81% of the scaffold distance. The observation that 35% of the genome consisted of mobile elements was determined using multiple repeat predictors. The Class I LTR retrotransposons accounted for 29% of the genome size. Multiple transcript libraries were developed and sequenced using Illumina or 454 technologies. This was combined with publicly available transcript data for gene modeling purposes using standard gene prediction software. A total of 66,634 transcripts were discovered to define 30,491 protein-coding gene models.

The publication of these two papers highlights the applications of a genome sequence to better understand a species. One application especially pertinent to crops, population genomics, investigates the history of a species from its wild state, through its landrace state, and finally to an improved cultivar. Because it has two gene pools that were derived from a single ancestral gene pool and underwent independent domestication events that occurred at a great distance, common bean can be a model for crop evolution. Schmutz et al. (2014) determined that the wild Andean gene pool was the result of a bottleneck that lasted 76,000 years and reduced its variation 75%, while the wild Middle American gene pool did not experience a bottleneck. The two resulting wild gene pools were highly differentiated (F_{ST} = 0.34). Further, domestication within these two gene pools reduced the variation ~15% relative to that found in the derived wild gene pools. As noted by Gaut (2015), common bean is well suited for the study of the dynamics of domestication because the two gene pools can serve as replicates of evolutionary events that took place at distant geographic locations. An important question for crop genomics is whether there is a single path to domestication, or put another way, is there a specific set of genes that must express a specific, genome-wide haplotype state for a plant to become domesticated. Or alternatively, are there multiple genomic paths to becoming a domesticated species. Schmutz et al. (2014) show that the later was the case, there are multiple paths to domestication. In their analysis, out of the 1875 Mesoamerican and 748 Andean genes determined to have undergone selection during domestication, only 59 were shared between the two gene pools. This demonstrates that bean domestication was most likely a relaxed event that followed multiple genomic paths.

Whole-genome sequencing also allows the investigation of speciation processes within a genus further up the evolutionary tree. Whole-genome duplications and gene duplications are major evolutionary events traceable within all higher plants. These duplication events occurred within ancestral seed plants [~340–320 million years ago (MYA)] and ancestral angiosperms (~235–190 MYAR) with additional independent duplications in the dicot and monocot lineages (~125 MYA, each) (Jiao et al. 2011). And further, from a common bean perspective, a legume-specific duplication (~59 MYA, Schmutz et al. 2010) must also be considered. Dating duplicates both within and between genome can assign specific genes to specific eras along the duplication time line, and clusters of genes can date-specific events. Neither Schmutz et al. (2014) nor Vlasova et al. (2016) found any evidence of a whole-genome duplication in common bean after the legume duplication. The dating of paralogs dated the *Phaseolus* (G19833 data) and *Glycine* divergence at ~19 MYA. For BAT 93, those paralogs duplicated most recently were typically expressed in the same tissues, while those duplicated earlier were expressed more variably from tissue and expression level perspectives. 79% of the BAT 93 genes were orthologous to a gene in the G19833 genome, and among these orthologs, divergent genes (<95% identity) were enriched for defense genes. Common bean is striking for its large number of clusters of disease-resistant genes. These clusters are shared between the two genomes demonstrating that the appearance of the clusters predates the Mesoamerican/Andean split. Also, specific BAT 93 gene expansions were noted for the ubiquitin pathway, seed development, and membrane-bound receptors. Seed size varies between the two gene pools, and the results from the Iberoamerican team shows that gene expansions involved in seed development differ between the two gene pools, and these different sets of genes may underlie the large difference in seed size between the Andean and Mesoamerican gene pools. As elegantly noted by the Iberoamerican team, multiple expansions of specific gene families may be responsible for the widespread adaptation of common bean to such very diverse environments as the highland desert of Mexico and the tropical regions of Central and South America. They further noted that this broad adaptation may have made it an ideal species for domestication as a crop.

6.3.2 Molecular Marker Development

A common utility of a crop genome sequence is the development of new markers that can be used for genetic studies and marker-assisted selection (MAS). The first markers to result from the USA bean genome project were the development of an extensive set of indel markers (Moghaddam et al. 2014). This development effort was funded by the BeanCAP project. The markers actually predate the V1.0 assembly and were developed from an early assembly (20× scaffolds) of the 454 reads. The project developed 2687 indel markers distributed across the genome, and the markers were selected to be diagnostic not only for intergene pool populations, but also useful for scoring intra-gene pool, and even within market class variation. Markers from this collection were instrumental in locating the WM7.1 and WM8.3 white mold tolerance QTL to narrow genomic regions (Mamidi et al. 2016) and providing gel-based markers for the *Co-1* anthracnose gene (Zuiderveen et al. 2016). As additional resources become available for large-scale resequencing of multiple bean genotypes, new indel collections will be regularly discovered from the resulting sequence information.

The richest marker set for any species are single nucleotide polymorphisms (SNP). The first SNP resource developed from USA sequence was the Illumina Infinium assay BARCBean6K_3 BeadChip. This tool was also funded by the USDA BeanCAP project. Two previous Infinium BeanChip were designed using early scaffold assemblies, and from those chips, the BARCBean6K_3 BeadChip was optimized to study variation among all levels of population classification (gene pool, race, and market class). This diagnostic SNP tool has found utility to study the genetics of population structure (Cichy et al. 2015), nitrogen fixation (Heilig et al. 2016), disease resistance (Nakedde et al. 2016), and pod characteristics in snap beans (Hagerty et al. 2016). A richer SNP data set was created using low-pass sequencing (Schröder et al. 2016) to generate a set of $\sim 150{,}000$ SNPs for the BeanCAP Mesoamerican Diversity Panel (MDP) (Moghaddam et al. 2016). The MDP is a collection ($n = \sim 300$) of historical and modern cultivars representative of USA, Canadian, and Latin America breeding programs. This SNP collection was used to define the population structure of the MDP, determine that linkage disequilibrium varies among Mesoamerican races, chromosomes and even among various regions within a chromosome, and evaluate the genetic architecture of several agronomic traits (Moghaddam et al. 2016). Other SNP sets are also being developed such as the one reported by Ariani et al. (2016) that discriminates wild and cultivated bean genotypes. The richest SNP data sets, short of a complete genome sequence, are the product of resequencing. Mamidi et al. (2016) applied resequencing at the level of $\sim 2\times$ per individual within pools of tolerant and susceptible white mold lines. Approximately 1.5 million SNPs distinguished the two pools, and these SNPs were used to map tolerance to vary narrow intervals. This approach, introgression mapping, can be applied to any mapping population, and the low-cost of pooled, rather than individual, resequencing makes this an attractive approach to rapid mapping to narrow intervals.

6.3.3 Transcriptomics

Another genome-wide application that is enabled by a reference genome sequence is a characterization of the transcribed portion of the genome or the transcriptome. A whole-genome perspective of the transcriptome, a gene atlas, was first published by O'Rourke et al. (2014) using samples from 24 different stages from leaves, stems, flowers, pods, seeds, roots, and nodules. The cultivar Jamapa, a race Mesoamerica black bean, was used for the analysis. Approximately 85% of the G19833 gene models were expressed in at least one tissue at some developmental stage. Across all gene models, 35% were differentially regulated when different tissue sources were compared. In general, transcription factors (TF) were underexpressed in seeds relative to other tissues, and 26 TF families showed differential expression among tissues. O'Rourke et al. (2014) monitored the transcriptome of seeds because of their importance as a food source. As a general observation, ten times more genes decreased expression as the seed developed compared to those that increased. The seed transcriptome was enriched for carbohydrate metabolism genes that as a group were highly expressed in the tissue. It was observed, conversely, that for soybean, carbohydrate metabolism genes are underexpressed while fatty acid metabolism genes are upregulated. This is consistent with the nutrient composition of the seeds of the two species, where carbohydrates are a main component of the modern common bean seed, whereas the modern soybean seed is rich in oils.

The Iberoamerican team also performed an in-depth study of the BAT 93 transcriptome (Vlasova et al. 2016). Notable observations include the following: (1) approximately 40% of the genes were expressed in all organs sampled (root, leaf, seed, pod, steam, flower, axial meristem); (2) 64% of the genes, on average, were expressed within any organ; (3) approximately 10% of the genes could be considered housekeeping genes; and (4) only a small percentage of the soybean orthologs of common bean housekeeping genes were classified as housekeeping genes. As an organism transitions between the various phases of development, the transcriptome also changes. Those changes were also monitored by this team where they discovered that when two successive stages of development were compared, the earlier stage had a greater number of differentially expressed genes. This was most pronounced for the early stages of the development of an organ. Further, the greatest number of differentially expressed genes was associated with flowers, pods, and seeds. It was also observed that the expression of only 2% of the genes is limited to a single developmental stage, and the putative functions of the developmental-stage-specific genes differ as plant tissues mature. Finally, from a co-expression analysis, it was noted that the densest hub was enriched for genes associated with photosynthesis and NADP metabolism.

6.3.4 Comparative Genomics

From a genomics perspective, a key question in evolutionary biology is the relationship between multiple species in a taxonomic lineage with regards to the nature of individual genes and gene families. This was first considered in common bean by O'Rourke et al. (2014) where they observed that the number of transcription factors within each TF family in soybean was twice that observed in common bean. This is consistent with the observation that soybean underwent a genome duplication following its divergence from the shared ancestor with common bean (Schmutz et al. 2010). This is not the case, though, for genes encoding nucleotide-binding site, leucine-rich repeat containing proteins (NLR). The NLR family members have consistently been shown to be associated with disease resistance and are often termed "resistance genes." Using similar bioinformatics approaches, Meziadi et al. (2016) identified 376 NLR genes in common bean, while Kang et al. (2012) determined the soybean genome consisted of 319 NLR genes. For both genomes, these NLR genes were clustered, but to a much greater extent in common bean where a cluster of 35 NLRs genes is located on the proximal end of chromosome Pv04, and 60 NLR genes are found on the distal end of Pv11. By contrast, the expected 2 to 1 relationship was observed for other gene families associated with disease defense, receptor-like kinases, receptor-like proteins, and LysM-containing protein families (McClean et al., unpublished data). So an important question from an evolutionary perspective is why is there such a discrepancy in the relative number of NLR genes between the two species, especially in light of the observation that additional resistance genes have a metabolic cost in terms of growth (Tian et al. 2003). It may be that the wide adaptation of common bean to various niches has exposed the species to more pathogens for which a larger number of resistance mechanisms evolved relative to soybean. As observed by Vlasova et al. (2016), the size of the common bean NLR repertoire was established before the split into gene pools suggesting the ancestor of the gene pools occupied a wide range that contained multiple pathogens that drove the disproportionate expansion of this gene family. Conversely, the energy cost of a complete set of duplicate NLR genes may have been too large to bear for soybean, and as its genome moved toward the ancestral diploid number of genes, the NLR genes were preferentially purged from the genome.

Genes of agronomic importance have also been considered from a comparative genomics perspective. *E1*, one of the ten soybean flowering time or maturity genes, encodes a bipartite nuclear localizing signal. The common bean ortholog of *E1*, *Phvul.009G204600*, is the closest non-soybean gene ortholog. Overexpression of *Phvul.009G204600* gave the same phenotype as the native soybean gene suggesting that it is also a flowering gene in common bean (Zhang et al. 2016). While this results points to *Phvul.009G204600* as a functional ortholog of *E1*, it was not located near flowering or maturity genome-wide association study (GWAS) peaks (Moghaddam et al. 2016). Functional conservation was also noted for the auxin response factor (ARF) TF family in both common bean and soybean (Le et al. 2016). This gene family is subdivided into five groups, and within each phylogenic group, the common bean member always groups with a pair or singleton soybean ortholog rather than with another common bean group member. This demonstrates that the ARF gene family was established prior to split of common bean and soybean in the Phaseoleae. Expansion though can occur at different times within a gene or gene family lineage and can occur independently in different lineages. For example, albumin 1b peptides (A1b) are insecticidal peptides, and as with other plant albumins, they are restricted to legume plants. An analysis revealed a major gene expansion through tandem duplications of this protein family in *Medicago truncatula* (n = 44) that involved seven of the eight chromosomes (Karaki et al. 2016). No such expansion was noted for other

Galegoid group legumes. Within the Phaseoleae family, which diverged from the Galegoid group ~54.3 MYA (Lavin et al. 2005), the only major expansion occurred in common bean. And this expansion only occurred on the distal end of Pv11 where all 21 members of the family are located in tandem. This clearly is an example of lineage-specific expansion in common bean via tandem duplication, whereas the expansion in *Medicago* was genome-wide. Clearly, expansions can occur by multiple modes in legumes. As the functions of more and more genes are defined, these types of comparative analyses across the legume lineage will provide hints as to the functions of uncharacterized genes in common bean.

6.4 Future Direction

6.4.1 Population Genomics: Deep Resequencing of Diversity Panels…A Requirement for More Gene-Based Genetic Analyses

Diversity panels that capture variation among a defined population are essential for the discovery of the genome-wide effects that control-specific phenotypes. Early panels include those developed and extensively utilized in poplar (Slavov et al. 2012), maize (Thornsberry et al. 2001), and *Arabidopsis* (Nordborg et al. 2005). Poplar would be a good example species to follow. Once the genome was sequenced (Tuskan et al. 2006), a common set of genotypes was selected for analysis across the geographic distribution of the species (Slavov et al. 2012). These were established as cuttings and were distributed as a common garden for phenotyping. The lines were genotyped initially with a 34K Infinium chip (Geraldes et al. 2013), and more recently, the SNP data was shared with others who adopted the common garden for their research. Currently, that panel has been resequenced to a minimum depth of 15× which enables the discovery of SNPs tightly linked to or within candidate genes (Muchero et al. 2015).

Common bean breeders and geneticists have recently developed three diversity panels: Mesoamerican Diversity Panel (Moghaddam et al. 2016); the Andean Diversity Panel (ADP, n = ~330; Cichy et al. 2015); and the Durango Diversity Panel (DDP, n = 122 originally, now n = 184; Soltani et al. 2016). These panels consist of modern genotypes (1930 onwards) used by growers in production fields. They are quite useful because the individual genotypes are adapted to modern production systems, so they can be evaluated under field conditions in which allelic variation critical to line success in those systems can be evaluated. Those populations have already been SNP genotyped with between 150k and 250k SNPs. While that marker depth is useful for marker-trait association discovery, in some cases it may not be granular enough to map genes in populations that provide high-level genetic resolution. That would require genotyping at a higher depth. Obviously, as any genotyping technology should be cost-effective, several approaches are possible. One that has high utility is to develop a SNP chip that is at the depth found in multiple humans SNP chip, ~500k. In humans, different chips representing different SNP sets are used collectively to genotypes individual in >100 clinics around the world, and coupled with phenotypic data for those individuals and imputation of the SNP data, very large effective mapping populations (n > 30,000) are developed to map candidate genes for complex human traits (Naj et al. 2011). For common bean, one goal would be to develop a community-wide chip with four SNPs per exon (~450k SNPs) and add additional intergenic SNPs to provide a genotyping platform of ~600k SNPs. This community tool would be useful to screen the panels described above or newer panels that would be for important agronomic phenotyping efforts. Importantly, the output from chip genotyping is SNP calls. This tool would be extremely useful since much of the common bean research community can manage genotype data but is challenged by the bioinformatics requirements to work with resequencing data. Whereas resequencing of complete panels would provide more data, including detection of more exotic SNPs that are not represented in the array, the challenging bioinformatics analysis would hinder utilization, next to the costs, which are currently favoring chip genotyping over WGS.

De novo assembly of key genetic resources and resequencing of diversity panels at high coverage has the potential to distinguish key differences between individuals and to uncover more SNPs than those assigned to a chip. Recent sequencing chemistries are making it easy to generate near whole-genome sequences at a depth that enables diploid de novo genome assembly for each individual. The 10× Genomics (http://www.10xgenomics.com/) sequencing approach generates localized tag sequences that can then be de novo assembled using the Supernova assembler. One application would be to generate de novo diploid assemblies of all individuals within a pathogen differential set. Proper phasing will then elucidate the critical difference between a susceptible line and a line carrying a single resistance gene which will facilitate accurate marker development. Currently, this approach would be prohibitively expensive for most common bean researchers to be used on a broad scale. An alternative would be to perform less expensive resequencing. The DDP has already been resequenced at an 8× depth, and ~700k SNPs have been identified (McClean et al., unpublished data). While resequencing at this depth would be helpful, a deeper sequencing depth would prove useful as the search for causative mutations, which most often fall within genes, moves forward. The effectiveness of in-depth resequencing to map phenotypic variation and discover putative causative variants to individual candidate genes has recently been shown for Poplar (Muchero et al. 2015). A set of 56 resequenced common bean samples (new and previously published) has been assembled and analyzed by Lobaton et al. (unpublished data), creating a common SNP set used to map intergene pool introgressions in varieties and land races.

From a genotypic perspective, deep resequencing of wild genotypes is currently lacking. That common bean consists of two gene pool diverged from a common ancestor about 100k years ago is well established (Mamidi et al. 2013). While a significant loss of variation was coupled with the appearance of Andean gene pool, the Mesoamerican gene pool contains essentially the same degree of variation as the ancestral pool (Mamidi et al. 2013, Schmutz et al. 2014). This lends credence to thoughts that the species arose in Mesoamerica, a hypothesis that is strongly supported by recent results that place the speciation of *P. vulgaris* in Mesoamerica and most likely Mexico (Bitocchi et al. 2012). The dual gene pool feature of common bean is an important distinguishing feature of the species and makes it useful to study early events associated with gene pool differentiation. This is best studied using population genomics approaches that consider both diversity and differentiation (Weigel and Nordborg 2015). These types of efforts would best be accomplished at the depth provided by resequencing data (8×–35×). With such data, SNP variation can initially be correlated with altitude, latitude, and longitude, or any available long-term environmental data, to begin a search for loci associated with local adaptation. Initial studies that adopted this approach have been reported, but with a very limited set of SNPs (Rodriguez et al. 2016). These considerations can also be addressed if performance data from controlled field trials could be collected. The near universal presence of photoperiod sensitivity, though, will limit such ecological-based trials to a narrow latitudinal range.

Even more than wild relatives, domesticated sister species of the secondary and tertiary gene pool of common bean (*Phaseolus acutifolius, Phaseolus coccineus, Phaseolus dumosus*) have been utilized by breeders to introgress valuable alleles into common bean breeding lines, e.g., for superior abiotic stress tolerance (Butare et al. 2011) or diseases resistance (Singh et al. 2001). Mapping rates against the G19833 reference drop below 50% in these species (Lobaton et al., unpublished), indicating large sequence and/or structural polymorphisms. Copy number variation (CNV) and other structural polymorphisms are obviously most pronounced in more distantly related germplasm. In cultivated common bean, they are probably most prominent in fast diverging disease-resistance clusters, which are of high interest to researchers and breeders. De novo assemblies using new technologies will deliver information on structural and CNV variation. Whereas such projects are currently very cost-intensive, new assemblies of the more diverse lines in the direct and extended breeding pools would represent very valuable genomics tools to understand and fully utilize these resources.

6.4.2 A Breeding Perspective

A stated output from all crop sequencing projects is the development of variants, SNPs, or indels, which can be applied to plant breeding purposes. A few examples include the discovery of SNPs linked to the *I* gene, a critical factor in resistance to the bean common mosaic virus (Drijfhout 1978). Using a combination of SNP genotyping of the MDP and additional Andean cultivars coupled with BCMV resistance/susceptibility phenotyping of these plant materials, Bello et al. (2014) described an approach called "in silico bulk segregant analysis" to discover codominant SNPs tightly linked to the *I* gene. This simple approach only requires the development of pools consisting of individuals with contrasting phenotypes followed by SNP or indel genotyping of the pools. These SNPs are then confirmed using an individual SNP testing protocol such as the Kompetitive Allele Specific PCR (KASP) assay (https://www.lgcgroup.com). This is a powerful approach that can then be applied to the MDP, ADP, and DDP if sufficient phenotypic data is available. Ferreira et al (2016) used genotyping by sequencing (GBS) sequence tags mapped against the reference to visualize backcross introgression of disease-resistant loci into fabada-type beans, demonstrating the near-isogenic line method for identification of *I* gene and *bc-3* resistance loci for virus resistance.

Of course, mapping with standard bi-parental populations or association panels followed by population screening is also possible with the SNP tools already available. For this purpose, the BARCBean6K_3 BeadChip has proven useful to map common bacterial blight (Viteri et al. 2014), halo blight (Miklas et al. 2014), anthracnose (Zuiderveen et al. 2016), leaf hopper (Brisco et al. 2014), and Fusarium root rot (Hagerty et al. 2015; Nakedde et al. 2016) resistance loci. Lobaton et al. (unpublished) adapted WGS data to develop improved molecular markers with superior specificity to resistance sources. A brief example of identifying an angular leaf spot resistance locus was recently published (Keller et al. 2015). WGS can also improve the power of new breeding tools like genomic selection. If complex populations are used that have a limited number of parental lines (e.g., 10), WGS data of these parental lines can be used to impute all SNPs (millions) based on lower coverage GBS data of the whole training/evaluation populations. Similarly, WGS of parental lines of complex or RIL populations can add valuable information to genetic analysis.

This BeanCAP research tool is an example of a community tool of great utility. The next generation of genomics tools for breeding programs should focus on loci of critical production importance. For common bean, with such a wide array of diseases that affect production, a "Resistance Chip" would be quite valuable. This chip could be constructed with the SNP loci that are linked to resistance genes, gene clusters, or QTL (best based on resequencing a larger number of resistance sources) and verified in confirmation populations or panels. Molecular marker research in bean suggests that the verification step might require that neighboring SNPs be discovered that are diagnostic in other market classes or gene pools than the one in which the original linked SNP was discovered (Miklas et al. 1996). Those additional linked SNPs would need to be included in the chip design to ensure its widest applicability. Given the long history of applying molecular markers for breeding purposes in common bean, this tool would be welcomed (Miklas et al. 2006).

Understanding the genotype-by-environment interaction is critical to the success of all plant breeding program because of the subtle differences between the regions that beans must be adapted. This concept of local adaptation, while a major academic research question in evolution (Kawecki and Ebert 2004), is also critical to the success of plant breeding. *Arabidopsis* research has defined a method that takes a genome-wide approach to identify loci closely associated with successful growth in specific environments (Fournier-Level et al. 2011; Hancock et al. 2011). The Cooperative Dry Bean Nursery (Singh 2000) is a historical effort that has been grown for >65 years and can be mined for performance data over many locations. The nursery was initiated and continues to provide value multisite performance data for plant breeders as advanced breeding lines from their program are considered for release. A recent survey showed over 500 genotypes have been evaluated, many over multiple years, and efforts are underway to collect seed and isolate DNA from as many lines as possible (MacQueen et al., unpublished). Those lines will then be genotyped using SNP technology. Environmental data from all site/year combinations is being collected, and the goal is to map the genetic variation relative to the environmental data to discover climatic conditions that are positive predictors of performance in specific environments. Those mapping results can then be used to identify markers for genetic regions associated with environmental variables unique to specific growing conditions. The effectiveness of those markers can then be tested in subsequent years of the trial across multiple locations. This is an example how the principles adopted by the field of population genomics (Weigel and Nordborg 2015) can be tested to determine how useful they are to crop improvement efforts in common bean.

6.5 Conclusion

Common bean is an important societal crop that needs a major research focus from the perspectives of food security and nutrition. Given its broad geographic production range, the focus by necessity must be diverse. Therefore, all appropriate tools must be employed for crop improvement efforts. The publication of genome sequences from the two gene pools will have translational significance beyond understanding the basic structure of the genome and how it has evolved over time. It is critical that the genomic focus switches to understanding how the crop successfully manages biotic and abiotic stresses across the many environments in which it is grown. This will require a gene-by-gene approach that successfully interacts with a more genome-wide perspective. These global approaches must be an interdisciplinary cooperative research effort that successfully incorporates all aspects of the biology of the plant. From that perspective, building up from the genome sequence to understand the full gene repertoire and how the interaction of genes is manifested in phenotype should be a central theme in common bean research. This is analogous to human research that is aimed at collecting a broad sweep of knowledge about human biology with the end goal of a healthy population. Understanding how bean genes work together will translate into healthy bean production and subsequently a more food-secure world.

References

Ariani A, Teran JCBM, Gepts P (2016) Genome-wide identification of SNPs and copy number variation in common bean (*Phaseolus vulgaris* L.) using genotyping-by-sequencing (GBS). Mol Breeding 36:87

Bello MH, Moghaddam SM, Massoudi M, McClean PE, Cregan PB, Miklas PN (2014) Application of in silico bulked segregant analysis for rapid development of markers linked to bean common mosaic virus resistance in common bean. BMC Genom 15:903

Beebe S, Skroch PW, Tohme J, Duque MC, Pedraza F, Nienhuis J (2000) Structure of genetic diversity among common bean landraces of Middle American origin based on correspondence analysis of RAPD. Crop Sci 40:264–273

Beebe S, Ramírez J, Jarvis A, Rao IM, Mosquera G, Bueno JM, Blair MW (2011) Genetic improvement of common beans and the challenges of climate change. In: Yadav SS, Redden R, Hatfield JL, Lotze-Campen H, Hall A (eds) Crop adaptation to climate change. Wiley-Blackwell, Oxford, UK, pp 356–369

Berdegué JA, Fuentealba R 2011. Latin America: the state of smallholders in agriculture. A paper presented at the IFAD conference on new directions for smallholder agriculture 24–25 Jan, 2011, Rome: IFAD

Bitocchi E, Nanni L, Bellucci E, Rossi M, Giardini A, Zeuli PS, Logozzo G, Stougaard J, McClean PE, Attene G, Papa R (2012) Mesoamerican origin of the common bean (*Phaseolus vulgaris* L.) is revealed by sequence data. Proc Natl Acad Sci USA 109:E788–E796

Blair MW, Pedraza F, Buendia HF, Gaitán-Solís E, Beebe SE, Gepts P, Tohme J (2003) Development of a genome-wide anchored microsatellite map for common bean (*Phaseolus vulgaris* L.). Theor Appl Genet 107:1362–1374

Brisco EI, Porch TG, Cregan PB, Kelly JD (2014) Quantitative trait loci associated with resistance to *Empoasca* in common bean. Crop Sci 54:2509–2519

Butare L, Rao I, Lepoivre P, Cajiao C, Polania J et al (2011) Phenotypic evaluation of interspecific recombinant inbred lines (RILs) of *Phaseolus* species for aluminium resistance and shoot and root growth response to aluminium-toxic acid soil. Euphytica 186:715–730

Cichy KA, Porch TG, Beaver JS, Cregan P, Fourie D, Glahn RP, Grusak MA, Kamfwa K, Katuuramu DN, McClean P, Mndolwa E, Nchimbi-Msolla S, Pastor-Corrales MA, Miklas PN (2015) A diversity panel for Andean bean improvement. Crop Sci 55:2149–2160

Cook OF (1919) Milpa agriculture, a primitive tropical system. Annual Report of the Board of Regents of the Smithsonian Institute pp 307–326. Government Printing Office, Washington, D.C.

Córdoba JM, Chavarro C, Schlueter JA, Jackson SA, Blair MW (2010) Integration of physical and genetic maps of common bean through BAC-derived microsatellite markers. BMC Genom 11:436

Debouck DG, Toro O, Paredes OM, Johnson WC, Gepts P (1993) Genetic diversity and ecological distribution of *Phaseolus vulgaris* (Fabaceae) in northwestern South America. Econ Bot 47:408–423

Drijfhout E (1978) Genetic interaction between *Phaseolus vulgaris* and bean common mosaic virus with implications for strain identification and breeding for resistance. Verslagen van Landbouwkundige Onderzoekingen No. 872:1–89

Ferreira JJ, Murube E, Campa A (2016) Introgressed genomic regions in a set of near-isogenic lines of common bean revealed by genotyping-by-sequencing. Plant Genome 9. doi:10.3835/plantgenome2016.08.0081

Fournier-Level A, Korte A, Cooper MD, Nordborg M, Schmitt J, Wilczek AM (2011) A map of local adaptation in *Arabidopsis thaliana*. Science 334:86–89

Freyre R, Skroch PW, Geffroy V, Adam-Blondon AF, Shirmohamadali A, Johnson WC, Llaca V, Nodari RO, Pereira PA, Tsai S-M, Tohme J, Dron M, Nienhuis J, Vallejos CE, Gepts P (1998) Towards an integrated linkage map of common bean. 4. Development of a core linkage map and alignment of RFLP maps. Theor Appl Genet 97:847–856

Gaut BS (2015) Evolution is an experiment: assessing parallelism in crop domestication and experimental evolution (Nei Lecture, SMBE 2014, Puerto Rico). Mol Biol Evol 32:1661–1671

Geraldes A, Difazio SP, Slavov GT, Ranjan P, Muchero W, Hannemann J, Gunter LE, Wymore AM, Grassa CUJ, Farzaneh N, Porth I, McKown AD, Skyba O, Li E, Fujita M, Klapste J, Martin J, Schackwitz W, Pennacchio C, Rokshar D, Friedmann MC, Wasteneys GO, Guy RD, El-Kassaby YA, Mansfield SD, Cronk QCB, Ehlting J, Douglas CJ, Tuskan GA (2013) A 34K SNP genotyping array for *Populus trichocarpa*: design, application to the study of natural populations and transferability to other *Populus* species. Mol Ecol Resour 13:306–323

Gourdji S, Läderach P, Valle AM, Martinez CZ, Lobell DB (2015) Historical climate trends, deforestation, and maize and bean yields in Nicaragua. Agric Forest Meteorol 200:270–281

Graham RD, Welch RW, Saunders DA, Ortiz-Monasterio I, Bouis HE, Bonierbale M, de Haan S, Bugos G, Thiele G, Liria R, Meisner AA, Beebe SE, Potts MJ, Kadian M, Hobbs PR, Gupta RK, Twomlow S (2007) Nutritious subsistence food systems. Adv Agron 92:1–74

Grisi MCM, Blair MW, Gepts P, Brondani C, Pereira PAA, Brondani RPV (2007) Genetic mapping of a new set of microsatellite markers in a reference common bean (*Phaseolus vulgaris*) population BAT93 × Jalo EEP558. Genet Mol Res 6:691–706

Hagerty CH, Cuesta-Marcos A, Cregan P, Song Q, McClean P, Myers JR (2016) Mapping snap bean pod and color traits, in a dry bean × snap bean recombinant inbred population. J Am Soc Hort Sci 141:131–138

Hagerty CH, Cuesta-Marcos A, Cregan PB, Song Q, McClean P, Noffsinger S, Myers JR (2015) Mapping and root rot resistance and root architecture quantitative trait loci in common bean. Crop Sci 55:1969–1977

Hancock AM, Brachi B, Faure N, Horton MW, Jarymowycz LB, Sperone FG, Toomajian C, RouxF Bergelson J (2011) Adaptation to climate across the *Arabidopsis thaliana* genome. Science 334:83–86

Heilig JA, Beaver JS, Wright EM, Song Q, Kelly JD (2016) QTL analysis of symbiotic nitrogen fixation in a black bean population. Crop Sci. doi:10.2135/cropsci2016.05.0348

Hyten DL, Song Q, Fickus EW, Quigley CV, Lim JS, Choi IY, Hwang EY, Pastor-Corrales M, Cregan PB (2010) High-throughput SNP discovery and assay development in common bean. BMC Genom 11:475

Jiao Y, Wickett NJ, Ayyampalayam S, Chanderbali AS, Landherr L, Ralph PE, Tomsho PP, Hu Y, Liang H, Soltis PS, Soltis DE, Clkifton SW, Schlarbaum SE, Schuster SC, Ma H, Leebens-Mack J, de Pamphilis CW (2011) Ancestral polyploidy in seed plants and angiosperms. Nature 473:97–100

Kami J, Poncet V, Geffroy V, Gepts P (2006) Development of four phylogenetically-arrayed BAC libraries and sequence of the APA locus in *Phaseolus vulgaris*. Theor Appl Genet 112:987–998

Kami J, Velásquez VB, Debouck DG, Gepts P (1995) Identification of presumed ancestral DNA sequences of phaseolin in *Phaseolus vulgaris*. Proc Natl Acad Sci USA 92:1101–1104

Kang YJ, Kim KH, Shim S, Yoon MY, Sun S, Kim MY et al (2012) Genome-wide mapping of NBS-LRR genes and their association with disease resistance in soybean. BMC Plant Biol 12:139

Karaki L, Da Silva P, Rizk F, Chouabe C, Chantret N, Eyraud V, Gressent F, Sivignon C, Rahioui I, Kahn D, Brochier-Armanet CB, Rahbe Yl, Royer C (2016) Genome-wide analysis identifies gain and loss/change of function within the small multigenic insecticidal Albumin 1 family of *Medicago truncatula*. BMC Plant Biol 16:63

Katungi E, Farrow A, Chianu J, Sperling L, Beebe S (2009) Common bean in Eastern and Southern Africa: a situation and outlook analysis. International Centre for Tropical Agriculture, Cali, Colombia

Katungi E, Sperling L, Karanja D, Farrow A, Beebe S (2011) Relative importance of common bean attributes and variety demand in the drought areas of Kenya. J Develop Agric Econ 3:411–422

Kawecki TJ, Ebert D (2004) Conceptual issues in local adaptation. Ecol Lett 7:1225–1241

Keller B, Manzanares C, Jara C, Lobaton JD, Studer B, Raatz B (2015) Fine-mapping of a major QTL controlling angular leaf spot resistance in common bean (*Phaseolus vulgaris* L.). Theor Appl Genet 128:813–826

Kelly JD, Gepts P, Miklas PN, Coyne DP (2003) Tagging and mapping of genes and QTL and molecular marker-assisted selection for traits of economic importance in bean and cowpea. Field Crops Res 82:135–154

Lavin M, Herendeen PS, Wojciechowski MF (2005) Evolutionary rates analysis of Leguminosae implicates a rapid diversification of lineages during the Tertiary. Syst Biol 54:575–594

Le B, Nawaz MA, Rehman HM, Le T, Yang SH, Golokhvast KS, Son E, Chung G (2016) Genome-wide characterization and expression pattern of auxin response factor (ARF) gene family in soybean and common bean. Genes Genom 38:1165–1178

Mamidi S, Miklas PN, Trapp J, Felicetti E, Grimwood J, Schmutz J, Lee R, McClean PE (2016) Sequence-based introgression mapping identifies candidate white mold tolerance genes in common bean. Plant Genome 9. doi:10.3835/plantgenome2015.09.0092

Mamidi S, Rossi M, Annam D, Moghaddam S, Lee R, Papa R, McClean P (2011) Investigation of the domestication of common bean (*Phaseolus vulgaris*) using multilocus sequence data. Funct Plant Biol 38:953–967

Mamidi S, Rossi M, Moghaddam SM, Annam D, Lee R, Papa R, McClean PE (2013) Demographic factors shaped diversity in the two gene pools of wild common bean *Phaseolus vulgaris* L. Heredity 110:267–276

McClean PE, Lee RK (2007) Genetic architecture of chalcone isomerase non-coding regions in common bean (*Phaseolus vulgaris* L.). Genome 50:203–214

McClean PE, Lee RK, Miklas PN (2004) Sequence diversity analysis of dihydroflavonol 4-reductase intron 1 in common bean. Genome 47:266–280

McClean PE, Mamidi S, McConnell M, Chikara S, Lee R (2010) Synteny mapping between common bean and soybean reveals extensive blocks of shared loci. BMC Genom 11:184

McClean P, Cannon S, Gepts P, Hudson M, Jackson S, Rokhsar D, Schmutz J, Vance C (2008) Towards a whole genome sequence of common bean, (*Phaseolus vulgaris* L.): background, approaches, applications. http://bic.css.msu.edu/_pdf/Bean_Genomics_Status_2008.pdf

McClean PE, Lee RK, Otto C, Gepts P, Bassett MJ (2002) Molecular and phenotypic mapping of genes controlling seed coat pattern and color in common bean (*Phaseolus vulgaris* L.). J Hered 93:148–152

McConnell M, Mamidi S, Lee R, Chikara S, Rossi M, Papa R, McClean P (2010) Syntenic relationships among legumes revealed using a gene-based genetic linkage map of common bean (*Phaseolus vulgaris* L.). Theor Appl Genet 121:1103–1116

Melotto M, Coelho MF, Pedrosa-Harand A, Kelly JD, Camargo LEA (2004) The anthracnose resistance locus Co-4 of common bean is located on chromosome 3 and contains putative disease resistance-related genes. Theor Appl Genet 109:690–699

Melotto M, Monteiro-Vitorello CB, Bruschi AG, Camargo LE (2005) Comparative bioinformatic analysis of genes expressed in common bean (*Phaseolus vulgaris* L.) seedlings. Genome 48:562–570

Meziadi C, Richard MM, Derquennes A, Thareau V, Blanchet S, Gratias A, Pflieger S, Geffroy V (2016) Development of molecular markers linked to disease resistance genes in common bean based on whole genome sequence. Plant Sci 242:351–357

Michael TP, Jackson S (2013) The first 50 plant genomes. Plant Genome 6. doi:10.3835/plantgenome2013.03.0001in

Miklas PN, Afanador L, Kelly JD (1996) Recombination-facilitated RAPD marker-assisted selection for disease resistance in common bean. Crop Sci 36:86–90

Miklas PN, Fourie D, Trapp J, Davis J, Myers JR (2014) New loci including conferring resistance to halo bacterial blight on chromosome Pv04 in common bean. Crop Sci 54:2099–2108

Miklas PN, Kelly JD, Beebe SE, Blair MW (2006) Common bean breeding for resistance against biotic and abiotic stresses: from classical to MAS breeding. Euphytica 147:105–131

Moghaddam SM, Mamidi S, Osorno JM, Lee R, Brick M, Kelly J, Miklas P, Urrea C, Song Q, Cregan P, Grimwood J, Schmutz J, McClean PE (2016) Genome-wide association study identifies candidate loci underlying agronomic traits in a Middle American diversity panel of common bean. Plant Genome 9. doi:10.3835/plantgenome2016.02.0012

Moghaddam SM, Song Q, Mamidi S, Schmutz J, Lee R, Cregan P, Osorno J, McClean PE (2014) Developing market class specific InDel markers from next generation sequence data in *Phaseolus vulgaris* L. Front Plant Sci 5:185

Muchero W, Guo J, Di Fazio SP, Chen JG, Ranjan P, Slavov GT, Gunter LE, Jawdy S, Bryan AC, Sykes R, Ziebell A, Klapste J, Porth I, Skyba O, Unda F, El-Kassaby YA, Douglas CA, Mansfield SD, Martin J, Schackwitz W, Evans LM, Czarnecki O, Tuskan GA (2015) High-resolution genetic mapping of allelic variants associated with cell wall chemistry in *Populus*. BMC Genom 16:1

Munishi LK, Lema AA, Ndakidemi PA (2015) Decline in maize and beans production in the face of climate change at Hai District in Kilimanjaro Region, Tanzania. Int J Climate Change Strat Manag 7:17–26

Naj AC, Jun G, Beecham GW, Wang L-S, Narayan B et al (2011) Common variants at MS4A4/MS4A6E, CD2AP, CD33 and EPHA1 are associated with late-onset Alzheimer's disease. Nat Gene 43:436–441

Nakedde T, Ibarra-Perez FJ, Mukankusi C, Waines JG, Kelly JD (2016) Mapping of QTL associated with Fusarium root rot resistance and root architecture traits in black beans. Euphytica 212:51–63

Nodari RO, Tsail SM, Gilbertson RL, Gepts P (1993) Towards an integrated linkage map of common bean 2. Development of an RFLP-based linkage map. Theor Appl Genet 85:513–520

Nordborg M, Hu TT, Ishino Y, Jhaveri J, Toomajian C, Zheng H, Bakker E, Calabrese P, Gladstone J, Goyal R, Jakobsson M, Kim S, Morozov Y, Padhukasahasram B, Plagnol V, Rosenberg NA, Shah C, Wall JD, Wang J, Zhao K, Kalbfleisch T, Schlz V, Kreitman M, Bergelson J (2005) The pattern of polymorphism in *Arabidopsis thaliana*. PLoS Biol 3: e196

O'Rourke JA, Iniguez LP, Fu F, Bucciarelli B, Miller SS, Jackson SA, McClean PE, Li J, Dai X, Zhao PX, Hernandez G, Vance CP (2014) An RNA-Seq based gene expression atlas of the common bean. BMC Genom 15:866

Papa R, Gepts P (2003) Asymmetry of gene flow and differential geographical structure of molecular diversity in wild and domesticated common bean (*Phaseolus vulgaris* L.) from Mesoamerica. Theor Appl Genet 106:239–250

Ramírez M, Graham MA, Blanco-López L, Silvente S, Medrano-Soto A, Blair MW et al (2005) Sequencing and analysis of common bean ESTs. Building a foundation for functional genomics. Plant Physiol 137:1211–1227

Rodriguez M, Rau D, Bitocchi E, Bellucci E, Biagetti E, Carboni A, Gepts P, Nanni L, Papa R, Attene G (2016) Landscape genetics, adaptive diversity and population structure in *Phaseolus vulgaris*. New Phytol 209:1781–1794

Salami A, Kamara AB, Brixiova Z (2010) Smallholder agriculture in East Africa: trends, constraints and opportunities. Working Papers Series No. 105. African Development Bank, Tunis, Tunisia

Schlueter JA, Goicoechea JL, Collura K, Gill N, Lin JY, Yu Y, Kudrna D, Zuccolo A, Vallejos CE, Munoz-Torres M, Blair MW, Tohme J, Tomkins J, McClean P, Wing RA, Jackson SA (2008) BAC-end sequence analysis and a draft physical map of the common bean (*Phaseolus vulgaris* L.) genome. Trop Plant Biol 1:40–48

Schmutz J, Cannon SB, Schlueter J, Ma J, Mitros T, Nelson W, Hyten DL, Song Q, Thelen JJ, Cheng J, Xu D, Hellsten U, May GD, Yu Y, Sakurai T, Umezawa T, Bhattacharyya MK, Sandhu D, Valliyodan B, Lindquist E, Peto M, Grant D, Shu S, Goodstein D, Barry K, Futrell-Griggs M, Abernathy B, Du J, Tian Z, Zhu L, Gill N, Joshi T, Libault M, Sethuraman A, Zhang X-C, Shinozaki K, Nguyen HT, Wing RA, Cregan P, Specht J, Grimwood J, Rokshar D, Stacey G, Shoemaker RC, Jackson SA (2010) Genome sequence of the palaeopolyploid soybean. Nature 463:178–183

Schmutz J, McClean P, Mamidi S, Wu GA, Cannon SB, Grimwood J, Jenkins J, Shu S, Song Q, Chavarro C, Torres-Torres M, Geffroy V, Moghaddam SM, Gao D, Abernathy B, Barry K, Blair M, Brick MA, Chovatia M, Gepts P, Goodstein DM, Gonzales M, Hellsten U, Hyten DL, Jia G, Kelly JD, Kudrna D, Lee R, Richard MMS, Miklas PN, Osorno JM, Rodrigues J, Thareau V, Urrea CA, Want M, Yu Y, Zhang M, Wing RA, Cregan PB, Rokhsar DS, Jackson SA (2014) A reference genome for common bean and genome-wide analysis of dual domestications. Nat Genet 46:707–713

Schröder S, Mamidi S, Lee R, McKain MR, McClean PE, Osorno JM (2016) Optimization of genotyping by sequencing (GBS) data in common bean (*Phaseolus vulgaris* L.). Mol Breeding 36:1–9

Singh SP (2000) 50 years of the cooperative dry bean nursery. Annu Rep Bean Improv Cooperative 43: 110–111

Singh SP, Gepts P, Debouck DG (1991) Races of common bean (*Phaseolus vulgaris*, Fabaceae). Econ Bot 45:379–396

Singh SP, Munoz CG, TeraN H (2001) Registration of common bacterial blight resistant dry bean germplasm VAX 1, VAX 3, and VAX 4. Crop Sci 41:275–276

Slavov GT, Di Fazio SP, Martin J, Schackwitz W, Muchero W, Rodgers-Melnick E, Lipphardt MF, Pennacchio CP, Hellsten U, Pennacchio LA, Gunter LE, Ranjan P, Vining K, Pomraning KR, Wilhelm LJ, Pellegrini M, Mockler TC, Freitag M, Geraldes A, El-Kassaby YA, Mansfield SD, Cronk QCB, Douglas CJ, Strauss SH, Rokhsar D, Tuskan GA (2012) Genome resequencing reveals multiscale geographic structure and extensive linkage disequilibrium in the forest tree *Populus trichocarpa*. New Phytol 196:713–725

Soltani A, Bello M, Mndolwa E, Schroder S, Moghaddam SM, Osorno JM, Miklas PN, McClean PE (2016) Targeted analysis of dry bean growth habit: interrelationship among architectural, phenological, and yield components. Crop Sci 56:3005–3015

Song Q, Jia G, Hyten DL, Jenkins J, Hwang EY, Schroeder SG, Osorno JM, Schmutz J, Jackson SA, McClean PE, Cregan PB (2015) SNP Assay Development for linkage map construction, anchoring whole-genome sequence, and other genetic and genomic applications in common bean. G3: Genes Genom Genet 5:2285–2290

Thibivilliers S, Joshi T, Campbell KB, Scheffler B, Xu D, Cooper B, Nguyen HT, Stacey G (2009) Generation of *Phaseolus vulgaris* ESTs and investigation of their regulation upon *Uromyces appendiculatus* infection. BMC Plant Biol 9:46

Thornsberry JM, Goodman MM, Doebley J, Kresovich S, Nielsen D, Buckler ES (2001) Dwarf8 polymorphisms associate with variation in flowering time. Nat Genet 28:286–289

Tian D, Traw MB, Chen JQ, Kreitman M, Bergelson J (2003) Fitness costs of R-gene-mediated resistance in *Arabidopsis thaliana*. Nature 423:74–77

Tuskan GA, Difazio S, Jansson S, Bohlmann J, Grigoriev I, Hellsten U, Putnam N, Ralph S, Rombauts S, Salamov A, Schein J, Sterck I, Aerts A, Bhalerao RR, Bhalerao RP, Blaudez D, Boerjan W, Brun A, Brunner A, Busov V, Campbell M, Carlson J, Chalot M, Chapman J, Chen G–L, Cooper D, Coutinho PM, Couturier J, Covert S, Cronk Q, Cunningham R, Davis J, Degroeve S, Dejardin A, dePamphilis C, Detter J, Dirks B, Dubchak I, Duplessis, Ehlting J, Ellis B, Gendler K, Goodstein D, Gribskov M, Grimwood J, Groover A, Gunter L, Hamberger B, Heinze B, Helariutta Y, Henrissat B, Holligan D, Holt R, Huang W, Islam-Faridi N, Jones S, Jones-Rhoades M, Jorgensen R, Joshi C, Kangasjarvi J, Karlsson K, Kelleher C, Kirkpatrick R, Kirst M, Kohler A, Kalluri U, Larimer F, Leebens-Mack J, Leple L–C, Locascio P, Lou Yl, Lucas S, Martin F, Montanini B, Napoli C, Nelson DR, Nelson C, Nieminen K, Nilsson O, Pereda V, Peter G, Philippe R, Pilate G, Poliakov A, Razumovskaya J, Rihcardson P, Rinaldi C, Ritland L, Rouze P, Ryaboy D, Schmutz J, Scharader J, Segerman B, Shin H, Siddiqui A, Sterky F, Terry A, Tsai C, Uberbacher E, Unneberg P, Vahala J, Wall K, Wessler S, Yang G, Yin T, Douglas C, Marra M, Sandber G, Van de Peer Y, Rokshar D (2006) The genome of black cottonwood, *Populus trichocarpa* (Torr. & Gray). Science 313:1596–1604

Ugen M, Musoni A, Cheminingwa G, Kimani P, Mucharo M, Katafire M (2012) Utilization of common bean for improved health and nutrition in eastern and central Africa. Association for Strengthening Agricultural Research in Eastern and Central Africa, Entebbe, Uganda

Viteri DM, Cregan PB, Trapp JJ, Miklas P, Singh SP (2014) A new common bacterial blight resistance QTL in VAX 1 common bean and interaction of the new QTL, SAP6, and SU91 with bacterial strains. Crop Sci 54:1598–1608

Vlasova A, Capella-Gutiérrez S, Rendón-Anaya M, Hernández-Oñate M, Minoche AE, Erb I, Camara F, Prieto-Barja P, Corvelo A, Sanseverino W, Westergaard G, Dohm J, Pappas GJ Jr, Saburido-Alvarez S, Kedra D, Gonzalez I, Cozzuto L, Gomez-Garrido J, Aguilar-Moron MA, Andreu N, Aguilar OM, Garcia-Mas J, Zehnsdorf M, Vazquez MP, Delgado-Salinas A, Delaye L, Lowy E, Mentaberry A, Vianello-Brondani RP, Garcia JL, Alioto T, Sanchez F, Himmelbauer H, Santalla M, Notredame C, Gabaldon T, Herrera-Estrella A, Guigo R (2016) Genome and transcriptome analysis of the Mesoamerican common bean and the role of gene duplications in establishing tissue and temporal specialization of genes. Genome Biol 17:32

Weigel D, Nordborg M (2015) Population genomics for understanding adaptation in wild plant species. Annu Rev Genet 49:315–338

Zhang X, Zhai H, Wang Y, Tian X, Zhang Y, Wu H, Lu S, Yang G, Li Y, Wang L, Hu B, Bu Q, Xia Z (2016) Functional conservation and diversification of the soybean maturity gene *E1* and its homologs in legumes. Scien Rep 6:29548

Zuiderveen GH, Padder BA, Kamfwa K, Song Q, Kelly JD (2016) Genome-wide association study of anthracnose resistance in Andean beans (*Phaseolus vulgaris*). PLoS ONE 11:e0156391

Organelle Genomes in *Phaseolus* Beans and Their Use in Evolutionary Studies

7

Maria I. Chacón Sánchez

Abstract

Chloroplasts and mitochondria originated from separate endosymbiotic events that occurred about 1.5 billion years ago. In plants, the quadripartite nature of the chloroplast genome is a conserved feature with very little size variation among species. In contrast, size of mitochondrial genomes varies greatly in plants with high rate of rearrangements in angiosperms, although highly conserved in sequence. Sequencing of organelle genomes has increased in the last years as new technologies developed, and today, the Organelle Genome Resources of GenBank contains about 1717 and 275 records of complete chloroplast and mitochondrial genomes for plants, 73 and six of them for legumes, respectively. In plants, plastid genomes have been very useful for phylogenetic and population genetics studies. In *Phaseolus* beans, polymorphisms in the plastid genome have been used in several studies to unravel the evolutionary history of the common bean and Lima bean in the wild and to pinpoint domestication places. However, all these studies have explored very few genomic regions of the plastid genome. Therefore, new genome resources need to be developed for *Phaseolus* beans. The sequencing of the plastid genome of the common bean in the year 2007 was a good start, but since then no new organelle genome sequences have been reported in this genus. The goal of this review is to stimulate the development of more organelle genomes resources in the genus *Phaseolus*, which will allow a better understanding of the rates and patterns of evolution and the dynamics of expression patterns of these genomes. Third-generation sequencing technologies and additional tools offer an opportunity to do so, and in the near future, we should see more developments in this direction.

M. I. Chacón Sánchez (✉)
Departamento de Agronomía, Universidad Nacional de Colombia—Bogotá—Facultad de Ciencias Agrarias, Carrera 30 no 45-03—Edificio 500, Bogotá, D.C Código 111321, Colombia
e-mail: michacons@unal.edu.co

M. Pérez de la Vega et al. (eds.), *The Common Bean Genome*,
Compendium of Plant Genomes, https://doi.org/10.1007/978-3-319-63526-2_7

Keywords
Chloroplast genome · Mitochondrial genome · Common bean · Lima bean · Genotyping by sequencing · Next-generation sequencing technologies · Population genetics

7.1 Introduction

The origin of mitochondria and chloroplasts has been attributed to separate endosymbiosis events, one of them more than 1.5 billion years ago between an α-proteobacterium and a unicellular organism of Archaea or eukaryotic nature that gave rise to mitochondria, and another one about 1.2–1.5 billion years ago between a cyanobacterium and an eukaryotic organism (already containing mitochondria) that gave rise to chloroplasts (reviewed in Dyall et al. 2004). There has been controversy about whether plastids originated from a primary cyanobacterial/eukaryotic endosymbiosis event or from more than one event, although much of the evidence seem to favor a monophyletic origin of plastids (see Palmer 2003 for a discussion). After the primary endosymbiosis event, plastids presumably spread across the major groups of algae through an unknown number of secondary eukaryote/eukaryote endosymbiosis events (Archibald and Keeling 2002; Palmer 2003). One of the most convincing evidence that supports the primary endosymbiosis hypotheses is the fact that plastids and mitochondria contain a small genome whose genes keep relationship with those of bacteria. The genomes of chloroplast and mitochondria are single chromosomes, and the predominant view has been that they occur in a stable circular form (Kolodner and Tewari 1979); however, recent evidence indicates that their structure may be more plastic than previously believed. These genomes may also occur as linear forms, and both circular and linear forms may show various sizes and may contain one or more genome equivalents, or even incomplete genome equivalents, as revealed by fiber-FISH experiments (Lilly et al. 2001; Burger et al. 2003; Bendich 2004, 2007)

As soon as it was evident that the mitochondria and chloroplasts contained their own genomes, studies started to unravel the structure and dynamics of these genomes, first based on denaturation/renaturation kinetics (DRK) and then on restriction fragment length polymorphisms (RFLP) of the whole molecule, PCR-RFLP, PCR-sequencing of small portions of the molecule, and lately sequencing of whole organelle genomes. In the 1970s, experiments based on DRK and restriction enzyme mapping showed that the plastid genome structure consisted of two inverted repeats (IRs) separated by a small single copy (SSC) region and a large single copy (LSC) region (Bedbrook and Kolodner 1979; Kolodner and Tewari 1979). Later, researches showed that this quadripartite structure of the plastid genome is a conserved feature in land plant evolution, except for some legume species (e.g., alfalfa, pea, broad bean, chickpea, wisteria) and two conifers (Douglas fir and radiata pine) that have lost one entire inverted repeat, showing a tripartite structure (Palmer et al. 1987; Strauss et al. 1988). The IR contains mainly ribosomal genes and in flowering plants may experience small and apparently random expansions which make the IR-LSC junction lie at different places even between closely related species (Goulding et al. 1996). Another relatively conserved feature of the plastid genome is its size, which usually falls in a narrow range between 120 and 160 Kb, except for some green algae groups with genomes under 100 Kb or over 200 Kb (Palmer 1985). The size variation of the plastid genome has been explained in part by the loss of one IR and in part by gene losses, especially in the angiosperms where about 27 protein-coding genes have been lost in at least one lineage during evolution (Magee et al. 2010). In contrast to the plastid genome, researches on the mitochondrial genome have shown that its size varies greatly in plants, with a range between 200 Kb and 2500 Kb (reviewed in Levings III and Brown 1989), and in angiosperms usually show internal rearrangements, even among closely related species, although it is quite conserved in primary sequence as a consequence of a low point mutation rate (Palmer and Herbon 1988). This high rate of rearrangement in angiosperms is in sharp contrast with the high level of synteny usually found in early land plants (Bryophytes), even among species that have diverged hundreds of millions of years ago (reviewed in Liu et al. 2012). It has been found that the extensive DNA rearrangements and size differences seen in mitochondrial genomes (or chondromes) of angiosperms are prompted by frequent recombination, trans-splicing of disrupted introns, the presence of sequences of chloroplast and nuclear origin (including retrotransposons), and ongoing transfer of genes to the nuclear genome (Kubo and Mikami 2007; Kubo and Newton 2008; Knoop et al. 2011; Knoop 2012).

Sequencing projects of organelle genomes started in the 1980s with the plastid genomes and in the 1990s with the mitochondrial genomes and have lately increased with the development of more robust and efficient sequencing and bioinformatics tools. Today, in the Organelle Genome Resources of GenBank, there are about 1717 records of complete plant chloroplast genomes, 96 of them for green algae and 1602 for land plants, and of these, about 73 are for legume species, including *Acacia ligulata, Cicer arietinum* (chickpea), *Glycine max* (soybean) and other nine *Glycine* species, *Lathyrus sativus* and other 12 *Lathyrus* species, *Lens culinaris* (Lentil), *Lotus japonicas* (Lotus), *Lupinus luteus, Medicago hybrida, Medicago papillosa, Medicago truncatula* (barrel medic), *Milletia pinnata, Phaseolus vulgaris* (common bean), *Pisum sativum* (pea), *Vicia sativa, Vigna angularis* (adzuki bean), *Vigna radiata* (mungbean), *Vigna unguiculata, Trifolium subterraneum* and other four *Trifolium* species. For the mitochondrial genome, there are 275 records for plants, of these, 53 are for green algae and 213 for land plants, and the only legume species are mungbean (Alverson et al. 2011), adzuki bean (Naito et al. 2013), *Vicia faba* (Negruk 2013), soybean (Chang et al. 2013), *Millettia pinnata* (Kazakoff et al. 2012), and *Lotus* (Organelle Genome Resources of GenBank, http://www.ncbi.nlm.nih.gov/genomes/GenomesHome.cgi?taxid=2759&hopt=html, accessed on September 26, 2016).

From these projects, we have learned that the average plastid genome of seed plants contains about 101–118 genes, and of these, 66–82 are protein-coding genes involved mainly in photosynthesis and functions related to gene expression (Jansen and Ruhlman 2012). On the other hand, the average mitochondrial genome contains about 40–50 genes, which code mainly for protein components of complexes I to IV of electron transport chain and ribosomal proteins of the mitochondrion itself (Lang et al. 1997).

Although a lot of effort has been put in the research of organelle genomes, the genomic resources developed for seed plants and in specific for legume species are in general relatively poor, which limits the undertaking of comparative studies leading to understanding the organization of these genomes throughout time and the evolutionary mechanisms involved, as well as the use of these molecules in comprehensive phylogenetic analyses, especially at deep levels. For evolutionary studies in plants, the plastid genome is quite useful for deep level phylogenetic analyses as well as for intraspecific studies (Palmer et al. 1988; Gielly and Taberlet 1994; Jansen et al. 2007; Byrne and Hankinson 2012; Dong et al. 2012) due to several characteristics. First, it does not show such a high rate of rearrangements as the mitochondrial genome; therefore, results are easier to interpret. Second, the rate of point mutations is higher in the chloroplast genome than in the mitochondrial genome, providing variation for deep as well as for shallow level phylogenetic and population genetics analyses. And third, plastid genomes (as well as mitochondrial genomes) are usually maternally inherited and may, therefore, trace different evolutionary histories when compared to the nuclear genome, which is biparentally inherited and normally exposed to hybridization and introgression events.

The goal of this review is to stimulate the development of more genomic resources for organelle genomes in legumes, with an emphasis in the genus *Phaseolus*, to address a diversity of questions especially concerning phylogenetics and population genetics. In this chapter, the use of organelle genomes for evolutionary studies in the genus *Phaseolus* is reviewed, with an emphasis in the plastid genome for the reasons exposed above. In the first part, the structure and gene content of the recently sequenced plastid genome of the common bean is revised. In the second part, a summary of how successfully this molecule has been used in evolutionary studies in *Phaseolus* beans, in wild and domesticated populations, is presented. The review ends with a brief overview of the tools that can be applied to carry out analyses of organelle genomes in the genus *Phaseolus* and other genera, for population level and phylogenetic studies.

7.2 Structure and Gene Content of the Common Bean Plastid Genome

The number of genes in current chloroplast genomes varies from about 100–250 (Gao et al. 2010), which is only a small fraction compared to the about 3000 genes that can be found in a free-living cyanobacteria, and therefore, chloroplasts depend on the products of nuclear-encoded genes to function (Bock 2007). In the plastome, genes have been classified into functional classes by several authors (Bock 2007; Wicke et al. 2011), which are summarized below. (1) Genes (approximately 62) that encode components of the expression machinery of the chloroplast; for example, genes for DNA- and RNA-processing enzymes, the gene *mat*K (a group IIA intron maturase), four ribosomal RNAs (ribosomal RNA *rrn23* for the large subunit, ribosomal RNAs *rrn16, rrn5,* and *rrn4.5* for the small subunit), 30 different tRNAs, 12 ribosomal proteins for the small subunit of the ribosome, and nine ribosomal proteins for the large subunit of the ribosome. (2) Genes ($\sim$50) coding for protein subunits involved in photosynthetic light and dark reactions and cytochrome C biogenesis. These include genes encoding subunits of a protochlorophyllide reductase, genes encoding proteins for CO_2 uptake, the *rbcL* gene encoding the large subunit of the RuBisCo enzyme, genes encoding proteins of the thylakoid membrane involved in photosynthetic light reactions, among them are protein subunits of the photosystems I (5 subunits) and II (15 subunits) where light is harvested, chaperons that help in the assembly of the two photosystems, six subunits of the cytochrome b_6f complex that connects the two photosystems by electron transfer, six subunits of the ATP synthase complex, and 11 subunits of the plastid NAD(P)H-complex involved in electron recycling. (3) Proteins on other metabolic pathways such as fatty acid synthesis (Acetyl-CoA carboxylase, *accD* gene) and sulfur metabolism mainly in liverworts, and open reading frames of unknown function (for example, *ycf*1 and *ycf*2).

Guo et al. (2007) reported the sequence for the plastid genome of the common bean cv. Negro Jamapa (GenBank accession DQ886273.1). To obtain the sequence, the authors extracted DNA from isolated chloroplasts, then fragmented the DNA by nebulization, and finally, cloned and sequenced the fragments by Sanger technology. The study showed that the size of the common bean chloroplast genome is approximately 150,826 bp, of which around 79,824 bp make up the LSC, 17,610 make up the SSC, and 26,426 bp make up each of the two IR (IRA and IRB). In previous experiments using restriction enzymes, it was established that the chloroplast genome of the common bean exists in two circular isomers that are different in the relative orientation of the two single copy regions, presumably as a result of flip-flop recombination among the IR (Palmer 1983; Marechal and Brisson 2010).

According to Guo et al. (2007), the content of non-coding regions (40%) of the common bean plastid genome is similar to that of tobacco, with a little bias toward AT content (around 64–68%), as reported in other plastomes. In the coding regions, the authors predicted 127 genes, with 19 of them duplicated in the inverted repeats. Among the 108 single copy genes, there are 4 rRNAs and 30 tRNAs similar to the tobacco plastome, and 75 protein-coding genes involved in photosynthesis and gene expression. A complete list of the protein-coding genes in the chloroplast genome of common bean (GenBank accession number DQ886273.1) can be found in Table 7.1, and the physical map can be seen in Fig. 7.1a. Among the particular features that Guo et al. (2007) found are the lack of the genes *rpl*22 and *inf*A as has been reported in legumes and other plants, the presence of two pseudogenes (*rps*16 and *rpl*33), a 51 Kb inversion in the LSC that goes from *rbc*L to *rps*16 that is characteristic of legume species in the Papilionoideae subfamily (Doyle et al. 1996), and an inversion at the junctions between *trn*H-*rpl*14 and *rps*8-*rps*19 absent in other legumes. Guo et al. (2007) confirmed the presence of these two pseudogenes and this last inversion by PCR amplification in other ten common bean varieties. Guo et al. (2007) also compared amino acid substitution rates of all protein-coding genes in the chloroplast genomes of common bean and soybean, taking different species as reference (*Arabidopsis thaliana*, lotus, barrel medic), and concluded that the common bean chloroplast genome evolves at faster substitution rates than soybean. Among the genes with faster substitution rates are *acc*D and *mat*K.

There is a second chloroplast genome for common bean reported in GenBank under accession number EU196765.1 (Moore et al. 2007). When comparing both reported genomes, only minor differences are found; for example, the genome reported by Moore et al. (2007) contains 229 bp in the IR annotated as *ycf*15 (pseudogene) between the genes *ycf*2 and *ndh*B that was not annotated in the other genome. The physical map of this genome can be seen in Fig. 7.1b.

7.3 Use of Organelle Genomes in Population Genetics

The predominant pattern of inheritance of organelle genomes in plants, including early land pants, is uniparental, mostly maternal. In spite of being a general feature, few attempts have been made to explain the origin of this type of inheritance. Some authors state that uniparental inheritance arose as a consequence of the high level of cytological differentiation among female and male germ lines; this differentiation being itself a consequence of selection in the germ line (Godelle and Reboud 1995). Basically, the organelles are maternally inherited because there are different transmission rates among female and male gametes, being the conditions of replication more difficult in the male germ line. Other authors argue that because the replication of nuclear DNA and organelle DNA is uncoupled, intracellular conflicts (nuclear DNA–organelle DNA or organelle–organelle DNA) can arise if natural selection, acting on both kinds of DNA molecules, does not favor the same characters (Eberhard 1980). These incompatibilities may lead to uniparental inheritance of organelle DNA by the elimination of the products from one parent. Another reason that has been invoked is that in order to prevent "infection" of the egg by "foreign" or "pathogenic DNA" during the fusion of gametes, there are degradation mechanisms in the egg that would degrade not only foreign DNA but also entering organelle DNA (Coleman 1982). However, not all plant species show uniparental plastid inheritance, many angiosperm species show or at least have the potential for biparental inheritance (Zhang 2010). One common method to inspect for potential biparental plastid inheritance (PBPI) is to look for the presence of plastid DNA in the male gamete by means of epifluorescence microscopy (Corriveau and Coleman 1991). Cytological studies have revealed that many legume species

Table 7.1 List of protein-coding genes found in the chloroplast genome of common bean cv. Negro Jamapa sequenced by Guo et al. (2007), their position in the large single copy (LSC), small single copy (SSC) or inverted repeat (IRA, IRB), and their classification into three gene categories

Gene name	Description	Start position	End position	Region	Gene category
*rps*3	ribosomal protein S3	25	675	LSC	Plastid genetic machinery
*rpl*16	ribosomal protein L16	829	2253	LSC	Plastid genetic machinery
*rpl*14	ribosomal protein L14	2379	2747	LSC	Plastid genetic machinery
*psb*A	photosystem II protein D1	3253	4314	LSC	Photosynthesis genes
*mat*K	maturase K	4964	6505	LSC	Plastid genetic machinery
*rbc*L	ribulose-1,5-bisphosphate carboxylase/oxygenase large subunit	8093	9523	LSC	Photosynthesis genes
*atp*B	ATP synthase CF1 beta subunit	10,293	11,789	LSC	Photosynthesis genes
*atp*E	ATP synthase CF1 epsilon subunit	11786	12187	LSC	Photosynthesis genes
*ndh*C	NADH dehydrogenase subunit 3	13451	13813	LSC	Photosynthesis genes
*ndh*K	NADH dehydrogenase subunit K	13857	14495	LSC	Photosynthesis genes
*ndh*J	NADH dehydrogenase subunit J	14597	15073	LSC	Photosynthesis genes
*rps*4	ribosomal protein S4	18027	18632	LSC	Plastid genetic machinery
*ycf*3	photosystem I assembly protein Ycf3	19209	21200	LSC	Photosynthesis genes
*psa*A	photosystem I P700 chlorophyll a apoprotein A1	21951	24203	LSC	Photosynthesis genes
*psa*B	photosystem I P700 chlorophyll a apoprotein A2	24229	26433	LSC	Photosynthesis genes
*rps*14	ribosomal protein S14	26578	26880	LSC	Plastid genetic machinery
*psb*Z	photosystem II protein Z	27952	28140	LSC	Photosynthesis genes
*psb*C	photosystem II CP43 chlorophyll apoprotein	28804	30225	LSC	Photosynthesis genes
*psb*D	photosystem II protein D2	30173	31234	LSC	Photosynthesis genes
*psb*M	photosystem II protein M	34282	34386	LSC	Photosynthesis genes
*pet*N	cytochrome b6/f complex subunit VIII	34964	35053	LSC	Photosynthesis genes
*rpo*B	RNA polymerase beta subunit	36566	39778	LSC	Plastid genetic machinery

(continued)

Table 7.1 (continued)

Gene name	Description	Start position	End position	Region	Gene category
*rps*2	ribosomal protein S2	47250	47960	LSC	Plastid genetic machinery
*atp*I	ATP synthase CF0 A subunit	48206	48949	LSC	Photosynthesis genes
*atp*H	ATP synthase CF0 C subunit	49943	50188	LSC	Photosynthesis genes
*atp*F	ATP synthase CF0 B subunit	50566	51830	LSC	Photosynthesis genes
*atp*A	ATP synthase CF1 alpha subunit	51903	53435	LSC	Photosynthesis genes
*psb*I	photosystem II protein I	55554	55664	LSC	Photosynthesis genes
*psb*K	photosystem II protein K	56132	56317	LSC	Photosynthesis genes
*acc*D	acetyl-CoA carboxylase beta subunit	57913	59337	LSC	Other genes
*psa*I	photosystem I subunit VIII	59604	59708	LSC	Photosynthesis genes
*cem*A	envelope membrane protein	61005	61688	LSC	Other genes
*pet*A	cytochrome f	61879	62841	LSC	Photosynthesis genes
*psb*J	photosystem II protein J	63317	63439	LSC	Photosynthesis genes
*psb*L	photosystem II protein L	63596	63712	LSC	Photosynthesis genes
*psb*F	photosystem II protein VI	63735	63854	LSC	Photosynthesis genes
*psb*E	photosystem II protein V	63864	64115	LSC	Photosynthesis genes
*pet*L	cytochrome b6/f complex subunit VI	65029	65124	LSC	Photosynthesis genes
*pet*G	cytochrome b6/f complex subunit V	65288	65401	LSC	Photosynthesis genes
*psa*J	photosystem I subunit IX	65991	66125	LSC	Photosynthesis genes
*rps*18	ribosomal protein S18	66899	67207	LSC	Plastid genetic machinery
*rpl*20	ribosomal protein L20	67442	67801	LSC	Plastid genetic machinery
*rps*12	ribosomal protein S12	68655	68768	LSC	Plastid genetic machinery
*clp*P	clp protease proteolytic subunit	68985	71036	LSC	Plastid genetic machinery
*psb*B	photosystem II 47 kDa protein	71441	72967	LSC	Photosynthesis genes

(continued)

Table 7.1 (continued)

Gene name	Description	Start position	End position	Region	Gene category
*psb*T	photosystem II protein T	73127	73234	LSC	Photosynthesis genes
*psb*N	photosystem II protein N	73293	73424	LSC	Photosynthesis genes
*psb*H	photosystem II protein H	73538	73759	LSC	Photosynthesis genes
*pet*B	cytochrome b6	73891	75330	LSC	Photosynthesis genes
*pet*D	cytochrome b6/f complex subunit IV	75525	76720	LSC	Photosynthesis genes
*rpo*A	RNA polymerase alpha subunit	76960	77961	LSC	Plastid genetic machinery
*rps*11	ribosomal protein S11	78036	78452	LSC	Plastid genetic machinery
*rpl*36	ribosomal protein L36	78768	78881	LSC	Plastid genetic machinery
*rps*8	ribosomal protein S8	79367	79771	LSC	Plastid genetic machinery
*rps*19	ribosomal protein S19	80396	80674	IRB	Plastid genetic machinery
*rpl*2	ribosomal protein L2	80729	82213	IRB	Plastid genetic machinery
*rpl*23	ribosomal protein L23	82232	82513	IRB	Plastid genetic machinery
*ycf*2	Hypothetical protein rf2	83119	89985	IRB	Other genes
*ndh*B	NADH dehydrogenase subunit 2	91242	93411	IRB	Photosynthesis genes
*rps*7	ribosomal protein S7	93761	94228	IRB	Plastid genetic machinery
*rps*12	ribosomal protein S12	94283	95069	IRB	Plastid genetic machinery
*ndh*F	NADH dehydrogenase subunit 5	106264	108516	SSC	Photosynthesis genes
*rpl*32	ribosomal protein L32	108943	109104	SSC	Plastid genetic machinery
*ccs*A	cytochrome c biogenesis protein	109540	110526	SSC	Photosynthesis genes
*ndh*D	NADH dehydrogenase subunit 4	110787	112283	SSC	Photosynthesis genes
*psa*C	photosystem I subunit VII	112409	112654	SSC	Photosynthesis genes
*ndh*E	NADH dehydrogenase subunit 4L	112939	113244	SSC	Photosynthesis genes
*ndh*G	NADH dehydrogenase subunit 6	113447	113977	SSC	Photosynthesis genes

(continued)

Table 7.1 (continued)

Gene name	Description	Start position	End position	Region	Gene category
*ndh*I	NADH dehydrogenase subunit I	114254	114739	SSC	Photosynthesis genes
*ndh*A	NADH dehydrogenase subunit 1	114823	117172	SSC	Photosynthesis genes
*ndh*H	NADH dehydrogenase subunit 7	117174	118355	SSC	Photosynthesis genes
*rps*15	ribosomal protein S15	118457	118729	SSC	Plastid genetic machinery
*ycf*1	Hypothetical protein RF1	119085	124364	SSC/IRB	Other genes
*rps*12	ribosomal protein S12	135040	135826	IRA	Plastid genetic machinery
*rps*7	ribosomal protein S7	135881	136348	IRA	Plastid genetic machinery
*ndh*B	NADH dehydrogenase subunit 2	136698	138867	IRA	Photosynthesis genes
*ycf*2	Hypothetical protein RF2	140124	146990	IRA	Other genes
*rpl*23	ribosomal protein L23	147596	147877	IRA	Plastid genetic machinery
*rpl*2	ribosomal protein L2	147896	149380	IRA	Plastid genetic machinery
*rps*19	ribosomal protein S19	149435	149713	IRA	Plastid genetic machinery

show PBPI such as chickpea (*C. arietinum*), *Lathyrus japonicus*, *L. orodatus*, *Medicago sativa*, *Melilotus alba*, *M. indica*, *M. officinalis*, pea (*Pisum sativum*), and *Wisteria sinensis* (Smith et al. 1986; Corriveau and Coleman 1988, 1991). Among the *Phaseolus* species analyzed by epifluorescence microscopy are the common bean and *Phaseolus aureus*, both with maternally inherited plastid DNA (Corriveau and Coleman 1988). Studying in detail the mode of inheritance of organelle DNA is important for the use of these molecules in evolutionary studies, especially because this will affect the effective population size.

The plastid genome has been very useful for evolutionary and domestication studies in *Phaseolus* beans (common bean and Lima bean). The earliest studies detected polymorphisms in the molecule at the intraspecific level by means of RFLP and PCR-RFLP, and later studies used PCR-sequencing. Chacón et al. (2005, 2007) investigated the evolutionary history of wild common bean, a species that is widely distributed from Mexico to Argentina, and the domestication areas by applying PCR-RFLP in 322 accessions of wild, weedy, and domesticated common bean. Seven non-coding regions (Table 7.2) were amplified and then restricted by a set of enzymes. A total of 32 point mutations and two indels were found by sequencing a small sample of wild beans, and of these, 16 point mutations were detectable by restriction digestion. These 16 point mutations defined 14 haplotypes, and the phylogenetic analysis of these haplotypes along with those of sister species (*P. dumosus* and *P. costaricensis*) suggested that the origin of wild common bean was probably Mesoamerica. Genetic relationships among wild and domesticated accessions suggested southern Peru as a possible area for a single domestication of the

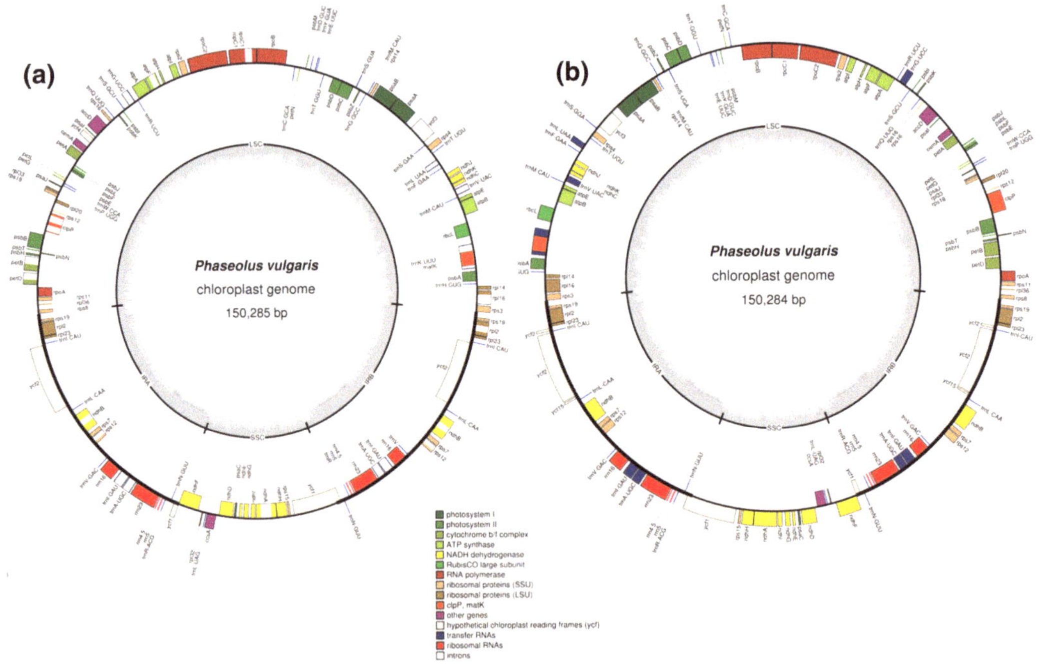

Fig. 7.1 Physical maps of the two plastid genomes that are available for the common bean. **a** GenBank accession number DQ886273.1. **b** GenBank accession number EU196765.1. Legends show color-coding for gene categories. GC content is shown in the inner circle in darker gray. Physical maps were drawn using complete genome sequences in the program OGDRAW (Lohse et al. 2013)

Andean landraces of common bean and pointed Mexico as a possible area for multiple domestication events that would have given rise to the Mesoamerican landraces.

For studying the biological origin of Lima bean (*Phaseolus lunatus* L.), a species that is also widely distributed from Mexico to Argentina in the wild, Fofana et al. (1999) analyzed chloroplast DNA polymorphisms by means of PCR-RFLP in a sample of Lima bean accessions and allied Mesoamerican and Andean species. The authors extracted genomic DNA, then amplified a set of six intergenic spacers (Table 7.2), and finally digested the PCR products with ten restriction enzymes. The authors also performed Southern blotting using as probe the chloroplast DNA purified from one accession of Lima bean. The phylogenetic analysis confirmed a monophyletic origin for the Andean and Mesoamerican gene pools in Lima bean, and an Andean origin of the species by the fact of a closer phylogenetic relationship of Lima bean with Andean allied species (*P. augusti*, *P. pachyrrhizoides* and *P. bolivianus*). Later, Fofana et al. (2001) studied the genetic diversity of wild and domesticated Lima beans by means of RFLP using two cpDNA probes in a set of 152 wild and domesticated accessions. In total, 89 bands were scored in the wild accessions, and about 55% of them resulted polymorphic. In their analyses, the authors evidenced the presence of three groups, the Mesoamerican, the Andean gene pools, and a third group called "transition group" with accessions from Peru, Ecuador, Colombia, and Costa Rica. They also suggested two places of domestication for Lima beans: one in the Andes of Ecuador and Peru, and a second one in the region Mexico–Guatemala.

In the last years, our research group has been investigating the biological origin and domestication areas of Lima bean in the Americas by sequencing two intergenic spacers of the chloroplast DNA (*trn*L_UAA-*trn*F_GAA and *atp*B-*rbc*L), both located in the large single copy

Table 7.2 Non-coding regions of plastid DNA that has been used for evolutionary studies in *Phaseolus* beans

Species	Intron	Intergenic spacers	Source
Phaseolus vulgaris	*trn*L intron *rpl*16 intron *ndh*A intron	*trn*T-*trn*L *trn*L-*trn*F *rps*14-*psa*B *acc*D-*psa*I	Chacón et al. (2005, 2007)
Phaseolus lunatus	–	*atp*B-*rbc*L *rps*14-*psa*B *pet*A-*psb*E *psb*C-*tRNA*ser *tRNA*ser-*tRNA*fmet *tRNA*thr-*tRNA*phe	Fofana et al. (1999)
Phaseolus lunatus	–	*trn*L-*trn*F *atp*B-*rbc*L	Motta-Aldana et al. (2010), Serrano-Serrano et al. (2010), Andueza-Noh et al. (2013)

of the molecule. A total of 1324 base pairs were sequenced from both spacers, and 33 polymorphic sites were found in a sample of 59 wild accessions representing the widespread distribution of the species, and a set of 12 allied species (Serrano-Serrano et al. 2010). These polymorphic sites defined 23 haplotypes, each one of them restricted to a single geographic region, except for one haplotype that was shared between Mesoamerica and South America. These data confirmed an Andean origin of wild Lima bean during Pleistocene and the presence of three gene pools, two Mesoamerican (MI and MII) and one Andean. Gene pool MI was distributed mainly from northern Mexico to Oaxaca, on the Pacific side. Gene pool MII was widely distributed on the plains of the Gulf of Mexico, Peninsula of Yucatan, Chiapas, in Central America from Costa Rica to Guatemala, and in South America on the eastern slope of the Andes in Venezuela, Colombia, Ecuador, Peru, and Argentina. The Andean gene pool was restricted to the Andes of Ecuador and northern Peru, on the western slope of the Andes. Motta-Aldana et al. (2010) and Andueza-Noh et al. (2013) tried to pinpoint areas of domestication for the Mesoamerican landraces using these two intergenic spacers and an increased sample of 262 individuals (40 haplotypes) and proposed two possible areas, one in central-western Mexico from gene pool MI, and another one in Guatemala–Costa Rica from gene pool MII. As it can be seen, these two intergenic spacers have been very useful for phylogeographic studies in Lima bean, but in order to have greater geographic resolution to pinpoint domestication places within Mexico with the use of plastid DNA, more polymorphisms need to be identified outside these two spacers. The new sequencing technologies provide an opportunity to deepen this aspect.

Another source of polymorphisms in the *Phaseolus* plastid genome has been the microsatellite sequences (SSR). Angioi et al. (2009) evaluated the usefulness of 39 SSR loci in several legume species, including common bean, *P. coccineus,* and Lima bean. The authors tested 36 primer pairs already reported for legumes, redesigned one primer pair based on previous reports, and designed two new primer pairs on the basis of available chloroplast genome sequences. In their study, 16 SSR loci were polymorphic for common bean and *P. coccineus* and 18 loci were polymorphic in 59 accessions of the *Phaseolus* genus (common bean, *P. coccineus*, *P. acutifolius*, *P. dumosus,* and Lima bean). Angioi et al. (2009) found that in the genus *Phaseolus*, the number of alleles per locus ranged from 2 to 12 and a total of 56 haplotypes were observed. As it can be seen, plastid SSR loci are more polymorphic than PCR-RFLP or RFLP of the chloroplast genome and therefore are good markers for intraspecific studies; however, as the authors noted, the possibility of homoplasy in these loci should be taken into account because different alleles not always originate from expansion/contraction of the repeat motif but from other causes such as indels and the presence of more than one type of SSR motif.

The SSR loci reported by Angioi et al. (2009) have been used to study the origin of the wild common bean (Desiderio et al. 2012). The authors extracted genomic DNA from 109 accessions of wild common bean and amplified 17 microsatellite loci by PCR. In the sample of wild beans, these loci showed between 2 and 12 alleles, with a total of 86 observed alleles, and genetic diversity values ranged from 0.13 to 0.85. In their study, the Mesoamerican population showed higher genetic diversity ($H_e = 0.54$) than the Andean population ($H_e = 0.40$), and this along with the population structure analyses indicated Central America as the place of origin of common bean.

In summary, the chloroplast genome has proved useful for evolutionary studies in *Phaseolus* beans for phylogenetic and population genetics analyses; however, some limitations are that these studies have analyzed polymorphisms in a small number of its regions (few non-coding regions and few SSR loci). As it can be seen below, new technologies offer an opportunity to carry out more comprehensive genomic analyses of plastid DNA.

7.4 Tools for the Analysis of Organelle Genomes

Among the earliest tools available to analyze organelle genomes was the use of restriction enzymes and probes for Southern blotting, to detect polymorphisms in the length of restriction fragments (Kochert et al. 1996; Bukhari et al. 1999). The first chloroplast genomes that were sequenced were those of *Nicotiana tabaccum* (representing angiosperms) (Shinozaki et al. 1986) and *Marchantia polymorpha* (representing bryophytes) (Ohyama 1996). Since the 1980s, the number of chloroplast genomes sequenced has increased, with more than 180 plastid genomes by the year 2010, 137 of them corresponding to land plants and of these, five genomes sequenced for the bryophytes (basal clades in the phylogeny of land plants), three for the lycophytes (the earliest divergent clade in vascular plants), four plastid genomes for the ferns/monilophytes, 17 of gymnosperms, and 108 angiosperms (Gao et al. 2010). The availability of whole plastid genome sequences soon provided the opportunity for the design of the so-called universal primers that anneal at conserved regions within coding sequences and can be used to amplify intervening and more variable introns and intergenic spacers (Taberlet et al. 1991). These primers allowed the amplification of these regions by PCR for downstream applications such as PCR-RFLP (Vekemans et al. 1998; Chacón et al. 2005) and PCR-sequencing (Motta-Aldana et al. 2010). The direct detection of DNA polymorphisms by sequencing of PCR products has been the tool of choice for the last 23 years. A list of available primers for amplification of chloroplast DNA regions can be found on the Internet at: http://bfw.ac.at/rz/bfwcms2.web?dok=4977.

The area of molecular systematics has benefited by the use of DNA sequencing of single or sets of coding and non-coding sequences for resolving phylogenetic relationships in several plant groups. In special, the use of non-coding regions, such as intron and intergenic spacers, has provided informative characters to resolve relationships at lower taxonomic levels. The plastid genome of tobacco taken as model indicates that about 60% of the genome corresponds to coding regions, about 40% to non-coding regions, with about 12% of the total length corresponding to introns. Borsch and Quandt (2009) reviewed what is known about the molecular evolution of the most commonly used non-coding regions of the plastid genome for phylogenetic analyses. The authors argue that it is important to understand the evolution of these regions in order to make a correct use of their information content in plant evolutionary studies. Among the regions with higher content of parsimony informative characters are the group II introns in the genes *pet*D, *rpl*16, *rps*16, (lost in various legume species, Doyle et al. 1995; Jansen et al. 2008) and *trn*K, and the intergenic spacers *trn*S-*trn*G (that includes a group II intron in *trn*G), *trn*T-*trn*F (that includes the *trn*T-*trn*L spacer, the group I intron in *trn*L, and the transcribed spacer *trn*L-*trn*F), *psb*A-*trn*H, and *atp*B-*rbc*L. However, the authors also point that the amount of variability is not the only characteristic that should be taken into account for a phylogenetic marker, because introns and spacers in the chloroplast genomes are generally found to be a "mosaic" of elements (e.g., base substitutions and microstructural mutations), and their molecular evolution should be better understood before being used as markers in any specific taxonomic group, either in deep phylogenies or at the species level.

The use of new technologies for sequencing in parallel millions of reads (NGS for Next-generation sequencing) has made possible the use of whole chloroplast genome sequences to carry out phylogenetic and phylogeographic studies at low taxonomic levels (Parks et al. 2009; Vachon and Freeland 2011). This approach has been useful for studies at the population level, which are usually limited by both the lack of informative loci and sequencing costs. Parks et al. (2009) carried out a phylogenomics study by applying massively parallel sequencing in the genus *Pinus* with the method described by Cronn et al. (2008), to unravel relationships left unresolved in previous studies. In this method, genomic DNA was enriched for plastid DNA by using 35 primer pairs designed from two *Pinus* reference genomes to amplify by PCR the entire plastid DNA molecule in amplicons of an average size of 3.6 Kb. After amplification, amplicons from each individual were pooled using approximately 10–30 ng of each and adapters containing unique 3 bp-tags were linked to the amplicons, and after library preparation, the amplicons were sequenced by Illumina. The short reads were assembled de novo and the contigs were aligned to a reference genome. With this approach, the authors obtained 67–98% of contigs aligned to the reference genome, covering between 78 and 94% of the reference genome. The authors estimated that the genomes sequenced were about 88–94% complete with a sequence depth of 55× to 186×, making of this a robust approach for phylogenetics and population genetics analyses.

A different plastid enrichment approach to sequence whole chloroplast genomes was recently developed by Stull et al. (2013). In this strategy, enrichment was carried out by using a set of plastid oligonucleotide (RNA) probes (approximately 55,000 RNA probes) of broad phylogenetic coverage. The probes were designed on the basis of 22 previously sequenced eudicot plastomes. In this method, libraries for Illumina sequencing were constructed, but before sequencing, these libraries were enriched using the RNA probes. After sequencing 24 species by Illumina in a single lane, the reads were assembled in de novo contigs and these were aligned against their closest plastid reference genome. In average, 59% of the total reads per sample were plastid sequences with a mean coverage of 717×, thus increasing significantly the number of plastid genomes that can be sequenced in parallel. A disadvantage that was observed by the authors is that the high coverage was biased toward coding sequences; therefore, studies at the population level should use an approach that targets longer insert lengths.

The methods described above use an enrichment approach and have the disadvantage of being sequence-dependent and laborious and time-consuming. McPherson et al. (2013) developed an alternative approach that uses whole genomic DNA with no enrichment procedures to assemble whole chloroplast genomes using the Illumina platform for paired ends, which promises to be cost-effective for routine applications. For this, total genomic DNA was extracted from four individuals of *Toona ciliata* from each of two locations in Australia; the DNA was normalized and then pooled into two libraries. These two libraries (multiplexed with seven more libraries from other studies) were sequenced with paired-end runs (100 bp fragments) in a single lane of an Illumina Genome Analyzer (GAIIx). Reads were assembled *de novo* into contigs using two strategies, and contigs containing chloroplast sequences were identified by blasting them against a database containing whole chloroplast genomes (134 in total); contigs with E value of zero were used for assembly. The chloroplast genome of the closest relative available (*Citrus sinensis*) was used as a scaffold for contig mapping. The authors compared the Illumina results (average coverage obtained was 214×) with the sequencing of the chloroplast genome of a single individual of *T. ciliata* from isolated chloroplasts on the 454 platform. The authors validated the observed SNPs by Sanger sequencing. The authors mentioned that their strategy was efficient to assemble chloroplast genomes without a reference genome and to discover SNPs in a population sample. They also stated that the quality of the sequencing using Illumina was comparable to that of 454, and because whole-genomic DNA was used, this strategy is appropriate for large-scale studies in plants. A similar method to sequence and assemble whole chloroplast genomes was reported by Ferrarini et al. (2013) who used the PacBio sequencing technology combined with the Illumina platform to sequence the chloroplast genome of *Potentilla micrantha*. With the PacBio long reads, the authors were able to assemble the whole chloroplast genome into a single contig without the need of a reference genome, with 320× coverage, no bias in coverage of GC-rich regions and with an increase in sequence accuracy provided by the Illumina short reads. The authors showed the usefulness of combining long and short read sequencing platforms for de novo assembly of chloroplast genomes, and recently, other researchers have adopted this same strategy (Li et al. 2014; Wu et al. 2014; Chen et al. 2015; Jackman et al. 2016).

From the above, we can see that in the last years, new methods to carry out phylogeographic and population genetics analyses using the plastid genome have become available. These methods are everytime more robust, easier to implement, and more cost-effective, and therefore, we are now at the right time to undertake more ambitious projects that seek to understand plastid and in general organelle genome diversity in plant populations. As more plastid genome sequences become available, it will be possible by means of comparative analyses to understand the rates and patterns of evolution of different regions in the plastome (coding and non-coding) to make a correct use of these in phylogenetics and population genetics analyses. In this regards, Sanitá Lima et al. (2016) called the attention to the fact that 97% of the studies reported on organelle genomes between 2010 and 2015 made only use of DNA-sequencing technologies and bioinformatics tools to characterize the chromosomes of organelle genomes, without the application of additional tools to understand the expression (transcription and translation) and structure of these genomes, in a similar way as it is done with nuclear genome projects. The authors highlight the importance to get insights in the near future of the dynamic structures of these genomes, posttranscriptional processes such as RNA editing, trans-splicing, transcriptional cleavage and polyadenylation, mutations rates, and cytonuclear interactions, in order to get a complete understanding of these genomes.

7.5 Polymorphisms in *Phaseolus* Chloroplast Genomes

As mentioned before, NGS tools allow the rapid identification of single-nucleotide polymorphisms (SNPs) for population-level studies, which have become the markers of choice for addressing many evolutionary questions. Here, I wanted to evaluate the usefulness of genotyping by sequencing (GBS) data generated from whole genomic DNA as a tool to discover SNPs in the chloroplast DNA. For this purpose, I used data derived from a GBS study in Lima bean from whole genomic DNA (unpublished data) to try to identify SNPs between the plastid genome of the common bean and the GBS reads of Lima bean. The GBS data were produced in a sample of 95 wild and domesticated accessions of Lima bean from the Americas by restricting whole genomic DNA with the *Ape*KI enzyme (G↓CTG↑C), which in the chloroplast genome of common bean has 111 recognition sites. All 95 samples were analyzed in a single lane of Illumina HiSeq 2000/2500 (100 bp, single-end reads). The GBS data from all 95 individuals were mapped against the chloroplast genome of the common bean reported by Guo et al. (2007; GenBank number DQ886273.1) using the program NGSEP (Duitama et al. 2014). Only high-quality SNPs with a minimum distance of five bases among variants were retained.

A total of 64 high-quality SNPs were observed (Table 7.3) among the plastid genome of common bean and aligned GBS reads from Lima bean, and 32 of them were located, according to the common bean plastid genome, in coding regions of 15 genes; nine were found in the introns of the genes *rpo*C1, *clp*P, *pet*B, and *ndh*A genes, two SNPs were found in the pseudogene *rps*16, and 21 SNPs were located in ten spacers. Therefore, non-coding regions harbor twice as much polymorphism as coding regions, as expected in these types of comparisons (among closely related species). The gene that showed more SNPs was *acc*D (7 SNPs), which agrees with the observation of Guo et al. (2007) as this is the fastest evolving gene in common bean. The spacer that showed more variation was *atp*B-*rbc*L with five SNPs. Of the 64 SNPs, only seven of them were polymorphic within Lima bean, four were found in just one spacer (*atp*B-*rbc*L), and the other three were found in coding regions. As shown in the previous section, the *atp*B-*rbc*L spacer has shown to be very informative for evolutionary and domestication studies in Lima bean.

Although the GBS method was useful to discover SNPs between these two closely related species, a large amount of missing data was observed. In this GBS experiment, the coverage obtained was around 15×, which is suitable for analyses of SNPs from the nuclear genome but not for GBS analyses of the plastid genome. This is probably due to the tiny size of organelle

Table 7.3 Position of the 64 SNPs found among the chloroplast genomes of common bean and aligned GBS reads from Lima bean. Position is given in relation to the sequence reported for the chloroplast genome of common bean GenBank accession DQ886273.1

Position	Region	Name of region	Allele in *P. vulgaris*	Allele in *P. lunatus*
2911	spacer	*rpl*14-*trn*H	C	A
2922	spacer	*rpl*14-*trn*H	A	C
3034	spacer	*psb*A-*trn*H	T	G
3050	spacer	*psb*A-*trn*H	G	C
3235	spacer	*psb*A-*trn*H	G	T
9819	spacer	*atp*B-*rbc*L	A	A/C
9867	spacer	*atp*B-*rbc*L	T	T/G
9885	spacer	*atp*B-*rbc*L	A	G
9923	spacer	*atp*B-*rbc*L	C	A/C
9963	spacer	*atp*B-*rbc*L	A	A/T
10403	coding	*atp*B	G	A
10475	coding	*atp*B	G	T
11807	coding	*atp*B	C	T
14240	coding	*ndh*K	G	A
15390	spacer	*ndh*J-*trn*F	A	G
30976	coding	*psb*D	A	C
34448	spacer	*psb*M-*pet*N	T	G

(continued)

Table 7.3 (continued)

Position	Region	Name of region	Allele in *P. vulgaris*	Allele in *P. lunatus*
36796	coding	*rpo*B	T	G
40536	intron	*rpo*C1	A	T
40704	intron	*rpo*C1	A	C
40716	intron	*rpo*C1	T	A
52589	coding	*atp*A	G	A/G
52638	coding	*atp*A	C	T
55254	spacer	*trn*G-*trn*S	T	C
55281	spacer	*trn*G-*trn*S	G	T
55388	coding	*trn*S	C	A
55434	spacer	*trn*S-*psb*I	C	A
55896	spacer	*psb*I-*psb*K	T	C
55915	spacer	*psb*I-*psb*K	G	A
56069	spacer	*psb*I-*psb*K	G	A
57346	pseudogene	*rps*16	A	C
57402	pseudogene	*rps*16	G	T
57411	spacer	*rps*16-*acc*D	C	A
57420	spacer	*rps*16-*acc*D	A	G
58744	coding	*acc*D	G	A
58757	coding	*acc*D	C	T
58769	coding	*acc*D	A	G
58818	coding	*acc*D	C	T
58851	coding	*acc*D	T	C
58935	coding	*acc*D	T	G
58978	coding	*acc*D	T	A
69944	intron	*clp*P	T	C
70043	coding	*clp*P exon 2	C	A
70093	coding	*clp*P exon 2	G	T
71272	spacer	*clp*P-psbB	T	C
72032	coding	*psb*B	G	A
72052	coding	*psb*B	G	T/G
72151	coding	*psb*B	C	A
72160	coding	*psb*B	C	T/C
72171	coding	*psb*B	C	T
74282	intron	*pet*B	G	T
77290	coding	*rpo*A	C	A
97949	coding	*rrn*16	G	C
112560	coding	*psa*C	C	T
112580	coding	*psa*C	T	A
116406	intron	*ndh*A	A	C

(continued)

Table 7.3 (continued)

Position	Region	Name of region	Allele in *P. vulgaris*	Allele in *P. lunatus*
116435	intron	*ndh*A	C	G
116444	intron	*ndh*A	A	C
116450	intron	*ndh*A	G	T
117714	coding	*ndh*H	A	C
117795	coding	*ndh*H	G	A
119955	coding	*ycf*1	T	G
119966	coding	*ycf*1	C	A
126051	coding	*rrn*4.5	G	C

genomes compared to the nuclear genome, thus requiring a larger coverage. In GBS experiments, multiplexing fewer individuals in a single Illumina lane can easily increase coverage.

7.6 Conclusions

Organelle genomic resources for *Phaseolus* beans have just started to develop. One big step undoubtedly has been the sequencing in the year 2007 of the chloroplast genome of the common bean, providing a reference for future analyses in this species. However, ten years later and with new sequencing technologies available, genomic resources for the plastid genome of many other important species of *Phaseolus* are still lacking. Current efforts for the development of *Phaseolus* genomics resources are centered in the nuclear genome; therefore, it is of urgent need that the scientific community interested in *Phaseolus* beans makes a greater effort to advance in the field of organelle genomics of beans.

References

Alverson AJ, Zhuo S, Rice DW, Sloan DB, Palmer JD (2011) The mitochondrial genome of the legume *Vigna radiata* and the analysis of recombination across short mitochondrial repeats. PLoS ONE 6:e16404

Andueza-Noh RH, Serrano-Serrano ML, Chacón Sánchez MI, Sánchez del Pino I, Camacho-Pérez L, Coello-Coello J, Mijangos Cortes J, Debouck DG, Martínez-Castillo J (2013) Multiple domestications of the Mesoamerican gene pool of lima bean (*Phaseolus lunatus* L.): evidence from chloroplast DNA sequences. Genet Resour Crop Evol 60:1069–1086

Angioi SA, Desiderio F, Rau D, Bitocchi E, Attene G, Papa R (2009) Development and use of chloroplast microsatellites in *Phaseolus* spp. and other legumes. Plant Biol 11:598–612

Archibald JM, Keeling PJ (2002) Recycled plastids: a 'green movement' in eukaryotic evolution. Trends Genet 18:577–584

Bedbrook JR, Kolodner R (1979) The structure of chloroplast DNA. Annu Rev Plant Physiol 30:593–620

Bendich AJ (2004) Circular chloroplast chromosomes: the grand illusion. Plant Cell 16:1661–1666

Bendich AJ (2007) The size and form of chromosomes are constant in the nucleus, but highly variable in bacteria, mitochondria and chloroplasts. BioEssays 29:474–483

Bock R (2007) Structure, function, and inheritance of plastid genomes. In: Bock R (ed) Cell and molecular biology of plastids. Springer, pp 29–63

Borsch T, Quandt D (2009) Mutational dynamics and phylogenetic utility of noncoding chloroplast DNA. Plant Syst Evol 282:169–199

Bukhari YM, Koivu K, Tigerstedt PMA (1999) Phylogenetic analysis of *Acacia* (Mimosaceae) as revealed from chloroplast RFLP data. Theor Appl Genet 98:291–298

Burger G, Gray MW, Lang BF (2003) Mitochondrial genomes: anything goes. Trends Genet 19:709–716

Byrne M, Hankinson M (2012) Testing the variability of chloroplast sequences for plant phylogeography. Aust J Bot 60:569–574

Chacón SMI, Pickersgill B, Debouck DG (2005) Domestication patterns in common bean (*Phaseolus vulgaris* L.) and the origin of the Mesoamerican and Andean cultivated races. Theor App Genet 110:432–444
Chacón SMI, Pickersgill B, Debouck DG, Arias JS (2007) Phylogeographic analysis of the chloroplast DNA variation in wild common bean (*Phaseolus vulgaris* L.) in the Americas. Plant Syst Evol 266:175–195
Chang S, Wang Y, Lu J, Gai J, Li J, Chu P, Guan R, Zhao T (2013) The mitochondrial genome of soybean reveals complex genome structures and gene evolution at intercellular and phylogenetic levels. PLoS ONE 8: e56502
Chen X, Li Q, Li Y, Qian J, Han J (2015) Chloroplast genome of *Aconitum barbatum* var. *puberulum* (Ranunculaceae) derived from CCS reads using the PacBio RS platform. Front Plant Sci 6:42
Coleman AW (1982) Sex is dangerous in a world of potential symbionts or the basis of selection for uniparental inheritance. J Theor Biol 97:367–369
Corriveau JL, Coleman AW (1988) Rapid screening method to detect potential biparental inheritance of plastid DNA and results for over 200 angiosperm species. Am J Bot, 1443–1458
Corriveau JL, Coleman AW (1991) Monitoring by epifluorescence microscopy of organelle DNA fate during pollen development in five angiosperm species. Develop Biol 147:271–280
Cronn R, Liston A, Parks M, Gernandt DS, Shen R, Mockler T (2008) Multiplex sequencing of plant chloroplast genomes using Solexa sequencing-by-synthesis technology. Nucl Acids Res 36:e122
Desiderio F, Bitocchi E, Bellucci E, Rau D, Rodriguez M, Attene G, Papa R, Nanni L (2012) Chloroplast microsatellite diversity in *Phaseolus vulgaris*. Front Plant Sci 3:312
Dong W, Liu J, Yu J, Wang L, Zhou S (2012) Highly variable chloroplast markers for evaluating plant phylogeny at low taxonomic levels and for DNA barcoding. PLoS ONE 7:e35071
Doyle JJ, Doyle JL, Ballenger JA, Palmer JD (1996) The distribution and phylogenetic significance of a 50 kb chloroplast DNA inversion in the flowering plant family Leguminosae. Mol Phylogenet Evol 5:429–438
Doyle JJ, Doyle JL, Palmer JD (1995) Multiple independent losses of two genes and one intron from legume chloroplast genomes. Syst Bot 20:272–294
Duitama J, Quintero JC, Cruz DF, Quintero C, Hubmann G, Foulquié-Moreno MR, Verstrepen KJ, Thevelein JM, Tohme J (2014) An integrated framework for discovery and genotyping of genomic variants from high-throughput sequencing experiments. Nucl Acids Res 42:e44–e44
Dyall SD, Brown MT, Johnson PJ (2004) Ancient invasions: from endosymbionts to organelles. Science 304:253–257
Eberhard WG (1980) Evolutionary consequences of intracellular organelle competition. Quart Rev Biol 55:231–249
Ferrarini M, Moretto M, Ward JA, Šurbanovski N, Stevanović V, Giongo L, Viola R, Cavalieri D, Velasco R, Cestaro A, Sargent DJ (2013) An evaluation of the PacBio RS platform for sequencing and *de novo* assembly of a chloroplast genome. BMC Genom 14:670–670
Fofana B, Baudoin JP, Vekemans X, Debouck DG, Du Jardin P (1999) Molecular evidence for an Andean origin and a secondary gene pool for the Lima bean (*Phaseolus lunatus* L.) using chloroplast DNA. Theor Appl Genet 98:202–212
Fofana B, Du Jardin P, Baudoin JP (2001) Genetic diversity in the Lima bean (*Phaseolus lunatus* L.) as revealed by chloroplast DNA (cpDNA) variations. Genet Resour Crop Evol 48:437–445
Gao L, Su Y-J, Wang T (2010) Plastid genome sequencing, comparative genomics, and phylogenomics: current status and prospects. J Syst Evol 48:77–93
Gielly L, Taberlet P (1994) The use of chloroplast DNA to resolve plant phylogenies: noncoding versus *rbcL* sequences. Mol Biol Evol 11:769–777
Godelle B, Reboud X (1995) Why are organelles uniparentally inherited? Proc Roy Soc B-Biol Sci 259:27–33
Goulding SE, Wolfe KH, Olmstead RG, Morden CW (1996) Ebb and flow of the chloroplast inverted repeat. Mol Gen Genet 252:195–206
Guo X, Castillo-Ramirez S, Gonzalez V, Bustos P, Fernandez-Vazquez JL, Santamaria RI, Arellano J, Cevallos MA, Davila G (2007) Rapid evolutionary change of common bean (*Phaseolus vulgaris* L.) plastome, and the genomic diversification of legume chloroplasts. BMC Genom 8:228
Jackman SD, Warren RL, Gibb EA, Vandervalk BP, Mohamadi H, Chu J, Raymond A, Pleasance S, Coope R, Wildung MR, Ritland CE, Bousquet J, Jones SJM, Bohlmann J, Birol I (2016) Organellar genomes of white spruce (*Picea glauca*): assembly and annotation. Genome Biol Evol 8:29–41
Jansen RK, Cai Z, Raubeson LA, Daniell H, Depamphilis CW, Leebens-Mack J, Muller KF, Guisinger-Bellian M, Haberle RC, Hansen AK, Chumley TW, Lee SB, Peery R, McNeal JR, Kuehl JV, Boore JL (2007) Analysis of 81 genes from 64 plastid genomes resolves relationships in angiosperms and identifies genome-scale evolutionary patterns. Proc Natl Acad Sci USA 104:19369–19374
Jansen RK, Ruhlman TA (2012) Plastid genomes of seed plants. In: Bock R, Knoop V (eds) Genomics of Chloroplasts and Mitochondria. Springer, pp 103–126
Jansen RK, Wojciechowski MF, Sanniyasi E, Lee S-B, Daniell H (2008) Complete plastid genome sequence of the chickpea (*Cicer arietinum*) and the phylogenetic distribution of *rps12* and *clpP* intron losses among legumes (Leguminosae). Mol Phylogenet Evol 48:1204–1217
Kazakoff SH, Imelfort M, Edwards D, Koehorst J, Biswas B, Batley J, Scott PT, Gresshoff PM (2012) Capturing the biofuel wellhead and powerhouse: the chloroplast and mitochondrial genomes of the

leguminous feedstock tree *Pongamia pinnata*. PLoS ONE 7:e51687

Knoop V (2012) Seed plant mitochondrial genomes: complexity evolving. In: Bock R, Knoop V (eds) Genomics of Chloroplasts and Mitochondria. Springer, Netherlands, pp 175–200

Knoop V, Volkmar U, Hecht J, Grewe F (2011) Mitochondrial genome evolution in the plant lineage. In: Kempken F (ed) Plant Mitochondria. Springer, New York, pp 3–29

Kochert G, Stalker HT, Gimenes M, Galgaro L, Lopes CR, Moore K (1996) RFLP and cytogenetic evidence on the origin and evolution of allotetraploid domesticated peanut, *Arachis hypogaea* (Leguminosae). Am J Bot 83:1282–1291

Kolodner R, Tewari KK (1979) Inverted repeats in chloroplast DNA from higher plants. Proc Natl Acad Sci USA 76:41–45

Kubo T, Mikami T (2007) Organization and variation of angiosperm mitochondrial genome. Physiol Plantarum 129:6–13

Kubo T, Newton KJ (2008) Angiosperm mitochondrial genomes and mutations. Mitochondrion 8:5–14

Lang BF, Burger G, O'Kelly CJ, Cedergren R, Golding GB, Lemieux C, Sankoff D, Turmel M, Gray MW (1997) An ancestral mitochondrial DNA resembling a eubacterial genome in miniature. Nature 387:493–497

Levings CS III, Brown GG (1989) Molecular biology of plant mitochondria. Cell 56:171–179

Li Q, Li Y, Song J, Xu H, Xu J, Zhu Y, Li X, Gao H, Dong L, Qian J, Sun C, Chen S (2014) High-accuracy *de novo* assembly and SNP detection of chloroplast genomes using a SMRT circular consensus sequencing strategy. New Phytol 204:1041–1049

Lilly JW, Havey MJ, Jackson SA, Jiang J (2001) Cytogenomic analyses reveal the structural plasticity of the chloroplast genome in higher plants. Plant Cell 13:245–254

Liu Y, Wang B, Li L, Qiu Y-L, Xue J (2012) Conservative and dynamic evolution of mitochondrial genomes in early land plants. In: Bock R, Knoop V (eds) Genomics of Chloroplasts and Mitochondria. Springer, Netherlands, pp 159–174

Lohse M, Drechsel O, Kahlau S, Bock R (2013) OrganellarGenomeDRAW—a suite of tools for generating physical maps of plastid and mitochondrial genomes and visualizing expression data sets. Nucl Acids Res 41:W575–W581

Magee AM, Aspinall S, Rice DW, Cusack BP, Semon M, Perry AS, Stefanovic S, Milbourne D, Barth S, Palmer JD, Gray JC, Kavanagh TA, Wolfe KH (2010) Localized hypermutation and associated gene losses in legume chloroplast genomes. Genome Res 20:1700–1710

Marechal A, Brisson N (2010) Recombination and the maintenance of plant organelle genome stability. New Phytol 186:299–317

McPherson H, van der Merwe M, Delaney SK, Edwards MA, Henry RJ, McIntosh E, Rymer PD, Milner ML, Siow J, Rossetto M (2013) Capturing chloroplast variation for molecular ecology studies: a simple next generation sequencing approach applied to a rainforest tree. BMC Ecol 13:8

Moore MJ, Triplett EW, Broughton WJ, Soltis PS, Soltis DE (2007) Complete nucleotide sequence of the plastid genome of the common bean, *Phaseolus vulgaris*. EU196765.1

Motta-Aldana JR, Serrano-Serrano ML, Hernández--Torres J, Castillo-Villamizar G, Debouck DG, Chacón SMI (2010) Multiple origins of lima bean landraces in the Americas: evidence from chloroplast and nuclear DNA polymorphisms. Crop Sci 50:1773–1787

Naito K, Kaga A, Tomooka N, Kawase M (2013) *De novo* assembly of the complete organelle genome sequences of azuki bean (*Vigna angularis*) using next-generation sequencers. Breeding Sci 63:176–182

Negruk V (2013) Mitochondrial genome sequence of the legume *Vicia faba*. Front Plant Sci 4:128

Ohyama K (1996) Chloroplast and mitochondrial genomes from a liverwort, *Marchantia polymorpha*–gene organization and molecular evolution. Biosci Biotechnol Biochem 60:16–24

Palmer JD (1983) Chloroplast DNA exists in two orientations. Nature 301:92–93

Palmer JD (1985) Comparative organization of chloroplast genomes. Annu Rev Genet 19:325–354

Palmer JD (2003) The symbiotic birth and spread of plastids: how many times and whodunit? J Phycol 39:4–12

Palmer JD, Herbon LA (1988) Plant mitochondrial DNA evolves rapidly in structure, but slowly in sequence. J Mol Evol 28:87–97

Palmer JD, Jansen RK, Michaels HJ, Chase MW, Manhart JR (1988) Chloroplast DNA variation and plant phylogeny. Ann Missouri Botan Garden 75:1180–1206

Palmer JD, Osorio B, Aldrich J, Thompson WF (1987) Chloroplast DNA evolution among legumes: loss of a large inverted repeat occurred prior to other sequence rearrangements. Curr Genet 11:275–286

Parks M, Cronn R, Liston A (2009) Increasing phylogenetic resolution at low taxonomic levels using massively parallel sequencing of chloroplast genomes. BMC Biol 7:84

Sanitá Lima M, Woods LC, Cartwright MW, Smith DR (2016) The (in)complete organelle genome: exploring the use and nonuse of available technologies for characterizing mitochondrial and plastid chromosomes. Mol Ecol Resour. doi:10.1111/1755-0998.12585

Serrano-Serrano ML, Hernández-Torres J, Castillo-Villamizar G, Debouck DG, Chacón Sánchez MI (2010) Gene pools in wild Lima bean (*Phaseolus lunatus* L.) from the Americas: evidences for an Andean origin and past migrations. Mol Phylogenet Evol 54:76–87

Shinozaki K, Ohme M, Tanaka M, Wakasugi T, Hayashida N, Matsubayashi T, Zaita N, Chunwongse J, Obokata J, Yamaguchi-Shinozaki K, Ohto C, Torazawa K, Meng BY, Sugita M, Deno H,

Kamogashira T, Yamada K, Kusuda J, Takaiwa F, Kato A, Tohdoh N, Shimada H, Sugiura M (1986) The complete nucleotide sequence of the tobacco chloroplast genome: its gene organization and expression. EMBO J 5:2043–2049

Smith SE, Bingham ET, Fulton RW (1986) Transmission of chlorophyll deficiencies in *Medicago sativa*: evidence for biparental inheritance of plastids. J Hered 77:35–38

Strauss SH, Palmer JD, Howe GT, Doerksen AH (1988) Chloroplast genomes of two conifers lack a large inverted repeat and are extensively rearranged. Proce Natl Acad Sci USA 85:3898–3902

Stull GW, Moore MJ, Mandala VS, Douglas NA, Kates H-R, Qi X, Brockington SF, Soltis PS, Soltis DE, Gitzendanner MA (2013) A targeted enrichment strategy for massively parallel sequencing of angiosperm plastid genomes. Appl Plant Sci 1:1200497

Taberlet P, Gielly L, Pautou G, Bouvet J (1991) Universal primers for amplification of three non-coding regions of chloroplast DNA. Plant Mol Biol 17:1105–1109

Vachon N, Freeland JR (2011) Phylogeographic inferences from chloroplast DNA: quantifying the effects of mutations in repetitive and non-repetitive sequences. Mol Ecol Resour 11:279–285

Vekemans X, Hardy O, Berken B, Fofana B, Baudoin J-P (1998) Use of PCR-RFLP on chloroplast DNA to investigate phylogenetic relationships in the genus *Phaseolus*. Biotechnol Agron Soc Environ 2:128–134

Wicke S, Schneeweiss GM, dePamphilis CW, Muller KF, Quandt D (2011) The evolution of the plastid chromosome in land plants: gene content, gene order, gene function. Plant Mol Biol 76:273–297

Wu Z, Gui S, Quan Z, Pan L, Wang S, Ke W, Liang D, Ding Y (2014) A precise chloroplast genome of *Nelumbo nucifera* (Nelumbonaceae) evaluated with Sanger, Illumina MiSeq, and PacBio RS II sequencing platforms: insight into the plastid evolution of basal eudicots. BMC Plant Biol 14:289

Zhang Q (2010) Why does biparental plastid inheritance revive in angiosperms? J Plant Res 123:201–206

Phytic Acid Biosynthesis and Transport in *Phaseolus vulgaris*: Exploitation of New Genomic Resources

8

Eleonora Cominelli, Gregorio Orozco-Arroyo and Francesca Sparvoli

Abstract

Although common bean has a good content in essential minerals, it also accumulates significant amounts of compounds that reduce its nutritional value by lowering nutrient bioavailability. Phytic acid, which is the major form of phosphorus stored in the seed, is one of such compounds, as, during gastro-intestinal passage, it binds trace elements and reduces their absorption, leading, under certain dietary circumstances, to mineral (mostly Fe, Zn, Ca) deficiencies. A major goal for grain crop improvement is the reduction of phytic acid content in the seed to improve micronutrient bioavailability, through the identification of *low phytic acid* (*lpa*) mutants. In common bean only one of such mutants has been described so far. Genes involved in phytic acid pathway and transport have been described in different species, including common bean. Recently, new genomic resources have become available for the common bean research community, thanks to the release of two whole genome sequences: the Andean G19833 and the Mesoamerican BAT93 genotypes. In this chapter we use the two common bean reference genomes to compare the sequences of genes involved or putatively involved in phytic acid synthesis and transport, some of them never reported in this species. Moreover, we discuss transcriptomic data of these genes, reported in different organs at different developmental stages for the Mesoamerican genotype. Finally, we discuss alternatives on how to exploit these new genomic resources to study and eventually manipulate phytic acid pathway and transport.

E. Cominelli (✉) · G. Orozco-Arroyo ·
F. Sparvoli (✉)
Institute of Agricultural Biology and Biotechnology, CNR, Via Bassini 15, 20133 Milan, Italy
e-mail: cominelli@ibba.cnr.it

F. Sparvoli
e-mail: sparvoli@ibba.cnr.it

G. Orozco-Arroyo
e-mail: gorozco@unam.mx

M. Pérez de la Vega et al. (eds.), *The Common Bean Genome*,
Compendium of Plant Genomes, https://doi.org/10.1007/978-3-319-63526-2_8

Keywords
Inositols · *lpa* · *Myo*-inositol · Phosphorous · Phytic acid · SNPs

8.1 Introduction

Common bean (*Phaseolus vulgaris* L.) is a highly important source of nutrients for more than 300 million people in developing countries, mainly in Eastern Africa, Central and South America, representing 65% of total protein consumed and 32% of energy (Broughton et al. 2003; Blair et al. 2010). In spite of the high iron concentration in common bean (average 55 μg/g), severe iron and zinc deficiencies are quite common in countries where beans are preponderant on the diet, these deficiencies convey to stunted growth, decreased immune function and anaemia. In the last years research on nutritional improvement of common bean has focused in increasing micronutrient content and to decrease antinutritional factors. In fact, the Harvest Plus programme, an initiative of the Consultative Group for International Agricultural Research (CGIAR), chose common bean as one of the target species to be iron biofortified (HarvestPlus website, Iron beans). Exploiting natural variation, bean varieties with iron concentration ranging from 92 to 99 μg/g have been developed through breeding, as recently reviewed by Petry et al. (2015). However, the main constriction against bean iron biofortification is the permanence of low levels of iron absorption, mainly attributed to the presence of high concentrations of phytic acid (*myo*-inositol-1,2,3,4,5,6-hexakisphosphate, PA). In plants, PA plays a major role for storage of phosphorous and minerals in the seed, which become available to the seedling immediately after seed germination, due to the activity of phytases. PA is highly negatively charged at physiological pH, it binds important mineral cations such as iron, zinc, potassium, calcium, and magnesium and easily precipitates in the form of phytate salts. Non-ruminants poorly digest PA due to the lack of phytases in their digestive tract. High molar ratios between PA and mineral cations are ascribed as one of the most important causes of mineral deficiencies in those populations whose diet is largely based on staple crops (Raboy 2001; Schlemmer et al. 2009). In this way, PA decreases the nutritional value of the seeds by limiting mineral bioavailability for human nutrition. Moreover, the inability of non-ruminant livestock to use PA as a source of phosphorous, necessitates supplementation of grain-based feeds with phosphorous and/or phytases. Consequent excretion of undigested PA and PA-derived phosphorus in animal waste may contribute to the pollution of ground and surface waters (Leytem and Maguire 2007).

In mature seeds, phytate is stored into spherical inclusions called globoids, which are in turn found within protein bodies. Phytate deposits are also observed to occur transiently in various tissues and subcellular compartments during grain development (Greenwood et al. 1984; Otegui et al. 2002).

In common bean, PA is mainly accumulated in the cotyledons (95–98%) and only a small portion is located in the embryo (1–3%) and the seed coat (0.5–4%), reaching 3% of total seed weight (Blair et al. 2012; Petry et al. 2015). During common bean seed development, PA is accumulated between 17 and 26 days after flowering (DAF), remaining rather constant until maturation (Coelho et al. 2005). PA concentration in common bean vary in wild or cultivated germplasm as well as in segregating populations (Guzman-Maldonado et al. 2003; Blair et al. 2009) and ranges from 4 to 26 mg/g, with a mean content of about 10 mg/g and a PA: iron molar ratio ranging from 6:1 to 33:1 (Petry et al. 2015). The PA level depends on the bean variety, but also on phosphorous concentration in the soil (Blair et al. 2012).

The development of *low phytic acid* (*lpa*) seeds in different crops has been proposed as a strategy that could offer both nutritional and environmental benefits (Raboy 2007, 2009). Different *lpa* mutants have been isolated in different crops and in model species, through forward and reverse genetics approaches (Sparvoli and Cominelli 2015). Unfortunately many of these mutants are of a limited value to breeders, because they show negative pleiotropic effects, such as low germination rates, reduced seed development and weight, stunted vegetative growth (Raboy 2009). The reason for these phenotypic alterations relies on the key role played by PA and its precursors in different plant developmental and physiological processes, including signal transduction, photomorphogenesis, sugar signalling, phosphorus homeostasis, hormone signalling (auxin, abscisic acid, jasmonic acid), membrane trafficking, abiotic and biotic stress response, chromatin modification and remodelling, and mRNA nuclear export (Sparvoli and Cominelli 2015).

However, some concerns with the use of biofortified crops exist such as the evidence that PA is a broad-spectrum antineoplastic agent acting in different steps of cancer development and progression, considered as a great promise in strategies for cancer prevention and therapy. So far, the PA protective role in cancer prevention and development has been only tested in vitro in different cancer cell models or in animal trials in which purified PA has been administered (Norazalina et al. 2010) and no study has been conducted using diets based on foods containing different concentrations of PA.

In *P. vulgaris*, only one *lpa* mutant was reported so far. It is affected in a PA ABC-transporter and does not show the negative pleiotropic effects, described for other *lpa* mutants (Campion et al. 2009; Panzeri et al. 2011). Moreover, it was used in a study on volunteers showing that iron absorption from the *lpa* beans is significantly higher than from non-mutant beans with normal PA levels (Petry et al. 2013), highlighting the potentiality of the common bean as a vehicle for iron biofortification (Petry et al. 2015).

Either any manipulation of PA content in the seed, directed towards increasing or lowering PA, requires knowledge of the key enzymes involved in its biosynthetic pathway. The majority of the genes of the PA pathway in *P. vulgaris* were identified and sequenced from the cv. Taylor's Horticultural (Asgrow), and from the breeding line 905 (Mesoamerican gene pool) and mapped on the common bean reference genetic map of McClean (NDSU) 2007, using an in silico mapping strategy against the soybean genome (Fileppi et al. 2010). In another study, these genes were mapped through a mapping population derived from a cross between an Andean (G19833) and a Mesoamerican (DOR364) parents, and some molecular markers for seed PA and phosphorous content were developed (Blair et al. 2012). Moreover, two common bean PA-ABC family transporters have been described (Panzeri et al. 2011).

8.2 Lipid-Dependent and Lipid-Independent Pathways

As shown in Fig. 8.1, PA may be synthesized through two different routes: (1) the lipid-independent pathway that consists in the sequential phosphorylation of the 6-carbon cyclic alcohol *myo*-inositol (Ins) and soluble inositol phosphates (InsPs); and (2) the lipid-dependent pathway that uses precursors that include phosphatidylinositol (PtdIns) and PtdIns phosphates. This last route is present in most eukaryotic cells, including plant vegetative tissues, while the first pathway appears to be predominant in seeds (Raboy 2009). These two pathways only differ in their early intermediate steps leading from Ins to $InsP_3$. In the lipid-dependent pathway a phosphatidylinositol synthase (PtdIS) is required to convert Ins to PtdIns. Sequential phosphorylations of the headgroup of PtdIns by phosphatidylinositol kinases produce PtdIns(4,5)P_2. A PtdIns-specific phospholipase C uses this molecule as substrate and releases Ins(1,4,5)P_3, a central molecule to signal transduction. The lipid-independent pathway consists of sequential

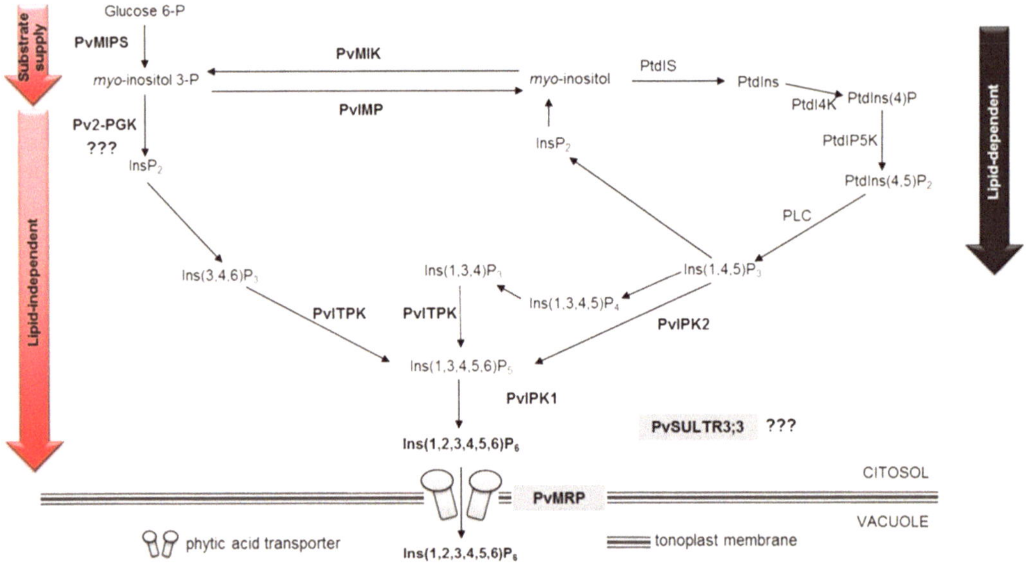

Fig. 8.1 Schematic representation of PA synthesis and transport in common bean. The substrate supply, lipid-independent and lipid-dependent sub-pathways for PA synthesis are indicated. PvMIPS, *myo*-inositol-3-phosphate synthase; PvIMP, bifunctional enzyme: *myo*-inositol-phosphate monophosphatase and galactose-1-phosphate phosphatase; PvMIK, *myo*-inositol kinase; Pv2-PGK, 2-phosphoglycerate kinase; PvIPK2, inositol 1,4,5-tris-phosphate kinase; PvITPK, inositol 1,3,4-triphosphate 5⁄6-kinase; PvIPK1, inositol 1,3,4,5,6 penta*kis*phosphate 2-kinase; PtdIS, phosphatidyl inositol phosphate synthase; PtdI4 K, phosphatidyl inositol 4-kinase; PtdI5 K phosphatidyl inositol 5-kinase, PtdIns phosphatidyl inositol; PtdIns(4)P_1, phosphatidyl inositol 4-phosphate; PtdIns(4,5)P_2, phosphatidyl inositol 4,5-biphosphate; PLC, phospholipase; PvMRP1, multidrug-resistance-associated protein PA transporter; PvSULTR3;3, sulfate transporter class 3;3. Question marks indicate that the function for these proteins is hypothetical

phosphorylation of the Ins ring to $InsP_6$, through the action of a number of specific inositol phosphate kinases. Both pathways can be split in three phases: an early phase or substrate supply pathway consisting in the production of Ins(3)P_1, Ins, or PtdIns; an early intermediate phase that generates $InsP_3$; and the common late inositol polyphosphate pathways which convert $InsP_3$ to PA (Raboy 2003; Rasmussen et al. 2010). Mutations in the different genes of the pathway may affect different aspects of PA metabolism, depending on the action timing of the different enzymes. Two main types of mutants are normally recognized: mutants affected in genes coding for 1D-*myo*-inositol-3-phosphate synthase (*MIPS*), *myo*-inositol kinase (*MIK*) and *myo*-inositol-phosphate monophosphatase (*IMP*), involved in the early and intermediate steps of the pathway, are characterized by a decrease in PA content, accompanied by a molar increase in free phosphate; meanwhile mutants affected in genes coding for enzymes involved in the late pathway, such as 2-phosphoglycerate kinase (*2-PGK*), inositol 1,4,5-tris-phosphate kinase (*IPK2*), inositol 1,3,4-triphosphate 5⁄6-kinase (*ITPK*) and inositol 1,3,4,5,6 penta*kis*phosphate 2-kinase (*IPK1*), show decreased PA content accompanied by a low increase in free phosphate and increase content of lower $InsP_s$ (Sparvoli and Cominelli 2015).

Interestingly, PA is synthesized in two cellular compartments: the nucleus and the cytoplasm (Raboy 2009). The cytoplasmic pathway probably contributes most to net seed PA synthesis, as shown through the cytoplasmic-targeted overexpression of a phytase active during seed development that abolishes seed PA accumulation (Bilyeu et al. 2008). The nuclear pool may mainly contribute to the regulatory roles of PA and its precursors. Thus, different enzymes of the pathway are targeted to the cytoplasm or the nucleus or to both compartments (Xia et al. 2003). However, it is not clear the relative contributions of the two pathways and their distribution between cell compartments (Raboy 2009).

As the lipid-independent route is of main interest for the manipulation of PA synthesis in seed, we will describe in details common bean genes involved in this pathway, with the exception of *IPK2* for which a specific role in PA synthesis in the seed was clearly highlighted only for the model species *Arabidopsis thaliana* (Stevenson-Paulik et al. 2005).

For all the genes involved in PA synthesis and transport, already described in common bean (Fileppi et al. 2010; Panzeri et al. 2011), we identified and analysed the corresponding sequences in the Andean G19833 genome sequence (*P. vulgaris* v1.0), deposited on Phytozome v11.0 portal (Phytozome v11.0 website) and in the Mesoamerican BAT93 genome (*P. vulgaris* BAT93), available at the CoGe database (CoGe website) and through BLAST search in PhylomeDB (PhylomedDB website), where the complete collection of evolutionary histories of bean genes was reconstructed (Vlasova et al. 2016). Moreover, through TBLASTN analysis on both genomes and through Phylome DB, using as queries proteins already identified in common bean, we confirmed the copy number for each class of enzymes and transporters or found new members of the group. In order to find the common bean genes encoding for 2-PGK and for the sulfate transporter class 3;3 (SULTR3;3), not yet described before, we performed TBLASTN analysis, using as queries protein sequences described in other species. The correspondence of all the genes in both genomes is reported in Table 8.1.

8.2.1 *PvMIPSs* and *PvMIPSv*

PA biosynthesis needs the de novo production of Ins through a highly conserved reaction, shared by all living organisms, in which the enzyme 1D-*myo*-inositol 3-phosphate synthase, also known as 1L-*myo*-inositol-1-phosphate synthase (MIPS), converts D-glucose-6-phosphate to 1D-*myo*-inositol-3-phosphate, designated also as 1L-*myo*-inositol-1-phosphate, depending on the counterclockwise or clockwise numbering of carbon atoms in the ring, or simply Ins(3)P_1, where the numbering is counterclockwise, starting from 1D position (Loewus and Murthy 2000). Here we will use this last simplified nomenclature for different InsPs. The reaction catalyzed by MIPS is the rate-limiting step in the synthesis of all inositol-containing compounds (Raboy 2009; Rasmussen et al. 2010).

The copy number of *MIPS* genes and the expression pattern in different plant organs varies in different species, and consequently varies the possibility that a mutation in one of these loci may confer a *lpa* phenotype to the seeds (Sparvoli and Cominelli 2015). For example, in rice only one *MIPS* gene was described and transgenic seeds expressing its antisense showed *lpa* phenotype (Kuwano et al. 2009). In *A. thaliana* genome, three *MIPS* genes are present and no single knock out mutant has *lpa* seeds, suggesting a redundant function of these three genes in this species (Kim and Tai 2011).

In common bean two different *MIPS* genes have been described: *PvMIPSs* (*MIPS* seed), expressed at early stages of pod development and during seed development, and *PvMIPSv* (*MIPS* vegetative) highly expressed in vegetative tissues and only at low levels in seeds (Fileppi et al. 2010).

We confirmed the existence of only these two copies of *MIPS* genes on both bean reference genomes with a similar genomic structure, consisting in 10 exons and 9 introns. The main difference between *PvMIPSs* and *PvMIPSv* genes consists in the length of the third intron, longer in *PvMIPSv* than in *PvMIPSs*. *PvMIPSs* and *PvMIPSv* coding sequences (CDS) as well as the corresponding proteins show very high sequence identity, 88 and 93%, respectively. *PvMIPSs* CDS of the two reference genotypes differ only for five single nucleotide polymorphisms (SNPs), three silent and two neutral at protein level, while intron sequences are more different, particularly in the fourth and sixth introns. Since natural

Table 8.1 Transcript names of genes from G19833 and BAT93 genomes, used in this study

Gene name	G19833	BAT93
PvMIPSs	Phvul.001G251000.1	PHASIBEAM10F002132T1
PvMIPSv	Phvul.002G261700.1	PHASIBEAM10F005171T6 (T1)
PvIMPα	Phvul.006G142400.1	PHASIBEAM10F001370T5 (T2)
PvIMPβ	Phvul.003G084500.1	PHASIBEAM10F010830T2
PvMIK	Phvul.008G261400.1	PHASIBEAM10F008479T1
Pv2PGK	Phvul.011G082700.1	PHASIBEAM10F002348T7 (T1)
PvIPK2	Phvul.011G214900.1	PHASIBEAM10F026266T1
PvITPKα1	Phvul.003G200900.1	PHASIBEAM10F025221T1 (T2)
PvITPKα2	Phvul.003G284100.1	PHASIBEAM10F004219T1
PvITPKα3	Phvul.002G139700.1	n.i.
PvITPKβ1	Phvul.009G113600.1	PHASIBEAM10F013400T1
PvITPKβ2	Phvul.007G047700.1	PHASIBEAM10F012326T2
PvITPKγ	Phvul.001G232000.1	PHASIBEAM10F007821T1
PvIPK1α	Phvul.009G060100.1	PHASIBEAM10F027025T10
PvIPK1β	Phvul.001G012200.1	PHASIBEAM10F010299T3 (T11)
PvMRP1	Phvul.001G165500.1	PHASIBEAM10F011179T3 (T2)
PvMRP2	Phvul.007G153800.1	PHASIBEAM10F003020T1
PvSULTR3;3α	Phvul.002G095300.1	PHASIBEAM10F020374T2
PvSULTR3;3β	Phvul.002G095200.1	PHASIBEAM10F020373T1

For each gene we identified the primary transcript in the G19833 genome. Among the different PCGs annotated in the BAT93 genome, we identified the ones that, when translated, are the most similar to the corresponding G19833 proteins. In those cases in which the BAT93 sequence versions, obtained after the TBLAST on Phylome Database, have not the highest similarity with the G19833 homologous sequence, the different versions are indicated in parenthesis

variability in PA levels is controlled by *PvMIPSs*, as shown in a QTL study (Blair et al. 2012), the identification of the SNPs and of the two insertions/deletions on the intronic regions between the two reference genotypes, may be useful to the development of molecular markers in order to distinguish the two gene pools (Fig. 8.2).

The *PvMIPSv* CDS from both reference genomes differs only for 5 SNPs, while the proteins are identical. Some more differences are present in the intron sequences, particularly the two sequences for the third intron share only 85% of identity (Fig. 8.2).

Gene expression data showed that *PvMIPSs* gene is highly expressed in young pods (from 4 and 6 days after flowering, DAF) and in cotyledons at 12 DAF, similarly to what observed in other species like rice (Yoshida et al. 1999) and soybean (Chiera and Grabau 2007), then its expression level decreased (Fileppi et al. 2010). Therefore, its highest expression level preceded the synthesis and accumulation of inositol phosphates ($InsP_3$ to $InsP_5$) that starts around 12 DAF and peaks at 21 DAF (Coelho et al. 2005). *PvMIPSs* resulted to be expressed at low level also in vegetative tissues, where *PvMIPSv* was highly expressed (Fileppi et al. 2010). Transcriptomic data derived from RNA-seq analysis, performed on the BAT93 genotype (Fig. 8.3), are generally in accordance with this expression pattern and show that *PvMIPSs* is the gene of the pathway expressed to the highest levels, particularly in the pods at different developmental stages. Interestingly, *PvMIPSs* is highly expressed not only in the developing seed, but also in the developing pod, as revealed by the analysis

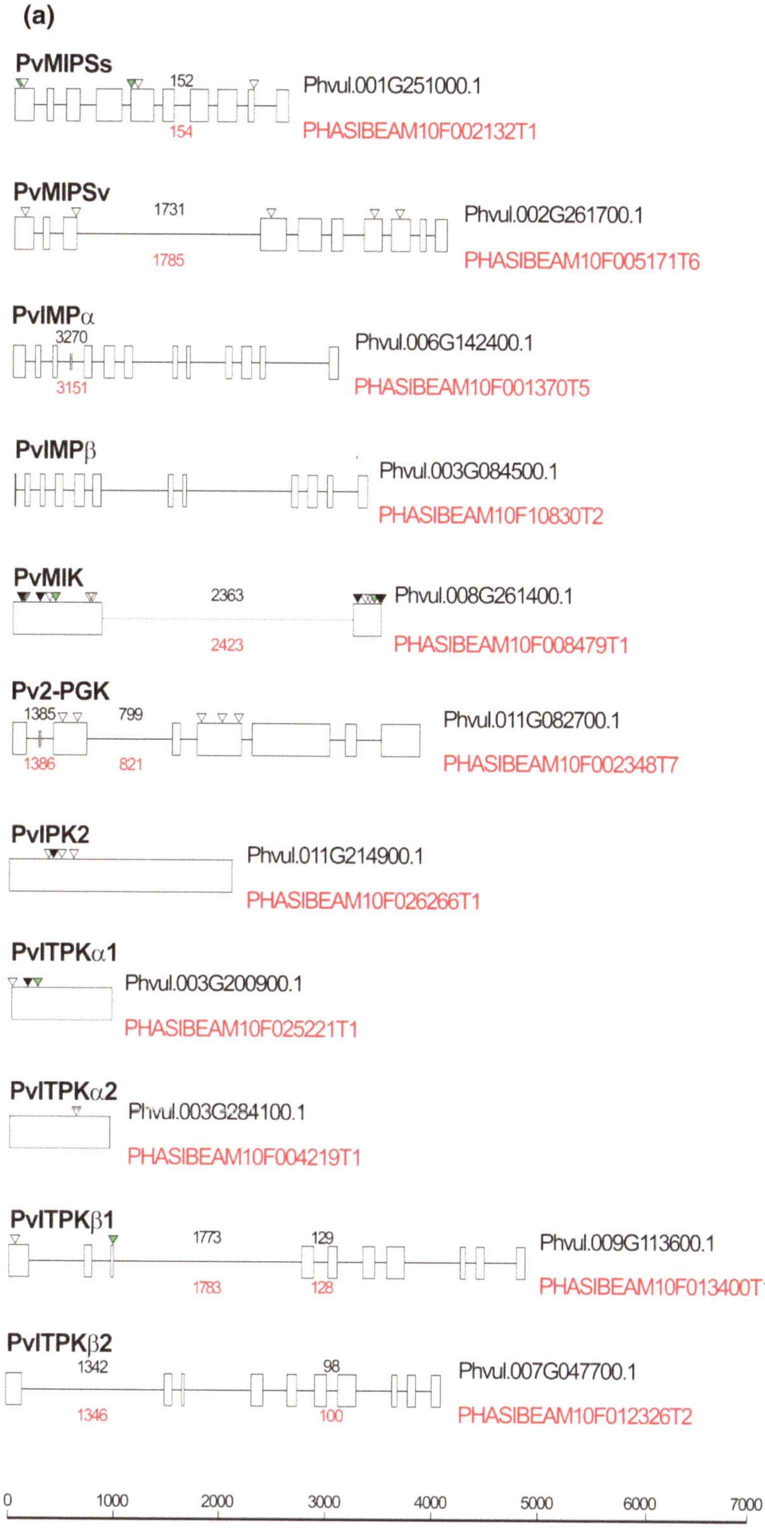
(a)
PvMIPSs
152
154
Phvul.001G251000.1
PHASIBEAM10F002132T1
PvMIPSv
1731
1785
Phvul.002G261700.1
PHASIBEAM10F005171T6
PvIMPα
3270
3151
Phvul.006G142400.1
PHASIBEAM10F001370T5
PvIMPβ
Phvul.003G084500.1
PHASIBEAM10F10830T2
PvMIK
2363
2423
Phvul.008G261400.1
PHASIBEAM10F008479T1
Pv2-PGK
1385
799
1386
821
Phvul.011G082700.1
PHASIBEAM10F002348T7
PvIPK2
Phvul.011G214900.1
PHASIBEAM10F026266T1
PvITPKα1
Phvul.003G200900.1
PHASIBEAM10F025221T1
PvITPKα2
Phvul.003G284100.1
PHASIBEAM10F004219T1
PvITPKβ1
1773
129
1783
128
Phvul.009G113600.1
PHASIBEAM10F013400T1
PvITPKβ2
1342
98
1346
100
Phvul.007G047700.1
PHASIBEAM10F012326T2
0
1000
2000
3000
4000
5000
6000
7000

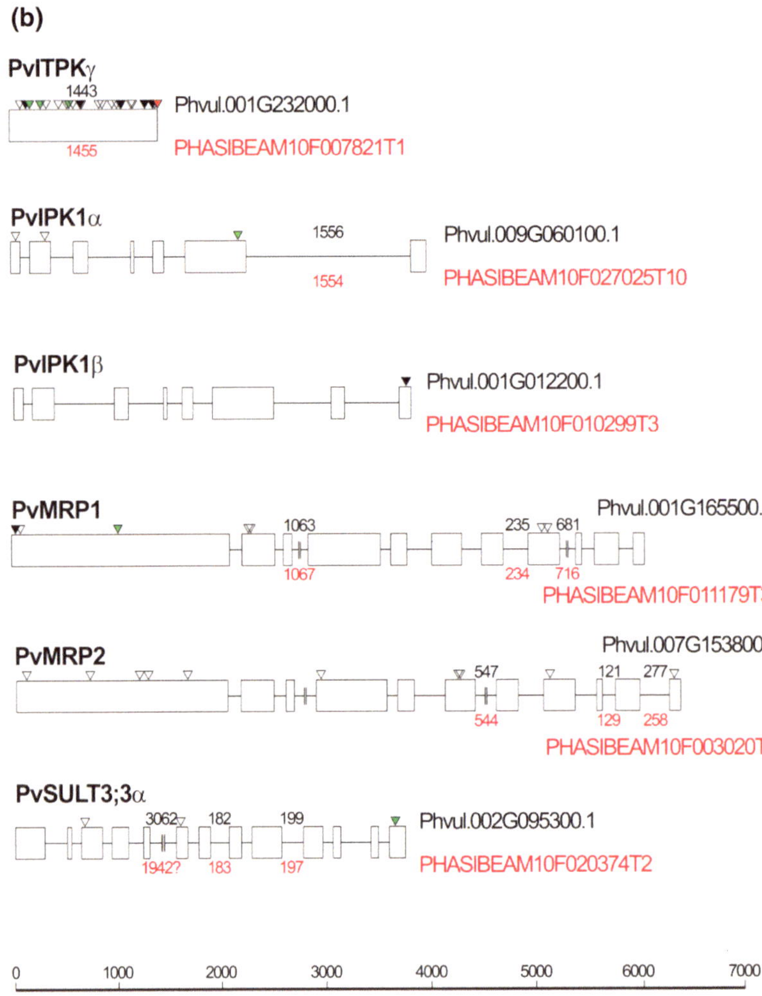

Fig. 8.2 (continued)

◄ **Fig. 8.2** **a**, **b** Schematic representation of exon-intron structure of the PA biosynthesis and transport related genes in G19983 (above) and BAT93 (below) common bean genotypes. Exons and introns are represented by white boxes and thin black lines, respectively. For the gene names see Fig. 8.1 legend. The SNPs between both sequences are indicated by white triangles if the coded amino acid on the protein sequence remained unchanged (silent substitution), by green triangles if the coded amino acid changed for other with similar biochemical properties (neutral substitution), and with black triangles if it changed the amino acid for another with different biochemical properties (missense). Red triangle indicates the presence of a stop codon. Changes in the size of introns and exons between genotypes are indicated with the number of nucleotides at the corresponding positions. A double bar is used to indicate the schematic shortening of long introns (i.e. those introns are not in scale). When no indications are made, the length of the intron or exon is the same for both genotypes. The exon and intron sizes (in base pairs) can be estimated using the scale at the bottom

performed on mature pods without seeds (MPWS_R9) and on immature seeds (IS_R9) at the same developmental stages (79 days of plant growth, corresponding to 19 DAF). This suggests an important role of the pod itself in addition to the seed in the first steps of PA biosynthesis.

	PvMIPSs	PvMIPSv	PvIMPα	PvIMPβ	PvMIK	Pv2-PGK	PvIPK2	PvITPKα1	PvITPKα2	PvITPKβ1	PvITPKβ2	PvITPKγ	PvIPK1α	PvIPK1β	PvMRP1	PvMRP2	PvSULTR3;3α
EC_V0	24.9	5.2	64.1	8.3	3.4	3.5	1.7	4.9	2.4	6.0	14.7	4.2	2.5	5.4	5.6	1.9	6.8
RD_V0	77.7	23.0	32.7	14.5	5.7	7.4	3.2	8.5	2.9	8.9	5.0	5.4	5.4	6.2	5.7	4.0	4.7
PR_V1	116.4	21.0	12.0	14.1	6.8	11.7	1.9	8.6	7.7	54.0	5.1	8.8	8.6	11.2	9.2	12.8	11.0
NR_V3	11.1	2.4	3.0	11.6	8.1	9.1	4.5	5.3	3.9	30.3	14.4	4.7	3.5	5.4	2.3	7.1	2.0
R_V3	18.7	0.4	3.9	10.5	6.5	9.1	4.3	8.7	3.9	45.1	15.9	5.5	2.4	4.8	2.4	6.4	1.6
R_R5	29.3	0.1	10.0	8.3	8.8	13.2	5.8	9.9	4.6	31.4	10.6	10.8	7.9	13.3	3.5	16.1	8.7
C_V1	15.2	1171.7	6.3	2.0	3.8	4.5	0.8	3.8	3.4	14.9	59.3	8.3	1.6	4.1	2.1	3.1	22.6
PL_V1	94.7	616.9	6.0	10.8	5.7	19.0	0.9	6.8	4.8	31.6	42.3	7.7	2.3	4.0	6.8	7.6	8.2
PL_V2	231.7	58.1	44.2	14.5	2.2	3.4	0.5	2.7	3.3	17.3	8.2	2.9	1.5	5.2	0.9	1.5	50.2
FTL_V2	316.4	61.1	55.0	18.1	2.7	9.3	0.4	3.2	5.6	55.9	9.8	3.4	2.0	6.2	1.9	3.7	36.6
TL_V4b	346.8	65.5	57.8	18.4	2.6	5.2	0.6	2.3	2.6	22.1	7.9	2.9	1.8	7.3	1.7	2.5	67.9
TL_R5	117.6	123.6	29.4	15.4	4.0	10.2	0.8	3.2	2.2	15.7	9.7	5.9	3.5	8.1	3.3	3.4	27.8
HYP_V0	46.5	14.9	22.1	11.6	7.1	10.7	1.1	5.2	6.2	62.0	3.7	4.0	4.4	4.7	8.1	4.7	1.2
HYP_V2	25.2	13.2	21.4	5.0	2.8	1.6	1.4	3.3	1.7	30.7	9.7	3.5	4.4	7.1	1.0	0.9	10
HYP_V4a	12.6	1.6	7.7	6.5	6.7	7.1	1.2	2.1	3.0	56.4	32.9	1.6	2.8	8.6	1.7	4.3	6.0
HYP_V1	85.1	167.3	8.5	6.0	5.6	4.3	1.7	6.3	2.6	38.0	6.7	4.7	5.5	6.1	4.8	6.3	3.2
EPI_V1	55.4	277.3	8.2	10.3	9.1	15.9	1.4	6.4	6.1	162.9	13.1	4.9	5.6	7.7	7.0	6.2	3.8
ST_V4b	59.4	13.9	16.2	13.0	4.1	6.9	2.1	3.3	2.0	27.0	8.9	3.7	6.6	9.7	4.8	2.5	9.0
SN_R5	18.3	163.2	13.7	6.7	3.7	3.2	1.5	2.6	2.0	19.3	4.5	3.2	4.5	6.6	1.7	1.9	14.9
AM_V4b	65.0	49.4	44.1	20.5	4.5	20.5	1.6	4.3	2.9	14.0	9.7	10.1	6.2	11.3	14.6	7.5	14.5
AM_R5	71.9	75.2	22.0	13.4	4.6	12.5	1.9	3.8	3.1	9.9	6.7	4.8	3.8	5.5	6.3	4.6	9.3
F_R6	209.3	15.8	15.7	2.7	11.5	2.5	6.1	7.1	6.9	5.9	1.9	2.0	1.7	1.9	2.1	3.1	3.6
SP_R7	3511.6	235.6	120.1	31.8	5.6	16.4	2.5	6.6	2.6	15.7	8.7	9.8	10.5	15.1	14.9	3.5	22.0
MEP_R8	3372.8	22.6	47.6	25.0	6.6	10.1	2.6	7.9	2.2	17.1	5.9	7.0	11.7	11.8	9.0	2.4	11.8
IS_R9	596.8	1.1	23.1	17.7	31.8	19.9	8.0	14.5	8.4	21.4	3.0	17.3	5.1	11.5	19.1	4.0	40.3
MPWS_R9	998.8	19.9	27.9	18.8	5.4	9.1	2.4	5.7	2.2	30.0	4.9	5.7	12.7	11.8	9.2	3.8	16.7
MP_R9	5188.9	214.0	81.8	31.0	6.0	11.6	2.7	8.6	2.2	16.3	8.9	12.4	14.4	15.6	12.7	2.0	10.0

Fig. 8.3 Expression values (*RPKM* normalization) for the genes described in the text, obtained from RNA-Seq analysis on BAT93 genotype, as reported by Vlasova et al. (2016). The average signal expression value of genes are represented with a colour scale in which yellow, orange, red and dark red indicate low, medium, high and very high level of expression, respectively. For the gene names see Fig. 8.1 legend. Sample names are the same used by Vlasova et al. (2016) and correspond to the following organs at the indicated developmental stages: EC_V0, Embryo + Cotyledons V0; RD_V0, Radicle V0; PR_V1, Primary root; NR_V3, Neck of the root V3; R_V3, Root V3; R_R5, Root R5; C_V1, Cotyledons V1; PL_V1, Primary leaf V1; PL_V2, Primary leaf V2; FTL_V2, First trifoliate leaf; TL_R5, trifoliate leaf R5; HYP_V0, Hypocotyl V0; HYP_V4a, Hypocotyl V4a; HYP_V1, Hypocotyl V1; EPI_V1, Epicotyl V1; ST_V4b, Stem V4b; SN_R5, Stem node R5; AM_V4b, Axial meristem V4b; AM_R5, Axial meristem R5; F_R6, Flower R6; SP_R7, Small pod R7; MEP_R8, Medium pod R8; IS_R9, Immature seed R9; MPWS_R9, Mature pod without seeds R9; MP_R9, Mature pod R9. Stages of the vegetative and reproductive developmental phases with collection times in parenthesis are: V0 = germination (48 h), V1 = emergence (6 days), V2 = primary leaves (10 days), V3 = first trifoliate leaf (14 days), V4 = third trifoliate leaf (a, 29 days; b, 35 days), R5 = pre-flowering (43 days), R6 = flowering (53 days), R7 = pod formation (60 days), R8 = pod filling (64 days), and R9 = maturity (79 days, except MP_R9 at 86 days)

The *PvMIPSv* gene is expressed at lower levels than *PvMIPSs* in the seeds, but also in other vegetative organs at different developmental stages and reaches the highest expression levels in cotyledons and primary leaf at 6 days of plant growth.

8.2.2 PvIMP

A specific Mg^{2+}-dependent *myo*-inositol 1-phosphatase (IMP) is required to dephosphorylate Ins(3)P_1 to generate free Ins, which is used in numerous metabolic pathways, including PtIns biosynthesis. Interestingly, the IMP enzyme has a dual activity as it hydrolyses L-galactose 1-phosphate (L-Gal 1-P), a precursor of ascorbic acid synthesis (Laing et al. 2004; Torabinejad et al. 2009).

We identified two putative PvIMP proteins from both reference genotypes, both identical in the two reference genomes, hereinafter called PvIMPα and PvIMPβ, sharing an identity degree of 79% (Table 8.1 and Fig. 8.2); and two putative PvIMPL (IMP-like), hereinafter called PvIMPL1 and PvIMPL2. PvIMPL1 is encoded by loci *Phvul.009G205700* and *PHASIBEAM10F006139*, and is a putative ortholog of AtIMPL1; PvIMPL2 is encoded by loci *Phvul.009G005100* and *PHASIBEAM10F025125* and is a putative ortholog of AtIMPL2. A phylogenetic tree of common bean IMP and IMPL proteins is represented in Fig. 8.4a. As no involvement in phytic acid biosynthesis has been described for IMP-like proteins, we did not consider PvIMPLs for further analysis. PvIMP sequence, reported by Fileppi et al. (2010), corresponds to a partial version of PvIMPα, lacking the first 34 aminoacids. PvIMPα and PvIMPβ share an overall aminoacid identity of 68 and 77% respectively with AtIMP, also known as AtVTC4 (Torabinejad et al. 2009). The coding sequences of both genes are identical in the two reference genomes, and only 6 SNPs are present in the introns of *PvIMPβ*, while some more differences are present in the intron sequences of *PvIMPα*. Particularly, the third intron, the longest one, that results not completely sequenced in the G19833 genome (100 unidentified nucleotides are present) in a region highly enriched in TG and TA dinucleotides, in the BAT93 genome shows two deleted sequences of 65 and 125 base pairs, respectively (Fig. 8.2). In the case of *PvIMPα* two alternative transcripts were annotated in the Andean genotype, differing only in the 5′ UTR length.

PvIMPα expression is higher at early stages of common bean seed development, and then declines to undetectable levels before the start of PA accumulation (Fileppi et al. 2010). These expression profiles were confirmed by transcriptomic data, showing a decreased *PvIMPα* expression levels during pod development with an expression peak in medium pods (MP_R9; Fig. 8.3). *PvIMPα* is expressed at higher levels than *PvIMPβ* in the majority of the analysed organs.

8.2.3 PvMIK

The activity of *myo*-inositol kinase (MIK) reverses the reaction catalysed by IMP. The importance of MIK in seed PA metabolism has been demonstrated by a number of mutations in the *MIK* gene in different species (Shi et al. 2005; Kim et al. 2008b), although Ins(3)P_1 may be produced directly from glucose 6-phosphate by MIPS. Probably MIK could provide more substrate diversity for the generation of $InsP_2$ to feed the lipid-independent pathway, since it is able to produce multiple inositol monophosphates (Shi et al. 2005).

In common bean, as in the majority of the species in which PA pathway has been studied so far, one *MIK* gene has been identified for which only a partial cDNA sequence was previously reported (Fileppi et al. 2010). The presence of a single copy for this gene, consisting in two exons and one intron, was confirmed on both reference genomes (Fig. 8.2). The *PvMIK* coding sequences differ for 15 SNPs, the majority of these substitutions have no effect at protein level, consisting in nine silent, two neutral and four missense substitutions. The last SNPs are in positions not highly conserved in the MIK

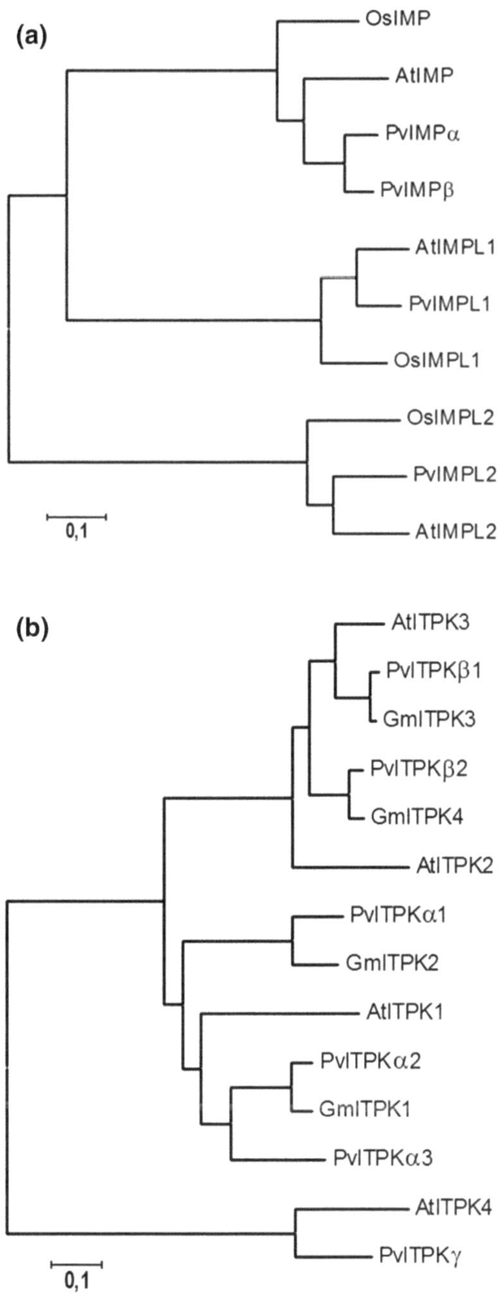

Fig. 8.4 Phylogenetic relationships among the common bean, *A. thaliana* and *O. sativa* or *G. max* of the characterised *PvIMP/PvIMPL* (**a**) and *PvITPK* genes (**b**). The phylogenies were inferred using the Neighbor-Joining method (Saitou and Nei 1987) with the amino acid sequences. Evolutionary analyses were conducted in MEGA6 (Tamura et al. 2013). The trees were drawn to scale with branch lengths in the same units as those of the evolutionary distances. The evolutionary distances were computed using the Poisson correction method (Zuckerkandl and Pauling 1965) and are in the units of the number of amino acid substitutions per site. Protein accession numbers or corresponding transcripts are: PvIMPα (*Phvul.003G084500.1*), PvIMPβ (*Phvul.003G084500.1*), PvIMPL1 (*Phvul.009G205700.1*), PvIMPL2 (Phvul.009G005100.1), AtIMP (At3g02870.1), AtIMPL1 (*At1g31190.1*), AtIMPL2 (*At4g39120.1*), OsIMP (*ABF97360.1*), OsIMPL1 (*XP_015623340.1*), OsIMPL2 (*XP_015646976.1*), PvITPKα1 (*Phvul.003G200900.1*), PvITPKα2 (*Phvul.003G284100.1*), PvITPKα3 (*Phvul.002G139700.1*), PvITPKβ1 (*Phvul.009G113600.1*), PvITPKβ2 (*Phvul.007G047700.1*), PvITPKγ (*Phvul.001G232000.1*), AtITPK1 (*At5g16760.1*), AtITPK3 (*At4g08170.1*), AtITPK4 (*At2g43980.1*), GmITPK-1 (*ABU93831*), GmITPK-2 (*ABU93832*), GmITPK-3 (*ABU93833*), GmITPK-4 (*ABU93834*)

proteins in different species, thus probably not affecting the protein function. When comparing the *PvMIK* BAT93 sequence with the *PvMIK* G19833, we observed that in the intron there are quite numerous SNPs and various short insertions (the longest of 39 nucleotides). *PvMIK* is expressed at similar levels in different organs at different developmental stages and shows the highest transcript accumulation in immature seeds (IS_R9; Fig. 8.3).

8.2.4 Putative *Pv2-PGK*

The production of $InsP_2$ from $InsP_1$ requires a monophosphate kinase activity. A good candidate to perform this step is a homolog of 2-phosphoglycerate kinase (2-PGK), which catalyses the production of 2,3-6 bisphosphoglycerate from 2-phosphoglycerate in Archaeobacteria (Raboy 2009). In contrast to other enzymes of the pathway, characterized in different species, enzymes that probably catalyse the synthesis of $InsP_2$ from $InsP_1$ were described only in rice and in *Arabidopsis* (Kim et al. 2008a; Kim and Tai 2010; Tagashira et al. 2015). In the genome of both species there are two putative *2-PGK* homologs, namely *Oslpa-1* and *LOC_Os09g39870* in rice and *At5g60760* and *At3g45090* in *Arabidopsis*. Most likely, only *Oslpa-1*, also named *OsPGK1*, and *At5g60760*, expressed during seed development, play a major role in PA biosynthesis in developing seeds, because the mutations of these genes resulted in a severe *lpa* seed phenotype (Kim et al. 2008a; Kim and Tai 2010). Moreover, *OsPGK1* overexpression produced a significant increase in total P content in rice seeds, due to increased levels of PA and inorganic phosphates (Tagashira et al. 2015). The second putative *2-PGK* gene present in *Arabidopsis* and in rice is poorly expressed in the seed and a knock out mutation in *At3g45090* does not affect seed PA content (Kim et al. 2008a; Kim and Tai 2010).

Through TBLASTN analysis performed on the two common bean reference genomes, using as queries the two characterised seed 2-PGK protein sequences from *Arabidopsis* and rice, we identified *Phvul.011G082700* and *PHASIBEAM10F002348* loci, in the Andean and Mesoamerican genotype, respectively, hereinafter called *Pv2-PGK* (Table 8.1 and Fig. 8.2), coding for identical proteins of 729 aminoacids. These putative Pv2-PGK proteins show 65% identity with At5g60760 and 59% with OsPGK1, respectively. The genomic structure of these loci, consisting of 8 exons and 7 introns (one intron is present in the 5'UTR), is very similar to the ones of the putative *Arabidopsis* and rice paralogs (Kim et al. 2008a; Kim and Tai 2010). In the Andean genotype two alternative splicing variants were annotated, differing only in 5′ UTR length. The comparison of the coding sequences in the two reference genomes revealed 5 SNPs with no effect at protein level, while numerous mismatches and also some short insertions in the Mesoamerican genotype, compared to the Andean one, were identified in the intron sequences.

Interestingly, TBLASTN analysis performed on the two reference genomes, using as queries the proteins encoded by *At3g45090* and *LOC_Os 09g39870*, identified again the Pv2-PGK protein, suggesting that a single *2-PGK* gene is present in the *P. vulgaris* genome. A search through Phylome DB supports the single copy hypothesis.

Pv2-PGK gene shows moderate differences in gene expression in different organs at different developmental stages, being mainly expressed in immature seeds (IS_R9; Fig. 8.3)

8.2.5 PvIPK2

Inositol 1,4,5-tris-phosphate kinase, or more appropriately inositol polyphosphate multikinase (IPK2/IPMK), has a 3-/6-kinase activity required to phosphorylate Ins(1,4,5)P_3 to generate Ins (1,3,4,5,6)P_5 (Fig. 8.1; Raboy 2009). Therefore, IPK2 kinase is specific for the lipid-dependent pathway, which is not the major route to PA synthesis in the seed (Raboy 2009). However, the *atipk2β* mutant, an *Arabidopsis* mutant with a *lpa* seed phenotype with a reduction of the PA content by 35–48%, evidences that the lipid-dependent pathway is active in the seed (Stevenson-Paulik et al. 2005). The presence of an intronless *PvIPK2* gene was confirmed in both bean reference genomes, as previously identified by Fileppi et al. (2010). We identified 4 SNPs between the two coding sequences in the two reference genomes and only one, occurring in a region not highly conserved that changes the encoded amino acid, a valine in the G19833 sequence is substituted for alanine in the BAT93 sequence, another aliphatic amino acid (Fig. 8.2).

PvIPK2 transcript accumulation was assessed during pod (4 and 6 DAF) and seed development (from 12 to 24 DAF) and showed a peak of expression between 12 and 16 DAF, consistent with its role in the late part of the pathway (Fileppi et al. 2010). Moderate differences in expression levels were found in the different analysed organs (Fig. 8.3).

8.2.6 PvITPKs

In the lipid-independent pathway, there is a missing link between the synthesis of $InsP_3$ from $InsP_2$. However, the two following steps of phosphorylation are catalysed by the activity of the inositol 1,3,4-trisphosphate 5-/6-kinases (ITPK), which belong to the family of ATP-grasp fold proteins (Fig. 8.1; Raboy 2009). This class of enzymes may have different substrates (Rasmussen et al. 2010): $InsP_3$s, such as Ins(1,3,4)P_3, Ins(3,5,6)P_3 and/or Ins(3,4,6)P_3; $InsP_4$s like Ins(3,4,5,6)P_4, Ins(1,3,4,5)P_4 and/or Ins(1,2,5,6)P_4. Moreover, some ITPKs show phosphatase activity (Rasmussen et al. 2010). In different species a variable number of *ITPK* genes have been reported, ranging from one in maize (Shi et al. 2003) to six in rice (Suzuki et al. 2007), all of them showed different expression profiles. The ITPK proteins cluster in three phylogenetic subgroups (α, β and γ); nonetheless this separation does not usually match with their tissue-specific expression pattern (Rasmussen et al. 2010).

In common bean, two *ITPK* genes were reported: *PvITPKα* and *PvITPKβ* (hereinafter called *PvITPKα1* and *PvITPKβ1*) clustering in the α and β subgroups, respectively, although Southern blot hybridization suggested the presence of at least a second member of the β subgroup (Fileppi et al. 2010). It was previously suggested that PvITPKα1 is important in the control of phytate content on a per-seed basis, as it was somewhat associated with a QTL related to this trait (Blair et al. 2012). TBLASTN analysis, using these proteins as queries against the G19833 and BAT93 genomes, revealed that the reported sequence of PvITPKβ1 was partial, and the existence of four more PvITPK proteins that throughout a phylogenetic analysis were classified as: PvITPKα2, PvITPKβ2 and PvITPKγ (Table 8.1; Figs. 8.2 and 8.4b). Only for the Andean genotype we found also a third PvITPKα, called PvITPKα3, but we were not able to identify the corresponding sequence in the Mesoamerican genotype even with different approaches (TBLASTN, BLAST analyses, search in the PhylomeDB database). It was reported that some genes are specifically lost in BAT93 genome (Vlasova et al. 2016), and *PvITPKα3* seems to be one of them.

The *PvITPKα1* and *PvITPKα2* genes are intronless and encode for proteins of 317 and 341 amino acids, respectively, while *PvITPKα3* consists of two exons and one intron and encodes a polypeptide of 304 amino acids. *PvITPKα1* CDS in the two reference genomes show three SNPs corresponding at protein level to a missense change in a position not highly conserved, a neutral and a silent one. In the case of *PvITPKα2* only a silent mismatch is present comparing the two reference genome sequences.

The two *PvITPKβ* genes have a similar genomic structure with 10 exons and 9 introns, and code for respectively polypeptides of 350 and 339 amino acids in the reference genomes. For *PvITPKβ2* also an alternative splicing variant with 9 coding exons (plus 2 exons and a part of the third exon constituting the 5′ UTR) was annotated in the G19833 genome coding for a 276 amino acids long protein. However, this last polypeptide is quite different from the other proteins belonging to the same subgroup and was not considered for further analyses. Comparison of *PvITPKβ1* CDS sequences in the two reference genomes showed only two SNPs, causing one silent and one neutral change. A few mismatches are present in the intron sequences of this gene. Comparison of the two reference genome sequences reveals a 100% identity for *PvITPKβ2* CDS sequences, while many mismatches are present in the first intron.

PvITPKγ gene consists in 2 exons and one intron, present inside the 5′ UTR or between the 5′ UTR and the translational start codon, depending on two splicing variants, annotated in the Andean genotype, both coding for the same polypeptide of 480 amino acids. Comparison between the CDS in the two genomes reveals 23 SNPs, 14 of them silent, 3 neutral and 5 missense, at protein level (all in variable regions of the protein among different orthologues), and one non-sense causing a reduction in length for the last 4 amino acids in the Andean genotype. This C-terminal region of the protein is not highly conserved in different orthologues in other species.

Among the different *PvITPK* genes, *PvITPKβ1* shows the highest expression levels in different organs, including pods at different developmental stages. It is mainly expressed in epicotyl (EPI_V1). *PvITPKα1*, *PvITPKα2* and *PvITPKγ* show their highest transcript accumulation in immature seed (IS_R9; Fig. 8.3), suggesting an important role of these genes in the seed. From data reported for the Andean genotype, *PvITPKα3* gene is barely expressed in all analysed samples, compared to all the other *PvITPKs* (data not shown).

8.2.7 PvIPK1

The last step of PA biosynthesis is catalysed by the inositol polyphosphate 2-kinase (IPK1), which specifically phosphorylates InsPs in the 2-position of Ins. A partial *PvIPK1* CDS was previously reported (Fileppi et al. 2010). In both Andean and Mesoamerican genome sequences we identified two *PvIPK1* genes, one on chromosome 9, corresponding to the already identified partial gene, and a second one on chromosome 1, hereinafter called *PvIPK1α* and *PvIPK1β*, respectively (Table 8.1 and Fig. 8.2). Two *IPK1* genes were also described in maize and *Arabidopsis* (Sweetman et al. 2006; Sun et al. 2007; Kim and Tai 2011). *PvIPK1α* genomic structure consists in 9 exons and 8 introns (one intron inside the 5′ UTR and another inside the 3′ UTR) and the gene codes for a protein of 453 amino acids. *PvIPK1β* has a similar genomic structure (one intron is between the stop codon and the 3′ UTR) and two splicing variants of the last intron were annotated in the Andean genotype. They encode for proteins of 489 and 466 amino acids, respectively. The overall amino acid sequence identity shared by *PvIPK1α* and *PvIPK1β* is of 65 or 68%, depending on the splicing variant for *PvIPK1β* used for the alignment analysis.

PvIPK1α and *PvIPK1β* CDS sequences show only three and one SNPs, respectively, in the two reference genomes; in the case of *PvIPK1α* two of the SNPs are silent and one neutral at protein level, while the single mismatch in *PvIPK1β* sequence is of missense type although outside the functional kinase domain. Very few mismatches were found comparing the intron sequences for both genes in the two genomes. In fact, *PvIPK1α* was considered impossible to be mapped, due to a very low polymorphism between the Mesoamerican DOR364 and the Andean G19833 genotypes of the mapping population (Blair et al. 2012).

PvIPK1β is generally more expressed than *PvIPK1α*, except in pods where the two genes are expressed at similar levels (Fig. 8.3). However, *PvIPK1β* expression is double, compared to *PvIPK1α* in immature seeds (IS_R9; Fig. 8.3). A more detailed analysis of both gene expressions during seed development may help in understanding if they may have a redundant function in the seeds.

8.3 Phytic Acid and Transport

PA is stored in the form of globoids inside the storage vacuoles. *lpa* mutants affected in PA transport are characterized by strong decrease in PA content accompanied by a molar increase in free phosphate. Two classes of *lpa* mutants affected in transporters were isolated: mutants in genes coding for multidrug-resistance-associated proteins (MRP), belonging to the ABCC cluster of plant ABC transporters (Nagy et al. 2009; Sparvoli and Cominelli 2014); and mutants in putative sulfate transporters, orthologues of the *Arabidopsis* AtSULTR3;3 protein (Takahashi et al. 2011; Ye et al. 2011).

8.3.1 *PvMRP1* and *PvMRP2*

In common bean two putative tonoplast PA ABCC transporters, PvMRP1 and PvMRP2, were described (Panzeri et al. 2011), showing an overall identity sequence of 78 and 77%, respectively with AtMRP5 protein for which this function was demonstrated (Nagy et al. 2009).

In common bean, the only *lpa* mutation (*lpa1*, also known as *lpa280-10*) reported so far affects the PvMRP1 transporter (Panzeri et al. 2011). The *lpa1* seeds display a very strong phenotype, consisting in a 90% reduction of PA accumulation, together with a decrease of raffinose containing sugars by 25% and *myo*-inositol by 30% (Campion et al. 2009; Panzeri et al. 2011). These seeds are characterized by a seven times increase of free iron (Panzeri et al. 2011) and they are really biofortified, as iron absorption derived from *lpa1* beans ingestion is significantly higher than from beans with normal PA levels, as shown in a study on volunteers (Petry et al. 2013). The agronomic performance of the mutant does not significantly differ from the wild type genotype, in term of seedling emergence, seed yield and plant growth (Campion et al. 2009, 2013). Moreover, the *lpa1* seeds performed equally or even better than the wild type ones under stressful germination conditions (Campion et al. 2009). Mutations in maize and rice *PvMRP1* orthologous genes showed similar decrease in PA content in the seeds, but convey also negative pleiotropic effects on seed germination and plant development (Sparvoli and Cominelli 2014). A possible explanation for this difference between cereals and common bean is that in common bean a second gene is present, *PvMRP2*, paralog of *PvMRP1*, and it is most likely able to complement the absence of a functional *PvMRP1* in all the organs but in the seed (Panzeri et al. 2011). Quantitative expression analyses confirm that *PvMRP2* is expressed in almost all the organs in a similar fashion to *PvMRP1*, with the exception of developing seeds (unpublished results), where *PvMRP1* has higher expression, this is also evident from transcriptomic data reported in Fig. 8.3.

Sequences and map positions for *PvMRP1* and partially for *PvMRP2* gene were previously reported in the breeding line 905 of Mesoamerican origin (Panzeri et al. 2011). We identified these sequences in the G19833 and BAT93 genome sequences (Table 8.1 and Fig. 8.2). Both genes show a similar genomic structure consisting of 13 exons and 12 introns with 2 introns inside the 5′ UTR. *PvMRP1* and *PvMRP2* genes encode for proteins of 1538 and 1513 amino acids, respectively, sharing an overall sequence identity of 82%.

PvMRP1 and *PvMRP2* coding sequences are highly conserved in both reference genomes: five silent, one conservative and one missense SNPs were detected for *PvMRP1* and ten silent SNPs for *PvMRP2*. Some more differences are present in intron sequences, particularly in *PvMRP1* 8th intron, showing also the presence of two short insertions in the BAT93 gene compared to the G19833 one.

8.3.2 Putative PvSULTR3;3

In barley and rice mutations affecting *HvST* and *OsSULTR3;3*, coding for two putative sulfate transporters, belonging to the SULTR3;3 class of sulfate transporters (Takahashi et al. 2011), confer seed *lpa* phenotype (Zhao et al. 2008, 2016; Ye et al. 2011). Both mutants exhibit an increase in inorganic phosphate as other *lpa* mutants, but also a decrease in total phosphate in the seed, differently from other *lpa* mutants characterized so far (Raboy et al. 2014; Zhao et al. 2016). The role of these transporters is not very clear. It was recently reported that OsSULTR3;3 is involved in sulfate as well as in phosphate homeostasis, although, when expressed in heterologous system such as yeast or *Xenopus* oocytes, it is not able to transport neither sulfate, nor phosphate, nor PA precursors

(Zhao et al. 2016). However, it cannot be excluded that in plant system OsSULTR3;3 may transport these molecules as well as PA. In plant cells OsSULTR3;3 is localized in the endoplasmic reticulum (ER). Previous studies have suggested that the final steps of PA synthesis (from $InsP_3$ to $InsP_6$) take place in the ER (Otegui et al. 2002). Moreover, *ossultr3;3* mutation has great impact on metabolism, as it alters the concentrations of sugars involved in the close biosynthetic pathway leading to PA, but also levels of free fatty acids, sugar alcohols, organic alcohols, organic acids and γ-aminobutyric acid (Zhao et al. 2016).

To our knowledge the common bean orthologous gene has not yet been analysed. In both reference genotypes we identified two putative orthologous proteins, sharing 72% identity, through TBLASTN analysis, called hereinafter PvSULTR3;3α and PvSULTR3;3β, using as queries LOC_Os04g55800 or AtSULTR3;3 (*At1g23090*) protein sequences. The main difference between PvSULTR3;3α and PvSULTR3;3β consists in the loss of 23 amino acids inside the putative sulfate transporter domain. Interestingly, their genomic loci are in tandem, suggesting a duplication event. However, from expression analyses reported for both reference genomes, it results that *PvSULTR3;3β* gene is not expressed at all in the majority of samples and is only scarcely expressed in stem. These data strongly suggest it might be a pseudogene. For this reason, we did not consider it for further analyses. PvSULTR3;3α is 647 amino acid long, it differs for only one amino acid in the two reference genomes and shares 78% identity with AtSULTR3;3, and 70.5% with the rice orthologue. *PvSULTR3;3α* genomic structure consists in 13 exons. The comparison of *PvSULT3;3α* CDS in both reference genomes revealed a high degree of conservation with only three SNPs of which only one has effect at protein level, causing a conservative amino acid change. Very few mismatches are present in intron sequences, except in the long fourth intron for which a portion was not properly sequenced in the BAT93 genome, compromising our comparison analysis.

PvSULT3;3α gene is differentially expressed in the analysed organs and it reaches its highest expression levels in leaves and during pod and seed development (Fig. 8.3).

8.4 Conclusions and Future Perspectives

The recent sequencing of the genomes of an Andean and a Mesoamerican landrace of common bean (Schmutz et al. 2014; Vlasova et al. 2016) and the phylogenetic, genomic and transcriptomic data generated (Vlasova et al. 2016) provide invaluable resources for the study of this important legume species and offer new tools and methodologies to generate superior varieties (Vlasova et al. 2016). Here we exploited this in silico resources to identify differences in genes involved in PA synthesis and transport between the two reference genomes.

From our analyses it is clear that coding sequences of genes involved in PA pathway and transport are highly conserved in the Mesoamerican and Andean genetic pools, as in the two genomes they share more than 95% of sequence identity, as shown for 94.5% of common bean genes (Vlasova et al. 2016). On the other hand, we found that some intron showed numerous SNPs and differences in term of insertions/deletions between the two genomes. For the majority of the genes of the pathway it was already observed the usefulness of knowledge of intronic regions to develop molecular markers able to discriminate between the two genetic pools (Blair et al. 2012). We confirmed this possibility also for the other genes of the pathway and transport. Molecular markers that can be developed, based on these sequence data, may be useful to produce new common bean lines with reduced PA content.

For the analysed genes we summarised transcriptomic data reported for the Mesoamerican genotype (Vlasova et al. 2016), because a more complete set of samples was evaluated in this genotype than in the Andean one (data present for each gene on Phytozome v11.0 portal, https://phytozome.jgi.doe.gov/pz/portal.html).

However, for the organs analysed in both genotypes we found very similar expression patterns for the different genes. Interestingly, all the genes, except *PvITPKβ1*, *PvITPKβ2* and *PvMRP2* (for which its major role in vegetative organs were already discussed) reach their highest expression level in one of the pod developmental stages (SP_R7, MEP_R8, MP_R9) or in immature seeds (IS_R9; Fig. 8.3), suggesting their major role in the synthesis/transport of PA with phosphorous storage function in the seed. On the other side, all the genes are also expressed in vegetative organs, suggesting also a role in the synthesis of PA with a more focused on regulatory than storage processes. A more detailed expression analysis on different seed developmental stages, separated from pods, may help in understanding their principal role in PA synthesis and transport.

The availability of these new common bean genomic resources would be also extremely useful to obtain new *lpa* mutants. Only the *lpa1-1* mutant, affecting the *PvMRP1* gene, has been isolated so far, as previously discussed, through the screening of an EMS mutagenized collection and the following mapping of the mutation following a forward genetic approach (Campion et al. 2009; Panzeri et al. 2011). The knowledge of the whole genome sequence allows the use of reverse genetic approaches, such as Targeting Induced Local Lesions in Genome (TILLING), also in its high throughput version of TILLING by sequencing, or genome editing, in order to isolate new mutants. At the moment a population of 3000 M2 mutant lines from the genotype BAT93 is available for TILLING screening (Porch et al. 2009). However, it is desirable to increase this population, as over 5000 mutant lines are required for adequate genome coverage and an effective TILLING approach (Porch et al. 2009). Genome editing approaches, aimed to targeted genome modifications, were developed in plants in recent years (Schaart et al. 2015). One of the first strategies used in plants, through zinc-finger nucleases (ZFNs) that induce a double-stranded break at their target locus, was successfully applied in maize to generate a *lpa* mutation in the *IPK1* gene (Shukla et al. 2009). Genome editing strategies require transformation approaches. Common bean, like other legumes, is generally considered recalcitrant to *Agrobacterium*-mediated transformation, due to poor regeneration in tissue culture (Arellano et al. 2009). However, the molecular characterisation of the first transgenic commercial common bean immune to bean golden mosaic virus was recently carried out, demonstrating the stability of the transgene after eight self-pollinated generations (Aragao et al. 2013). Subsequently, common bean transformation may become a more routine strategy, increasing the possibility to use genome-editing approaches also in this species.

The common bean genomic resources may be also exploited to study regulatory aspects of PA pathway in this species. No transcription factors involved in the regulation of PA metabolism have been described to date. Only very recently transcriptomic data showed that more than 300 differentially expressed genes in two maize lines with opposite PA content encode for transcription factors (Zhang et al. 2016). Moreover, studies of promoter or of other regulatory sequences, such as UTRs, of different genes involved in PA synthesis and transport are possible, taking advantage from these genomic resources. For example, it was recently shown that there is a correlation between the number of repeats of a dinucleotide in the 5′ UTR and the transcription level of the *CaIMP* gene from *Cicer arietinum* L. A shorter 5′ UTR confers higher *CaIMP* expression and consequently higher PA content and increased drought-tolerance than a longer 5′ UTR sequence (Joshi-Saha and Reddy 2015).

As it was shown that PA is the major determinant of iron bioavailability in common bean, we consider the new genomic resources as important tools to develop strategies aimed to reduce its content in this important legume crop.

Acknowledgements We thank Dr. Walter Sanseverino (CRAG, Spain) and Dr. Salvador Capella-Gutiérrez (BIST, Spain; UPF, Spain; CBS Fungal Biodiversity Centre, The Netherlands) for their help with the use of database developed for BAT93 genome and Dr. Roberto Bollini for critical reading of the manuscript. This work was partially supported by CERES—"NewPearl" project, jointly funded by Fondazione Cariplo and Agropolis Fondation.

References

Aragao F, Nogueira E, Tinoco M et al (2013) Molecular characterization of the first commercial transgenic common bean immune to the *Bean golden mosaic virus*. J Biotechnol 166:42–50

Arellano J, Fuentes S, Castillo-Espana P et al (2009) Regeneration of different cultivars of common bean (*Phaseolus vulgaris* L.) via indirect organogenesis. Plant Cell Tissue Org 96:11–18

Bilyeu KD, Zeng P, Coello P et al (2008) Quantitative conversion of phytate to inorganic phosphorus in soybean seeds expressing a bacterial phytase. Plant Physiol 146:468–477

Blair M, Herrera A, Sandoval T et al (2012) Inheritance of seed phytate and phosphorus levels in common bean (*Phaseolus vulgaris* L.) and association with newly-mapped candidate genes. Mol Breed 30:1265–1277

Blair M, Medina J, Astudillo C et al (2010) QTL for seed iron and zinc concentration and content in a Mesoamerican common bean (*Phaseolus vulgaris* L.) population. Theor Appl Genet 121:1059–1070

Blair M, Sandoval T, Caldas G et al (2009) Quantitative trait locus analysis of seed phosphorus and seed phytate content in a recombinant inbred line population of common bean. Crop Sci 49:237–246

Broughton W, Hernandez G, Blair M et al (2003) Beans (*Phaseolus* spp.)—model food legumes. Plant Soil 252:55–128

Campion B, Glahn R, Tava A et al (2013) Genetic reduction of antinutrients in common bean (*Phaseolus vulgaris* L.) seed, increases nutrients and in vitro iron bioavailability without depressing main agronomic traits. Field Crops Res 141:27–37

Campion B, Sparvoli F, Doria E et al (2009) Isolation and characterisation of an *lpa* (*low phytic acid*) mutant in common bean (*Phaseolus vulgaris* L.). Theor Appl Genet 118:1211–1221

Chiera J, Grabau E (2007) Localization of myo-inositol phosphate synthase (GmMIPS-1) during the early stages of soybean seed development. J Exp Bot 58:2261–2268

Coelho C, Tsai S, Vitorello V (2005) Dynamics of inositol phosphate pools (tris-, tetrakis- and pentakisphosphate) in relation to the rate of phytate synthesis during seed development in common bean (*Phaseolus vulgaris*). J Plant Physiol 162:1–9

CoGe website. https://genomevolution.org/CoGe/SearchResults.pl?s=20365. Accessed 29 April 2016

Fileppi M, Galasso I, Tagliabue G et al (2010) Characterisation of structural genes involved in phytic acid biosynthesis in common bean (*Phaseolus vulgaris* L.). Mol Breed 25:453–470

Greenwood J, Gifford D, Bewley J (1984) Seed development in *Ricinus communis* cv Hale (astor bean). II. Accumulation of phytic acid in the developing endosperm and embryo in relation to the deposition of lipid, protein, and phosphorus. Can J Bot 62:255–261

Guzman-Maldonado S, Martinez O, Acosta-Gallegos J et al (2003) Putative quantitative trait loci for physical and chemical components of common bean. Crop Sci 43:1029–1035

HarvestPlus website, Iron beans. http://www.harvestplus.org/content/iron-beans. Accessed 29 April 2016

Joshi-Saha A, Reddy KS (2015) Repeat length variation in the 5'UTR of *myo-inositol monophosphatase* gene is related to phytic acid content and contributes to drought tolerance in chickpea (*Cicer arietinum* L.). J Exp Bot 66:5683–5690

Kim S, Andaya C, Goyal S et al (2008a) The rice *OsLpa1* gene encodes a novel protein involved in phytic acid metabolism. Theor Appl Genet 117:769–779

Kim S, Andaya C, Newman J et al (2008b) Isolation and characterization of a *low phytic acid* rice mutant reveals a mutation in the rice orthologue of maize MIK. Theor Appl Genet 117:1291–1301

Kim S, Tai T (2010) Genetic analysis of two *OsLpa1*-like genes in *Arabidopsis* reveals that only one is required for wild-type seed phytic acid levels. Planta 232:1241–1250

Kim S, Tai T (2011) Identification of genes necessary for wild-type levels of seed phytic acid in *Arabidopsis thaliana* using a reverse genetics approach. Mol Genet Genomics 286:119–133

Kuwano M, Mimura T, Takaiwa F et al (2009) Generation of stable '*low phytic acid*' transgenic rice through antisense repression of the *1D-myo-inositol 3-phosphate synthase* gene (*RINO1*) using the 18-kDa oleosin promoter. Plant Biotechnol J 7:96–105

Laing W, Bulley S, Wright M et al (2004) A highly specific L-galactose-1-phosphate phosphatase on the

path to ascorbate biosynthesis. Proc Natl Acad Sci USA 101:16976–16981
Leytem AB, Maguire RO (2007) Environmental implications of inositol phosphates in animal manures. In: Turner BL, Richardson AE, Mullaney EJ (eds) Inositol phosphates: linking agriculture and the environment. CAB International, Wallingford, Oxfordshire, pp 150–168
Loewus FA, Murthy PPN (2000) *myo*-Inositol metabolism in plants. Plant Sci 150:1–19
Nagy R, Grob H, Weder B et al (2009) The *Arabidopsis* ATP-binding cassette protein AtMRP5/AtABCC5 is a high affinity inositol hexakisphosphate transporter involved in guard cell signaling and phytate storage. J Biol Chem 284:33614–33622
Norazalina S, Norhaizan M, Hairuszah I et al (2010) Anticarcinogenic efficacy of phytic acid extracted from rice bran on azoxymethane-induced colon carcinogenesis in rats. Exp Toxicol Pathol 62:259–268
Otegui M, Capp R, Staehelin L (2002) Developing seeds of *Arabidopsis* store different minerals in two types of vacuoles and in the endoplasmic reticulum. Plant Cell 14:1311–1327
Panzeri D, Cassani E, Doria E et al (2011) A defective ABC transporter of the MRP family, responsible for the bean *lpa1* mutation, affects the regulation of the phytic acid pathway, reduces seed myo-inositol and alters ABA sensitivity. New Phytol 191:70–83
Petry N, Boy E, Wirth JP et al (2015) Review: the potential of the common bean (*Phaseolus vulgaris*) as a vehicle for iron biofortification. Nutrients 7:1144–1173
Petry N, Egli I, Campion B et al (2013) Genetic reduction of phytate in common bean (*Phaseolus vulgaris* L.) seeds increases iron absorption in young women. J Nutr 143:1219–1224
PhylomedDB website. http://www.phylomedb.org/. Accessed 29 April 2016
Phytozome v11.0 website. https://phytozome.jgi.doe.gov/pz/portal.html. Accessed 29 April 2016
Porch T, Blair M, Lariguet P, Galeano C, Pankhurst C, Broughton W (2009) Generation of a mutant population for TILLING common bean genotype BAT 93. J Am Soc Hortic Sci 134:348–355
Raboy V (2001) Seeds for a better future: 'low phytate', grains help to overcome malnutrition and reduce pollution. Trends Plant Sci 6:458–462
Raboy V (2003) myo-Inositol-1,2,3,4,5,6-hexakisphosphate. Phytochem 64:1033–1043
Raboy V (2007) The ABCs of low-phytate crops. Nat Biotechnol 25:874–875
Raboy V (2009) Approaches and challenges to engineering seed phytate and total phosphorus. Plant Sci 177:281–296
Raboy V, Cichy K, Peterson K et al (2014) Barley (*Hordeum vulgare* L.) *low phytic acid* 1-1: an endosperm-specific, filial determinant of seed total phosphorus. J Hered 105:656–665
Rasmussen S, Ingvardsen C, Torp AM (2010) Mutations in genes controlling the biosynthesis and accumulation of inositol phosphates in seeds. Biochem Soc T 38:689–694
Saitou N, Nei M (1987) The neighbor-joining method: a new method for reconstructing phylogenetic trees. Mol Biol Evol 4:406–425
Schaart JG, van de Wiel CC, Lotz LA et al (2015) Opportunities for products of new plant breeding techniques. Trends Plant Sci 21:438–449
Schlemmer U, Frølich W, Prieto RM et al (2009) Phytate in foods and significance for humans: food sources, intake, processing, bioavailability, protective role and analysis. Mol Nutr Food Res 53(Suppl 2):S330–S375
Schmutz J, McClean PE, Mamidi S et al (2014) A reference genome for common bean and genome-wide analysis of dual domestications. Nat Genet 46:707–713
Shi J, Wang H, Hazebroek J et al (2005) The maize *low-phytic acid 3* encodes a myo-inositol kinase that plays a role in phytic acid biosynthesis in developing seeds. Plant J 42:708–719
Shi J, Wang H, Wu Y et al (2003) The maize low-phytic acid mutant *lpa2* is caused by mutation in an *inositol phosphate kinase* gene. Plant Physiol 131:507–515
Shukla V, Doyon Y, Miller J et al (2009) Precise genome modification in the crop species *Zea mays* using zinc-finger nucleases. Nature 459: 437-U156
Sparvoli F, Cominelli E (2014) Phytate transport by MRPs. In: Geisler M (ed) Plant ABC transporters. Springer, Berlin, Germany, vol 22, pp 19–38
Sparvoli F, Cominelli E (2015) Seed biofortification and phytic acid reduction: a conflict of interest for the plant? Plants 4:728–755
Stevenson-Paulik J, Bastidas R, Chiou S et al (2005) Generation of phytate-free seeds in *Arabidopsis* through disruption of inositol polyphosphate kinases. Proc Natl Acad Sci USA 102:12612–12617
Sun Y, Thompson M, Lin G et al (2007) Inositol 1,3,4,5,6-pentakisphosphate 2-kinase from maize: molecular and biochemical characterization. Plant Physiol 144:1278–1291
Suzuki M, Tanaka K, Kuwano M et al (2007) Expression pattern of inositol phosphate-related enzymes in rice (*Oryza sativa* L.): Implications for the phytic acid biosynthetic pathway. Gene 405:55–64
Sweetman D, Johnson S, Caddick S et al (2006) Characterization of an *Arabidopsis* inositol 1,3,4,5,6-pentakisphosphate 2-kinase (AtIPK1). Biochem J 394:95–103
Tagashira Y, Shimizu T, Miyamoto M et al (2015) Overexpression of a gene involved in phytic acid biosynthesis substantially increases phytic acid and total phosphorus in rice seeds. Plants 4:196–208
Takahashi H, Buchner P, Yoshimoto N et al (2011) Evolutionary relationships and functional diversity of plant sulfate transporters. Front Plant Sci 2:119
Tamura K, Stecher G, Peterson D, Filipski A, Kumar S (2013) MEGA6: molecular evolutionary genetics analysis version 6.0. Mol Biol Evol 30:2725–2729
Torabinejad J, Donahue J, Gunesekera B et al (2009) VTC4 is a bifunctional enzyme that affects

myo-inositol and ascorbate biosynthesis in plants. Plant Physiol 150:951–961

Vlasova A, Capella-Gutiérrez S, Rendón-Anaya M et al (2016) Genome and transcriptome analysis of the Mesoamerican common bean and the role of gene duplications in establishing tissue and temporal specialization of genes. Genome Biol 17:32

Xia H, Brearley C, Elge S et al (2003) *Arabidopsis* inositol polyphosphate 6-/3-kinase is a nuclear protein that complements a yeast mutant lacking a functional ArgR-Mcm1 transcription complex. Plant Cell 15:449–463

Ye H, Zhang X, Broughton S et al (2011) A nonsense mutation in a putative sulphate transporter gene results in *low phytic acid* in barley. Funct Integr Genom 11:103–110

Yoshida K, Wada T, Koyama H et al (1999) Temporal and spatial patterns of accumulation of the transcript of myo-inositol-1-phosphate synthase and phytin-containing particles during seed development in rice. Plant Physiol 119:65–72

Zhang S, Yang W, Zhao Q et al (2016) Analysis of weighted co-regulatory networks in maize provides insights into new genes and regulatory mechanisms related to inositol phosphate metabolism. BMC Genomics 17:129

Zhao H, Liu Q, Ren X et al (2008) Gene identification and allele-specific marker development for two allelic *low phytic acid* mutations in rice (*Oryza sativa* L.). Mol Breed 22:603–612

Zhao H, Frank T, Tan Y et al (2016) Disruption of *OsSULTR3;3* reduces phytate and phosphorus concentrations and alters the metabolite profile in rice grains. New Phytol 211:926–939

Zuckerkandl E, Pauling L (1965) Evolutionary divergence and convergence in proteins. In: Bryson V, Vogel HJ (eds) Evolving genes and proteins. Academic Press, New York, USA, pp 97–166

What Is Present at Common Bean Subtelomeres? Large Resistance Gene Clusters, Knobs and *Khipu* Satellite DNA

9

Manon M. S. Richard, Vincent Thareau,
Nicolas W. G. Chen, Chouaib Meziadi,
Stéphanie Pflieger and Valérie Geffroy

Abstract

In plants, the largest class of resistance (*R*) gene encodes nucleotide-binding leucine-rich repeat (NB-LRR) proteins. This multigene family is often organized in clusters of tightly linked genes. In the common bean (*Phaseolus vulgaris* L.) genome, most of the well-characterized large *R* gene clusters are not randomly distributed since they are often located at the ends of the linkage groups (LG), suggesting that this location is favorable for *R* gene proliferation. In addition, terminal knobs (heterochromatic blocks) are present at most chromosome (Chr) ends of *P. vulgaris,* and we have identified a satellite DNA referred to as *khipu* that is a component of most of them. Plasticity of subtelomeres has been described in various organisms such as yeast and human but is not well documented in plants. In common bean, the *B4* cluster of *R* gene was shown to derive from the *Co-2 R* gene cluster through an ectopic recombination between non-homologous chromosomes in subtelomeric regions. These unusual features of common bean genome (subtelomeric localization of large NB-LRR sequences, the presence of terminal knobs, and plasticity of subtelomeres) have been

M. M. S. Richard · V. Thareau · N. W. G. Chen · C. Meziadi · S. Pflieger · V. Geffroy (✉)
Institut of Plant Sciences Paris-Saclay (IPS2), CNRS, INRA, Université Paris-Sud, Université d'Evry, Université Paris-Diderot, Sorbonne Paris-Cité, Université Paris-Saclay, Bâtiment 630, 91405 Orsay, France
e-mail: valerie.geffroy@u-psud.fr

N. W. G. Chen
Institut de Recherche en Horticulture et Semences, UMR 1345 INRA-AgroCampus-Ouest, 49045 Angers, France

M. Pérez de la Vega et al. (eds.), *The Common Bean Genome*,
Compendium of Plant Genomes, https://doi.org/10.1007/978-3-319-63526-2_9

investigated with the availability of the complete common bean genome sequence.

Keywords
Phaseolus vulgaris · Subtelomeres · *Khipu* · Gene cluster

9.1 Introduction

In the human genome, extensive cytogenetic and sequence analyses have revealed that subtelomeres, i.e., regions adjacent to telomeres, are hotspots of interchromosomal recombination and segmental duplications (Linardopoulou et al. 2005). This peculiar dynamic activity of subtelomeres has been reported in such diverse organisms as yeast, human, and the malaria parasites, *Plasmodium* ssp. (Louis 1995; Freitas-Junior et al. 2000, 2005). As expected for a plastic region of the genome subject to reshuffling through recombination events, subtelomeres exhibit extremely high levels of within-species structural and nucleotide polymorphism and they often contain fast-evolving genes involved in adaptive processes (Linardopoulou et al. 2005). In parasitic eukaryotes such as *Plasmodium* spp. and pathogenic fungi, many virulence genes reside at subtelomeres (Leech et al. 1984; Sunkin and Stringer 1996; Vanhamme et al. 2001; Barry et al. 2003; Kamoun 2007). For example, in *Magnaporthe oryzae*, the agent of rice blast, many avirulence genes are located in subtelomeric regions (Valent and Khang 2010; Farman 2007; Grandaubert et al. 2014). In plants, this plasticity of subtelomeres has not been identified in *Arabidopsis thaliana* and, to our knowledge, has not yet been investigated at a large scale for other plant species with full genome sequences available. In common bean, classical genetic analyses have suggested a particular organization of disease resistance (*R*) genes with most large clusters located at the end of the linkage groups. The availability of the complete genome sequence of *Phaseolus vulgaris* has offered a unique opportunity to investigate the peculiar features of common bean subtelomeres at the genome-wide level, in particular, concerning resistance genes and repeated sequences.

9.2 Disease Resistance Gene Clusters Against Various Pathogens Are Often Located at the End of the Linkage Groups in Common Bean Genome

In the common bean (*P. vulgaris* L.) genome, classical genetic analyses have shown that most of the well-characterized large resistance (*R*) gene clusters, containing resistance genes against various pathogens, are located at the ends (rather than the centers) of the linkage groups. For example, the *Co-x*, *I*, *B4*, *Co-4,* and *Co-2* loci have been mapped to the ends of B01, B02, B04, B08, and B11, respectively (Fig. 9.1). Genetic complexity of these clusters is illustrated by the *B4 R* gene cluster where *R* specificities and *R* quantitative trait loci (QTL) against various pathogens including the fungi *Colletotrichum lindemuthianum* (*Co* genes), *Uromyces appendiculatus* (*Ur* genes), *Pseudocercospora griseola* (*Phg* genes), and *Erysiphe diffusa* (*Pm* genes) and the bacteria *Pseudomonas syringae* (*Pse* or *Rpsar* genes) have been mapped (Geffroy et al. 1998, 1999; Miklas et al. 2006; Goncalves-Vidigal et al. 2013; Perez-Vega et al. 2013). Similarly, at the *Co-2* (end of LG B11) and *Co-4 R* locus (end of LG B08), *R* specificities against both the bacteria *P. syringae* and the fungi *C. lindemuthianum* have been mapped (Adam-Blondon 1994; Geffroy et al. 1999; David et al. 2008; Perez-Vega et al. 2013). The *I* cluster, located at one end of

LG B2, is another remarkable example of complex cluster of *R* gene mainly against viruses but also against fungi and bacteria (Fisher and Kyle 1994; Morales and Singh 1997; Miklas et al. 2006; Geffroy et al. 2008; Pflieger et al. 2014; Hart and Griffiths 2015). Whether these genes are tightly linked genes, alleles, or eventually the same gene is not known.

9.3 Genomic Distribution of NB-LRR (NL) Sequences in Common Bean: Case Study of the *B4* and the *Co-2* Resistance Gene Clusters

In plants, the prevalent class of *R* genes encodes proteins containing a nucleotide-binding (NB) and a C-terminal leucine-rich repeat (LRR) domain (NB-LRR). *R* genes belonging to this class have been identified in various plant species, in monocots as well as in dicots, and correspond to *R* genes effective against all types of pathogens and pests, including fungi, bacteria, viruses, nematodes, oomycetes, and insects (Dangl et al. 2013; Michelmore et al. 2013). This NB-LRR protein class can be divided into two subclasses on the basis of their amino-terminal sequence, corresponding to two ancient lineages (Bai et al. 2002; Meyers et al. 2003; Ameline-Torregrosa et al. 2008). One subclass is composed of toll-interleukin-1 receptor (TIR)-NB-LRR (TNL) encoding genes characterized by the TIR domain homologous to the *Drosophila* toll and mammalian interleukin-1 receptor (Hammond-Kosack and Jones 1997). The second subclass is composed of NB-LRR encoding genes without the TIR motif, which often includes a coiled-coil (CC) domain (Pan et al. 2000).

In common bean, molecular analysis of the *Co-2* and *B4* clusters based on BAC clone sequencing has revealed that these two complex *R* clusters consist of a tandem array of more than 40 CC-NB-LRR (CNL) sequences (Geffroy et al. 1998; Creusot et al. 1999) (Fig. 9.2a). Fluorescence in situ hybridization (FISH) analysis revealed a subtelomeric location for these two complex *R* clusters (Geffroy et al. 2009) (Fig. 9.2b, e).

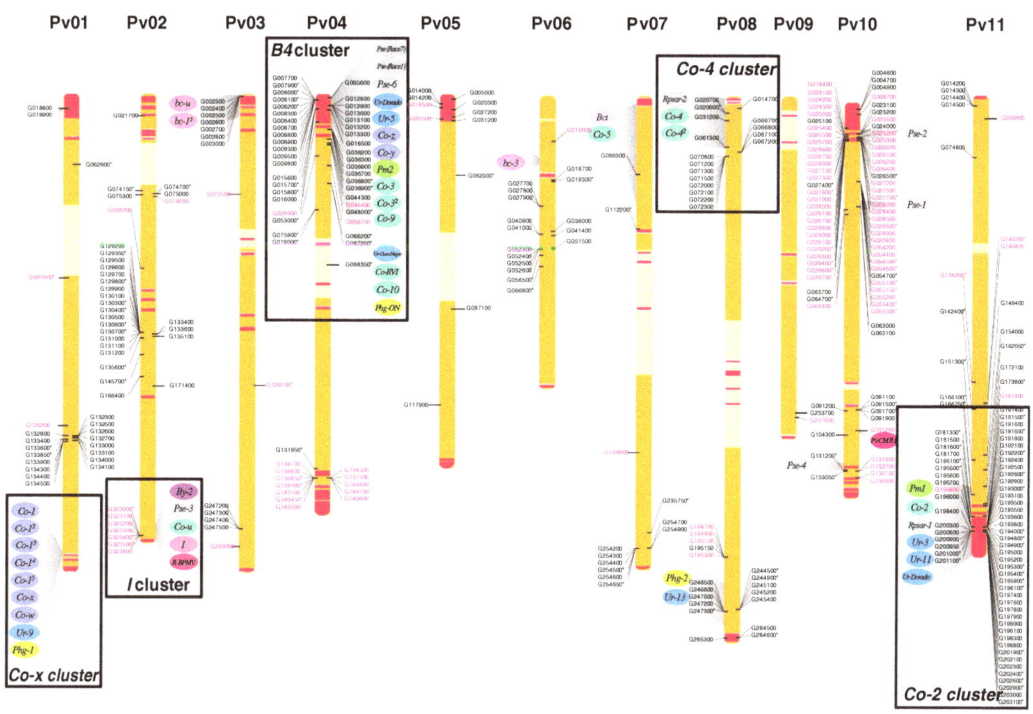

Conserved microsynteny was observed between the *P. vulgaris B4* locus and corresponding regions of *Medicago truncatula* (*Mt*) and *Lotus japonicus* (*Lj*) in chromosomes *Mt6* and *Lj2*, respectively (Fig. 9.2a). The notable exception was the CNL sequences, which were completely absent in these regions in *M. truncatula* and *L. japonicus*. The origin of the *P. vulgaris B4*-CNL sequences was investigated through phylogenetic analysis (Fig. 9.3), which reveals that, in the *P. vulgaris* genome, closely related CNL sequences (*B4* and *Co-2*-CNL) are shared among non-homologous chromosomes (4 and 11). Together, these results suggest that *P. vulgaris B4*-CNL was derived from CNL sequences from another cluster, the *Co-2* cluster, through an ectopic recombination event between non-homologous chromosomes in subtelomeric regions (David et al. 2009). Plasticity of subtelomeres has been described in various organisms but is not well documented in plants.

◄ **Fig. 9.1** Chromosomal distribution of common bean NB-LRR (NL) encoding genes, approximate locations of genetically characterized monogenic disease resistance genes against various pathogens, and *khipu* satellites. The relative map position of 377 NB-LRR (NL) encoding genes is shown on the individual pseudomolecules depicting the chromosomes Pv01–Pv11 as identified in Schmutz et al. (2014). Pseudomolecules Pv05, Pv07, Pv08, and Pv11 are in the same orientation than linkage groups (LG) B5, B7, B8, and B11, respectively, while pseudomolecules Pv01, Pv02, Pv03, Pv04, Pv06, Pv09, and Pv10 are in reverse orientation relative to LG B1, B2, B3, B4, B6, B9, and B10, respectively, according to genetic maps presented in Freyre et al. (1998), Geffroy et al. (2000), and Miklas et al. (2006). Each NB-LRR gene has a unique label representing the seven last informative digits from the annotation. For example, G132300 located on pseudomolecule 1 corresponds to the gene *Phvul.001G132300*. Genes encoded by the positive and negative DNA strands are depicted on the right and left sides of the chromosome, respectively. TNL sequences are presented in pink font and CNL sequences are presented in black font. NLs corresponding to pseudogenes are denoted by an asterisk (*) after their name. Centromeric positions on individual pseudomolecules were identified using centromere satellite repeats identified in Iwata et al. (2013) and are represented in light yellow bars on the pseudomolecules. To locate *R* gene on the physical map, sequences of available linked markers were used as query in BLASTN analysis against G19833 genome (http://phytozome.jgi.doe.gov/pz/portal.html). Approximate locations of monogenic disease resistance genes are indicated by gray area connecting colored bubbles (*R* genes) to candidate location on the pseudomolecules. The *Co* are anthracnose (*Colletotrichum lindemuthianum*) resistance loci (Goncalves-Vidigal and Kelly 2006; Miklas et al. 2006; David et al. 2008; Geffroy et al. 2008; Campa et al. 2009; Vallejo and Kelly 2009; Goncalves-Vidigal et al. 2011; Richard et al. 2014; Oblessuc et al. 2015). Andean and Mesoamerican anthracnose resistance loci are represented by lavender and turquoise bubbles, respectively. *Ur* rust (*U. appendiculatus*) resistance loci (*Ur-Dorado*, *Ur-Ouro Negro*, and *Ur-BAC 6* refer to the line source of unnamed genes) (Miklas et al. 2006; de Souza et al. 2011; Souza et al. 2014) are represented by blue bubbles, *Phg* angular leaf spot (*P. griseola*) resistance loci (Goncalves-Vidigal et al. 2011, 2013) are represented by light green bubbles, *Pm* powdery mildew (*E. diffusa*) resistance loci (Perez-Vega et al. 2013) are represented by green bubbles, and *Pse* or *Rpsar* halo blight (*P. syringae*) resistance loci (Chen et al. 2010; Miklas et al. 2014) are represented by gray bubbles. *R* genes against viruses are depicted by bubbles in various shades of pink. *I* is a resistance locus to BCMNV (*Bean common mosaic necrosis virus*), BCMV (*Bean common mosaic virus*), ZYMV (*Zucchini yellow mosaic virus*), BSMV (*Bean severe mosaic virus*), PWVK (*Passionfruit woodiness virus-K*), WMV (*Watermelon mosaic virus-2*), CabMV (*Cowpea aphid-borne mosaic virus*), ThPV (*Thaïland passiflora mosaic virus*) and SMV (*Soybean mosaic virus*) (Fisher and Kyle 1994, 1996; Morales and Singh 1997; Vallejos et al. 2000, 2006; Bello et al. 2014). *Bct* is a locus for resistance to BCTV (*Beet curly top virus*) (Miklas et al. 2006), *By-2* is an *R* gene for resistance to BYMV (*Bean yellow mosaic virus*), and CLYVV (*Clover yellow vein virus*) (Hart and Griffiths 2015), *PvCMR1* is an *R* gene (TNL) for resistance to CMV (*Cucumber mosaic virus*) positioned using sequences of specific PCR primers as query in BLASTN analysis against G19833 genome (http://phytozome.jgi.doe.gov/pz/portal.html) (Seo et al. 2006). *R-BPMV* is an *R* gene for resistance to BPMV (*Bean pod mottle virus*) (Pflieger et al. 2014), and *bc* are recessive genes for resistance to BCMNV and BCMV (Strausbaugh et al. 1999; Mukeshimana et al. 2005; Miklas et al. 2000; Hart and Griffiths 2013). The major subtelomeric *R* clusters *Co-x*, *I*, *B4*, *Co-4*, and *Co-2* are indicated with black frames. *Khipu* satellite sequences are indicated with pink bars on pseudomolecule structures according to Richard et al. (2013) with size of block proportional to number of *khipu* units

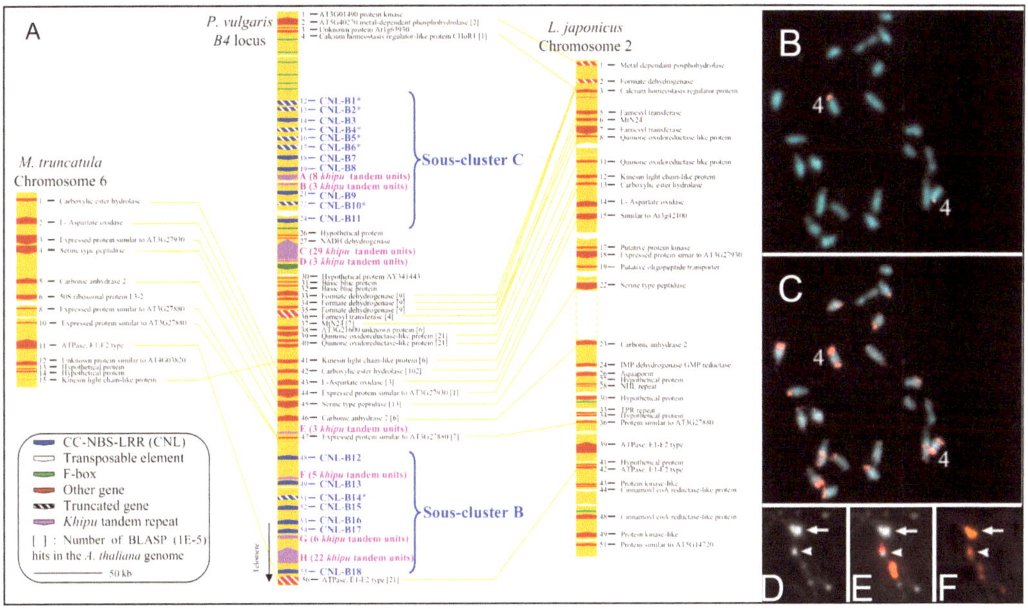

Fig. 9.2 **A** Sequence comparison between the *P. vulgaris* BAT93 B4 410-kb contig (center) and syntenic regions in *M. truncatula* chromosome 6 (left) and *L. japonicus* chromosome 2 (right). Yellow lines indicate significant homology matches between predicted genes. **B** and **C**. FISH to mitotic Pv BAT93 chromosomes using B4-CNL (**B**) and *khipu* (**C**) as probe. **D, E, F**. FISH to pachytene chromosomes (**D**) using B4-CNL (**E**) and *khipu* as probe (**F**). The major knob and the minor knob are indicated with an arrow and an arrowhead, respectively (David et al. 2009)

9.4 Genome-Wide Analysis of NB-LRR (NL) in the Bean Genome (G19833)

The availability of the common bean genome (genotype G19833) allowed genome-wide annotation of NB-LRR (NL) encoding genes (Schmutz et al. 2014); 377 NL genes were identified, of which 106 (28.2%) encoded TNL and 270 (71.8%) encoded CNL (Table 9.1). The majority of these NL sequences were physically organized in complex clusters, often located at the ends of chromosomes (Fig. 9.1). In particular, three large clusters were identified at the ends of chromosomes Pv04, Pv10, and Pv11 and contained more than 40 NL genes. Two of them correspond to the previously mentioned *B4* (Pv04) and *Co-2 R* (Pv11) clusters containing mainly CNL sequences. The later one, located at one end of Pv10, is enriched for TNL sequences. Approximate location of genetically characterized *R* genes against various pathogens in common bean is presented in Fig. 9.1. Most of the complex genetically characterized clusters of *R* genes are located at the end of the pseudomolecules and correspond to NL clusters. This is true for the previously mentioned *B4* and *Co-2 R* clusters where many *R* specificities against the fungi *C. lindemuthianum* (*Co* genes), *U. appendiculatus* (*Ur* genes), and *E. diffusa* (*Pm* genes) as well as against the bacteria *P. syringae* (*Pse* and *Rpsar* genes) co-localize with huge NL sequences clusters (Geffroy et al. 1998, 1999; David et al. 2008; Chen et al. 2010; de Souza et al. 2011; Goncalves-Vidigal et al. 2013; Perez-Vega et al. 2013; Miklas et al. 2014). Similarly, the multiresistance *I* cluster located at the end of Pv02 co-localizes with a cluster of TNL sequences. The same situation (complex distal *R* gene cluster co-locating with NL sequences) is also observed at the ends of Pv08 and Pv10 (Fig. 9.1). This strongly suggests that NL genes are the molecular basis of these

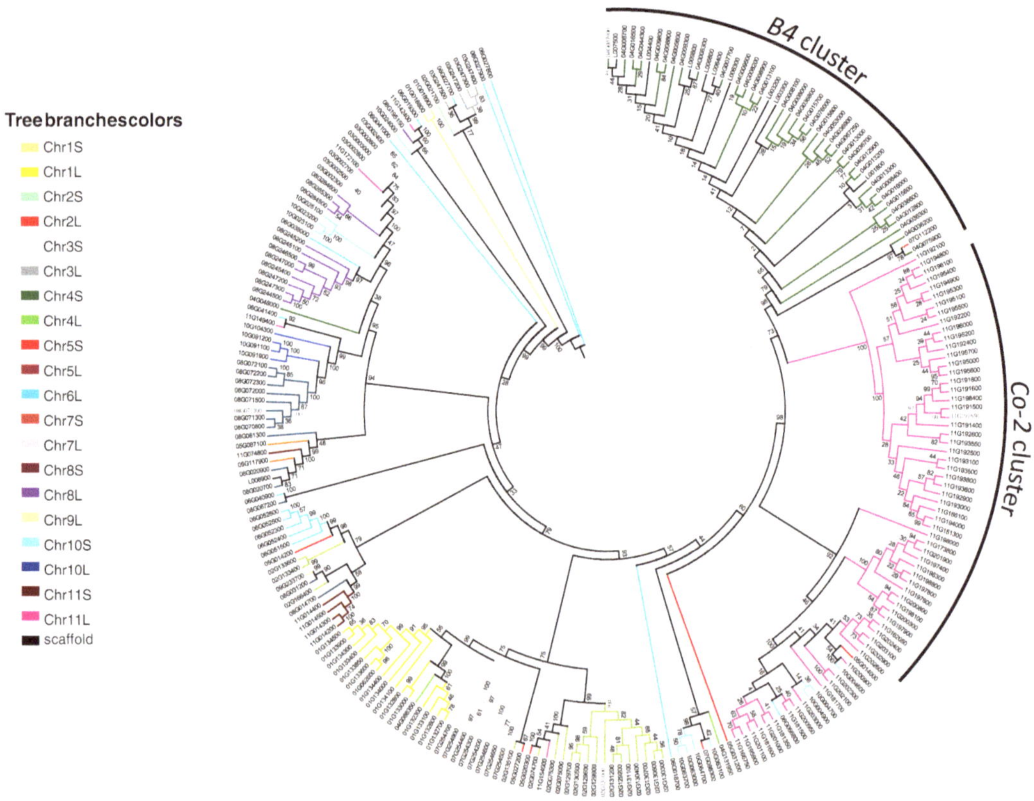

Fig. 9.3 Maximum likelihood phylogenetic analysis of the predicted common bean non-TNL genes. Phylogenetic tree of the 270 non-TNL genes from the G19833 genome sequence. Nucleic acid sequences of the region spanning from the P-loop to the MHD-conserved protein motifs were used to construct the phylogenetic tree. A color has been assigned to each chromosome arm, L (long arm) or S (short arm), and consequently, each gene from a same chromosome arm has branches with the same color in the tree. Non-TNL sequences from *B4* and *Co-2* clusters are indicated with brackets

Table 9.1 Number and classification of predicted NB-LRR encoding *R* genes in *Phaseolus vulgaris* genome

	#Full length genes	#Pseudogenes	#Total	Percentage
TNL type	*82*	*24*	*106*	*28.1*
TIR-NB-LRR	73	20	93	
TIR-NB	9	4	13	
non-TNL type	*186*	*85*	*271*	*71.9*
CC-NB	3	1	4	
NB	5	2	7	
NB-LRR	92	64	156	
CC-NB-LRR	86	18	104	
#Total	268	109	377	

R specificities. However, a different situation is observed for the resistance specificities *Co-1*, *Co-1*^{2}, *Co-1*^{3}, *Co-1*^{4}, *Co-1*^{5}, *Co-x*, *Co-w*, *Ur-9*, and *Phg-1* at the end of Pv01 since they are located in a region that does not contain NL sequences, suggesting that these *R* genes might correspond to atypical *R* genes (Richard et al. 2014). Recessive *R* genes against viruses have been shown to correspond to *eIF4e* genes in various plant species such as pepper, lettuce, and *Arabidopsis* (Robaglia and Caranta 2006), and similar results were observed in common bean for *bc-3* gene on Pv06 (Hart and Griffiths 2013).

If the NL sequences represent the prevalent class of *R* genes in all plant species, the preferential location of NL cluster at the end of the pseudomolecules was not observed for other plant species such as *Arabidopsis* (Meyers et al. 2003), rice (Luo et al. 2012), or *Medicago* (Young et al. 2011). However, similar distal distribution seems to be present in the potato genome (Jupe et al. 2012). Phylogenetic analysis of CNL sequences confirms that *B4*-CNL sequences are more similar to *Co-2*-CNL sequences than to any other CNL sequences (Fig. 9.3).

9.5 Terminal Knobs Containing *Khipu* Satellite DNA

Satellite DNA can be defined as highly reiterated noncoding DNA sequences, organized as long arrays of head-to-tail linked repeats of 150–180bp or 300–360bp monomers located in the constitutive heterochromatin (Plohl et al. 2008). During the bioinformatic analysis of the *B4 R* gene cluster, a new 528 bp satellite repeat, referred to as *khipu*, was identified between the CNL sequences (David et al. 2009). In order to determine the pattern of *khipu* distribution at a greater resolution, FISH was performed on meiotic pachytene chromosomes because they are less condensed than somatic chromosomes. Terminal knobs (heterochromatic blocks) of different sizes were visible at most chromosome ends of *P. vulgaris*, and *khipu* tandem repeats were present on 17 chromosome ends (Fig. 9.2c), mostly corresponding to cytologically visible terminal knobs (Fig. 9.2d, f). Southern experiments on a range of legume species using *khipu* as a probe have shown that *khipu* is specific to the *Phaseolus* genus (David et al. 2009). The presence of *khipu* on 17 chromosome ends, combined with its specificity to the *Phaseolus* genus, suggests the existence of frequent ectopic recombination events in *Phaseolus* subtelomeric regions.

The availability of the complete genome sequence of the common bean genotype G19833 allows investigating the genomic organization and the evolution of *khipu*. This genome-wide analysis led to the identification of 2460 *khipu* units located, as expected based on the FISH results, at most distal ends of the sequenced regions (Richard et al. 2013) (Fig. 9.1; Table 9.2). *khipu* sequences are arranged in discrete blocks of at least 2–55 units and are heterogeneously distributed among the different chromosome ends of G19833 (from 0 to 555 *khipu* units per chromosome end) (Table 9.2). Notably, chromosome ends containing the highest number of *khipu* units (short arms of chromosome 4 and long arms of chromosomes 10 and 11, with 349, 435, and 555 *khipu* units, respectively) are also those harboring the largest NL clusters in common bean genome (Fig. 9.1; Table 9.2). Comparison between phylogeny and physical distribution of *khipu* repeats reveals that phylogenetically related *khipu* units are spread between numerous chromosome ends, confirming the existence of frequent exchanges between non-homologous subtelomeres in common bean (Richard et al. 2013). However, most subclades contain numerous *khipu* units from only one or few chromosome ends indicating that local duplication is also driving *khipu* expansion. Altogether, these data suggest extensive sequence exchanges in subtelomeres between non-homologous chromosomes in common bean and confirm that subtelomeres represent one of the most dynamic and rapidly evolving regions in eukaryotic genomes. Interestingly, co-localization of *khipu* sequences with large NL clusters was observed on Pv04, Pv10, and Pv11 suggesting that *khipu* sequences could have played a role in the amplification on NL sequences through unequal crossing over. However, not all large NL clusters co-localized with *khipu* sequences, as for example the *I* cluster located at one end of chromosome 2.

In contrast to large genome plant species, heterochromatin is largely restricted to pericentromeric regions in plant species with a small genome. For example, only two knobs have been

Table 9.2 Number of *khipu* units in each pseudomolecule of *Phaseolus vulgaris*

Pseudomolecules	Short arm	Centromere	Long arm	Total
Chr01	193	0	19	212
Chr02	46	0	39	85
Chr03	60	3	17	80
Chr04	349	13	261	623
Chr05	169	0	33	202
Chr06	0	0	17	17
Chr07	17	7	14	38
Chr08	7	55	51	113
Chr09	0	2	10	12
Chr10	71	1	435	507
Chr11	16	0	555	571
Total		81		2460

reported in the compact *A. thaliana* genome (125 Mbp), while in the large maize (*Zea mays*) genome (2671 Mbp), numerous knobs have been reported (Peacock et al. 1981; Ananiev et al. 1998; Fransz et al. 1998). The relatively small genome of *P. vulgaris* does not seem to follow this tendency. Indeed, heterochromatic knobs have been detected in most *P. vulgaris* chromosome termini and *khipu* tandem repeats are components of most of them (David et al. 2009) (Fig. 9.2). The presence of terminal knobs in *P. vulgaris* is in sharp contrast with results from other legume species such as *L. japonicus* and *M. truncatula*, where most of the heterochromatin is localized at pericentromeric regions and no terminal heterochromatic blocks have been reported, except for the 45S rDNA cluster on *Lj2* (Kulikova et al. 2001; Pedrosa et al. 2002). Thus, the complexity of bean subtelomeres does not seem to be obviously related to its genome size, because with 587 Mbp, *P. vulgaris* genome is not significantly larger than *L. japonicus* (472 Mbp) or *M. truncatula* (500 Mbp) genomes (Sato et al. 2008; Young et al. 2011). In cereals, similar results were obtained for rye (*Secale cereale* L.) genome. Indeed, rye differs from phylogenetically related wheat (*Triticum*) and barley (*Hordeum*) in having large heterochromatin blocks in the subtelomeric regions of its chromosomes (Evtushenko et al. 2010; Evtushenko and Vershinin 2010).

9.6 Conclusions: Hypothesis Concerning the Huge Size of Subtelomeric NL Clusters in Common Bean

In the common bean genome, most of the *R* gene clusters are located at the ends of the chromosomes and present a huge size compared to other species (Fig. 9.1). For some of them, FISH analyses have revealed a subtelomeric location. This is the case for the *B4*, *Co-4,* and *Co-2* clusters on Pv04, Pv08, and Pv11, respectively (David et al. 2009; Geffroy et al. 2009; Melotto et al. 2004). With regard to their huge size, it is particularly impressive for the NL clusters located at the end of Pv04, Pv10, and Pv11 each containing more than 40 NL sequences. Microsynteny analyses of the *Co-2* and *B4* clusters between soybean and common bean have shown an impressive amplification of NL sequences in common bean since only one and ~20 NL were identified in the corresponding regions of soybean for the *B4* and *Co-2* cluster, respectively (Innes et al. 2008; David et al. 2009). This suggests that these subtelomeric regions in common bean are favorable "niches" for *R* gene proliferation.

R gene expression may be associated with a fitness cost on the host plant (Purrington 2000;

Tian et al. 2003). Consequently, *R* gene expression must be carefully regulated, especially in plants species containing many NL genes. Different mechanisms are emerging as implicated in the regulation of NL sequences including miRNA modulation of NL expression (Zhai et al. 2011; Gonzalez et al. 2015). However, because this process is not specific to common bean, since it has been identified in various plant species from tomato to spruce, it cannot explain the huge size of NL clusters in common bean specifically (Kallman et al. 2013; Li et al. 2012; Shivaprasad et al. 2012). One element that could explain this impressive expansion of NL sequences is their peculiar genomic location at common bean subtelomeres. First, distal regions of the chromosomes are known to be highly recombinant compared to pericentromeric regions (Gore et al. 2009; Schmutz et al. 2010, 2014) and are consequently favorable to promote NL amplification through unequal crossing over. In addition, the identification of *khipu* tandem repeats tightly linked to the huge NL clusters on Pv04, Pv10, and Pv11 strongly suggests that amplification of NL sequences could have been promoted by unequal crossing over involving *khipu* sequences. Secondly, peculiar features of common bean genome include the presence of terminal knobs (heterochromatic blocks), most of them containing the *khipu* satellite DNA (David et al. 2009; Richard et al. 2013). At least two common bean NL clusters have been reported to be in vicinity of these terminal knobs, *B4* and *Co-2* clusters (David et al. 2009; Chen et al., in preparation). Epigenetic silencing of genes in proximity to heterochromatic regions and to repeated sequences through spreading of methylation is a known process (Vitte et al. 2014; Talbert and Henikoff 2006). Consequently, we propose that this peculiar genomic environment may favor the proliferation of large *R* gene clusters due to, not only increased recombination, but also to some form of silencing (Yi and Richards 2007), allowing a large amplification of NL sequences without fitness cost in common bean.

References

Adam-Blondon AF (1994) Cartographie génétique de *Phaseolus vulgaris L.* localisation de gènes de résistance à l'anthracnose. INAPG, Paris, France, p 160

Ameline-Torregrosa C, Wang BB, O'Bleness MS, Deshpande S, Zhu HY, Roe B, Young ND, Cannon SB (2008) Identification and characterization of nucleotide-binding site-leucine-rich repeat genes in the model plant *Medicago truncatula*. Plant Physiol 146:5–21. doi:10.1104/pp.107.104588

Ananiev EV, Phillips RL, Rines HW (1998) Complex structure of knob DNA on maize chromosome 9: retrotransposon invasion into heterochromatin. Genetics 149:2025–2037

Bai JF, Pennill LA, Ning JC, Lee SW, Ramalingam J, Webb CA, Zhao BY, Sun Q, Nelson JC, Leach JE, Hulbert SH (2002) Diversity in nucleotide binding site-leucine-rich repeat genes in cereals. Genome Res 12:1871–1884

Barry JD, Ginger ML, Burton P, McCulloch R (2003) Why are parasite contingency genes often associated with telomeres? Int J Parasit 33:29–45. doi:10.1016/s0020-7519(02)00247-3

Bello MH, Moghaddam SM, Massoudi M, McClean PE, Cregan PB, Miklas PN (2014) Application of in silico bulked segregant analysis for rapid development of markers linked to *Bean common mosaic virus* resistance in common bean. BMC Genom 15. doi:10.1186/1471-2164-15-903

Campa A, Giraldez R, Ferreira JJ (2009) Genetic dissection of the resistance to nine anthracnose races in the common bean differential cultivars MDRK and TU. Theor Appl Genet 119:1–11. doi:10.1007/s00122-009-1011-8

Chen NWG, Sevignac M, Thareau V, Magdelenat G, David P, Ashfield T, Innes RW, Geffroy V (2010) Specific resistances against *Pseudomonas syringae* effectors AvrB and AvrRpm1 have evolved differently in common bean (*Phaseolus vulgaris*), soybean (*Glycine max*), and *Arabidopsis thaliana*. New Phytol 187:941–956. doi:10.1111/j.1469-8137.2010.03337.x

Creusot F, Macadre C, Cana EF, Riou C, Geffroy V, Sevignac M, Dron M, Langin T (1999) Cloning and molecular characterization of three members of the NBS-LRR subfamily located in the vicinity of the *Co-2* locus for anthracnose resistance in *Phaseolus vulgaris*. Genome 42:254–264. doi:10.1139/gen-42-2-254

Dangl JL, Horvath DM, Staskawicz BJ (2013) Pivoting the plant immune system from dissection to deployment. Science 341:746–751. doi:10.1126/science.1236011

David P, Chen NWG, Pedrosa-Harand A, Thareau V, Sevignac M, Cannon SB, Debouck D, Langin T, Geffroy V (2009) A nomadic subtelomeric disease resistance gene cluster in common bean. Plant Physiol 151:1048–1065. doi:10.1104/pp.109.142109

David P, Sevignac M, Thareau V, Catillon Y, Kami J, Gepts P, Langin T, Geffroy V (2008) BAC end sequences corresponding to the *B4* resistance gene cluster in common bean: a resource for markers and synteny analyses. Mol Genet Genomics 280:521–533. doi:10.1007/s00438-008-0384-8

de Souza TLPO, Dessaune SN, Sanglard DA, Moreira MA, de Barros EG (2011) Characterization of the rust resistance gene present in the common bean cultivar Ouro Negro, the main rust resistance source used in Brazil. Plant Pathol 60:839–845. doi:10.1111/j.1365-3059.2011.02456.x

Evtushenko EV, Elisafenko EA, Vershinin AV (2010) The relationship between two tandem repeat families in rye heterochromatin. Mol Biol 44:1–7. doi:10.1134/s0026893310010012

Evtushenko EV, Vershinin AV (2010) Heterogeneous organization of a tandem repeat family in subtelomeric heterochromatin of rye. Russ J Genet 46:1074–1076. doi:10.1134/s1022795410090152

Farman ML (2007) Telomeres in the rice blast fungus *Magnaporthe oryzae*: the world of the end as we know it. FEMS Microbiol Lett 273:125–132. doi:10.1111/j.1574-6968.2007.00812.x

Fisher ML, Kyle MM (1994) Inheritance of resistance to potyviruses in *Phaseolus vulgaris* L. 3. Cosegregation of phenotypically similar dominant responses to 9 potyviruses. Theor Appl Genet 89:818–823

Fisher ML, Kyle MM (1996) Inheritance of resistance to potyviruses in *Phaseolus vulgaris* L. 4. Inheritance, linkage relations, and environmental effects on systemic resistance to four potyviruses. Theor Appl Genet 92:204–212. doi:10.1007/bf00223377

Fransz P, Armstrong S, Alonso-Blanco C, Fischer TC, Torres-Ruiz RA, Jones G (1998) Cytogenetics for the model system *Arabidopsis thaliana*. Plant J 13:867–876. doi:10.1046/j.1365-313X.1998.00086.x

Freitas-Junior LH, Bottius E, Pirrit LA, Deitsch KW, Scheidig C, Guinet F, Nehrbass U, Wellems TE, Scherf A (2000) Frequent ectopic recombination of virulence factor genes in telomeric chromosome clusters of *P. falciparum*. Nature 407:1018–1022

Freitas-Junior LH, Hernandez-Rivas R, Ralph SA, Montiel-Condado D, Ruvalcaba-Salazar OK, Rojas-Meza AP, Mancio-Silva L, Leal-Silvestre RJ, Gontijo AM, Shorte S, Scherf A (2005) Telomeric heterochromatin propagation and histone acetylation control mutually exclusive expression of antigenic variation genes in malaria parasites. Cell 121:25–36. doi:10.1016/j.cell.2005.01.037

Freyre R, Skroch PW, Geffroy V, Adam-Blondon AF, Shirmohamadali A, Johnson WC, Llaca V, Nodari RO, Pereira PA, Tsai SM, Tohme J, Dron M, Nienhuis J, Vallejos CE, Gepts P (1998) Towards an integrated linkage map of common bean. 4. Development of a core linkage map and alignment of RFLP maps. Theor Appl Genet 97:847–856

Geffroy V, Creusot F, Falquet J, Sévignac M, Adam-Blondon AF, Bannerot H, Gepts P, Dron M (1998) A family of LRR sequences in the vicinity of the *Co-2* locus for anthracnose resistance in *Phaseolus vulgaris* and its potential use in marker-assisted selection. Theor Appl Genet 96:494–502

Geffroy V, Macadre C, David P, Pedrosa-Harand A, Sevignac M, Dauga C, Langin T (2009) Molecular analysis of a large subtelomeric nucleotide-binding-site-leucine-rich-repeat family in two representative genotypes of the major gene pools of *Phaseolus vulgaris*. Genetics 181:405–419. doi:10.1534/genetics.108.093583

Geffroy V, Sevignac M, Billant P, Dron M, Langin T (2008) Resistance to *Colletotrichum lindemuthianum* in *Phaseolus vulgaris*: a case study for mapping two independent genes. Theor Appl Genet 116:407–415. doi:10.1007/s00122-007-0678-y

Geffroy V, Sevignac M, De Oliveira JCF, Fouilloux G, Skroch P, Thoquet P, Gepts P, Langin T, Dron M (2000) Inheritance of partial resistance against *Colletotrichum lindemuthianum* in *Phaseolus vulgaris* and co-localization of quantitative trait loci with genes involved in specific resistance. Mol Plant Microbe Interact 13:287–296. doi:10.1094/mpmi.2000.13.3.287

Geffroy V, Sicard D, de Oliveira JCF, Sevignac M, Cohen S, Gepts P, Neema C, Langin T, Dron M (1999) Identification of an ancestral resistance gene cluster involved in the coevolution process between *Phaseolus vulgaris* and its fungal pathogen *Colletotrichum lindemuthianum*. Mol Plant Microbe Interact 12:774–784

Goncalves-Vidigal MC, Cruz AS, Garcia A, Kami J, Vidigal PS, Sousa LL, McClean P, Gepts P, Pastor-Corrales MA (2011) Linkage mapping of the *Phg-1* and *Co-1*^{4} genes for resistance to angular leaf spot and anthracnose in the common bean cultivar AND 277. Theor Appl Genet 122:893–903. doi:10.1007/s00122-010-1496-1

Goncalves-Vidigal MC, Cruz AS, Lacanallo GF, Vidigal Filho PS, Sousa LL, Pacheco CMNA, McClean P, Gepts P, Pastor-Corrales MA (2013) Co-segregation analysis and mapping of the anthracnose *Co-10* and angular leaf spot *Phg-ON* disease-resistance genes in the common bean cultivar Ouro Negro. Theor Appl Genet 126:2245–2255. doi:10.1007/s00122-013-2131-8

Goncalves-Vidigal MC, Kelly JD (2006) Inheritance of anthracnose resistance in the common bean cultivar Widusa. Euphytica 151:411–419. doi:10.1007/s10681-006-9164-x

Gonzalez VM, Mueller S, Baulcombe D, Puigdomenech P (2015) Evolution of NBS-LRR gene copies among dicot plants and its regulation by members of the miR482/2118 superfamily of miRNAs. Mol Plant 8:329–331. doi:10.1016/j.molp.2014.11.013

Gore MA, Chia JM, Elshire RJ, Sun Q, Ersoz ES, Hurwitz BL, Peiffer JA, McMullen MD, Grills GS, Ross-Ibarra J, Ware DH, Buckler ES (2009) A first-generation haplotype map of maize. Science 326:1115–1117. doi:10.1126/science.1177837

Grandaubert J, Balesdent MH, Rouxel T (2014) Evolutionary and adaptive role of transposable elements in fungal genomes. In: Martin FM (ed) Fungi, vol 70. Adv Botan Res., pp 79–107. doi:10.1016/b978-0-12-397940-7.00003-3

Hammond-Kosack KE, Jones JDG (1997) Plant disease resistance genes. Ann Rev Plant Physiol Plant Mol Biol 48:575–607

Hart JP, Griffiths PD (2013) A series of *eIF4E* alleles at the *Bc-3* locus are associated with recessive resistance to *Clover yellow vein virus* in common bean. Theor Appl Genet 126:2849–2863. doi:10.1007/s00122-013-2176-8

Hart JP, Griffiths PD (2015) Genotyping-by-sequencing enabled mapping and marker development for the *By-2* potyvirus resistance allele in common bean. Plant Genome 8. doi:10.3835/plantgenome2014.09.0058

Innes RW, Ameline-Torregrosa C, Ashfield T, Cannon E, Cannon SB, Chacko B, Chen NWG, Couloux A, Dalwani A, Denny R, Deshpande S, Egan AN, Glover N, Hans CS, Howell S, Ilut D, Jackson S, Lai H, Mammadov J, del Campo SM, Metcalf M, Nguyen A, O'Bleness M, Pfeil BE, Podicheti R, Ratnaparkhe MB, Samain S, Sanders I, Segurens B, Sevignac M, Sherman-Broyles S, Thareau V, Tucker DM, Walling J, Wawrzynski A, Yi J, Doyle JJ, Geffroy V, Roe BA, Maroof MAS, Young ND (2008) Differential accumulation of retroelements and diversification of NB-LRR disease resistance genes in duplicated regions following polyploidy in the ancestor of soybean. Plant Physiol 148:1740–1759. doi:10.1104/pp.108.127902

Iwata A, Tek AL, Richard MMS, Abernathy B, Fonseca A, Schmutz J, Chen NWG, Thareau V, Magdelenat G, Li YP, Murata M, Pedrosa-Harand A, Geffroy V, Nagaki K, Jackson SA (2013) Identification and characterization of functional centromeres of the common bean. Plant J 76:47–60. doi:10.1111/tpj.12269

Jupe F, Pritchard L, Etherington GJ, MacKenzie K, Cock PJA, Wright F, Sharma SK, Bolser D, Bryan GJ, Jones JDG, Hein I (2012) Identification and localisation of the NB-LRR gene family within the potato genome. BMC Genom 13. doi:10.1186/1471-2164-13-75

Kallman T, Chen J, Gyllenstrand N, Lagercrantz U (2013) A Significant fraction of 21-nucleotide small RNA originates from phased degradation of resistance genes in several perennial species. Plant Physiol 162:741–754. doi:10.1104/pp.113.214643

Kamoun S (2007) Groovy times: filamentous pathogen effectors revealed. Curr Opin Plant Biol 10:358–365. doi:10.1016/j.pbi.2007.04.017

Kulikova O, Gualtieri G, Geurts R, Kim DJ, Cook D, Huguet T, de Jong JH, Fransz PF, Bisseling T (2001) Integration of the FISH pachytene and genetic maps of *Medicago truncatula*. Plant J 27:49–58

Leech JH, Barnwell JW, Miller LH, Howard RJ (1984) Identification of a strain-specific malarial antigen exposed on the surface of *Plasmodium falciparum* infected erythrocytes. J Exp Med 159:1567–1575. doi:10.1084/jem.159.6.1567

Li F, Pignatta D, Bendix C, Brunkard JO, Cohn MM, Tung J, Sun H, Kumar P, Baker B (2012) MicroRNA regulation of plant innate immune receptors. Proc Natl Acad Sci USA 109:1790–1795. doi:10.1073/pnas.1118282109

Linardopoulou EV, Williams EM, Fan YX, Friedman C, Young JM, Trask BJ (2005) Human subtelomeres are hot spots of interchromosomal recombination and segmental duplication. Nature 437:94–100

Louis EJ (1995) The chromosome ends of *Saccharomyces cerevisiae*. Yeast 11:1553–1573. doi:10.1002/yea.320111604

Luo S, Zhang Y, Hu Q, Chen J, Li K, Lu C, Liu H, Wang W, Kuang H (2012) Dynamic nucleotide-binding site and leucine-rich repeat-encoding genes in the grass family. Plant Physiol 159:197–210. doi:10.1104/pp.111.192062

Melotto M, Coelho MF, Pedrosa Harand A, Kelly JD, Camargo LEA (2004) The anthracnose resistance locus *Co-4* of common bean is located on chromosome 3 and contains putative disease resistance-related genes. Theor Appl Genet 109:690–699

Meyers BC, Kozik A, Griego A, Kuang HH, Michelmore RW (2003) Genome-wide analysis of NBS-LRR-encoding genes in *Arabidopsis*. Plant Cell 15:809–834

Michelmore RW, Christopoulou M, Caldwell KS (2013) Impacts of resistance gene genetics, function, and evolution on a durable future. Ann Rev Phytopathol 51:291–319. doi:10.1146/annurev-phyto-082712-102334

Miklas PN, Fourie D, Trapp J, Davis J, Myers JR (2014) New loci including *Pse-6* conferring resistance to halo bacterial blight on chromosome Pv04 in common bean. Crop Sci 54:2099–2108. doi:10.2135/cropsci2014.03.0213

Miklas PN, Kelly JD, Beebe SE, Blair MW (2006) Common bean breeding for resistance against biotic and abiotic stresses: from classical to MAS breeding. Euphytica 147:105–131

Miklas PN, Larsen RC, Riley R, Kelly JD (2000) Potential marker-assisted selection for *bc-1(2)* resistance to bean common mosaic potyvirus in common bean. Euphytica 116:211–219. doi:10.1023/a:1004006514814

Morales FJ, Singh SP (1997) Inheritance of the mosaic and necroses reactions induced by bean severe mosaic comoviruses in *Phaseolus vulgaris* L. Euphytica 93:223–226. doi:10.1023/a:1002902207395

Mukeshimana G, Paneda A, Rodriguez-Suarez C, Ferreira JJ, Giraldez R, Kelly JD (2005) Markers linked to the *bc-3* gene conditioning resistance to bean common mosaic potyviruses in common bean. Euphytica 144:291–299. doi:10.1007/s10681-005-7397-8

Oblessuc PR, Francisco C, Melotto M (2015) The *Co-4* locus on chromosome Pv08 contains a unique cluster of 18 *COK-4* genes and is regulated by immune response in common bean. Theor Appl Genet 128:1193–1208. doi:10.1007/s00122-015-2500-6

Pan QL, Liu YS, Budai-Hadrian O, Sela M, Carmel-Goren L, Zamir D, Fluhr R (2000) Comparative genetics of nucleotide binding site-leucine rich repeat resistance gene homologues in the genomes of two dicotyledons: tomato and *Arabidopsis*. Genetics 155:309–322

Peacock WJ, Dennis ES, Rhoades MM, Pryor AJ (1981) Highly repeated DNA-sequence limited to knob heterochromatin in maize. Proc Natl Acad Sci USA 78:4490–4494

Pedrosa A, Sandal N, Stougaard J, Schweizer D, Bachmair A (2002) Chromosomal map of the model legume *Lotus japonicus*. Genetics 161:1661–1672

Perez-Vega E, Trabanco N, Campa A, Ferreira JJ (2013) Genetic mapping of two genes conferring resistance to powdery mildew in common bean (*Phaseolus vulgaris* L.). Theor Appl Genet 126:1503–1512. doi:10.1007/s00122-013-2068-y

Pflieger S, Blanchet S, Meziadi C, Richard MMS, Thareau V, Mary F, Mazoyer C, Geffroy V (2014) The "one-step" *Bean pod mottle virus* (BPMV)-derived vector is a functional genomics tool for efficient overexpression of heterologous protein, virus-induced gene silencing and genetic mapping of *BPMV R*-gene in common bean (*Phaseolus vulgaris* L.). BMC Plant Biol 14:232. doi:10.1186/s12870-014-0232-4

Plohl M, Luchetti A, Mestrovic N, Mantovani B (2008) Satellite DNAs between selfishness and functionality: structure, genomics and evolution of tandem repeats in centromeric (hetero)chromatin. Gene 409:72–82. doi:10.1016/j.gene.2007.11.013

Purrington CB (2000) Costs of resistance. Curr Opin Plant Biol 3:305–308. doi:10.1016/s1369-5266(00)00085-6

Richard MM, Pflieger S, Sevignac M, Thareau V, Blanchet S, Li Y, Jackson SA, Geffroy V (2014) Fine mapping of *Co-x*, an anthracnose resistance gene to a highly virulent strain of *Colletotrichum lindemuthianum* in common bean. Theor Appl Genet 127:1653–1666. doi:10.1007/s00122-014-2328-5

Richard MMS, Chen NWG, Thareau V, Pflieger S, Blanchet S, Pedrosa-Harand A, Iwata A, Chavarro C, Jackson SA, Geffroy V (2013) The subtelomeric *khipu* satellite repeat from *Phaseolus vulgaris*: lessons learned from the genome analysis of the Andean genotype G19833. Front Plant Sci 4:109. doi:10.3389/fpls.2013.00109

Robaglia C, Caranta C (2006) Translation initiation factors: a weak link in plant RNA virus infection. Trends Plant Sci 11:40–45. doi:10.1016/j.tplants.2005.11.004

Sato S, Nakamura Y, Kaneko T, Asamizu E, Kato T, Nakao M, Sasamoto S, Watanabe A, Ono A, Kawashima K, Fujishiro T, Katoh M, Kohara M, Kishida Y, Minami C, Nakayama S, Nakazaki N, Shimizu Y, Shinpo S, Takahashi C, Wada T, Yamada M, Ohmido N, Hayashi M, Fukui K, Baba T, Nakamichi T, Mori H, Tabata S (2008) Genome structure of the legume, *Lotus japonicus*. DNA Res 15:227–239. doi:10.1093/dnares/dsn008

Schmutz J, Cannon S, Schlueter J, Ma J, Mitros T, Nelson W, Hyten D, Song Q, Thelen J, Cheng J, Xu D, Hellsten U, May G, Yu Y, Sakurai T, Umezawa T, Bhattacharyya M, Sandhu D, Valliyodan B, Lindquist E, Peto M, Grant D, Shu S, Goodstein D, Barry K, Futrell-Griggs M, Abernathy B, Du J, Tian Z, Zhu L (2010) Genome sequence of the palaeopolyploid soybean. Nature 14:178–183

Schmutz J, McClean PE, Mamidi S, Wu GA, Cannon SB, Grimwood J, Jenkins J, Shu S, Song Q, Chavarro C, Torres-Torres M, Geffroy V, Moghaddam SM, Gao D, Abernathy B, Barry K, Blair M, Brick MA, Chovatia M, Gepts P, Goodstein DM, Gonzales M, Hellsten U, Hyten DL, Jia G, Kelly JD, Kudrna D, Lee R, Richard MMS, Miklas PN, Osorno JM, Rodrigues J, Thareau V, Urrea CA, Wang M, Yu Y, Zhang M, Wing RA, Cregan PB, Rokhsar DS, Jackson SA (2014) A reference genome for common bean and genome-wide analysis of dual domestications. Nat Genet 46:707–713. doi:10.1038/ng.3008

Seo Y-S, Rojas MR, Lee J-Y, Lee S-W, Jeon J-S, Ronald P, Lucas WJ, Gilbertson RL (2006) A viral resistance gene from common bean functions across plant families and is up-regulated in a non-virus-specific manner. Proc Natl Acad Sci USA 103:11856–11861. doi:10.1073/pnas.0604815103

Shivaprasad PV, Chen HM, Patel K, Bond DM, Santos B, Baulcombe DC (2012) A microRNA superfamily regulates nucleotide binding site-leucine-rich repeats and other mRNAs. Plant Cell 24:859–874. doi:10.1105/tpc.111.095380

Souza T, Ragagnin VA, Dessaune SN, Sanglard DA, Carneiro JES, Moreira MA, Barros EG (2014) DNA marker-assisted selection to pyramid rust resistance genes in "carioca" seeded common bean lines. Euphytica 199:303–316. doi:10.1007/s10681-014-1126-0

Strausbaugh CA, Myers JR, Forster RL, McClean PE (1999) *Bc-1* and *bc-u* - two loci controlling bean common mosaic virus resistance in common bean are linked. J Am Soc Hort Sci 124:644–648

Sunkin SM, Stringer JR (1996) Translocation of surface antigen genes to a unique telomeric expression site in *Pneumocystis carinii*. Mol Microbiol 19:283–295. doi:10.1046/j.1365-2958.1996.375905.x

Talbert PB, Henikoff S (2006) Spreading of silent chromatin: inaction at a distance. Nat Rev Genet 7:793–803. doi:10.1038/nrg1920

Tian D, Traw MB, Chen JQ, Kreitman M, Bergelson J (2003) Fitness costs of *R*-gene-mediated resistance in *Arabidopsis thaliana*. Nature 423:74–77

Valent B, Khang CH (2010) Recent advances in rice blast effect or research. Curr Opin Plant Biol 13:434–441. doi:10.1016/j.pbi.2010.04.012

Vallejo V, Kelly JD (2009) New insights into the anthracnose resistance of common bean landrace G2333. Open Hort J 2:29–33. doi:10.2174/1874840600902010029

Vallejos CE, Astua-Monge G, Jones V, Plyler TR, Sakiyama NS, Mackenzie SA (2006) Genetic and molecular characterization of the *I* locus of *Phaseolus vulgaris*. Genetics 172:1229–1242

Vallejos CE, Malandro JJ, Sheehy K, Zimmermann MJ (2000) Detection and cloning of expressed sequences linked to a target gene. Theor Appl Genet 101:1109–1113

Vanhamme L, Lecordier L, Pays E (2001) Control and function of the bloodstream variant surface glycoprotein expression sites in *Trypanosoma brucei*. Int J Parasitol 31:523–531. doi:10.1016/s0020-7519(01)00143-6

Vitte C, Fustier M-A, Alix K, Tenaillon MI (2014) The bright side of transposons in crop evolution. Brief Funct Genom 13:276–295. doi:10.1093/bfgp/elu002

Yi H, Richards EJ (2007) A cluster of disease resistance genes in *Arabidopsis* is coordinately regulated by transcriptional activation and RNA silencing. Plant Cell 19:2929–2939. doi:10.1105/tpc.107.051821

Young ND, Debelle F, Oldroyd GED, Geurts R, Cannon SB, Udvardi MK, Benedito VA, Mayer KFX, Gouzy J, Schoof H, Van de Peer Y, Proost S, Cook DR, Meyers BC, Spannagl M, Cheung F, De Mita S, Krishnakumar V, Gundlach H, Zhou SG, Mudge J, Bharti AK, Murray JD, Naoumkina MA, Rosen B, Silverstein KAT, Tang HB, Rombauts S, Zhao PX, Zhou P, Barbe V, Bardou P, Bechner M, Bellec A, Berger A, Berges H, Bidwell S, Bisseling T, Choisne N, Couloux A, Denny R, Deshpande S, Dai XB, Doyle JJ, Dudez AM, Farmer AD, Fouteau S, Franken C, Gibelin C, Gish J, Goldstein S, Gonzalez AJ, Green PJ, Hallab A, Hartog M, Hua A, Humphray SJ, Jeong DH, Jing Y, Jocker A, Kenton SM, Kim DJ, Klee K, Lai HS, Lang CT, Lin SP, Macmil SL, Magdelenat G, Matthews L, McCorrison J, Monaghan EL, Mun JH, Najar FZ, Nicholson C, Noirot C, O'Bleness M, Paule CR, Poulain J, Prion F, Qin BF, Qu CM, Retzel EF, Riddle C, Sallet E, Samain S, Samson N, Sanders I, Saurat O, Scarpelli C, Schiex T, Segurens B, Severin AJ, Sherrier DJ, Shi RH, Sims S, Singer SR, Sinharoy S, Sterck L, Viollet A, Wang BB, Wang KQ, Wang MY, Wang XH, Warfsmann J, Weissenbach J, White DD, White JD, Wiley GB, Wincker P, Xing YB, Yang LM, Yao ZY, Ying F, Zhai JX, Zhou LP, Zuber A, Denarie J, Dixon RA, May GD, Schwartz DC, Rogers J, Quetier F, Town CD, Roe BA (2011) The *Medicago* genome provides insight into the evolution of rhizobial symbioses. Nature 480:520–524. doi:10.1038/nature10625

Zhai J, Jeong D-H, De Paoli E, Park S, Rosen BD, Li Y, Gonzalez AJ, Yan Z, Kitto SL, Grusak MA, Jackson SA, Stacey G, Cook DR, Green PJ, Sherrier DJ, Meyers BC (2011) MicroRNAs as master regulators of the plant NB-LRR defense gene family via the production of phased, transacting siRNAs. Genes Develop 25:2540–2553. doi:10.1101/gad.177527.111

Comparison of Gene Families: Seed Storage and Other Seed Proteins

10

Jaya Joshi, Sudhakar Pandurangan, Marwan Diapari
and Frédéric Marsolais

Abstract

Common bean (*Phaseolus vulgaris*) is an important source of protein and dietary fiber in human diets. Seed proteins, therefore, determine, at least in part, the nutritional value of common bean. From the very beginning of plant molecular biology, in the 1980s, common bean has been a prominent model plant to study seed storage proteins. The recent availability of several sequences for the common bean genome, coupled with seed transcriptomic and proteomic information, enables a comprehensive, in-depth view of seed protein genes in this organism. Comparisons between these sequences highlight interesting variation in lectin gene composition between the two centers of domestication. Alleles conferring storage protein deficiency may be used to improve the levels of essential sulfur amino acids and therefore protein quality. Some of the seed proteins represent anti-nutritionals, including some lectins, trypsin inhibitors, and lipoxygenases, and represent targets to be potentially removed from the genome. Other proteins have potential as bioproducts due to their biological activity against fungi or insects, including defensin D1 and albumin-1.

Keywords

Seed storage protein · Globulin · Albumin · Lectin · Legumin · Protein quality · Defensin

10.1 Introduction

Common bean (dry bean, *Phaseolus vulgaris* L.) is consumed primarily as a dietary source of protein and fiber (De Ron et al. 2015). Therefore, seed storage and other seed proteins determine in part its nutritional value. Seed proteins such as

J. Joshi · S. Pandurangan · F. Marsolais (✉)
Department of Biology, University of Western Ontario; and Genomics and Biotechnology, London Research and Development Centre, Agriculture and Agri-Food Canada, London, ON, Canada
e-mail: Frederic.Marsolais@agr.gc.ca

M. Diapari
Genomics and Biotechnology, London Research and Development Centre, Agriculture and Agri-Food Canada, London, ON, Canada

M. Pérez de la Vega et al. (eds.), *The Common Bean Genome*,
Compendium of Plant Genomes, https://doi.org/10.1007/978-3-319-63526-2_10

globulins may act as a storage source of organic nitrogen. Some of the seed proteins, including the lectins and cysteine-rich peptides, have biological activities that protect the seed against pathogens or herbivory, while others like the lipoxygenases may impact flavor. Some of these proteins may be considered anti-nutritional. This chapter will review the state of our knowledge on seed proteins in light of the recent release of publically available genomes, including the sequence of the reference Andean G19833 (Schmutz et al. 2014), and Mesoamerican BAT-93 (Vlasova et al. 2016) and OAC-Rex (Perry et al., unpublished results; https://mega.nz/#%21KU13nB7B%21LS1afOiDNDPp6koxlmJpGYa8Xnk2_5hCvAiFXyKTOMA) genomes. This will be discussed in relation with our current knowledge of the seed transcriptome and proteome (Yin et al. 2011; Marsolais et al. 2010; O′Rourke et al. 2014; Liao et al. 2012; Parreira et al. 2016; De La Fuente et al. 2011; Natarajan et al. 2013; Majumdar et al. 2015; Balsamo et al. 2015; López-Pedrouso et al. 2014b). The biological activities of legume seed proteins have been recently reviewed by Sparvoli et al. (2015).

10.2 7S Globulin Phaseolin

The 7S globulin, phaseolin, is the most abundant seed protein, accounting for up to 50% of total protein in commercial varieties (Vitale and Bollini 1995). The biochemistry and molecular biology of phaseolin have been reviewed previously by Hall et al. (1999). Phaseolin does not undergo post-translational cleavage other than its signal peptide but is N-glycosylated and phosphorylated (Sturm et al. 1987; López-Pedrouso et al. 2014a). Phaseolin electrophoretic types have been used as markers and provided the first evidence for distinct centers of domestication, Mesoamerican (S type phaseolin from Sanilac cultivar) and Andean (T type phaseolin from Tendergreen cultivar) (Gepts and Bliss 1986). Using the genetic diversity of phaseolin from wild accessions has been suggested to improve the protein quality of common bean, particularly the bioavailable methionine (Montoya et al. 2010). There are two types of genes encoding two phaseolin subtypes, α- and β-. Their coding sequence is distinguished by the presence of a 27-base pair repeat in *α-phaseolin* (Slightom et al. 1985; Kami and Gepts 1994). At the genomic level, the α- and *β-phaseolin* genes are more dissimilar in their promoters and 5′-untranslated regions (Slightom et al. 1983; Anthony et al. 1990). Phaseolins are encoded by a single complex locus predicted to contain multiple genes in tandem. Southern hybridization studies showed that the phaseolin multigene family consists of approximately seven members (Talbot et al. 1984). Nevertheless, the first assembly of the reference *P. vulgaris* genome (v1.0) contained unique functional copies of α- and *β-phaseolin* located on chromosome 7, whereas the most recently released version (v2.1) contains three copies of *α-phaseolin* and two copies of *β-phaseolin* (Fig. 10.1). They are within an interval of approximately 130 kb, and the α- and *β-phaseolin* genes are on opposite strands. At this locus, there is also a gene coding for an uncharacterized 7S globulin sharing 53% identity with α-phaseolin (*Phvul.007g598011*). As expected, the chromosome 7 segment containing the phaseolin locus is syntenic with a segment of chromosome 20 containing the conglycinin locus in soybean (Tsubokura et al. 2012; McClean et al. 2010; Schmutz et al. 2014). According to data from the atlas of gene expression in the Negro Jamapa black bean, a Mesoamerican genotype, expression of α- and *β-phaseolin* is seed-specific (Fig. 10.2). The amino acid sequence of phaseolins is devoid of cysteine residues, and although they have few methionine residues, the seed concentration of methionine is actually correlated with phaseolin accumulation in cultivated varieties (Gepts and Bliss 1984).

The copy number of phaseolin genes in the genome was examined by the quantitative polymerase chain reaction (qPCR) according to Ingham et al. (2001), in the genetic stock SARC1 which contains both α- and β-phaseolin (Osborn et al. 2003). As expected, the threshold cycle (C_q) value of *phaseolin* was low compared to the single-copy genes (*BSAS3.1*, *PDS*, and *Lox*) at any temperature of annealing (T_a) tested (Pandurangan and Marsolais 2017). The average C_q value for single-copy genes was similar

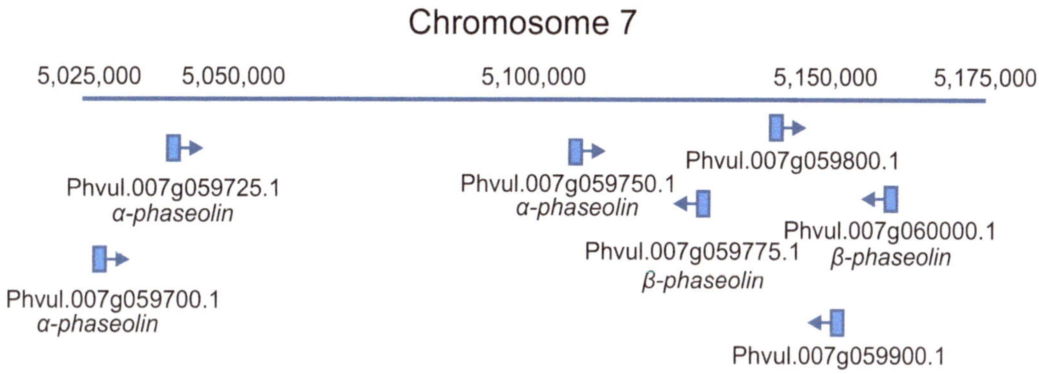

Fig. 10.1 Representation of the phaseolin locus from the reference genome G19833 v.2.1 in JBrowse

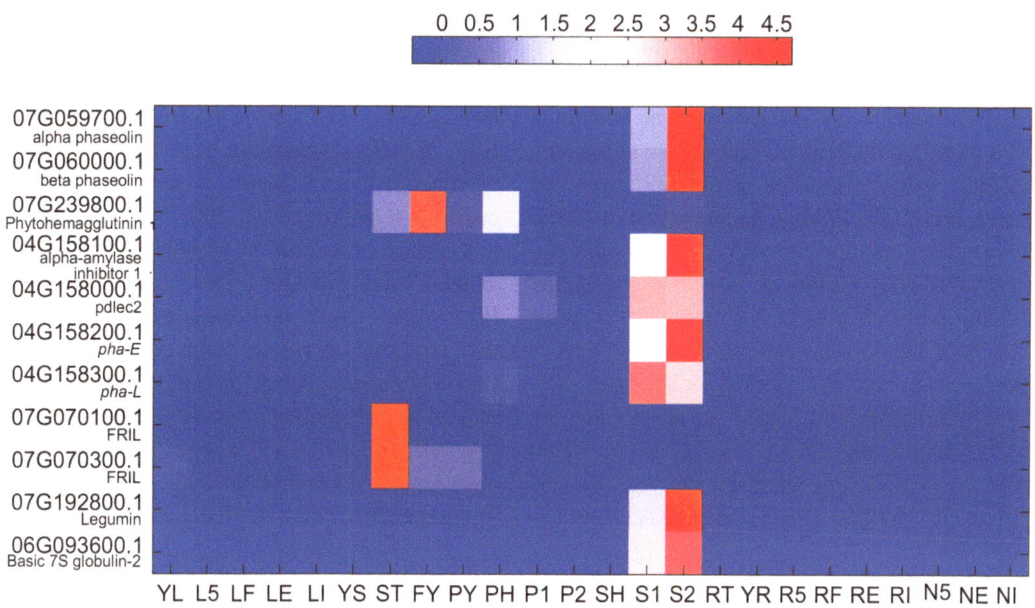

Fig. 10.2 Expression of seed protein transcripts in different tissues according to the gene atlas (O'Rourke et al. 2014). Accession numbers should be preceded by the prefix "Phvul.0-". Expression value is given as *Z*-score (Severin et al. 2010). Calculation of the *Z*-score is based on the reads per kilobase of transcript per million mapped reads (RPKM)-normalized $\log_2$-transformed transcript count data as follows: $Z = (X - \mu)/s$, where X is the transcript count of a gene for a specific tissue/timepoint, and μ is the mean transcript count of a gene across all tissues/developmental stages, and s is the transcript count standard deviation of a gene across all tissues/developmental stages. Therefore, the *Z*-score numerical value measures the number of standard deviations the expression level of a gene in a specific tissue deviates from the mean expression level in all tissues. *Z*-scores above 3.4–3.6 are considered highly tissue specific, and this has been validated in soybean. Heat maps were generated with MATLAB. Descriptions of relevant tissues are as follows: *ST* shoot tips; *FY* flowers; *PY* pods containing globular stage embryos (1–4 days after fertilization); *PH* pods containing heart stage seeds; *P1* pods associated with stage 1 seeds (pods only); *P2* pods associated with stage 2 seeds (pods only); *SH* heart stage seeds (ca. 7 mg); *S1* stage 1 seeds (ca. 50 mg); *S2* stage 2 seeds (ca. 150 mg)

(approximately 22.5 cycles). The value was intermediate for *β-phaseolin* (20 cycles) and lower for nonspecific phaseolin primers which amplify both *α*- and *β-phaseolin* (19 cycles). Based on exponential accumulation, the copy number of phaseolin genes is estimated from these values to be equal to eight, with four copies each of *β-phaseolin* and *α-phaseolin* genes. This is consistent with the above-noted hybridization results of Talbot et al. (1984), and this is close to the number of genes present in assembly v.2.1. A possible reason to explain the discrepancy between the number of genes in the genome assembly and that measured by empirical methods may be the presence of additional, nonfunctional gene copies.

Regulation of *β-phaseolin* gene expression has been extensively studied. Several functional regions within the promoter have been defined by deletion analyses (Bustos et al. 1991; van der Geest and Hall 1996; Chandrasekharan et al. 2003). They include four RY repeat motifs (5′-CATGC/TA-3′) (Bobb et al. 1997), a G-box binding motif (5′-CACGTG-3′), an E-box motif (5′-CACCTG-3′) (Kawagoe et al. 1994), CACA element (Li et al. 1999), vicilin box (5′-GCCACCTCAA-3′) (Chern et al. 1996a, b), ACGT motif, and CAAT box (Li et al. 1999) (Fig. 10.3). Deletion analyses showed that the G-box, RY motifs, E-box, and CAAT box are required for high-level expression of a reporter in transgenic *Arabidopsis thaliana* seed (Chandrasekharan et al. 2003). Binding of the B3-domain containing VP1/ABI3 member PvAlf transcription factor is necessary for expression (Bobb et al. 1995). Gene activation is a two-step process, requiring PvAlf and abscisic acid (Li et al. 1999). Each step is associated with specific chromatin modifications (Ng et al. 2006), resulting in nucleosomal displacement over phased TATA boxes (Li et al. 1998). PvAlf binds to the promoter via the RY repeat motifs (Carranco et al. 2004). ABI5, a bZIP transcription factor, acts downstream from abscisic acid, likely through the G-box (Ng and Hall 2008). Using a heterologous system, two additional essential transcriptional regulators have been identified, *RINGLET 2* (*RLT2*) and *AINTEGUMENTA-LIKE 5* (*AIL5*) (Sundaram et al. 2013). Analysis of genetic stocks differing in seed protein composition (Osborn et al. 2003) associated low levels of *β*-phaseolin accumulation in the bean genetic lines SMARC1-PN1 and SMARC1N-PN1 with a single nucleotide polymorphism in the proximal promoter converting a second G-box element present in Sanilac and SARC1 into an ACGT element (Pandurangan et al. 2016) (Fig. 10.3). This ACGT element does not influence levels of expression in the promoter from Tendergreen (Chandrasekharan et al. 2003).

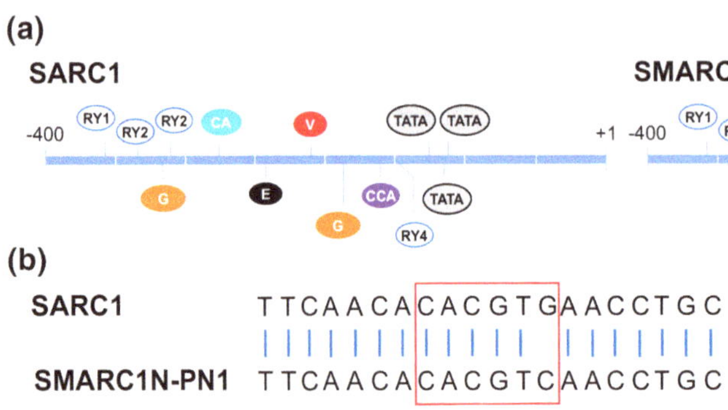

Fig. 10.3 **a** Diagram of *cis*-regulatory elements present in the promoter of *β*-phaseolin gene in SARC1 and Sanilac *versus* SMARC1N-PN1 and SMARC1-PN1. Regulatory motifs are designated as by Chandrasekharan et al. (2003). **b** A single nucleotide substitution converts a G-box in SARC1 and Sanilac to an ACGT motif in SMARC1N-PN1 and SMARC1-PN1. *G* G-box; *E* E-box; *C* CACA box; *V* vicillin box; *CCA* CCAAT box. Reproduced from Pandurangan et al. (2016)

10.3 Lectins

Lectins are the second most abundant seed proteins accounting for 5–10% of total protein (Vitale and Bollini 1995). They include erythroagglutinating phytohemaggutinin (Pha-E), leucoagglutinating phytohemagglutinin (Pha-L), and α-amylase inhibitor and α-amylase inhibitor-like protein (Mirkov et al. 1994). Arcelins are present in certain wild accessions and other *Phaseolus* species and protect against bruchid pests (Osborn et al. 1988; Zaugg et al. 2013; Mbogo et al. 2009; González Vélez et al. 2012). The lectins are usually glycoproteins and vary in subunit association and post-translational processing. α-Amylase inhibitor-1 is processed post-translationally into two subunits (Moreno and Chrispeels 1989). Most lectins are encoded at a single complex locus designated as the *arcelin/phytohemagglutinin/α-amylase inhibitor* (*APA*) locus on chromosome 4 (Freyre et al. 1998; Osborn et al. 1986). Exceptions include *mannose lectin FRIL* (*Flt3 receptor interacting lectin*) (Moore et al. 2000) and two other *lec* genes mapping to linkage group B7, *Lec-2,* and *Lec-3* (Kami et al. 2006; Gepts et al. 1993). This is supported by the presence of a phytohemagglutinin gene close to the end of chromosome 7 in the reference genome. In addition, there are two copies of *FRIL* in tandem on chromosome 7, approximately 1.2 Mb downstream from the phaseolin locus. The tissue profile of expression of these genes differs from those of lectin genes at the *APA* locus (Fig. 10.2). Gene expression of APA lectins is largely seed-specific. The phytohemagglutinin on chromosome 7 is expressed in shoot tips, flowers, and young pods, whereas *FRIL* expression is predominant at the shoot apex. This is interesting because *FRIL* has been reported as a protective factor of mammalian stem cells (Colucci et al. 1999; Moore et al. 2000). The mannose lectin *FRIL* gene is absent from several Mesoamerican navy bean cultivars, including Sanilac, BAT-93, and OAC-Rex. In the phaseolin deficient genotype SMARC1N-PN1, mannose lectin FRIL was derived from the wild accession G12882 (Pandurangan et al. 2016).

For analysis of gene diversity at the *APA* locus, genomic templates were annotated manually after blastn against NCBInr, and blastx of individual APA coding sequences against UniProt, based on highest sequence identity to a known lectin accession. The BAC-71F18 clone from an arcelin-5 genotype was included among genomic templates (Kami et al. 2006). The order and orientation of lectin genes are conserved across genotypes. However, the composition of

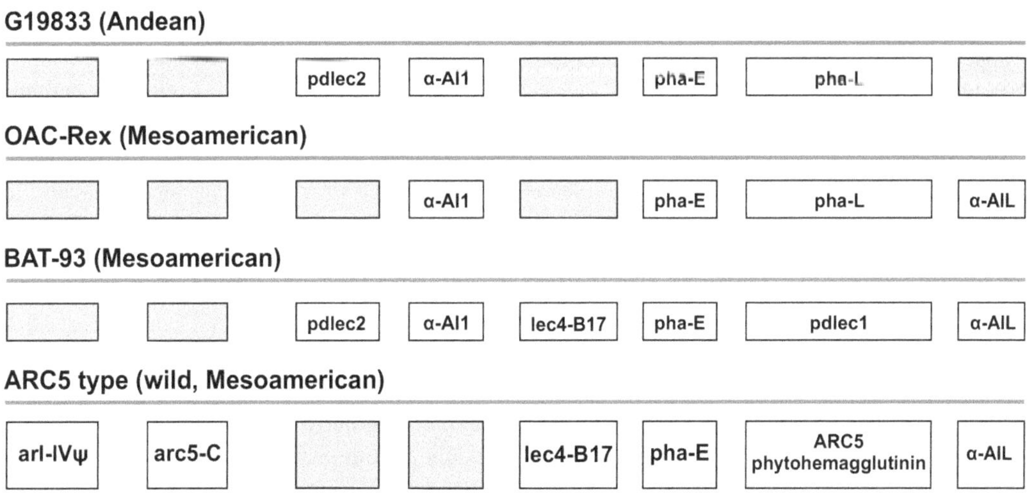

Fig. 10.4 Schematic representation of lectin genes at the arcelin-phytohemagglutinin-α-amylase inhibitor (APA) locus in different common bean genotypes

APA genes varied (Fig. 10.4). G19833 contains *pdlec2* encoding a leucoagglutinating phytohemagglutinin (Voelker et al. 1986), α-amylase inhibitor 1 (Moreno and Chrispeels 1989), and *pha-E* and *pha-L* (Hoffman and Donaldson 1985). OAC-Rex lacks *pdlec2* but harbors the α-amylase inhibitor-like protein gene, downstream from *pha-L* (Finardi-Filho et al. 1996; Wato et al. 2000). BAT-93 possesses all the lectin genes listed above, as well as *lec4-B17* upstream of *pha-E* (Lioi et al. 2003). However, *pha-L* is replaced by the pseudogene *pdlec1*, previously characterized in the pinto bean genotype Pinto UI 111 (Voelker et al. 1986). The *pdlec1* allele is characterized by a deletion of a single nucleotide, cytosine, after position 32 of the coding sequence, resulting in a premature stop codon at position 132. This polymorphism is associated with a lack of transcript accumulation (Voelker et al. 1990). In G02771, a wild arcelin-5 genotype, arcelin-5 phytohemagglutinin substitutes for *pha-L* (Kami et al. 2006). Arcelin-5 phytohemagglutinin is 99% identical to *pdlec1* but is a functional gene.

Genetic polymorphisms associated with differences in lectin composition in genetic lines SARC1, SMARC1-PN1, and SMARC1N-PN1 and their recurrent parent Sanilac were analyzed by aligning Illumina paired-end reads to BAT93 (Pandurangan et al. 2016). In SMARC1N-PN1, lectin deficiency was introgressed from Great Northern 1140. The genes *lec4-B17* and *pha-E* appeared to be absent from SMARC1N-PN1. This genotype also possesses the *pdlec1* allele. The similarity with Pinto UI 111 further extends to the presence of *pdlec2*. These results suggest that Great Northern 1140 and Pinto UI 111 share the same *APA* locus (Osborn and Bliss 1985). These genotypes are representative of market classes that both belong to race Durango, from the Mesoamerican center of domestication (Singh et al. 1991; Mensack et al. 2010). While the genetic relationship between the two genotypes is unknown, these results suggest that they share a common ancestor (McClean and Myers 1990). Bollini et al. (1985) had already hypothesized a unique origin of phytohemagglutinin deficiency, based on the similar patterns of DNA hybridization observed with different phytohemagglutinin-deficient cultivars probed with phytohemagglutinin cDNAs. SARC1 contains arcelin-1 introgressed from the wild accession G12882. Proteomic information and genomic PCR data indicated the presence of at least three functional arcelin genes in SARC1: *Arc1*, *arc3-II,* and *arc4-I* (Lioi et al. 2003). This is consistent with the detection of multiple arcelin variants differing in subunit structure and N-terminal sequence in this genotype (Hartweck et al. 1991). The large difference in α-amylase inhibitor-like protein accumulation between SARC1 and SMARC1N-PN1 is associated with multiple polymorphisms in the proximal promoter introducing several potential positive *cis*-regulatory elements in the latter genotype (Pandurangan et al. 2016). Although lectins are relatively low in sulfur amino acid residues, the accumulation of arcelin-like protein 4 and α-amylase inhibitor *β* subunit is elevated in response to sulfate fertilization (Pandurangan et al. 2015). The mature α-amylase inhibitor *β* subunit does not contain any sulfur amino acids. However, its polypeptide precursor contains four methionine residues at its N-terminal signal peptide which are absent from the mature protein (Prescott et al. 2005).

10.4 11S Globulin Legumin

Among grain legumes, *P. vulgaris* is characterized by low levels of the 11S globulin legumin, accounting approximately for 3% of seed protein (Muhling et al. 1997). Legumin is characterized by post-translational cleavage into an acidic (α) and a basic (*β*) subunit. There is a unique gene in the genome, coding for a predominant, high-molecular weight legumin, about 1 Mb upstream from the *APA* locus (Fig. 10.2). *Phaseolus vulgaris* legumin belongs to the subgroup of high-molecular weight legumins including group-2 glycinins from soybean, arachin-5 from peanut, and minor small legumin from pea (Yin et al. 2011). Members of this subgroup are characterized by an extended C-terminal half of the α-subunit arising from the expansion and mutation of a sequence repeat. The identity of repeats differs between *P. vulgaris* legumin, glycinin A5A4B3 from *G. max* and arachin 5 from peanut, indicating that expansion of repeats took place after the separation of the lineages leading to the three species. The 11S globulins are relatively rich in sulfur amino acids, and their subunits are held in place by one intermolecular and one intramolecular disulfide bonds (Staswick et al. 1984; Adachi et al. 2001). Mature *P. vulgaris* legumin subunits contain 0.9% of their residues as cysteine and 0.7% as methionine. Legumin contributes to the increased sulfur amino acid concentration in SMARC1N-PN1 as it is the most abundant seed protein in this genotype, accounting for 17% of total protein concentration (Marsolais et al. 2010). Another legumin-like protein (*Phvul.005g140300.1*) has been reported; however, its levels continuously decrease during seed development (Parreira et al. 2016). A closer examination of the genome reveals a number of other sequences with cupin domains representative of globulins; however, none of these proteins are likely to be very abundant in mature seed.

10.5 Basic 7S Globulin

Basic 7S globulin has been characterized from soybean (Yamauchi et al. 1984), also referred to as sulfur-rich protein (Monaghan et al. 2008), and as γ-conglutin in lupin (Duranti et al. 2008). It accounts for up to 5% of seed protein in soybean (Monaghan et al. 2008). The mature polypeptide is processed into a high-molecular weight and low-molecular weight subunit. It has structural similarity with xyloglucan-specific endo-β-1,4-glucanase inhibitors and aspartic proteases, although it does not possess either of these catalytic activities (Czubinski et al. 2015; Yoshizawa et al. 2011). Basic 7S globulin has insulin binding properties and glucose-lowering nutritional effects (Magni et al. 2004; Lovati et al. 2012; Hanada et al. 2003; Barbashov et al. 1991). The mature protein has five intra-chain and one interchain disulfide bonds. The predicted amino acid sequence of the basic 7S globulin-2 from common bean has 3% of its residues as cysteine and 1.9% as methionine. Its transcript levels are elevated in SMARC1N-PN1, likely contributing to the increased concentration of cysteine and methionine in this genotype (Liao et al. 2012). There is a single gene for basic 7S globulin, located on chromosome 6 (Fig. 10.2). However, the protein has not yet been reported or characterized.

10.6 Albumin-2

Albumin-2 was first characterized in pea (Higgins et al. 1987). It contains four hemopexin repeats involved in polyamine binding (Gaur et al. 2010). A pea genotype lacking albumin-2 has altered levels of polyamines (Vigeolas et al. 2008). Pea albumin-2 was shown to be resistant to gastrointestinal digestion in piglets, representing a potential allergenic and anti-nutritional protein (Le Gall et al. 2007), although true amino acid digestibility was found to be positively correlated with albumin-2 levels in broiler chickens (Gabriel et al. 2008). Interestingly, albumin-2 is absent from the soybean genome (Schmutz et al. 2010). The *P. vulgaris* protein (*Phvul.007g275800*) is relatively rich in sulfur amino acid residues with 1.3% of cysteine and 0.4% of methionine. Both the transcript and protein are elevated in SMARC1N-PN1 (Marsolais et al. 2010; Liao et al. 2012). The protein accumulation in seed positively responds to sulfate fertilization (Pandurangan et al. 2015), whereas in pea, the albumin-2 transcript shows a slight response to sulfur nutrition (Higgins et al. 1987). In common bean, the albumin-2 gene is part of a complex locus on chromosome 7 comprising a total of 13-related genes, spanning an interval of approximately 88 kb (Fig. 10.5). There are 5 additional albumin-2 genes on chromosome 7, including a cluster of three genes, and four other genes on chromosomes 2, 8, and 9. An examination of the data from the RNA-seq atlas suggests abundant expression of several albumin-2 in flower buds as well as seed or pod expression of additional family members (Fig. 10.5).

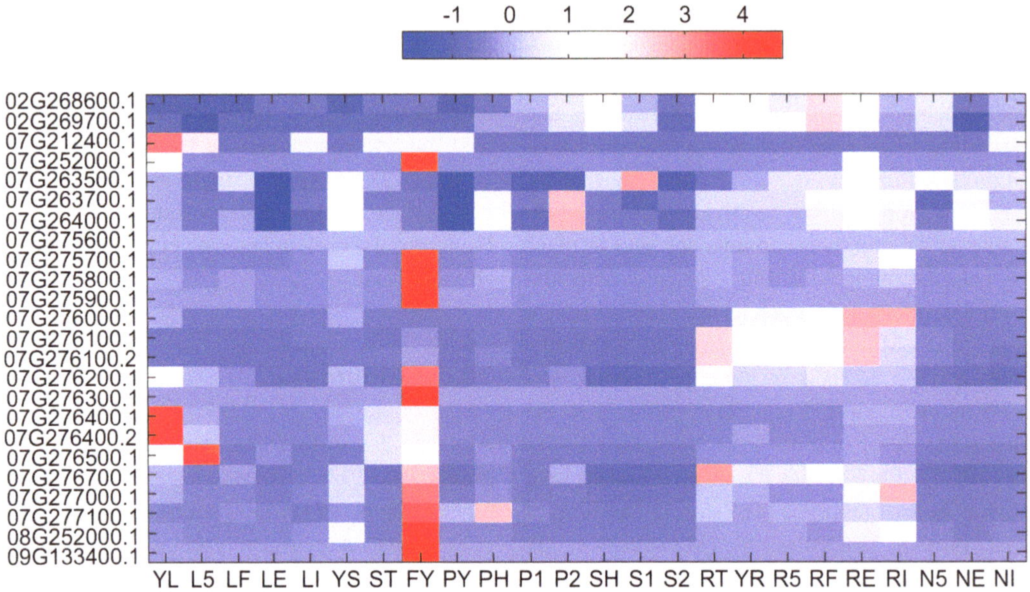

Fig. 10.5 Expression of albumin-2 transcripts in different tissues according to the gene atlas. Values are given as Z-score. See legend to Fig. 10.2 for explanation

10.7 Albumin-1

Albumin-1 is part of the cysteine-rich or knottin fold proteins. It is encoded by a single complex locus on chromosome 11, including 21 genes and spanning 285 kb (Fig. 10.6). Albumin-1 was originally characterized from pea (Higgins et al. 1986). The albumin-1 proteins have insecticidal activity, including the one from common bean (Louis et al. 2004), targeting the insect vacuolar ATPase (Muench et al. 2014). Soybean albumin-1, or leginsulin, binds to the 7S basic globulin and stimulates its autophosphorylation (Hanada and Hirano 2004). Comparative genomic analyses across legumes revealed a monophyletic origin of the common bean albumin-1 genes (Karaki et al. 2016), indicating that they evolved from a single gene through successive duplications, after the separation of lineages leading to the different legume crops. Albumin-1 identified from mature seed of common bean is present in methanol soluble extracts (Marsolais et al. 2010) and contributes to elevated cysteine concentration in SMARC1N-PN1. It may be encoded by two genes encoding an identical protein (*Phvul.011g205300* and *Phvul.011g205400*). The encoded protein has 7.9% of its residues as cysteine and 4.7% as methionine. Transcripts for these genes and a number of other albumin-1 transcripts were shown to be elevated in SMARC1N-PN1 by up to 10-fold as compared with SARC1 (Liao et al. 2012). These data and the RNA-seq atlas identify a group of genes as seed-specific, which also includes the adjacent *Phvul.011g205200*, *Phvul.011g205500*, and *Phvul.011g205600* (Fig. 10.6).

10.8 Trypsin Inhibitors

The Bowman–Birk serine protease inhibitor family has been characterized by cloning of a BAC clone which comprises all three genes of this family (Galasso et al. 2009). The Bowman–Birk inhibitors are cysteine-rich peptides with a low molecular mass (about 9 kDa) and are double-headed with two binding loops per molecule of inhibitor. The three genes encode inhibitors that have been well characterized at the protein level (Wilson and Laskowski 1973,

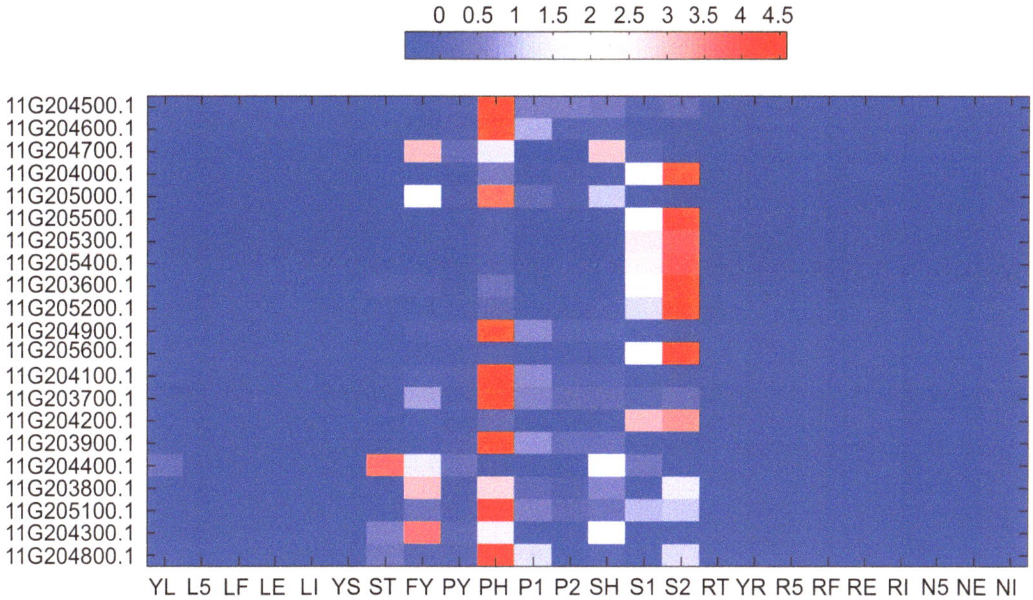

Fig. 10.6 Expression of albumin-1 transcripts in different tissues according to the gene atlas. Values are given as Z-score. See legend to Fig. 10.2 for explanation

1975). They encode Bowman–Birk type proteinase inhibitor 2, double-headed trypsin inhibitor, and trypsin/chymotrypsin inhibitor. They are present on chromosome 4 in an interval of 19 kb. The Bowman–Birk type proteinase inhibitor 2 has been detected in total protein extracts from mature seed. The protein is more abundant in SMARC1N-PN1 than SARC1 by about 2-fold (Marsolais et al. 2010). Each of the three transcripts is also elevated in SMARC1N-PN1, by 3- to 4-fold (Liao et al. 2012). Data from the RNA-seq atlas show that the three genes are expressed in a seed-specific manner (Fig. 10.7). Recently, a combined approach including TILLING and screening natural populations for genetic variation enabled the isolation of trypsin inhibitor null pea genotypes (Clemente et al. 2015). In soybean, an approach was used where an inactive Bowman–Birk inhibitor is overexpressed in seed, to compete with the expression of the native form (Livingstone et al. 2007). These genotypes may be advantageous when used as animal feed.

10.9 Kunitz Trypsin Inhibitors

The Kunitz trypsin inhibitors are structurally distinct from the Bowman–Birk type proteinase inhibitors. They encode larger proteins (ca. 23 kDa). The *P. vulgaris* genome contains 21 Kunitz trypsin inhibitors genes, including a complex locus with 15 genes on chromosome 4, spanning approximately 315 kb. At least eight of them show preferential expression in seed (Fig. 10.8). Transcript of one of the Kunitz trypsin inhibitor genes was elevated in SMARC1N-PN1 as compared with SARC1 (*Phvul.004g129900.1*) (Liao et al. 2012). The corresponding protein has 2.3% of its residues as cysteine and 0.9% as methionine. However, Kunitz trypsin inhibitors have not yet been characterized at the protein level from common bean. Although the Kunitz trypsin inhibitors are encoded by a multigene family in soybean, one gene encodes the prominent isoform present in seed (Jofuku and Goldberg 1989). A null allele

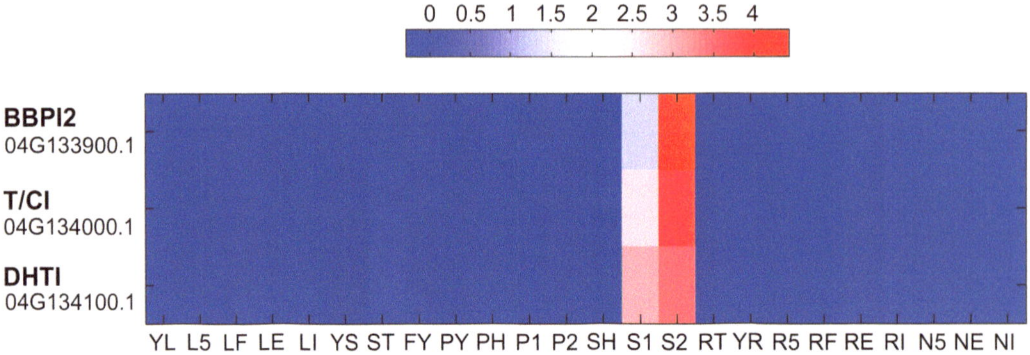

Fig. 10.7 Expression of trypsin inhibitor transcripts in different tissues according to the gene atlas. BBPI2: Bowman–Birk type proteinase inhibitor 2; *T/CI* trypsin/chymotrypsin inhibitor; *DHTI* double-headed trypsin inhibitor. Values are given as *Z*-score. See legend to Fig. 10.2 for explanation

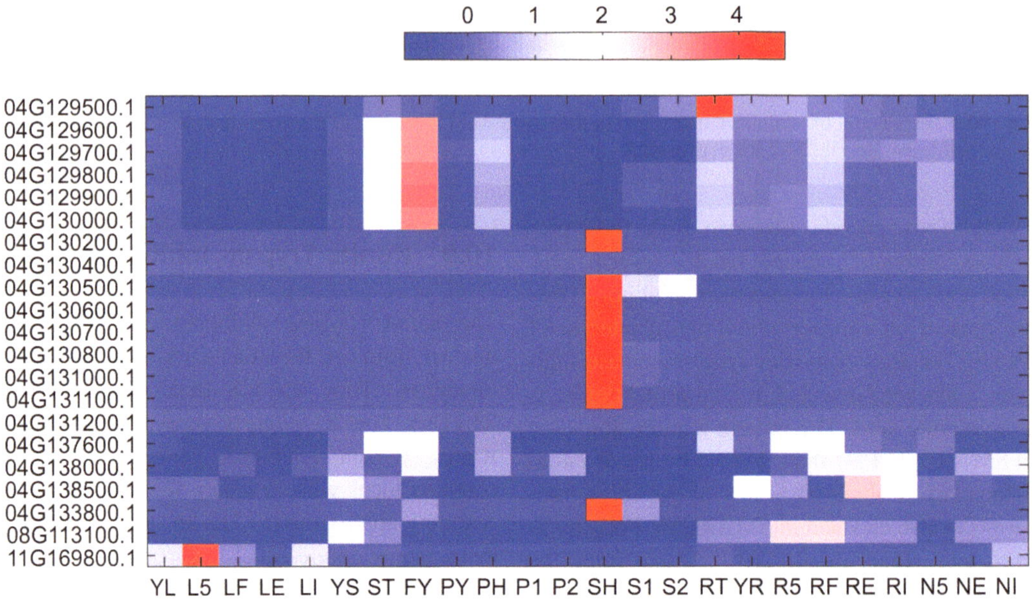

Fig. 10.8 Expression of Kunitz trypsin inhibitor transcripts in different tissues according to the gene atlas. Values are given as *Z*-score. See legend to Fig. 10.2 for explanation

of Kunitz trypsin inhibitor has been combined with two other null alleles (for P34 and agglutinin) to create a low allergenicity soybean cultivar (Schmidt et al. 2015). Recently, a new null allele has been identified and shown to have increased levels of Bowman–Birk inhibitor as a compensatory response (Gillman et al. 2015).

10.10 Defensins

The defensins are another group of cysteine-rich peptides involved in seed defense against fungi. The genome contains 12 defensin genes, six of which are highly expressed in seed (Fig. 10.9).

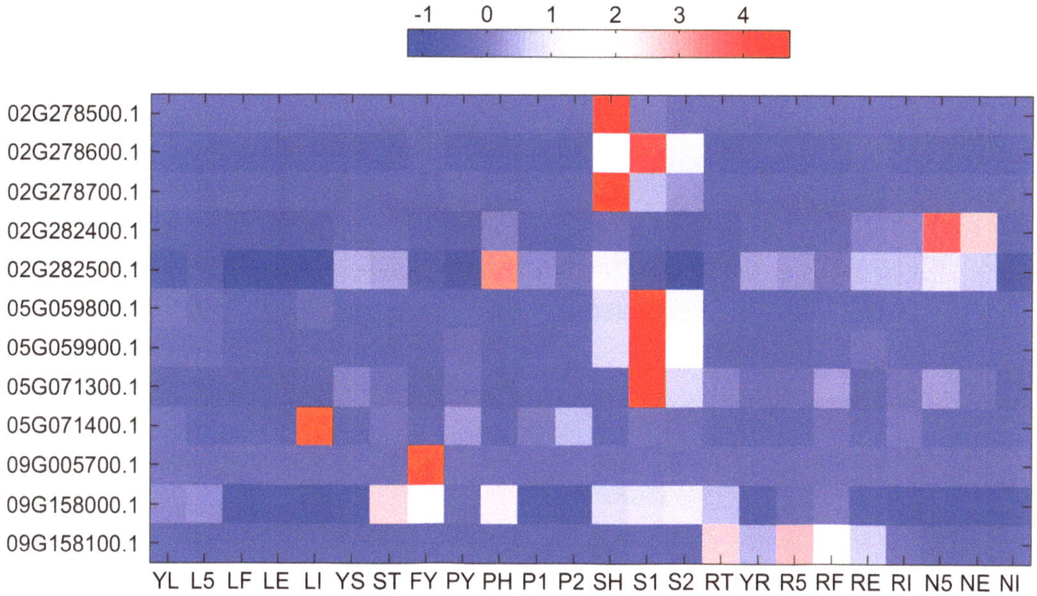

Fig. 10.9 Expression of defensin transcripts in different tissues according to the gene atlas. Values are given as *Z*-score. See legend to Fig. 10.2 for explanation

These include *defensin D1* (*Phvul.005G071300.1*) (Games et al. 2008). This defensin is present in propanol soluble extracts from mature seed, at higher levels in SMARC1N-PN1 than SARC1 (Marsolais et al. 2010). Defensin D1 has 12.2% of its residues as cysteine and 1.4% as methionine. Pea defensin D1 (Psd1) has a similar biological activity (Lacerda et al. 2014).

10.11 Lipoxygenases

Lipoxygenases are part of a large gene family comprising 32 members (Fig. 10.10). Data from the RNA-seq atlas highlight that two of them, located on chromosome 5, *Lox-2* (*Phvul.005g156700.1*) and *Lox-3* (*Phvul.005g157000.1*) have seed-specific expression (Fig. 10.10). Both have been detected in the mature seed proteome (Marsolais et al. 2010; Parreira et al. 2016). Lox-3 accumulation was shown to respond positively to sulfate nutrition in SMARC1N-PN1 but not in SARC1 (Pandurangan et al. 2015). Lox-3 is rich in sulfur amino acids with 1.7% of its residues as methionine and 0.5% as cysteine. Based on its similar molecular weight, it is likely that the sulfur-responsive albumin protein of 95 kDa identified in pea is actually Lox-3 (Higgins et al. 1987). Lipoxygenase is associated with an off-taste in food-grade soybeans. Therefore, researchers have sought to isolate lipoxygenase-free soybeans (Reinprecht et al. 2011; Lenis et al. 2010). A similar trait might be useful in common bean for food utilization of flour or fractions.

10.12 Other Proteins

Of note is the fact that the *P. vulgaris* genome lacks a gene coding for lunasin, a sulfur-rich peptide from soybean with anticancer properties (Hernández-Ledesma et al. 2013). Among the proteins present in mature seed, a set is related to responses to oxidative stress and dehydration and may be involved in seed longevity, including

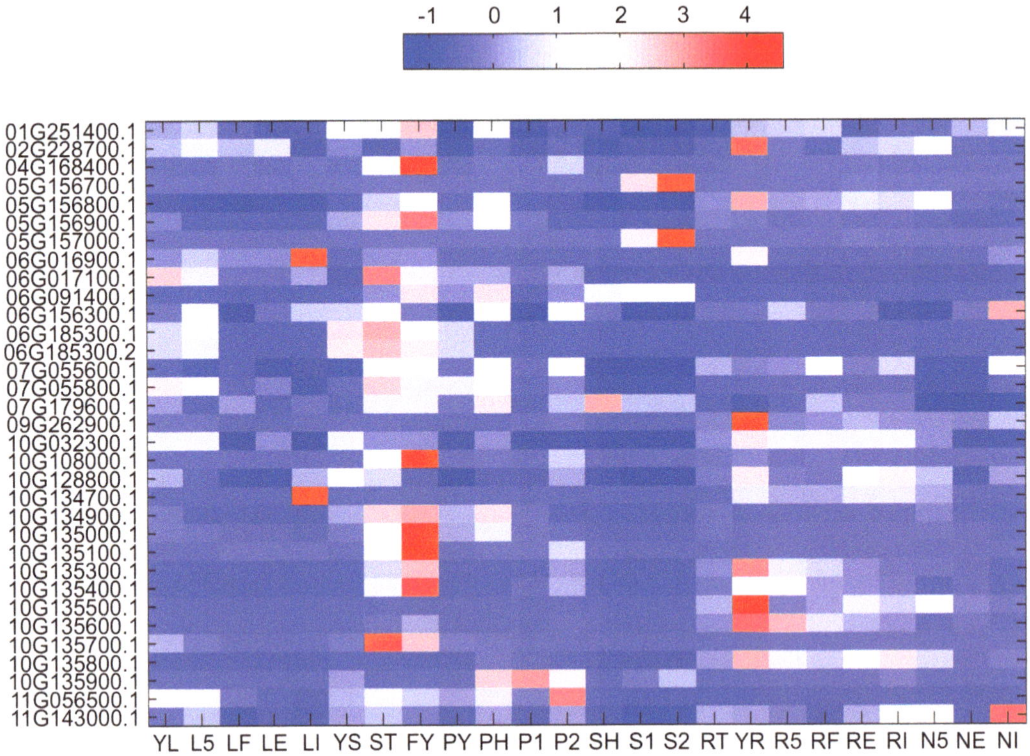

Fig. 10.10 Expression of lipoxygenase transcripts in different tissues according to the gene atlas. Values are given as *Z*-score. See legend to Fig. 10.2 for explanation

peroxiredoxin and group 3 late embryogenesis protein (Marsolais et al. 2010; Tolleter et al. 2007; Delahaie et al. 2013; Parreira et al. 2016).

Acknowledgements This work was supported by a grant from the Ontario Research Fund—Research Excellence program. We thank Gregory Perry, Seth Munholland, William Crosby, and Peter Pauls for sharing the OAC-Rex sequence. We are grateful to Alex Molnar for help with figures.

References

Adachi M, Takenaka Y, Gidamis AB, Mikami B, Utsumi S (2001) Crystal structure of soybean proglycinin A1aB1b homotrimer. J Mol Biol 305:291–305

Anthony JL, Vonder Haar RA, Hall TC (1990) Nucleotide sequence of an α-phaseolin gene from *Phaseolus vulgaris*. Nucleic Acids Res 18:3396

Balsamo GM, Valentim-Neto PA, Mello CS, Arisi ACM (2015) Comparative proteomic analysis of two varieties of genetically modified (GM) Embrapa 5.1 common bean (*Phaseolus vulgaris* L.) and their non-GM counterparts. J Agric Food Chem 63:10569–10577

Barbashov SF, Egorov TA, Kochkina VM (1991) Isolation and characteristics of insulin-binding proteins from soybeans. Bioorg Khim 17:421–423

Bobb AJ, Eiben HG, Bustos MM (1995) PvAlf, an embryo-specific acidic transcriptional activator enhances gene expression from phaseolin and phytohemagglutinin promoters. Plant J 8:331–343

Bobb AJ, Chern MS, Bustos MM (1997) Conserved RY-repeats mediate transactivation of seed-specific promoters by the developmental regulator PvALF. Nucleic Acids Res 25:641–647

Bollini R, Allavena A, Vitale A (1985) Genomic analysis of phytohemagglutinin-deficient *Phaseolus vulgaris* cultivars. Annu Rep Bean Improv Coop 37:82

Bustos MM, Begum D, Kalkan FA, Battraw MJ, Hall TC (1991) Positive and negative *cis*-acting DNA domains are required for spatial and temporal regulation of gene expression by a seed storage protein promoter. EMBO J 10:1469–1479

Carranco R, Chandrasekharan MB, Townsend JC, Hall TC (2004) Interaction of PvALF and VP1 B3 domains with the β-phaseolin promoter. Plant Mol Biol 55:221–237

Chandrasekharan MB, Bishop KJ, Hall TC (2003) Module-specific regulation of the β-phaseolin promoter during embryogenesis. Plant J 33:853–866
Chern MS, Bobb AJ, Bustos MM (1996a) The regulator of *MAT2* (ROM2) protein binds to early maturation promoters and represses PvALF-activated transcription. Plant Cell 8:305–321
Chern MS, Eiben HG, Bustos MM (1996b) The developmentally regulated bZIP factor ROM1 modulates transcription from lectin and storage protein genes in bean embryos. Plant J 10:135–148
Clemente A, Arques MC, Dalmais M, Le Signor C, Chinoy C, Olias R, Rayner T, Isaac PG, Lawson DM, Bendahmane A, Domoney C (2015) Eliminating anti-nutritional plant food proteins: the case of seed protease inhibitors in pea. PLoS ONE 10:e0134634
Colucci G, Moore JG, Feldman M, Chrispeels MJ (1999) cDNA cloning of FRIL, a lectin from *Dolichos lablab*, that preserves hematopoietic progenitors in suspension culture. Proc Natl Acad Sci USA 96:646–650
Czubinski J, Barciszewski J, Gilski M, Szpotkowski K, Debski J, Lampart-Szczapa E, Jaskolski M (2015) Structure of γ-conglutin: insight into the quaternary structure of 7S basic globulins from legumes. Acta Cryst D 71:224–238
De La Fuente M, Borrajo A, Bermúdez J, Lores M, Alonso J, López M, Santalla M, De Ron AM, Zapata C, Alvarez G (2011) 2-DE-based proteomic analysis of common bean (*Phaseolus vulgaris* L.) seeds. J Proteomics 74:262–267
De Ron AM, Papa R, Bitocchi E, González AM, Debouck DG, Brick MA, Fourie D, Marsolais F, Beaver J, Geffroy V, McClean P, Santalla M, Lozano R, Yuste-Lisbona FJ, Casquero PA (2015) Common bean. In: De Ron AM (ed) Grain Legumes. Handbook of Plant Breeding, vol 10. Springer, New York, pp 1–36
Delahaie J, Hundertmark M, Bove J, Leprince O, Rogniaux H, Buitink J (2013) LEA polypeptide profiling of recalcitrant and orthodox legume seeds reveals ABI3-regulated LEA protein abundance linked to desiccation tolerance. J Exp Bot 64:4559–4573
Duranti M, Consonni A, Magni C, Sessa F, Scarafoni A (2008) The major proteins of lupin seed: Characterisation and molecular properties for use as functional and nutraceutical ingredients. Trends Food Sci Technol 19:624–633
Finardi-Filho F, Mirkov TE, Chrispeels MJ (1996) A putative precursor protein in the evolution of the bean α-amylase inhibitor. Phytochemistry 43:57–62
Freyre R, Skroch PW, Geffroy V, Adam-Blondon AF, Shirmohamadali A, Johnson WC, Llaca V, Nodari RO, Pereira PA, Tsai SM, Tohme J, Dron M, Nienhuis J, Vallejos CE, Gepts P (1998) Towards an integrated linkage map of common bean. 4. Development of a core linkage map and alignment of RFLP maps. Theor Appl Genet 97:847–856
Gabriel I, Quillien L, Cassecuelle F, Marget P, Juin H, Lessire M, Sève B, Duc G, Burstin J (2008) Variation in seed protein digestion of different pea (*Pisum sativum* L.) genotypes by cecectomized broiler chickens: 2. Relation between in vivo protein digestibility and pea seed characteristics, and identification of resistant pea polypeptides. Livest Sci 113:262–273
Galasso I, Piergiovanni AR, Lioi L, Campion B, Bollini R, Sparvoli F (2009) Genome organization of Bowman-Birk inhibitor in common bean (*Phaseolus vulgaris* L.). Mol Breeding 23:617–624
Games PD, dos Santos IS, Mello EO, Diz MSS, Carvalho AO, de Souza-Filho GA, Da Cunha M, Vasconcelos IM, dos S. Ferreira B, Gomes VM (2008) Isolation, characterization and cloning of a cDNA encoding a new antifungal defensin from *Phaseolus vulgaris* L. seeds. Peptides 29:2090–2100
Gaur V, Qureshi IA, Singh A, Chanana V, Salunke DM (2010) Crystal structure and functional insights of hemopexin fold protein from grass pea. Plant Physiol 152:1842–1850
Gepts P, Bliss FA (1984) Enhanced available methionine concentration associated with higher phaseolin levels in common bean seeds. Theor Appl Genet 69:47–53
Gepts P, Bliss FA (1986) Phaseolin variability among wild and cultivated common beans (*Phaseolus vulgaris*) from Colombia. Econ Bot 40:469–478
Gepts P, Nodari R, Tsai SM, Koinange EMK, Llaca V, Gilbertson LA, Guzman P (1993) Linkage mapping in common bean. Annu Rep Bean Improv Coop 45: xxiv–xxxviii
Gillman JD, Kim W-S, Krishnan HB (2015) Identification of a new soybean Kunitz trypsin inhibitor mutation and its effect on Bowman—Birk protease inhibitor content in soybean seed. J Agric Food Chem 63:1352–1359
González Vélez A, Ferwerda F, Abreu E, Beaver JS (2012) Development of bean lines (*Phaseolus vulgaris* L.) resistant to BGYMV, BCMNV and bean weevil (*Acanthoselides obtectus* Say). Annu Rep Bean Improv Coop 55:89–90
Hall TC, Chandrasekharan MB, Li G (1999) Phaseolin: its past, properties, regulation and future. In: Shewry PR, Casey R (eds) Seed Proteins. Kluwer Academic Publishers, Dordrecht, The Netherlands, pp 209–240
Hanada K, Hirano H (2004) Interaction of a 43-kDa receptor-like protein with a 4-kDa hormone-like peptide in soybean. Biochemistry 43:12105–12112
Hanada K, Nishiuchi Y, Hirano H (2003) Amino acid residues on the surface of soybean 4-kDa peptide involved in the interaction with its binding protein. Eur J Biochem 270:2583–2592
Hartweck LM, Vogelzang RD, Osborn TC (1991) Characterization and comparison of arcelin seed protein variants from common bean. Plant Physiol 97:204–211
Hernández-Ledesma B, Hsieh CC, de Lumen BO (2013) Chemopreventive properties of peptide lunasin: a review. Protein Pept Lett 20:424–432
Higgins TJ, Chandler PM, Randall PJ, Spencer D, Beach LR, Blagrove RJ, Kortt AA, Inglis AS (1986) Gene structure, protein structure, and regulation of the synthesis of a sulfur-rich protein in pea seeds. J Biol Chem 261:11124–11130

Higgins TJV, Beach LR, Spencer D, Chandler PM, Randall PJ, Blagrove RJ, Kortt AA, Guthrie RE (1987) cDNA and protein sequence of a major pea seed albumin (PA 2: Mr ≈ 26 000). Plant Mol Biol 8:37–45

Hoffman LM, Donaldson DD (1985) Characterization of two *Phaseolus vulgaris* phytohemagglutinin genes closely linked on the chromosome. EMBO J 4:883–889

Ingham DJ, Beer S, Money S, Hansen G (2001) Quantitative real-time PCR assay for determining transgene copy number in transformed plants. Biotechniques 31:132–140

Jofuku KD, Goldberg RB (1989) Kunitz trypsin inhibitor genes are differentially expressed during the soybean life cycle and in transformed tobacco plants. Plant Cell 1:1079–1093

Kami JA, Gepts P (1994) Phaseolin nucleotide sequence diversity in *Phaseolus*. I. Intraspecific diversity in *Phaseolus vulgaris*. Genome 37:751–757

Kami J, Poncet V, Geffroy V, Gepts P (2006) Development of four phylogenetically-arrayed BAC libraries and sequence of the APA locus in *Phaseolus vulgaris*. Theor Appl Genet 112:987–998

Karaki L, Da Silva P, Rizk F, Chouabe C, Chantret N, Eyraud V, Gressent F, Sivignon C, Rahioui I, Kahn D, Brochier-Armanet C, Rahbé Y, Royer C (2016) Genome-wide analysis identifies gain and loss/change of function within the small multigenic insecticidal albumin 1 family of *Medicago truncatula*. BMC Plant Biol 16:1–19

Kawagoe Y, Campbell BR, Murai N (1994) Synergism between CACGTG (G-box) and CACCTG cis-elements is required for activation of the bean seed storage protein β-phaseolin gene. Plant J 5:885–890

Lacerda AF, Vasconcelos EAR, Pelegrini PB, Grossi de Sa MF (2014) Antifungal defensins and their role in plant defense. Front Microbiol 5:116

Le Gall M, Quillien L, Seve B, Gueguen J, Lalles JP (2007) Weaned piglets display low gastrointestinal digestion of pea (*Pisum sativum* L.) lectin and pea albumin 2. J Anim Sci 85:2972–2981

Lenis JM, Gillman JD, Lee JD, Shannon JG, Bilyeu KD (2010) Soybean seed lipoxygenase genes: molecular characterization and development of molecular marker assays. Theor Appl Genet 120:1139–1149

Li G, Chandler SP, Wolffe AP, Hall TC (1998) Architectural specificity in chromatin structure at the TATA box in vivo: nucleosome displacement upon β-phaseolin gene activation. Proc Natl Acad Sci USA 95:4772–4777

Li G, Bishop KJ, Chandrasekharan MB, Hall TC (1999) β-Phaseolin gene activation is a two-step process: PvALF-facilitated chromatin modification followed by abscisic acid-mediated gene activation. Proc Natl Acad Sci USA 96:7104–7109

Liao D, Pajak A, Karcz SR, Chapman BP, Sharpe AG, Austin RS, Datla R, Dhaubhadel S, Marsolais F (2012) Transcripts of sulphur metabolic genes are co-ordinately regulated in developing seeds of common bean lacking phaseolin and major lectins. J Exp Bot 63:6283–6295

Lioi L, Sparvoli F, Galasso I, Lanave C, Bollini R (2003) Lectin-related resistance factors against bruchids evolved through a number of duplication events. Theor Appl Genet 107:814–822

Livingstone D, Beilinson V, Kalyaeva M, Schmidt MA, Herman EM, Nielsen NC (2007) Reduction of protease inhibitor activity by expression of a mutant Bowman-Birk gene in soybean seed. Plant Mol Biol 64:397–408

López-Pedrouso M, Alonso J, Zapata C (2014a) Evidence for phosphorylation of the major seed storage protein of the common bean and its phosphorylation-dependent degradation during germination. Plant Mol Biol 84:415–428

López-Pedrouso M, Bernal J, Franco D, Zapata C (2014b) Evaluating two-dimensional electrophoresis profiles of the protein phaseolin as markers of genetic differentiation and seed protein quality in common bean (*Phaseolus vulgaris* L.). J Agric Food Chem 62:7200–7208

Louis S, Delobel B, Gressent F, Rahioui I, Quillien L, Vallier A, Rahbe Y (2004) Molecular and biological screening for insect-toxic seed albumins from four legume species. Plant Sci 167:705–714

Lovati MR, Manzoni C, Castiglioni S, Parolari A, Magni C, Duranti M (2012) Lupin seed gamma-conglutin lowers blood glucose in hyperglycaemic rats and increases glucose consumption of HepG2 cells. Br J Nutr 107:67–73

Magni C, Sessa F, Accardo E, Vanoni M, Morazzoni P, Scarafoni A, Duranti M (2004) Conglutin γ, a lupin seed protein, binds insulin in vitro and reduces plasma glucose levels of hyperglycemic rats. J Nutr Biochem 15:646–650

Majumdar S, Almeida IC, Arigi EA, Choi H, VerBerkmoes NC, Trujillo-Reyes J, Flores-Margez JP, White JC, Peralta-Videa JR, Gardea-Torresdey JL (2015) Environmental effects of nanoceria on seed production of common bean (*Phaseolus vulgaris*): a proteomic analysis. Environ Sci Technol 49:13283–13293

Marsolais F, Pajak A, Yin F, Taylor M, Gabriel M, Merino DM, Ma V, Kameka A, Vijayan P, Pham H, Huang S, Rivoal J, Bett K, Hernández-Sebastià C, Liu Q, Bertrand A, Chapman R (2010) Proteomic analysis of common bean seed with storage protein deficiency reveals up-regulation of sulfur-rich proteins and starch and raffinose metabolic enzymes, and down-regulation of the secretory pathway. J Proteomics 73:1587–1600

Mbogo KP, Davis J, Myers JR (2009) Transfer of the arcelin-phytohaemagglutinin-α amylase inhibitor seed protein locus from tepary bean (*Phaseolus acutifolius* A. Gray) to common bean (*P. vulgaris* L.). Biotechnology 8:285–295

McClean P, Myers J (1990) Pedigrees of dry bean cultivars, lines and PIs. Annu Rep Bean Improv Coop 33:xxv–xxx

McClean PE, Mamidi S, McConnell M, Chikara S, Lee R (2010) Synteny mapping between common bean and soybean reveals extensive blocks of shared loci. BMC Genom 11:184

Mensack MM, Fitzgerald VK, Ryan EP, Lewis MR, Thompson HJ, Brick MA (2010) Evaluation of diversity among common beans (*Phaseolus vulgaris* L.) from two centers of domestication using 'omics' technologies. BMC Genom 11:686

Mirkov TE, Wahlstrom J, Hagiwara K, Finardi-Filho F, Kjemtrup S, Chrispeels MJ (1994) Evolutionary relationships among proteins in the phytohemagglutinin-arcelin-α-amylase inhibitor family of the common bean and its relatives. Plant Mol Biol 26:1103–1113

Monaghan EK, Venkatachalam M, Seavy M, Beyer K, Sampson HA, Roux KH, Sathe SK (2008) Enzyme-linked immunosorbent assay (ELISA) for detection of sulfur-rich protein (SRP) in soybeans (*Glycine max* L.) and certain other edible plant seeds. J Agric Food Chem 56:765–777

Montoya CA, Lallès JP, Beebe S, Leterme P (2010) Phaseolin diversity as a possible strategy to improve the nutritional value of common beans (*Phaseolus vulgaris*). Food Res Int 43:443–449

Moore JG, Fuchs CA, Hata YS, Hicklin DJ, Colucci G, Chrispeels MJ, Feldman M (2000) A new lectin in red kidney beans called PvFRIL stimulates proliferation of NIH 3T3 cells expressing the Flt3 receptor. Biochim Biophys Acta 1475:216–224

Moreno J, Chrispeels MJ (1989) A lectin gene encodes the α-amylase inhibitor of the common bean. Proc Natl Acad Sci USA 86:7885–7889

Muench SP, Rawson S, Eyraud V, Delmas AF, Da Silva P, Phillips C, Trinick J, Harrison MA, Gressent F, Huss M (2014) PA1b inhibitor binding to subunits c and e of the vacuolar ATPase reveals its insecticidal mechanism. J Biol Chem 289:16399–16408

Muhling M, Gilroy J, Croy RRD (1997) Legumin proteins from seeds of *Phaseolus vulgaris* L. J Plant Physiol 150:489–492

Natarajan SS, Pastor-Corrales MA, Khan FH, Garrett WM (2013) Proteomic analysis of common bean (*Phaseolus vulgaris* L.) by two-dimensional gel electrophoresis and mass spectrometry. J Basic Appl Sci 9:424–437

Ng DWK, Hall TC (2008) PvALF and FUS3 activate expression from the phaseolin promoter by different mechanisms. Plant Mol Biol 66:233–244

Ng DW, Chandrasekharan MB, Hall TC (2006) Ordered histone modifications are associated with transcriptional poising and activation of the phaseolin promoter. Plant Cell 18:119–132

O'Rourke JA, Iniguez LP, Fu F, Bucciarelli B, Miller SS, Jackson SA, McClean PE, Li J, Dai X, Zhao PX, Hernandez G, Vance CP (2014) An RNA-Seq based gene expression atlas of the common bean. BMC Genom 15:866

Osborn TC, Bliss FA (1985) Effects of genetically removing lectin seed protein on horticultural and seed characteristics of common bean. J Am Soc Hortic Sci 110:484–488

Osborn TC, Blake T, Gepts P, Bliss FA (1986) Bean arcelin 2. Genetic variation inheritance and linkage relationships of a novel seed protein of *Phaseolus vulgaris* L. Theor Appl Genet 71:847–855

Osborn TC, Alexander DC, Sun SSM, Cardona C, Bliss FA (1988) Insecticidal activity and lectin homology of arcelin seed protein. Science 240:207–210

Osborn TC, Hartweck LM, Harmsen RH, Vogelzang RD, Kmiecik KA, Bliss FA (2003) Registration of *Phaseolus vulgaris* genetic stocks with altered seed protein compositions. Crop Sci 43:1570–1571

Pandurangan S, Marsolais F (2017) Analysis of phaseolin copy number by quantitative PCR. Annu Rep Bean Improv Coop 60:107–108

Pandurangan S, Sandercock M, Beyaert R, Conn KL, Hou A, Marsolais F (2015) Differential response to sulfur nutrition of two common bean genotypes differing in storage protein composition. Front Plant Sci 6:92

Pandurangan S, Diapari M, Yin F, Munholland S, Perry G, Chapman BP, Huang S, Sparvoli F, Bollini R, Crosby W, Pauls KP, Marsolais F (2016) Genomic analysis of storage protein deficiency in genetically related lines of common bean (*Phaseolus vulgaris*). Front Plant Sci 7:389

Parreira JR, Bouraada J, Fitzpatrick MA, Silvestre S, Bernardes da Silva A, Marques da Silva J, Almeida AM, Fevereiro P, Altelaar AFM, Araújo SS (2016) Differential proteomics reveals the hallmarks of seed development in common bean (*Phaseolus vulgaris* L.). J Proteomics 143:188–198

Prescott VE, Campbell PM, Moore A, Mattes J, Rothenberg ME, Foster PS, Higgins TJV, Hogan SP (2005) Transgenic expression of bean α-amylase inhibitor in peas results in altered structure and immunogenicity. J Agric Food Chem 53:9023–9030

Reinprecht Y, Luk-Labey SY, Yu K, Poysa VW, Rajcan I, Ablett GR, Peter Pauls K (2011) Molecular basis of seed lipoxygenase null traits in soybean line OX948. Theor Appl Genet 122:1247–1264

Schmidt MA, Hymowitz T, Herman EM (2015) Breeding and characterization of soybean Triple Null; a stack of recessive alleles of Kunitz trypsin inhibitor, soybean agglutinin, and P34 allergen nulls. Plant Breed 134:310–315

Schmutz J, Cannon SB, Schlueter J, Ma J, Mitros T, Nelson W, Hyten DL, Song Q, Thelen JJ, Cheng J, Xu D, Hellsten U, May GD, Yu Y, Sakurai T, Umezawa T, Bhattacharyya MK, Sandhu D, Valliyodan B, Lindquist E, Peto M, Grant D, Shu S, Goodstein D, Barry K, Futrell-Griggs M, Abernathy B, Du J, Tian Z, Zhu L, Gill N, Joshi T, Libault M, Sethuraman A, Zhang XC, Shinozaki K, Nguyen HT, Wing RA, Cregan P, Specht J, Grimwood J, Rokhsar D, Stacey G, Shoemaker RC, Jackson SA

(2010) Genome sequence of the palaeopolyploid soybean. Nature 463:178–183

Schmutz J, McClean PE, Mamidi S, Wu GA, Cannon SB, Grimwood J, Jenkins J, Shu S, Song Q, Chavarro C, Torres-Torres M, Geffroy V, Moghaddam SM, Gao D, Abernathy B, Barry K, Blair M, Brick MA, Chovatia M, Gepts P, Goodstein DM, Gonzales M, Hellsten U, Hyten DL, Jia G, Kelly JD, Kudrna D, Lee R, Richard MM, Miklas PN, Osorno JM, Rodrigues J, Thareau V, Urrea CA, Wang M, Yu Y, Zhang M, Wing RA, Cregan PB, Rokhsar DS, Jackson SA (2014) A reference genome for common bean and genome-wide analysis of dual domestications. Nat Genet 46:707–713

Severin AJ, Woody JL, Bolon YT, Joseph B, Diers BW, Farmer AD, Muehlbauer GJ, Nelson RT, Grant D, Specht JE, Graham MA, Cannon SB, May GD, Vance CP, Shoemaker RC (2010) RNA-seq atlas of *Glycine max*: a guide to the soybean transcriptome. BMC Plant Biol 10:160

Singh SP, Gepts P, Debouck DG (1991) Races of common bean (*Phaseolus vulgaris*, Fabaceae). Econ Bot 45:379–396

Slightom JL, Sun SM, Hall TC (1983) Complete nucleotide sequence of a French bean storage protein gene: phaseolin. Proc Natl Acad Sci USA 80:1897–1901

Slightom JL, Drong RF, Klassy RC, Hoffman LM (1985) Nucleotide sequences from phaseolin cDNA clones: the major storage proteins from *Phaseolus vulgaris* are encoded by two unique gene families. Nucleic Acids Res 13:6483–6498

Sparvoli F, Bollini R, Cominelli E (2015) Nutritional value. In: De Ron AM (ed) Grain Legumes. Handbook of Plant Breeding, vol 10. Springer Science+Business Media, New York, pp 291–325

Staswick PE, Hermodson MA, Nielsen NC (1984) Identification of the cystines which link the acidic and basic components of the glycinin subunits. J Biol Chem 259:13431–13435

Sturm A, Van Kuik JA, Vliegenthart JF, Chrispeels MJ (1987) Structure, position, and biosynthesis of the high mannose and the complex oligosaccharide side chains of the bean storage protein phaseolin. J Biol Chem 262:13392–13403

Sundaram S, Kertbundit S, Shakirov EV, Iyer LM, Juricek M, Hall TC (2013) Gene networks and chromatin and transcriptional regulation of the phaseolin promoter in Arabidopsis. Plant Cell 25:2601–2617

Talbot DR, Adang MJ, Slightom JL, Hall TC (1984) Size and organization of a multigene family encoding phaseolin, the major seed storage protein of *Phaseolus vulgaris* L. Mol Gen Genet 198:42–49

Tolleter D, Jaquinod M, Mangavel C, Passirani C, Saulnier P, Manon S, Teyssier E, Payet N, Avelange-Macherel M-H, Macherel D (2007) Structure and function of a mitochondrial late embryogenesis abundant protein are revealed by desiccation. Plant Cell 19:1580–1589

Tsubokura Y, Hajika M, Kanamori H, Xia Z, Watanabe S, Kaga A, Katayose Y, Ishimoto M, Harada K (2012) The β-conglycinin deficiency in wild soybean is associated with the tail-to-tail inverted repeat of the α-subunit genes. Plant Mol Biol 78:301–309

van der Geest AH, Hall TC (1996) A 68 bp element of the β-phaseolin promoter functions as a seed-specific enhancer. Plant Mol Biol 32:579–588

Vigeolas H, Chinoy C, Zuther E, Blessington B, Geigenberger P, Domoney C (2008) Combined metabolomic and genetic approaches reveal a link between the polyamine pathway and albumin 2 in developing pea seeds. Plant Physiol 146:74–82

Vitale A, Bollini R (1995) Legume storage proteins. In: Kigel J, Galili G (eds) Seed Development and Germination. Marcel Dekker, New York, pp 73–102

Vlasova A, Capella-Gutierrez S, Rendon-Anaya M, Hernandez-Onate M, Minoche AE, Erb I, Camara F, Prieto-Barja P, Corvelo A, Sanseverino W, Westergaard G, Dohm JC, Pappas GJ Jr, Saburido-Alvarez S, Kedra D, Gonzalez I, Cozzuto L, Gomez-Garrido J, Aguilar-Moron MA, Andreu N, Aguilar OM, Garcia-Mas J, Zehnsdorf M, Vazquez MP, Delgado-Salinas A, Delaye L, Lowy E, Mentaberry A, Vianello-Brondani RP, Garcia JL, Alioto T, Sanchez F, Himmelbauer H, Santalla M, Notredame C, Gabaldon T, Herrera-Estrella A, Guigo R (2016) Genome and transcriptome analysis of the Mesoamerican common bean and the role of gene duplications in establishing tissue and temporal specialization of genes. Genome Biol 17:32

Voelker TA, Staswick P, Chrispeels MJ (1986) Molecular analysis of two phytohemagglutinin genes and their expression in *Phaseolus vulgaris* cv. Pinto, a lectin-deficient cultivar of the bean. EMBO J 5:3075–3082

Voelker TA, Moreno J, Chrispeels MJ (1990) Expression analysis of a pseudogene in transgenic tobacco: a frameshift mutation prevents mRNA accumulation. Plant Cell 2:255–261

Wato S, Kamei K, Arakawa T, Philo JS, Wen J, Hara S, Yamaguchi H (2000) A chimera-like α-amylase inhibitor suggesting the evolution of *Phaseolus vulgaris* α-amylase inhibitor. J Biochem 128:139–144

Wilson KA, Laskowski M (1973) Isolation of three isoinhibitors of trypsin from garden bean, *Phaseolus vulgaris*, having either lysine or arginine at the reactive site. J Biol Chem 248:756–762

Wilson KA, Laskowski M (1975) The partial amino acid sequence of trypsin inhibitor II from garden bean, *Phaseolus vulgaris*, with location of the trypsin and elastase-reactive sites. J Biol Chem 250:4261–4267

Yamauchi F, Sato K, Yamagishi T (1984) Isolation and partial characterization of a salt-extractable globulin from soybean seeds. Agric Biol Chem 48:645–650

Yin F, Pajak A, Chapman R, Sharpe A, Huang S, Marsolais F (2011) Analysis of common bean expressed sequence tags identifies sulfur metabolic pathways active in seed and sulfur-rich proteins highly expressed in the absence of phaseolin and major lectins. BMC Genom 12:268

Yoshizawa T, Shimizu T, Yamabe M, Taichi M, Nishiuchi Y, Shichijo N, Unzai S, Hirano H, Sato M, Hashimoto H (2011) Crystal structure of basic 7S globulin, a xyloglucan-specific endo-β-1,4-glucanase inhibitor protein-like protein from soybean lacking inhibitory activity against endo-β-glucanase. FEBS J 278:1944–1954

Zaugg I, Magni C, Panzeri D, Daminati MG, Bollini R, Benrey B, Bacher S, Sparvoli F (2013) QUES, a new *Phaseolus vulgaris* genotype resistant to common bean weevils, contains the Arcelin-8 allele coding for new lectin-related variants. Theor Appl Genet 126:647–661

A Comparison of Phenylpropanoid Pathway Gene Families in Common Bean. Focus on P450 and C4H Genes

11

Yarmilla Reinprecht, Gregory E. Perry and K. Peter Pauls

Abstract

The focus of this chapter is on gene families encoding enzymes of phenylpropanoid pathway in common bean. The introductory section contains a short overview of the phenylpropanoid pathway. Section 11.2 introduces major gene families encoding enzymes of this pathway in common bean, soybean, and Arabidopsis in the current annotations of their complete genome sequences (*Phaseolus vulgaris* v1.0, *Glycine max* Wm82.a2.v1, and *Arabidopsis thaliana* TAIR10) deposited in Phytozome 10.2. For each of the 21 enzyme classes, their functional annotations were based on the commonly used Pfam and KOG databases, while the number of genes in each family was based on Phytozome and KEGG databases. Section 11.3 describes cytochrome P450s involved in the phenylpropanoid pathway with particular emphasis on ten families included in the general (central) phenylpropanoid pathway, C4H (family *CYP73A*), in the lignin/lignan branch, C3H (family *CYP98A*) and F5H (family *CYP84A*), in the flavonoid/anthocyanin/proanthocyanidin branch, F3′H (family *CYP75B*), F3′5′H (family *CYP75A*), and FNS (family *CYP93B*), and in the isoflavonoid branch IFS (family *CYP93C*), I2′H (family *CYP81E*), F6H (family *CYP71D*), and D6aH (family *CYP93A*). The availability of the complete genome sequences enabled a thorough inventory of putative P450 genes encoding enzymes of this metabolic pathway. The P450 gene sequences from common bean were compared to homologs from *Arabidopsis* and soybean and confirmed with the information published for both soybean and common bean genomes. Cinnamate 4-hydroxylase (C4H) is the first P450 enzyme in the phenylpropanoid pathway and is described in detail in Sect. 11.4. It belongs to the relatively small *CYP73A* gene family. Genome locations and gene structures including *cis*-regulatory regions in 5′UTRs (5′ regulatory sequences) are detailed for

Y. Reinprecht (✉) · G.E. Perry · K. Peter Pauls
Department of Plant Agriculture, University of Guelph, Guelph, ON N1G 2W1, Canada
e-mail: yreinpre@uoguelph.ca

M. Pérez de la Vega et al. (eds.), *The Common Bean Genome*,
Compendium of Plant Genomes, https://doi.org/10.1007/978-3-319-63526-2_11

this family in common bean. In addition, the expression patterns of these genes in different tissues (Phytozome 10.2) and syntenic relationships (Plant Genome Duplication Database) between common bean and soybean were examined. Finally, genes encoding the C4H enzyme in landrace G19833 (Andean gene pool, Phytozome 10.2) and in cultivar OAC Rex (Mesoamerican gene pool) were compared and searched for polymorphisms. These sequence differences can be used to develop C4H gene-based marker(s) to explore the roles of these genes in various processes such as lignin or anthocyanin biosynthesis.

Keywords
Common bean · Cytochrome P450 superfamily · *C4H* gene · Genome sequence · In silico · Phenylpropanoid pathway · Synteny

In this chapter:

- Common names in plants: Arabidopsis (*Arabidopsis thaliana*), soybean (*Glycine max*), common bean (*Phaseolus vulgaris*)
- Chromosome-based locus (gene model) identifier (Phytozome)

11.1 Introduction

As sessile organisms, plants produce numerous secondary metabolites to overcome biotic and abiotic stressors, to attract pollinators and nitrogen-fixing microorganisms, and to communicate with other plants (Koes et al. 2005; Noel et al. 2005; Moura et al. 2010; Agati et al. 2012, 2013; Baxter and Stewart 2013). Many of these compounds are synthesized by the phenylpropanoid pathway, which is likely one of the most studied pathways in plants. It is relatively well understood and was extensively reviewed (Goujon et al. 2003; Raes et al. 2003; Wang 2011; Falcone Ferreyra et al. 2012; Petrussa et al. 2013). Individual branches of the pathway have been thoroughly characterized. Most of the enzymes that catalyze individual steps of the pathway have been identified, and the genes coding for them have been isolated in a number of plant species, including *Arabidopsis* and soybean (Graham et al. 2008; Fraser and Chapple 2011).

The core (general or central) pathway consists of three steps, including (1) the conversion of the aromatic amino acid phenylalanine into *trans*-cinnamic acid, which is catalyzed by phenylalanine ammonia-lyase (PAL); (2) the conversion of *trans*-cinnamic acid into *p*-coumaric acid, catalyzed by cinnamate 4-hydroxylase (C4H); and (3) the transformation of *p*-coumaric acid into *p*-coumaroyl-CoA, catalyzed by 4-coumarate:CoA ligase (4CL). The compound *p*-coumaroyl-CoA serves as a starting point for several branches of the phenylpropanoid pathway leading to biosynthesis of lignin, lignans, coumarins, stilbenes, flavonoids, anthocyanin, condensed tannins (proanthocyanidins), and isoflavonoids (Vogt 2010; Cheynier et al. 2013). These products have important functions not only for plant survival, growth, and development but they could also be powerful supplements to the human diet. For example, lignans, stilbenes, and isoflavonoids have been associated with the reduced onset/development of certain chronic disease in humans, including some forms of cancer and heart diseases (Cassidy et al. 2000; Chen et al. 2006; Adlercreutz 2007; Xiao 2008; Brunetti et al. 2013) (Fig. 11.1).

Lignin biosynthesis is a two-step process. First, monolignol is synthesized through a series

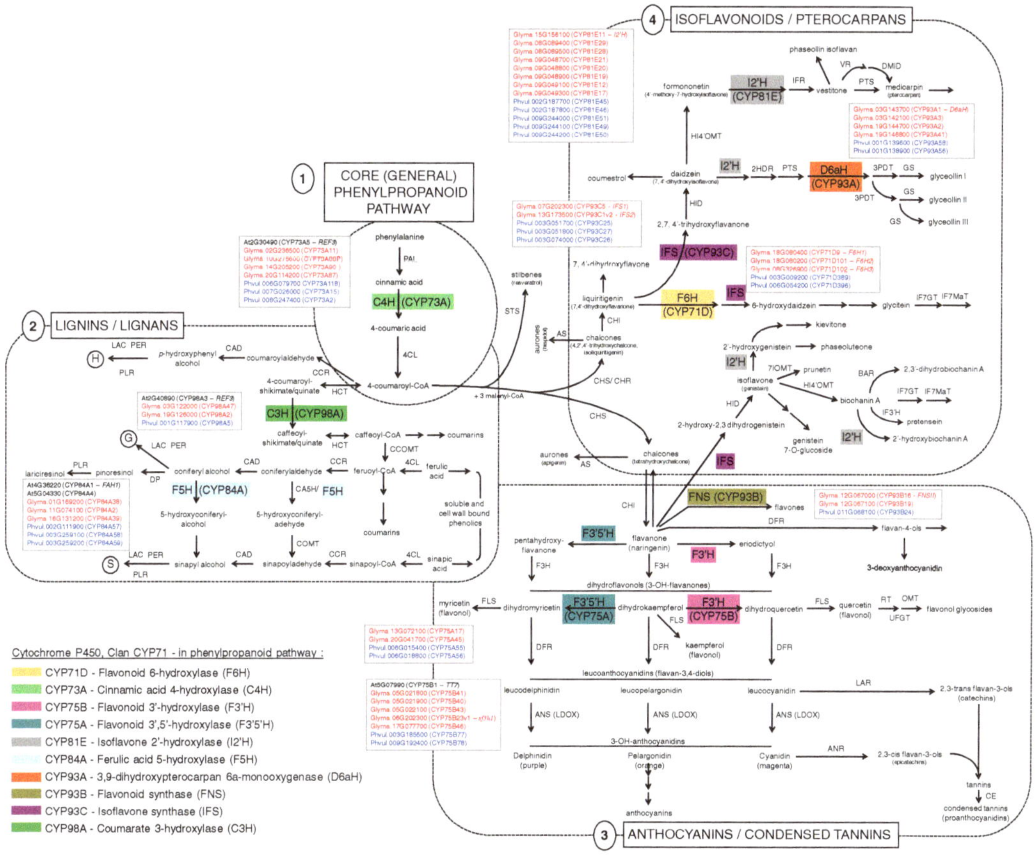

Fig. 11.1 Cytochrome P450s involved in the phenylpropanoid pathway. The positions of ten enzymes and locus (gene model) identifiers (https://phytozome.jgi.doe.gov/pz/portal.html) in the pathway in common bean (blue), soybean (red), and *Arabidopsis* (black) are indicated; 1. Core phenylpropanoid pathway: cinnamate 4-hydroxylase (C4H, CYP73A); 2. Lignin/lignans branch: coumarate 3-hydroxylase (C3H, CYP98A) and ferulic acid 5-hydroxylase (F5H, CYP84A); 3. Anthocyanins/condensed tannins branch: flavonoid 3′-hydroxylase (F3′H, CYP75B), flavonoid 3′,5′-hydroxylase (F3′,5′H, CYP75A) and flavone synthase (FNS, CYP93B); 4. Isoflavonoid branch: isoflavone synthase (IFS, CYP93C), isoflavone 2′-hydroxylase (I2′H, CYP81E), flavonoid 6-hydroxylase (F6H, CYP71D), and 3,9-dihydroxypterocarpan 6a-monooxygenase (D6aH, CYP93A)

of hydroxylations, *O*-methylations, and conversions of side-chain carboxyl into *p*-coumaryl, coniferyl, and sinapyl alcohols (Humphreys and Chapple 2002; Boerjan et al. 2003; Vanholme et al. 2010; Weng and Chaple 2010; Labeeuw et al. 2015). A second step involves monolignol polymerization by peroxidases (PER), laccases (LAC), and dirigent proteins (DP). In a reversible reaction, hydroxycinnamoyl-CoA:shikimate/quinate hydroxycinnamoyltransferase (HCT) converts *p*-coumaroyl-CoA and caffeoyl-CoA into their corresponding shikimate/quinate esters, which are then transformed by coumarate 3-hydroxylase (C3H) into their corresponding caffeoyl esters (Schoch et al. 2001). Caffeoyl-CoA *O*-methyltransferase (CCoAOMT) catalyzes methylation of caffeoyl-CoA to generate feruloyl-CoA. Cinnamoyl-CoA reductase (CCR) converts hydroxycinnamoyl-CoA esters into their corresponding aldehydes, and cinnamyl-alcohol dehydrogenase (CAD) catalyzes the conversion of cinnamyl aldehydes into their corresponding alcohols. Ferulic acid 5-hydroxylase (F5H) converts ferulic acid into 5-hydroxyferulic acid. F5H is also known as coniferaldehyde 5-hydroxylase (CAld5H), since the enzyme preferably transforms coniferaldehyde and/or coniferyl alcohol into synapaldehyde and/or sinapyl alcohol, respectively (Humphreys et al. 1999; Osakabe et al. 1999). Caffeic acid *O*-methyltransferase (COMT) converts 5-hydroxyconiferaldehyde and/or 5-hydroxyconiferyl alcohol into sinapaldehyde and/or sinapyl alcohol, respectively (Osakabe et al. 1999; Parvathi et al. 2001; Zubieta et al. 2002). COMT was previously thought to be a bifunctional enzyme, methylating caffeic and 5-hydroxyferulic acids.

Chalcone synthase (CHS) is the first enzyme in the flavonoid/anthocyanin branch of the phenylpropanoid pathway. It catalyzes the biosynthesis of chalcone from one molecule of *p*-coumaroyl-CoA with three molecules of malonyl-CoA. This basic flavonoid structure is then transformed by a set of various isomerases, reductases, hydroxylases, Fe^{2+}/2-oxoglutarate-dependent dioxygenases, and transferases into different flavonoids, including flavanones, flavones, flavonols, anthocyanins, and condensed tannins (Winkel-Shirley 2001; Ralston et al. 2005; Ferrer et al. 2008; Saito et al. 2013). CHS and chalcone isomerase (CHI) catalyze the two-step condensation, producing a colorless flavanone (naringenin), which is then oxidized by flavanone 3-hydroxylase (F3H) into the colorless dihydroflavonol (dihydrokaempferol). Subsequent hydroxylation of this compound (at the 3′ or 5′ position of the B-ring), catalyzed by flavonoid 3′-hydroxylase (F3′H) and flavonoid 3′,5′-hydroxylase (F3′5′H), produces dihydroquercetin and dihydromyricetin. These two enzymes (F3′H and F3′5′H) can also hydroxylate flavanone (naringenin) to produce eriodictyol and pentahydroxy-flavanone, which are then hydroxylated by F3H into dihydroquercetin and dihydromyricetin, respectively. The next step in the pathway is the conversion of the three dihydroflavonols (dihydroquercetin, dihydrokaempferol, and dihydromyricetin). These compounds can be transformed into flavonols (kaempferol, quercetin, and myricetin) by flavonol synthases (FLS). Dihydroflavonol 4-reductase (DFR) converts dihydroflavonols into leucoanthocyanidins (colorless flavan-3,4-diols: leucocyanidin, leucopelargonidin, and leucodelphinidin), which are then oxidized by anthocyanin synthase [ANS, also known as leucoanthocyanidin dioxygenase (LDOX)] into colored but unstable anthocyanidins [cyanidin (red-magenta), pelargonidin (orange), and delphinidin (purple-mauve)]. Stable anthocyanins (colored) are produced by glycosylation of these compounds by the UDP-glucose:flavonoid 3-*O*-glucosyl transferases (UFGT). Some anthocyanins (cyanidin-3-glucoside and delphinidin-3-glucoside) may be further methylated by methyltransferases (MTs) to produce peonidin-3-glucoside and petunidin- or malvidin-3-glucoside, respectively.

Condensed tannins are synthesized through two branches of the anthocyanin pathway. The reduction of leucocyanidin to catechin (2,3-*trans* flavan-3-ols) is catalyzed by leucoanthocyanidin reductase (LAR), and the conversion of cyanidin into epicatechin (2,3-*cis* flavan-3-ols) is driven by anthocyanidin reductase (ANR). The subsequent steps catalyzed by polyphenol oxidases and condensing enzymes possibly take place in vacuoles.

Legume-specific isoflavonoids are produced through two branches of the isoflavonoid pathway having major reactions in common. The branch leading to the isoflavone genistein uses the same naringenin intermediate, which is synthesized in the flavonoid/anthocyanin branch of the phenylpropanoid pathway by a two-step condensation catalyzed by CHS and CHI (common to majority of plants) (Lozovaya et al. 2007). On the other hand, isoflavone daidzein is synthesized through the co-action of CHS and legume-specific chalcone reductase (CHR), yielding isoliquiritigenin (trihydroxychalcone), which is then transformed into liquiritigenin (dihydroxyflavanone), a core intermediate of this branch of the isoflavonoid pathway (Austin and Noel 2003). Isoflavone synthase [IFS, also known as 2-hydroxyisoflavanone synthase (2-HIS)] converts flavanone (naringenin or isoliquiritigenin) into 2-hydroxyisoflavanones (through an aryl migration of the aromatic B-ring from C-2 to C-3 position and hydroxylation in position C-2) (Steele et al. 1999; Jung et al. 2000), which are then dehydrated (formation of a double bond between C-2 and C-3) to the corresponding isoflavones (genistein and daidzein) by 2-hydroxyisoflavanone dehydratase (HID) (Akashi et al. 2005; Shimamura et al. 2007). They are further modified by isoflavonoid-specific enzymes to produce major phytoalexins, including medicarpin, biochanin A, glyceollin, pisatin, and maackiain (Latunde-Dada et al. 2001; Lozovaya et al. 2007; Artigot et al. 2013).

Biosynthesis of lignin, flavonoids/anthocyanins/proanthocyanidins, and isoflavonoids is under complex regulation. The expression of the lignin biosynthetic genes is coordinately regulated by a number of transcription factors. The majority of these genes contain a common AC *cis*-element, which is required for their expression in cells undergoing lignification. NST1/2/3 (NAC secondary wall thickening promoting factor 1/2/3) and Myb26/Myb83 transcription factors act as master switches to regulate biosynthesis of major secondary wall components, including cellulose, xylan, and lignin in *Arabidopsis* (Zhong and Ye 2009; Zhao and Dixon 2011; Hao and Mohnen 2014; Yoon et al. 2015). In *Arabidopsis* flavonoid pathway, genes for early biosynthetic enzymes (CHS, CHI, F3H, and F3′H) are regulated by the three functionally redundant R2R3-MYB transcription factors (MYB11, MYB12, and MYB111), while the activation of late biosynthetic genes is controlled by the R2R3-MYB/bHLH/WD40 (MBW) complex (Grotewold 2005; Hartman et al. 2005; Ramsey and Glover 2005; Gonzalez et al. 2008; Gou et al. 2011; Petrussa et al. 2013; Li et al. 2014; Xu et al. 2014, 2015). Genes of legume-specific isoflavonoid branch of phenylpropanoid pathway are regulated by a different set of transcription factors. For example, GmMYB176, a R1 MYB transcription factor, regulates *CHS8* expression and isoflavonoid synthesis in soybean (Yi et al. 2010a, b; Dhaubhadel 2011). The constitutive over-expression of LjMYB14 was associated with the activation of dozen of genes coding for enzymes in the core phenylpropanoid pathway and isoflavonoid branch in *Lotus japonicus* (Shelton et al. 2012). At the same time, the expression of other transcription factors was altered resulting in coordinated down-regulation of the competing biosynthetic pathways.

Genes encoding the major enzymes of the phenylpropanoid pathway have been identified in a number of plant species (Tsai et al. 2006; Tohge et al. 2007; Xu et al. 2009). In most species, enzymes involved in the phenylpropanoid pathway are encoded by gene families of various sizes. For example, plants' CADs can reduce various aldehydes, including those expressed in response to pathogens (Barakat et al. 2010; Miedes et al. 2014). The nine putative *CAD* genes that were identified in *Arabidopsis* are split into three classes based on protein phylogenetic analysis (Raes et al. 2003). Using Southern hybridization of genomic DNA, Ryder et al. (1987) identified six to eight *CHS* genes in common bean, some of them tightly clustered, which represented different loci, not allelic variation. The soybean *CHS* gene family consists of nine members (*CHS1* to *CHS9*), some of which are clustered (Akada and Dube 1995; Yi et al. 2010a). They share a high degree of sequence similarity and play different roles in plant development and interactions with environment. Matsumura et al. (2005) mapped eight *CHS* genes on five linkage groups (A1, A2, B1, DIa, and K) in soybean. Duplicated *CHS1* gene was associated with the suppressed seed coat pigmentation in yellow soybean (Senda et al. 2002).

Gene families arise from interspecific hybridization, polyploidization, and local duplication. Genome duplication results in biased gene content (Freeling 2009) and non-random divergence in gene expression (Casneuf et al. 2006; Wang et al. 2012, 2013). After a duplication event, the new gene copy (or the original copy) can retain the same function (subfunctionalization), undergo neo-functionalization, or become non-functional (loss of function) (Lynch and Conery 2000; Hanada et al. 2011; Barker et al. 2012). Gene clusters formed by gene duplication have been frequently found in multigene families, including plant specialized metabolism (Nutzmann and Osbourn 2014, 2015). For example, clusters encoding enzymes of all steps in lignin biosynthesis have been identified in the *Eucalyptus grandis* EST libraries (Harakava 2005). The authors also predicted co-localization of several phenylpropanoid pathway enzymes including PAL, C4H, 4CL, C3H, and F5H on the endoplasmic reticulum (ER) membrane. This may suggest the existence of metabolons involving P450 multienzyme complexes and channeling of pathway intermediates without their release into the general metabolic pool (Hrazdina and Wagner 1985; Winkel-Shirley 1999; Ralston and Yu 2006; Bassard et al. 2012).

The availability of complete genome sequences enabled genome-wide analyses of the phenylpropanoid pathway genes in several species (Naoumkina et al. 2010). Shi et al. (2010) identified 95 genes (ten gene families) associated with phenylpropanoid pathway in *Populus trichocarpa* and identified functional redundancy at the transcript level for six lignin biosynthetic genes [*PAL, C4H, 4CL, HCT, CCoAOMT, CAld5H (F5H)*]. Using an in silico approach, Costa et al. (2003) analyzed the organization and function of phenylpropanoid pathway gene network in *Arabidopsis*, while Lucheta et al. (2007) focused on genes encoding key enzymes in the flavonoid pathway in *Citrus sinensis*. Hamberger et al. (2007) conducted genome-wide analysis of phenylpropanoid pathway gene families in poplar and compared them to homologs in *Arabidopsis* and rice. The focus of these studies was on the genes of the core pathway and the lignin branch. To explore the evolution of phenylpropanoid pathway diversity, Tohge et al. (2013) compared 65 gene families involved in the pathway among 23 species, including *Arabidopsis* and soybean. Another evolutionary study was focusing on the isoflavonoid pathway (Chu et al. 2014). The research examined nine major isoflavonoid genes in seven plant species, including *Arabidopsis*, soybean, and common bean. Genes coding for PAL, C4H, 4CL, CHS, and CHI were identified in all analyzed species, while for CHR, IFS, IOMT (isoflavonoid *O*-methyltransferase), and IFR (isoflavonoid reductase) were confirmed to be legume-specific. Divergent evolutionary patterns were observed among different gene copies of centrally located branch-point enzymes (4CL, CHS, and CHI) regardless of the level of polymorphism or the evolutionary rate.

However, information about this important pathway in common bean is still fragmentary. In our previous study (Reinprecht et al. 2013), 35 phenylpropanoid pathway genes were cloned and mapped in silico in common bean genome (annotation *Phaseolus vulgaris* v1.0). The work also identified syntenic regions containing phenylpropanoid pathway genes in common bean and soybean (annotation *Glycine max* v1.1) (Reinprecht et al. 2013). In another study, 22 phenylpropanoid pathway genes have been mapped in the Bat93 × Jalo EEP558 (a core mapping resource for *P. vulgaris*) and OAC Rex × SVM Taylor recombinant inbred line (RIL) populations (Yadegari 2013). Currently, work on identifying an association between these genes and different seed phenolics in common bean using an association mapping approach is underway. Cytochrome P450 gene family encodes several key enzymes in the phenylpropanoid pathway. Alber and Ehlting (2012) reviewed P450s involved in lignin biosynthesis. The availability of the complete common bean genome sequence allowed Kumar et al. (2015) to identify members of this gene family. The focus of our work was to study gene families encoding enzymes of phenylpropanoid pathway in common bean, using an in silico approach.

11.2 Gene Families Encoding Enzymes of Phenylpropanoid Pathway in Common Bean

Currently, complete genome sequences for 55 plant species, including common bean (Schmutz et al. 2014; current annotation *P. vulgaris* v1.0), are deposited in Phytozome 10.3 (a comparative genomic database, available at http://phytozome.jgi.doe.gov/pz/portal.html; accessed 16 Nov 2015; Goodstein et al. 2012). This allowed us to study the complete gene families encoding enzymes of phenylpropanoid pathway in common bean, thus extending our previous work

(Reinprecht et al. 2013). In particular, we examined their conservation and diversification through comparative analyses with previously sequenced soybean (Schmutz et al. 2010; current annotation *G. max* Wm82.a2.v1) and *Arabidopsis* (The Arabidopsis Genome Initiative 2000; Lamesch et al. 2012; current annotation *Arabidopsis thaliana* TAIR10) genomes. The basic information for the sequenced *Arabidopsis*, soybean, and common bean genomes is presented in Table 11.1.

Genome annotations for common bean (Schmutz et al. 2014), soybean (Schmutz et al. 2010), and *Arabidopsis* (The Arabidopsis Genome Initiative 2000) were obtained from Phytozome 10.2 (Goodstein et al. 2012). For each gene, identifiers and descriptions for all Pfam (**Pro**tein **fam**ilies), KEGG (**K**yoto **E**ncyclopedia of **G**enes and **G**enomes), GO (**G**ene **O**ntology), PANTHER (**P**rotein **AN**alysis **TH**rough **E**volutionary **R**elationships), and KOG (Eu**K**aryotic **O**rthologous **G**roups) classifications assigned to this gene can be found.

Table 11.2 contains the list and the number of putative genes in each of the major gene families encoding enzymes of the phenylpropanoid pathway in common bean, soybean, and *Arabidopsis* in the current annotations of their complete genome sequences (*P. vulgaris* v1.0, *G. max* Wm82.a2.v1, and *A. thaliana* TAIR10) deposited in Phytozome. For each of the 21 enzyme classes, their functional annotations were based on the Pfam and KOG databases (commonly used), while the number of genes in each family was based on Phytozome and KEGG databases. For example, with the KOG0222 search, four phenylalanine ammonia-lyase (*PAL*, EC:4.3.1.24) genes were identified in *Arabidopsis*, eight *PAL* genes were identified in soybean, and six *PAL* genes were found in common bean (Table 11.2). Several large gene families are involved in phenylpropanoid pathway, including the cytochrome P450 family.

11.3 The Role of Cytochrome P450 Superfamily in Phenylpropanoid Pathway

11.3.1 Cytochrome P450

Cytochromes P450 (CYPs) are ubiquitous monooxygenase enzymes involved in the oxidation of various substrates using oxygen and NADPH. Plant P450s play vital roles in metabolism and detoxification (Mizutani and Ohta 2010; Hamberger and Back 2013). They catalyze reactions in both primary metabolism and secondary metabolism and are involved in the biosynthesis of various metabolites, including fatty acids, sterols, hormones, phenylpropanoids, terpenoids, and signaling molecules. Chemical diversity across plant species is well correlated with the heterogeneity of the P450s (Mizutani and Sato 2011; Mizutani 2012; Sezutsu et al. 2013). They contain a heme cofactor, which

Table 11.1 Basic information for the sequenced genomes of *A. thaliana*, *G. max*, and *P. vulgaris*

Species	Genome				
	Version	Size (Mb)/ chromosomes	Protein coding loci	Data retrieval	Reference
Arabidopsis thaliana (*Arabidopsis* or thale cress)	TAIR10	135/5	27,416	TAIR[a]	The Arabidopsis Genome Initiative (2000)
Glycine max (soybean)	Wm82.a2.v1	978/20	56,044	JGI[b]	Schmutz et al. (2010)
Phaseolus vulgaris (common bean)	v1.0	521/11	27,197	JGI	Schmutz et al. (2014)

[a]TAIR, The Arabidopsis Information Resource [available at ftp://ftp.arabidopsis.org/home/tair/Genes/ (accessed 15 June 2015)]
[b]JGI, DOE Joint Genome Institute [available at http://phytozome.jgi.doe.gov/pz/portal.html (accessed 15 June 2015)]

Table 11.2 Major gene families encoding enzymes of phenylpropanoid pathway in *A. thaliana*, *G. max*, and *P. vulgaris*

Gene family		Number of genes		
Enzyme class (EC)	Functional annotation[a]	*A. thaliana* TAIR10	*G. max* Wm82. a2.v1	*P. vulgaris* v1.0
EC:1—Oxidoreductases				
1. Cytochrome P450[b]	PF00067 Cytochrome P450	249	443	264
	KOG0156 Cytochrome P450 CYP2 subfamily	150	213	134
2. Alcohol dehydrogenase	PF00107 Zinc-binding dehydrogenase	40	76	48
	KOG0023 Alcohol dehydrogenase, class V	9	16	11
2.1	K00083 Cinnamyl-alcohol dehydrogenase (CAD) EC:1.1.1.195	5/9[c]	8/12	5/11
3. Aldehyde dehydrogenase	PF00171 Aldehyde dehydrogenase family	15	62	28
	KOG2450 Aldehyde dehydrogenase	7	30	14
3.1	K12355 Coniferyl-aldehyde dehydrogenase (ALDH1A, *REF1*) EC:1.2.1.68	1	10/8	4
4. 2O-iron/ascorbate oxidoreductase	PF03171 2OG-Fe(II) oxygenase superfamily	120	276	145
	KOG0143 Iron/ascorbate family oxidoreductases	101	223	124
4.1	K00475 Naringenin 3-dioxygenase (F3H) EC:1.14.11.9	1	4	1
4.2	K05277 Leucoanthocyanidin dioxygenase (ANS, LDOX) EC:1.14.11.19	2	2	1
4.3	K05278 Flavonol synthase (FLS) EC:1.14.11.23	2/5	4/3	2/3
5. NmrA-like family	PF05368 NmrA-like family	15	44	26
	PTHR10366 NAD-dependent epimerase/dehydratase	79	174	99
5.1	K05291 Isoflavonoid reductase (IFR) EC:1.3.1.45	NA	NA	NA
5.2	K13081 Leucocyanidin reductase (LAR) EC:1.17.1.3	NA/0	NA/2	NA/1
5.3	K13266 Pterocarpan reductase (PTR) EC:1.23.1.-	NA	NA	NA
5.4	Pinoresinol/lariciresinol reductase (PLR) EC:1.23.1.2 1.23.1.1	2/NA	0/NA	0/NA
6. Flavonol reductase/cinnamoyl-CoA reductase	PF01370 NAD-dependent epimerase/dehydratase family	65	118	75
	KOG1502 Flavonolreductase/cinnamoyl-CoA reductase	27	48	38
6.1	K09753 Cinnamoyl-CoA reductase EC:1.2.1.44	2	4	2

(continued)

Table 11.2 (continued)

Gene family		Number of genes		
Enzyme class (EC)	Functional annotation[a]	*A. thaliana* TAIR10	*G. max* Wm82.a2.v1	*P. vulgaris* v1.0
6.2	K08695 Anthocyanidin reductase EC:1.3.1.77	1	2	2
6.3	K13082 Bifunctional dihydroflavonol 4-reductase/flavanone 4-reductase (DFR) EC:1.1.1.219 1.1.1.234	3/1	12/3	7/3
6.4	K13265 Vestitone reductase (VR) EC:1.1.1.348	NA/0	NA/1	NA/0
7. Multicopper oxidase	PF00394 Multicopper oxidase	41	98	51
	KOG1263 Multicopper oxidase	40	90	47
7.1	K05909 Laccase (LAC) EC:1.10.3.2	17	59/48	29/30
8. Polyphenol oxidase	PF12142 Polyphenol oxidase middle domain	0	17	3
8.1	K00422 Polyphenol oxidase (PPO) EC:1.10.3.1	0/NA	15/NA	3/NA
9. Peroxidase	PF00141 Peroxidase	82	208	100
9.1	K00430 Peroxidase (POX) EC:1.11.1.7	65/73	174/152	87/91
EC:2—Transferases				
Transferase class 2.1.1				
10. Methyltransferase	PF00891 *O*-methyltransferase	17	56	37
	KOG3178 Hydroxyindole-*O*-methyltransferase and related SAM-dependent methyltransferase	17	49	36
10.1	K13066 Caffeic acid 3-*O*-methyltransferase (COMT) EC:2.1.1.68	NA/1	NA/11	NA/16
10.2	K05279 Flavonol 3-*O*-methyltransferase (FOMT) EC:2.1.1.76	1/0	21/2	18/0
10.3	K13262 Isoflavone-7-*O*-methyltransferase (7IOMT) EC:2.1.1.150	NA/0	NA/1	NA/1
10.4	K13259 Isoflavone-4-*O*-methyltransferase (HI4OMT) EC:2.1.1.46	NA/0	NA/1	NA/1
11. Methyltransferase	PF01596 *O*-methyltransferase	7	17	8
	KOG1663 *O*-methyltransferase	7	14	7
11.1	K00588 Caffeoyl-CoA *O*-methyltransferase (CCoAOMT) EC:2.1.1.104	2/4	5/11	3/5

(continued)

Table 11.2 (continued)

Gene family		Number of genes		
Enzyme class (EC)	Functional annotation[a]	*A. thaliana* TAIR10	*G. max* Wm82. a2.v1	*P. vulgaris* v1.0
Transferase class 2.3.1				
12. Acyltransferases	PF02458 Transferase family	63	152	88
	PTHR20961 Glycosyltransferase	7	5	2
12.1	K13264 Isoflavone-7-*O*-beta-glucoside 6″-*O*-malonyltransferase (IF7MaT) EC:2.3.1.115	NA/0	NA/2	NA/0
12.2	K13065 Shikimate *O*-hydroxycinnamoyltransferase (HCT) EC:2.3.1.133	1	4/24	10/21
12.3	K12936 Anthocyanin 5-aromatic acyltransferase (5AT) EC:2.3.1.153	1/NA	NA	NA
13. Acyltransferases	PF00248 Aldo/keto reductase family	22	57	35
	KOG1577 Aldo/keto reductase family	10	26	19
13.1	Chalcone reductase (CHR)/6′-deoxychalcone synthase (DCHS) EC:2.3.1.170	NA	NA	NA
14. Acyltransferases	PF00195 Chalcone and stilbene synthases, N-terminal domain	4	24	16
	PTHR11877 Hydroxymethylglutaryl-CoA synthase	6	31	20
14.1	K00660 Chalcone synthase (CHS) EC:2.3.1.74	1	19/16	11/16
Transferase class 2.4.1				
15. Glycosyltransferases	PF00201 UDP-glucoronosyl and UDP-glucosyltransferase	115	249	174
	KOG1192 UDP-glucuronosyl and UDP-glucosyltransferase	117	226	172
15.1	K15787 Flavonol 3-*O* rhamnosyltransferase (UGT78D1) EC:2.4.1.-	NA/1	NA/0	NA/0
15.2	K13496 UDP-glucosyl transferase 73C (UGT-73C) EC:2.4.1.-	7/NA	23/NA	14/NA
15.3	K10757 Flavonol 3-*O*-glucosyltransferase (F3OGT) EC:2.4.1.91	22/1	54/0	41/0
15.4	K12356 Coniferyl-alcohol glucosyltransferase (CAGT/UGT72E) EC:2.4.1.111	4/3	9/3	4/3
15.5	K12930 Anthocyanidin 3-*O*-glucosyltransferase (A3OGT) EC:2.4.1.115	NA	NA	NA
15.6	K13068 Sinapate 1-glucosyltransferase (S1GT) EC:2.4.1.120	4/1	2/0	2/0
15.7	Flavonol-3-*O*-glucoside L-rhamnosyltransferase (RT) EC:2.4.1.159	NA	NA	NA

(continued)

Table 11.2 (continued)

Gene family		Number of genes		
Enzyme class (EC)	Functional annotation[a]	*A. thaliana* TAIR10	*G. max* Wm82. a2.v1	*P. vulgaris* v1.0
15.8	K13263 Isoflavone 7-*O*-glucosyltransferase (IF7GT) EC:2.4.1.170	NA/0	NA/1	NA/0
15.9	K13080 Flavanone 7-*O*-glucoside 2″-*O*-beta-L-rhamnosyltransferase EC:2.4.1.236	NA	NA	NA
15.10	Flavonol 7-*O*-glucosyltransferase (F7OGT) EC:2.4.1.237	NA	NA	NA
16. Transferase class 2.5.1	PF00043/PF02798 Glutathione S-transferase	49/52	93/91	50/45
	KOG0406/KOG0867 Glutathione S-transferase	31/17	61/16	29/14
16.1	K00799 Glutathione S-transferase (GST) EC:2.5.1.18	15/46	3/70	4/42
EC:3—Hydrolases				
Hydrolase class 3.2.1				
17a. Glycosidases	PF00232 Glycosyl hydrolase family 1 (GH1)	48	67	49
	KOG0626 β-glucosidase, lactase phlorizinhydrolase, and related protein	47	50	45
17a.1	K01188 β-glucosidase GH1 EC:3.2.1.21	7/30	37/39	21/38
17a.2	K05350 β-glucosidase GH1 (bglB) EC:3.2.1.21	3/4	7/9	5/8
17b. Glycosidases	PF00933/PF01915 Glycosyl hydrolase family 3 N-terminal domain/C-terminal domain (GH3)	15/16	30/30	15/16
17b.1	K05349 β-glucosidase GH3 (bglX) EC:3.2.1.21	1/6	4/12	2/7
EC:4—Lyases				
18. Lyases class 4.2.1	PF	NA	NA	NA
	KOG	NA	NA	NA
18.1	K132582-hydroxyisoflavanone dehydratase (HIDH) EC:4.2.1.105	NA/0	NA/1	NA/0
19. Lyases class 4.3.1	PF00221 Aromatic amino acid lyase	4	10	7
	KOG0222 Phenylalanine and histidine ammonia-lyase	4	8	6
19.1	K10775 Phenylalanine ammonia-lyase (PAL) EC:4.3.1.24	4	8	6/7

(continued)

Table 11.2 (continued)

Gene family		Number of genes		
Enzyme class (EC)	Functional annotation[a]	*A. thaliana* TAIR10	*G. max* Wm82.a2.v1	*P. vulgaris* v1.0
EC:5—Isomerases				
20. Isomerases class 5.5.1	PF02431 Chalcone-flavanone isomerase	7	12	9
20.1	K01859 Chalcone isomerase (CHI) EC:5.5.1.6	2	4	4
EC:6—Ligases				
21. Ligases class 6.2.1	PF00501 AMP-binding enzyme	45	93	48
	KOG1176 Acyl-CoA synthetase	29	53	30
21.1	K019044-Coumarate-CoA ligase (4CL) EC:6.2.1.12	6/7	16/14	8/9

NA not available
[a]Phytozome 10.2 (http://phytozome.jgi.doe.gov/pz/portal.html), where PF, Pfam; PTHR, Panther; KOG, KOG; K, KEGGORTH
[b]P450—details in Sect. 11.2
[c]Phytozome-first number, KEGG ORTHOLOGY (http://www.kegg.jp/)-second number; if same, a number of genes is presented as a single value

absorbs light at 450 nm, and are named for this trait (**P**igment absorbing at **450** nm), as well as their cellular localization. Plant P450s are typically membrane-bound to the cytoplasmic surface of the endoplasmic reticulum (ER) by a short N-terminal segment.

The P450s are one of the largest families of enzymes in plants and, in most of plant species, exist as a superfamily. The number of P450 genes is highly variable among plants (Nelson 2006) and represents 0.57–1.07% of the protein coding genes in various plant species [1.07% in *Arabidopsis* (246/23,000) (Nelson et al. 2004), 0.71% in soybean (332/46,500) (Guttikonda et al. 2010), and 0.78% in common bean (247/31,638) (Kumar et al. 2015)]. The large number of P450s in higher plants is due to gene duplication and diversification (Werck-Reichhart and Feyereisen 2000).

The P450 gene superfamily is characterized by enormous structural and functional diversity (Nelson et al. 2008; Nelson and Werck-Reinchhart 2011; Nagano 2014). Homology and phylogeny were used to group P450s into families (>40% amino acid sequence identity) and subfamilies (>55% amino acid sequence identity) (Nelson et al. 1996). Plant P450 proteins are numbered as CYP51, CYP71 to CYP99, and CYP701 to CYP772. They belong to ten clans (group of genes originated from a single ancestor), which are named by their lowest numbered member [six single-family clans (*CYP51, CYP74, CYP97, CP710, CYP711,* and *CYP727*) and four multiple-family clans (*CYP71, CYP72, CYP85,* and *CYP86*)] (Werck-Reichhart and Feyereisen 2000; Nelson et al. 2004; Schuler and Werck-Reinchhart 2003; Schuler et al. 2006). Following recommendations of a nomenclature committee (Nelson et al. 1996), the name of P450s consists of a CYP italicized root symbol, followed by a number of the family, a letter of the subfamily and ending by a number of the gene (e.g., *CYP71D9*—family 71, subfamily D, gene 9), which is determined by the order of identification regardless of the origin.

Initially, P450s were divided into a large A-type clade, which included members that are involved in secondary metabolism (clan CYP71) and several smaller, non-A-type clades, involved in primary metabolism (such as fatty acids and sterols) (Nelson 2006). The occurrence of large numbers of A-type P450s, compared to the non-A-type, suggests a rapid expansion of A-type P450 gene families in plants (Bak et al. 2011).

11.3.2 Clan CYP71—P450s Involved in the Phenylpropanoid Pathway

Based on the current genome annotations [Pfam:00067 (cytochrome P450) functional annotation at Phytozome 10.2; http://phytozome.jgi.doe.gov/pz/portal.html—accessed 26 June 2015], there are 249 P450 genes in *A. thaliana* TAIR10, 443 P450 genes in *G. max* Wm82.a.v1, and 264 P450 genes in *P. vulgaris* v1.0. However, the number of published P450s in these species is slightly different, 272 genes (including 28 pseudogenes) in *Arabidopsis* (Bak et al. 2011) and 247 genes (including 15 pseudogenes) in common bean (Kumar et al. 2015). P450s in common bean were classified into ten clans that contain 47 families. The largest *CYP71* clan (A-type) consists of 19 families with 144 genes. The majority of the genes (>70%) contain a single intron, but more than 20% of the genes have two introns and only a small number of genes (4%) are intronless. In addition, over 80% of the introns are of the zero phase (intron sequence inserted between two successive codons).

It was reported that over 16 P450s are involved in the synthesis and metabolism of phenylpropanoids (Werck-Reichhart 1995). They are placed at the several key positions in the phenylpropanoid pathway, and their roles in phenylpropanoid metabolism were extensively reviewed. For example, Ehlting et al. (2006) and Alber and Ehlting (2012) focused on P450s involved in the core phenylpropanoid pathway and lignin branch, Ayabe and Akashi (2006) in flavonoid metabolism, while Tanaka (2006) and Tanaka and Brugliera (2013) reviewed the role of P450s in flower color.

Seven gene families that encode P450 enzymes involved in phenylpropanoid pathway, as identified in the current genome annotations in common bean, soybean, and *Arabidopsis*, are listed in Table 11.3. It should be noted, however, that the number of genes in analyzed genomes may change as more work on annotations is done. For example, the *CYP71D* family in soybean had 81 genes (including 39 pseudogenes) in *G. max* v1.0 (Nelson 2009) and 52 genes (including 16 pseudogenes) in *G. max* Wm82.a2.v1. Eleven gene sequences did not correspond between the two genome annotations.

We used the standard nomenclature of chromosome-based locus (gene model) identifiers in plant genome annotations and assemblies (Phytozome), which consists of four segments:

- species [AT or At (*A. thaliana*), Glyma. (*G. max*), Phvul. (*P. vulgaris*)],
- chromosome number [1 to 5 (*A. thaliana*), 01 to 20 (*G. max*), 001 to 011 (*P. vulgaris*)],
- gene (G or g), and
- five-digit code [*A. thaliana—At2g37040* for phenylalanine ammonia-lyase 1 (*PAL1*)] or six-digit code [*G. max* (*Glyma.03g181700*, *PAL1*) and *P. vulgaris* (*Phvul.001g177800*, *PAL1*)], numbered from top to bottom of chromosome.

These gene families encode enzymes that catalyze various reactions in different branches of the phenylpropanoid pathway (Fig. 11.1), including

1. core phenylpropanoid pathway: cinnamate 4-hydroxylase (C4H, CYP73A),
2. lignin/lignan branch: coumarate 3-hydroxylase (C3H, CYP98A) and ferulic acid 5-hydroxylase (F5H, CYP84A),
3. anthocyanin/condensed tannin branch: flavonoid 3′-hydroxylase (F3′H, CYP75B), flavonoid 3′,5′-hydroxylase (F3′,5′H, CYP75A), and flavone synthase (FNS, CYP93B), and
4. isoflavonoid branch: isoflavone synthase (IFS, CYP93C), isoflavone 2′-hydroxylase

Table 11.3 Clan *CYP71* cytochrome P450 gene families encoding enzymes of the phenylpropanoid pathway in *A. thaliana*, *G. max*, and *P. vulgaris*

Clan *CYP71*				Loci encoding enzymes of phenylpropanoid pathway[c]					
Gene family		Number of genes		Gene	Functional annotation		Number of genes		
		G. max[a]	*P. vulgaris*[b]		KEGG-ORTH	EC number	*G. max* Wm82.a2.v1	*P. vulgaris* v1.0	*A. thaliana* TAIR10
CYP71	*CYP71D*	36 + 16P[d]	21 + 4P		K13267	1.14.11.-	3	1	0
CYP73	*CYP73A*	3 + 1P	3		K00487	1.14.13.11	3	3	1
CYP75	*CYP75A*	2 + 1P	2		K13083	1.14.13.88	2	2	0
	CYP75B	5 + 1P	2	*F3'H*	K05280	1.14.13.21	5	2	1
CYP81	*CYP81E*	12 + 4P	12 + 2P	*I2'H*	K13260	1.14.13.89	8	4	0
CYP84	*CYP84A*	3 + 1P	3	*F5H*	K09755	1.14.-.-	3	3	2
CYP93	*CYP93A*	8 + 2P	7 + 1P	*D6aH*	K13261	1.14.13.28	5	1	0
	CYP93B	2 + 1P	1	*FNS*	K13077	1.14.11.22	2	1	0
	CYP93C	2	3	*IFS*	K13257	1.14.13.136	2	3	0
CYP98	*CYP98A*	2	1	*C3'H*	K09754	1.14.13.-	2	1	1
	Total	75(+27P)	55(+7P)						5

[a]Soybean P450 Database; available at http://drnelson.uthsc.edu/soybean.html (Accessed: 6 Apr 2015); Nelson (2009). Gene Model Correspondence Lookup at SoyBase (http://www.soybase.org/) was used to connect different annotations of soybean genome; if there were no correspondence with the current soybean annotation (Wm82.a2.v1), the older annotations were checked/retrieved from the Phytozome 10.2 (http://phytozome.jgi.doe.gov/pz/portal.html)
[b]Kumar et al. (2015)
[c]Phytozome10.2
[d]P—Pseudogene

(I2′H, CYP81E), flavonoid 6-hydroxylase (F6H, CYP71D), and 3,9-dihydroxy pterocarpan 6a-monooxygenase (D6aH, CYP93A).

11.3.3 Gene Structure, Conserved Domains, and Motifs of P450s Involved in the Phenylpropanoid Pathway

Seven P450s families (clan *CYP71*) that encode enzymes in the phenylpropanoid pathway in common bean, soybean, and *Arabidopsis* contain 135 members, with one to 36 genes per family (Table 11.3). Most of these genes contain introns. Only one gene is intronless (*Phvul.009g244000, CYP81E51*). The number of introns ranges from one to four. The majority of the genes contain one (63%) or two introns (32%). The proteins that they encode range in size from 408 amino acids (*Phvul.001g139500, CYP93A57*) to 543 amino acids (*Phvul.002g014800, CYP81E44*). The protein sequences were aligned using Clustal Omega at EMBL-EBI (http://www.ebi.ac.uk/Tools/msa/clustalo/), and conserved regions were displayed with a sequence logo generated from the alignment using a Web-based WebLogo 3.4 (Crooks et al. 2004; available at http://weblogo.threeplusone.com/). All of the P450 sequences included the following domains: a heme-binding region (FxxGxRxCxG), a PERF motif (PERF/W), a K-helix region (KETRL) involved in defining the heme pockets and stabilizing the protein structure, and an I-helix region (AGxDT) involved in oxygen binding (Fig. 11.2).

11.3.4 Phylogenetic Analysis of P450s Involved in the Phenylpropanoid Pathway

The alignment and tree construction of 135 protein sequences (Table 11.3) from seven P450 gene families (clan *CYP71*) involved in the phenylpropanoid pathway were performed in MEGA6 (Tamura et al. 2013). These analyses were based on the full-length genes from the three genomes, with one nearly intact soybean *C4H* pseudogene included (indicated by P at the end of the CYP name—*CYP73A88**P***). A member from the soybean *CYP81E* family (*CYP81E220de1b, Glyma.16g149200*) is truncated (101 amino acids) and was not included in the tree construction.

The phylogenetic tree (Fig. 11.3) separates P450 protein sequences (clan 71) from the two species into seven families:

- *CYP71*—*CYP71D* is a legume-specific cluster and contains 36 genes in soybean (and 16 pseudogenes, not included) and 21 genes in common bean (and four pseudogenes, not included). A single flavonoid 6-hydroylase (F6H) in common bean was clustered with three F6H proteins in soybean.
- *CYP73*—*CYP73A* family contains four genes for cinnamic acid 4-hydroxylase (C4H) in soybean (including one pseudogene), three genes in common bean, and a single gene in *Arabidopsis*. The C4H cluster splits into class I and class II enzymes.
- *CYP75* family is split into two subfamilies. *CYP75A* consists of two genes for flavonoid 3′,5′-hydroxylase (F3′5′H) (and one pseudogene, not included) in soybean and two genes in common bean. There are no genes for F3′5′H in *Arabidopsis*. Subfamily *CYP75B* contains five genes for flavonoid 3′-hydroxylase (F3′H) (and one pseudogene, not included) in soybean, two genes in common bean, and a single gene for F3′H in *Arabidopsis*.
- *CYP81*—*CYP81E* is a legume-specific cluster and consists of 12 genes coding isoflavone 2′-hydroxylase-like (I2′H) genes (and four pseudogenes, not included) in soybean and 12 genes (and two pseudogenes, not included) in common bean.
- *CYP84*—*CYP84A* cluster contains three genes encoding ferulic acid 5-hydroxylase (F5H) (and one pseudogene, not included) in soybean, three genes in common bean, and two genes in *Arabidopsis*.

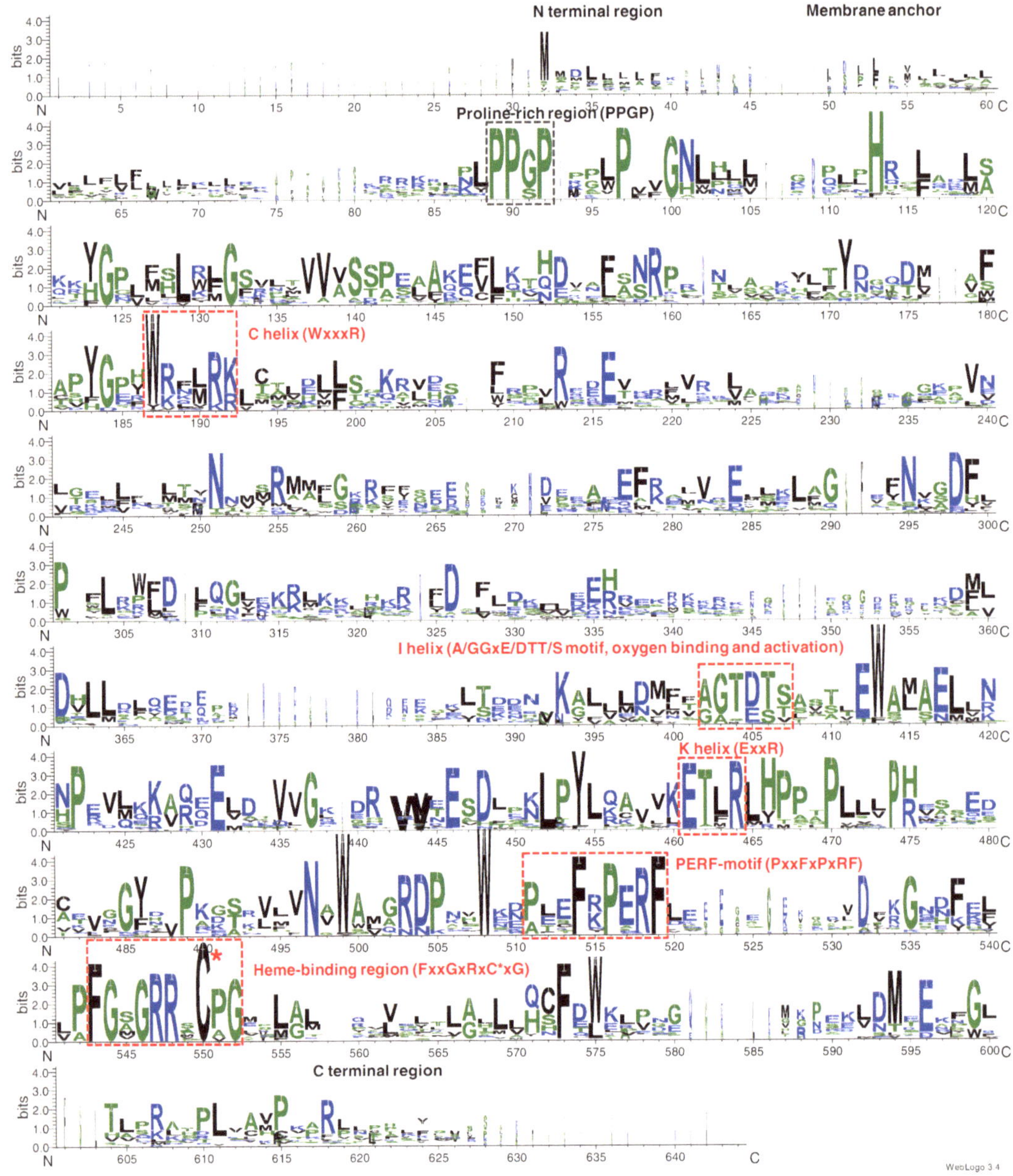

Fig. 11.2 Conserved domains and motif patterns of P450s, CYP71 clan involved in biosynthesis of various phenylpropanoids. P450 domains including a heme-binding region [cysteine (C*) residue is indicated by an asterisk (*)], PERF motif, K-helix and I-helix regions are indicated in red rectangles; the other regions (such as N-terminal region, proline-rich region, membrane anchor, and C-terminal region) are shown in black

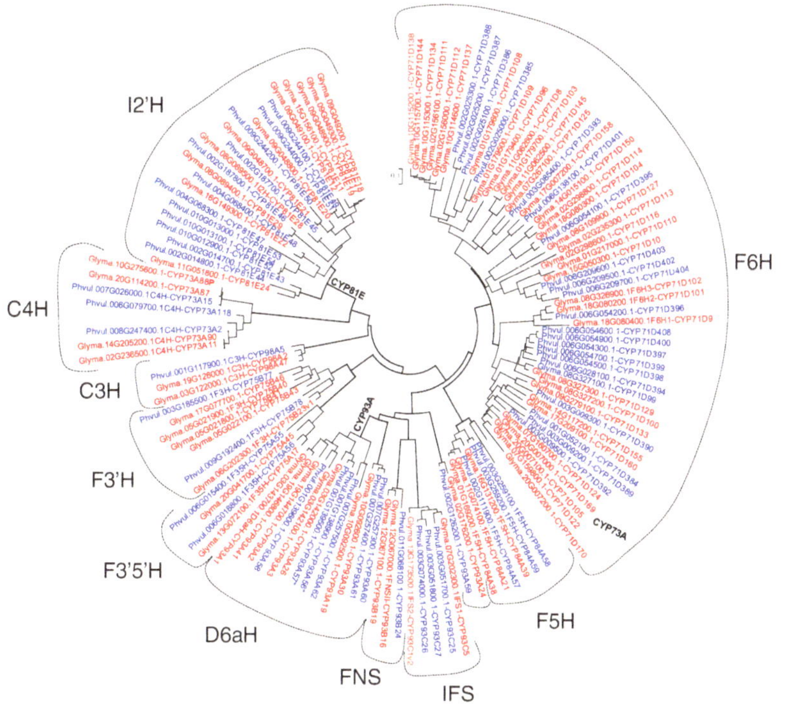

Fig. 11.3 Protein sequences of the seven gene families from the clan *CYP71* involved in the phenylpropanoid pathway in soybean and common bean. A neighbor-joining tree (Poisson model, complete deletion) was built using MEGA6. Soybean sequences are labeled in red, and common bean in blue; **P** at the end of CYP name indicates pseudogene (*Glyma.10G275600-CYP73A88**P***); shorter protein sequences are indicated by an asterisk (*); a truncated (101 amino acids) Glyma.16g149200-*CYP81E220del b* was excluded from the tree construction

- *CYP93*—The family is clustered into three subfamilies. *CYP93A* is a legume-specific subfamily. It consists of eight genes for 3,9-dihydropterocarpan 6a-monooxygenase (D6aH) (and two pseudogenes, not included) in soybean and seven genes (and one pseudogene, not included) in common bean. The *CYP93B* subfamily contains two genes encoding flavonoid synthase (FNS) in soybean and a single gene in common bean. There are no *FNS* genes in *Arabidopsis*. *CYP93C* is a legume-specific branch. It consists of two genes for isoflavone synthase (IFS) in soybean and three genes in common bean.
- *CYP98*—*CYP98A* cluster consists of two genes for coumarate 3-hydroxylase (C3H) in soybean and single genes in common bean and *Arabidopsis* genomes, respectively. There are two additional pollen-specific *CYP98A*s in *Arabidopsis* (*CYP98A8* and *CYP98A9*; Matsuno et al. 2009—not included in tree construction).

11.3.5 Genome Organization of the Clan CYP71 Gene Families Involved in Phenylpropanoid Pathway in Common Bean

A common bean in silico map that contained genes coding for enzymes of phenylpropanoid pathway, including nine P450s, was developed previously (Reinprecht et al. 2013). The map was created by BLASTing the genomic sequences of the phenylpropanoid pathway genes against the whole common bean genome (*P. vulgaris* v1.0, Phytozome) using the starting nucleotide positions of the resulting alignments with the chromosome as the map positions for each of the gene sequences.

A similar approach was used to develop a common bean P450-based in silico map, which contains 144 P450, clan *CYP71* genes. The mapping was initiated with 134 genes that were identified at Phytozome by searching for KOG0156 functional annotations (cytochrome P450 CYP2 subfamily). Selected gene sequences were BLASTed against the complete common bean genome sequence (Phytozome) to identify their locations. Gene identity was confirmed with the published common bean P450s (Kumar et al. 2015), and ten new sequences (not annotated as KOG0156 in Phytozome) were added to the map. Gene families involved in the phenylpropanoid pathway (shown in larger font, color-coded) were found throughout the common bean genome, except for chromosome Pv05 (Fig. 11.4).

Within the same family, P450s are usually grouped into clusters and the structure of the same P450 family is generally conserved (Nelson et al. 2004; Paquette et al. 2009). In the common bean genome, clustering of genes from the same family was noticed on the chromosomes Pv03 for family *CYP93C* (all three *IFS* genes) and Pv09 for family *CYP81E* (three *I2'H* genes). Some of the *CYP71* genes are tandem arranged with at least four genes from the same subfamily in a row. Many of these clustered genes are found in the same orientation on four chromosomes [Pv01 (four *CYP712B*, all forward), Pv02 (four *CYP71D*, all forward), Pv04 (ten *CYP82A*, all forward; five *CYP71AU*, all reverse; five *CYP736A*, all reverse) and Pv06 (eight *CYP71D*, all reverse; four *CYP79D*, all forward)] but in a different orientation on three chromosomes [Pv03 (four *CYP71D*), Pv04 (four *CYP81E*), and Pv06 (four *CYP71D*)]. However, members of the large *CYP71D* subfamily clustered in the same orientation on chromosomes Pv02 (four) and Pv06 (eight) but in a different orientation on the chromosomes Pv03 (four) and Pv06 (four). Therefore, the subfamily distribution may not follow a regular pattern. Due to clustered organization, the 144 *CYP71 P450* genes (Kumar et al. 2015) were not evenly distributed in the common bean genome. They ranged from two genes on the chromosome Pv05 to 25 genes on the chromosome Pv04 (Fig. 11.4).

11.4 Cinnamate 4-Hydroxylase (C4H, EC:1.14.13.11, *CYP73A*)

11.4.1 C4H Catalytic Reaction and Position in the Phenylpropanoid Pathway

Cinnamate 4-hydroxylase (*trans*-cinnamate 4-monooxygenase, C4H, EC:1.14.13.11, CYP73A) is the first P450 enzyme in the phenylpropanoid pathway. It is an ER membrane-bound P450 and belongs to the family of oxidoreductases that act on paired donors with incorporation of molecular oxygen. The enzyme catalyzes an irreversible (and rate-limiting) region-specific hydroxylation of the aromatic ring of *trans*-cinnamic acid (only at the 4-position or *para* position) to produce *p*-coumaric (hydroxycinnamic) acid (Fig. 11.5), a precursor for many phenylpropanoids including flavonoids, phytoalexins, and monolignols (Hahlbrock and Scheel 1989; Anterola and Lewis 2002; Lu et al. 2006). For activity, C4H requires molecular oxygen and a cytochrome P450 reductase (CPR).

Mizutani et al. (1997) isolated a cDNA and a genomic clone encoding cinnamate 4-hydroxylase from *Arabidopsis* (*CYP73A5*) and found its coordinated expression with *PAL* and *4CL* genes. Mutations in this gene affected phenylpropanoid metabolism, growth, and development (Schilmiller et al. 2009). The gene was mapped to the lower arm of chromosome 2 and was highly expressed in all *Arabidopsis* tissues, especially in roots and lignifying cells (Bell-Lelong et al. 1997). Genes targeted by the same transcription factors tend to show similar expression patterns, which usually suggest

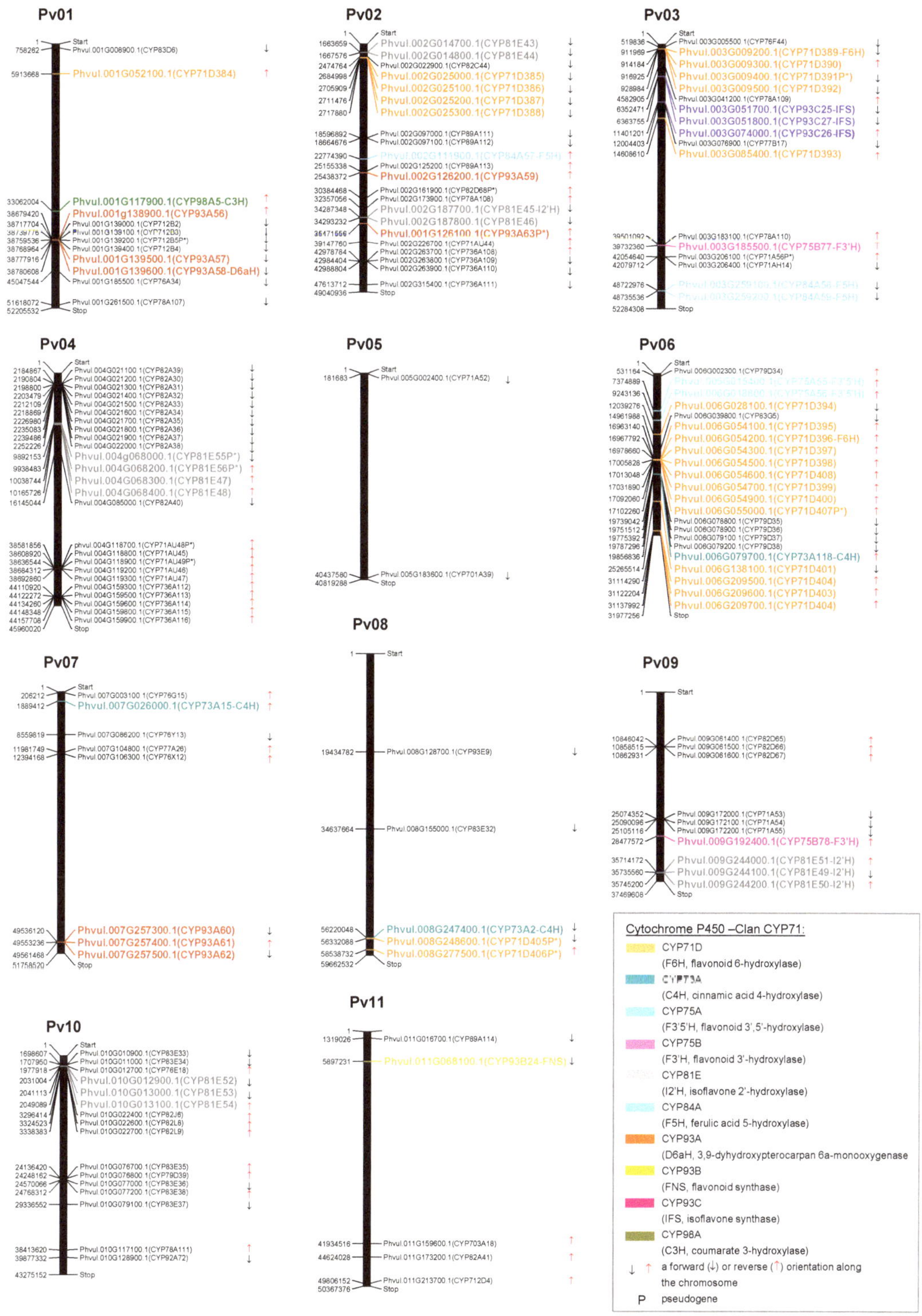

Fig. 11.4 Distribution of cytochrome P450—clan *CYP71* genes [locus (gene model) identifiers—Phytozome] in the common bean genome (identified on the right on bars). Genes belonging to families involved in the phenylpropanoid pathway are color-coded; **P** at the end of CYP name indicates a pseudogene; the orientation along the chromosome is indicated by a forward or reverse arrow. The starting nucleotide position of the resulting alignment with the chromosome was used as the map position for each P450 gene sequence (indicated on the left on bars)

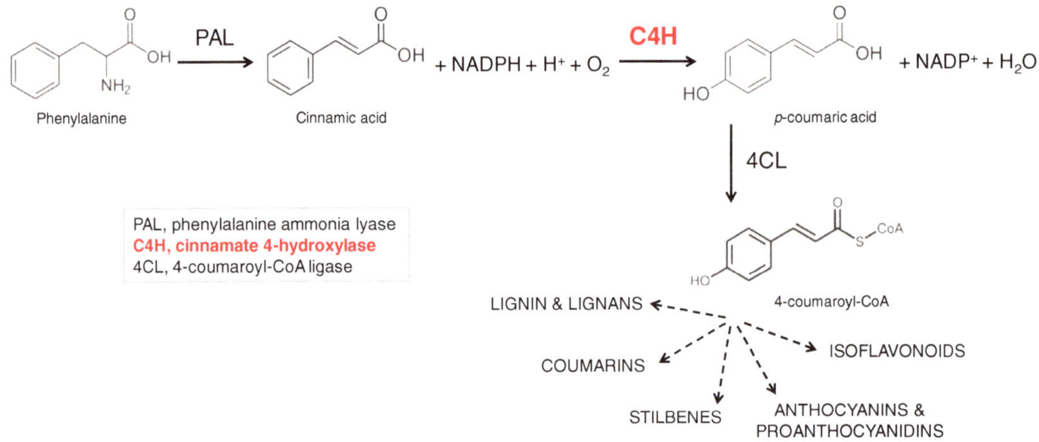

Fig. 11.5 Core (general) phenylpropanoid pathway and the catalytic reaction of cinnamate 4-hydroxylase (C4H, red). The enzyme catalyzes the first oxygenation step of the core phenylpropanoid pathway leading to synthesis of lignin, pigments, and phytoalexins

relationships among the genes. Down-regulation of genes coding for PAL and C4H was associated with reduced lignin content and altered lignin composition in transgenic tobacco (Sewalt et al. 1997). The position of C4H in the phenylpropanoid pathway protein network is shown in Fig. 11.6a. Highly connected proteins have a stable steady-state distribution of gene expression (Fig. 11.6b).

Separation of three common beans, four soybeans, and single *Arabidopsis* sequences into two groups (Fig. 11.7) confirmed earlier groupings of C4H into class I and class II proteins (Ehlting et al. 2006). This diversification occurred early in the evolution of vascular plants through gene duplication. Common bean and soybean have both classes of C4Hs, while *Arabidopsis* (Brassicaceae) contains only one gene encoding class I C4H. The alignment of C4H protein sequences (ClustalW2 at EMBL-EBI, available at http://www.ebi.ac.uk/Tools/msa/clustalw2/) revealed high conservation (60–98% identity) among the proteins (85–98% within five C4H class I proteins and 90% between two class II C4H proteins). However, when both monocots and dicots were compared, class I C4H was highly conserved (over 80% protein level), while class II C4Hs were more divergent (less than 70% protein level). This suggests that class I C4Hs "maintained an essential function that does not allow these genes to be lost or even changed much, and it is appealing to assume that this essential function is developmental lignification" (Alber and Ehlting 2012). Class II C4Hs are only present in some plant species, and the class seems to have more specialized functions.

The sequences of eight C4H proteins from common bean, soybean, and *Arabidopsis* were aligned using Clustal Omega (http://www.ebi.ac.uk/Tools/msa/clustalo/) and BoxShade (http://www.ch.embnet.org/software/BOX_form.html). The sequences were most divergent in their N-terminal membrane anchors. Conserved motifs found in plant P450s (Fig. 11.8, shown in bold) were present in all eight proteins, including proline-rich (PPGP) region, C helix (WrkmR), oxygen binding and activation I-helix (AAIETT), K-helix (EtlR), PERF motif (PeeFrPeRF), and heme-binding region (FgvGrRsCpG) at C-terminus. The only exception is soybean C4H (CYP73A88P) encoded by a pseudogene (*Glyma.10g275600*). It has truncated N-terminal region, and the generally highly conserved PERF motif has an arginine (R, Arg) to lysine (K, Lys) substitution (Fig. 11.8, highlighted).

Secondary structures of C4H proteins were predicted by programs GOR (**G**arnier-**O**sguthorpe-**R**obson), IV (Garnier et al. 1996; https://npsa-prabi.ibcp.fr/cgi-bin/npsa_automat.pl?page=npsa_gor4.html), and Phyre2 (**P**rotein **H**omology/

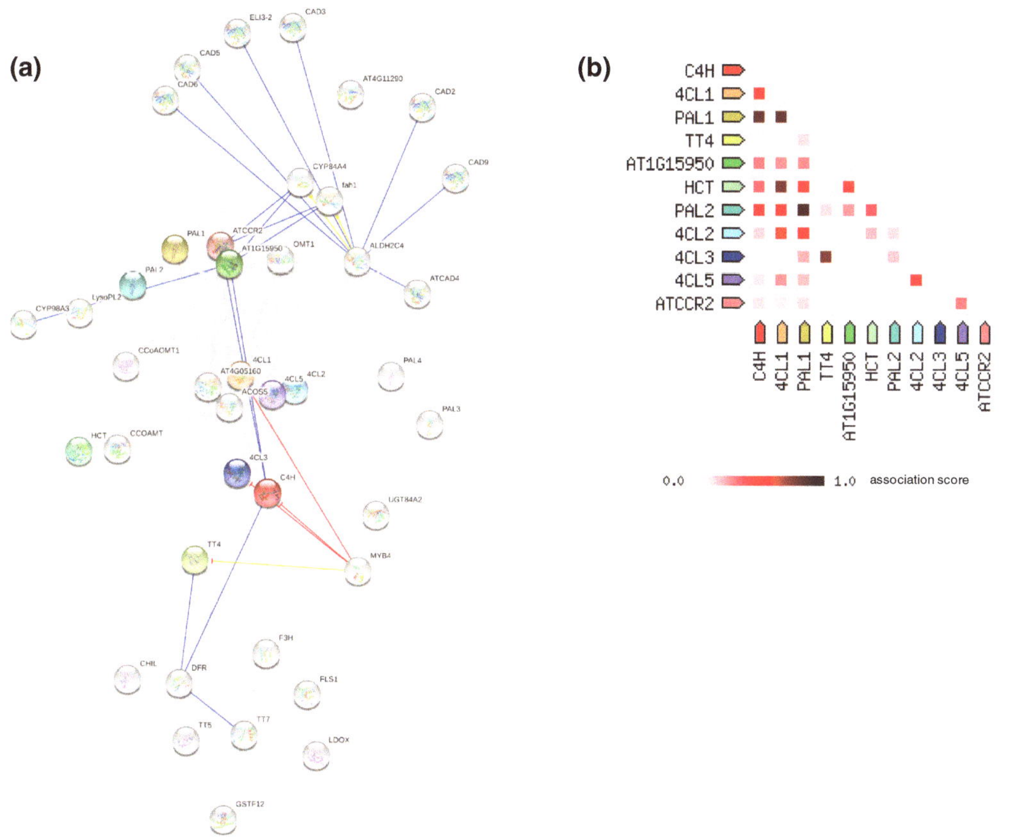

Fig. 11.6 Functional protein association network in *Arabidopsis* (action view) visualized on the STRING Web site (http://string-db.org/; accessed: 25 June 2015). **a** *C4H* is colored red, and modes of action are shown in different colors. Nodes directly linked to *C4H* are colored; **b** Co-expression of *C4H* with other phenylpropanoid pathway genes in *Arabidopsis*; locus AT1G15950 is a *CCR1* gene

analog**Y** **R**ecognition **E**ngine V 2.0) (Kelley et al. 2015; http://www.sbg.bio.ic.ac.uk/phyre2/html/page.cgi?id=index). Transmembrane helices were predicted by program TMHMM-2.0 (**T**rans**M**embrane prediction using **H**idden **M**arkov **M**odels; Krogh et al. 2001; http://www.cbs.dtu.dk/services/TMHMM-2.0/). All proteins have secondary structures similar to the previously published P450s (Graham and Peterson 1999) including alpha helices (blue), beta sheets (red), and random coils (pink) (Fig. 11.9a). They consist of 36–45% alpha helices, 14–18% extended (or beta) strands, and 40–46% random coils. There is a slight difference between the classes of common bean and soybean C4H proteins. Class I C4H proteins contain higher percentages of alpha helices, while class II C4H proteins were predicted to have higher percentages of extended (or beta) strands and random coils. Membrane anchors were predicted for all proteins except for soybean C4H (CYP73A88P) encoded by the pseudogene *Glyma.10g026000* (Fig. 11.9b). All C4H proteins are globular proteins as predicted by Phyre2 (Fig. 11.9c). Common bean and soybean C4Hs have tertiary structures similar to the previously identified CYP73A5 in *Arabidopsis* (*At2g30490*) and also contain an alpha-domain and a beta-domain (Rupasinghe et al. 2003).

Gene ontology (GO) annotations for C4H proteins (Table 11.4) were predicted using the protein function prediction (PFP), a sequence similarity-based protein function prediction server at Kihara Bioinformatics Laboratory (http://

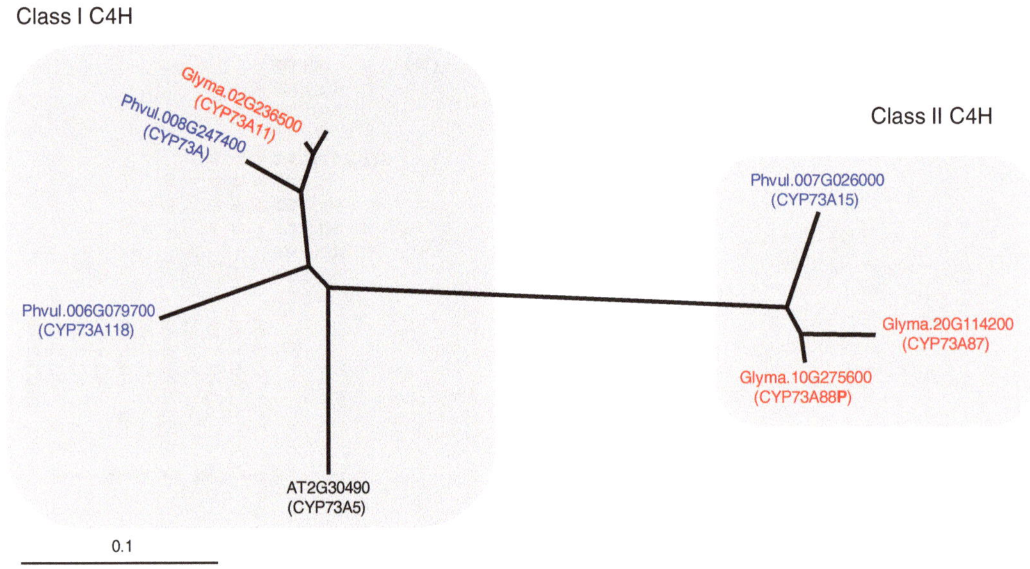

Fig. 11.7 Phylogenetic tree of class I and class II C4H proteins. Common bean sequences are labeled in blue, soybean in red, and *Arabidopsis* in black

kiharalab.org/; Hawkins et al. 2009). PFP takes into account weakly similar sequences as well as GO term associations observed in known annotations.

11.4.2 CYP73A Gene Family—Structure and Genome Location of C4H Genes

C4Hs are encoded by the relatively small *CYP73A* gene family. It consists of three genes in common bean {*Phvul.006g079700—CYP73A118, Phvul.007g026000—CYP73A15,* and *Phvul.008g247400—CYP73A* [this P450 was incorrectly named as *CYP73A2* in common bean (Kumar et al. 2015); however, *CYP73A2* was identified in mung bean (Mizutani et al. 1993), *Vigna radiata* (*previously Phaseolus aureus*; recently moved from the genus *Phaseolus* to *Vigna*)], four genes (including one pseudogene, *Glyma.10g275600—CYP73A88**P***) in soybean, and a single gene in *Arabidopsis* (*At2g30490*; *CYP73A5*; *REF3*).

The gene is well conserved in plants, including soybean and common bean. It contains the Pfam domain (PF00067), found as a "duplication-resistant" gene (Paterson et al. 2006). The first C4Hs were identified in Jerusalem artichoke (*Helianthus tuberosus—CYP73A1*, GenBank accession Z17369; Teutsch et al. 1993) and mung bean (*V. radiata—CYP73A2*, GenBank accession L07634; Mizutani et al. 1993). Soybean C4H (*CYP73A11*), a class I C4H enzyme, was identified as an elicitor-induced cytochrome P450, using differential display of mRNA (Schopfer and Ebel 1998). In contrast, common bean C4H (*CYP73A15*) was identified as a class II C4H enzyme, whose expression was associated with differentiation (Nedelkina et al. 1999).

The genes coding for C4H in common bean, soybean, and *Arabidopsis* differ in their exon/intron structures. The exons are conserved, while introns are more variable. Genes encoding class II proteins in common bean and soybean consist of two exons separated by an intron of moderate size (354 and 463 bp, respectively). Both exons are split, resulting in four exons, in the two genes encoding class I C4Hs in soybean. These genes are characterized by a long intron 3 (1499 and 1272 bp, respectively). The class I C4H gene in *Arabidopsis* and the two genes in common bean all have three exons (Fig. 11.10).

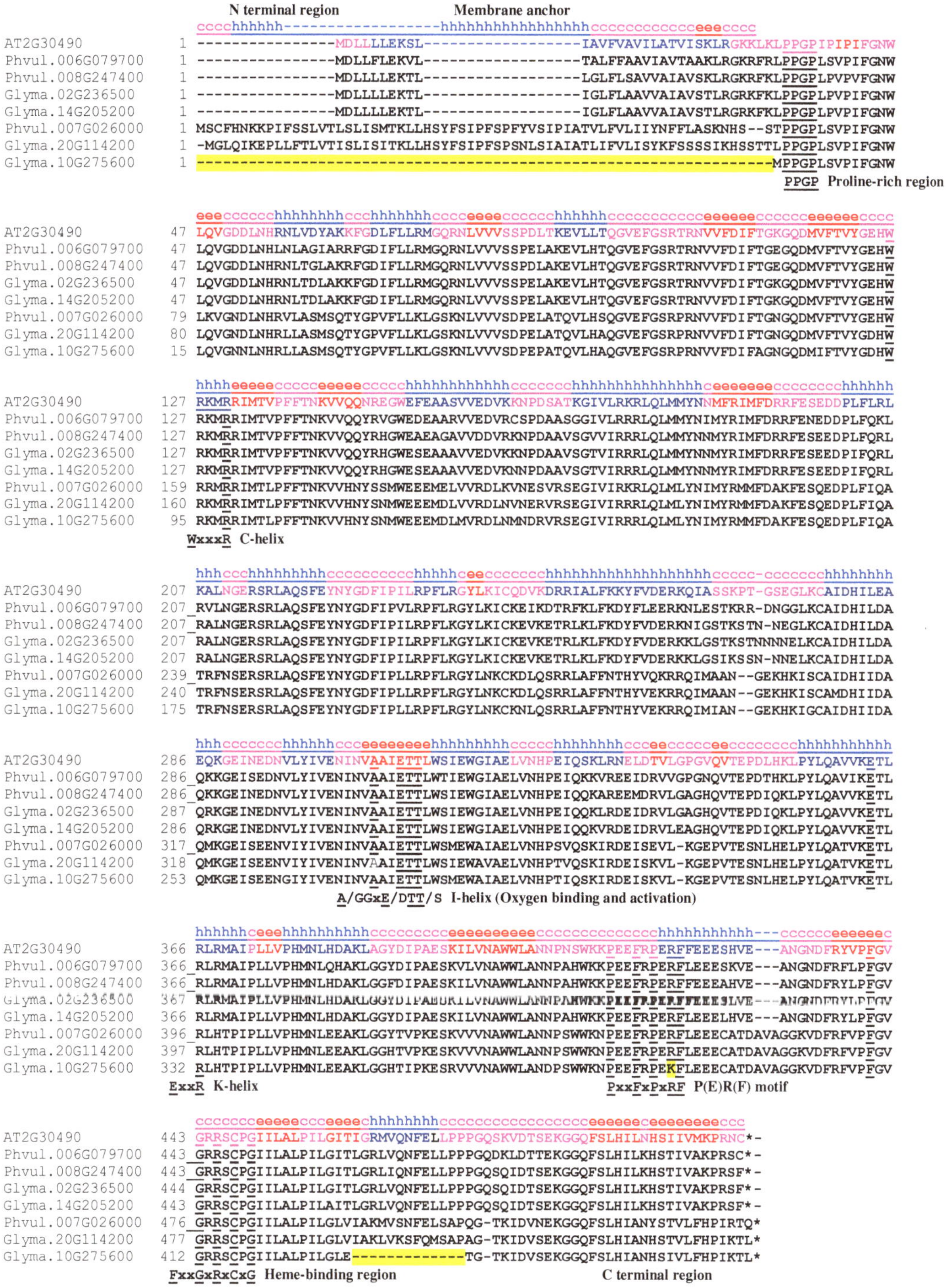

Fig. 11.8 Comparison of C4H protein sequences from common bean, soybean, and *Arabidopsis*. Conserved motifs and sequences are shown in bold. Secondary structures predicted for *Arabidopsis* C4H gene (*At2g30490*) are color-coded [shown at the top of sequences alignment, where H (blue) indicates alpha helices, E (red) represents extended (beta) strands, and C (pink) indicates random coils]

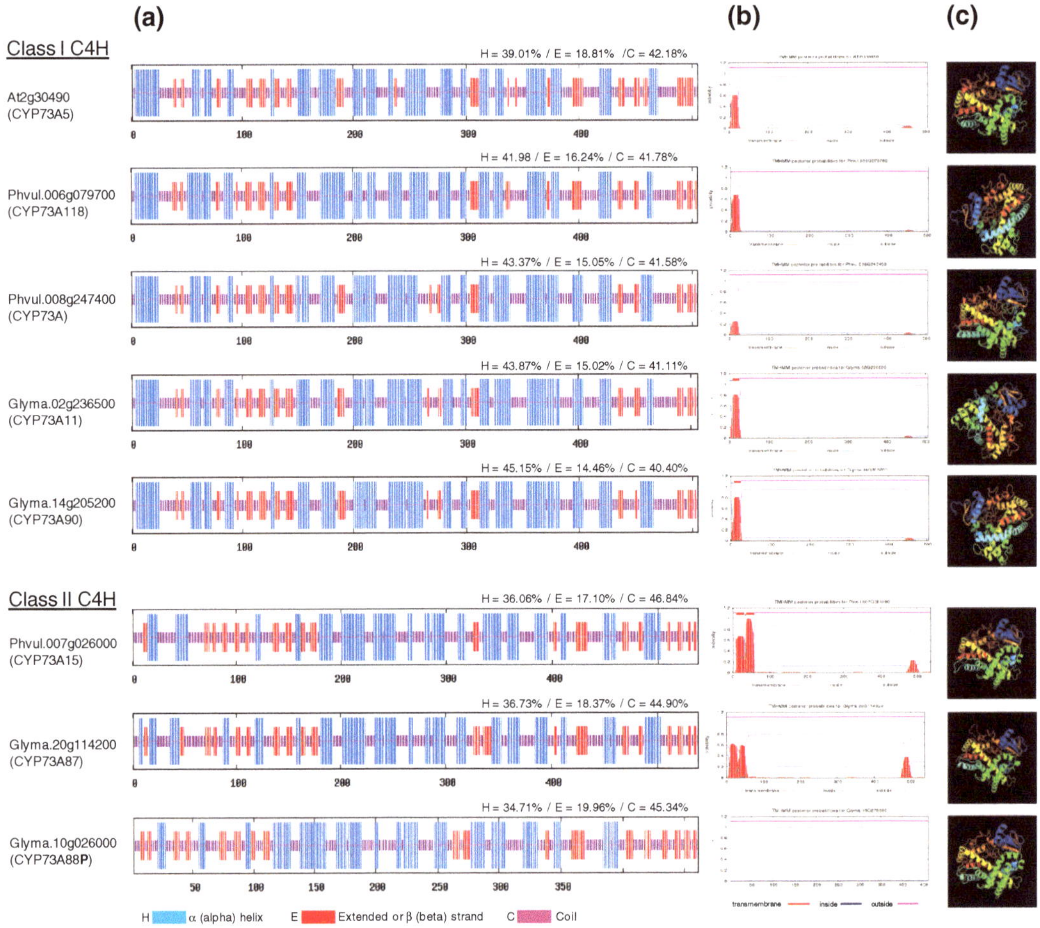

Fig. 11.9 Predicted structure of C4H class I and class II proteins in *Arabidopsis*, soybean, and common bean. **a** Secondary structures of C4H proteins (predicted by GOR IV); **b** Transmembrane helices of C4H proteins (predicted by TMHMM); **c** Tertiary structure of C4H proteins (predicted by Phyre2)

11.4.3 Tissue-Specific Expression of Genes Encoding C4Hs

Using publicly available microarray data, Ehlting et al. (2008) created a tool for co-expression analysis of P450s in *Arabidopsis*. RNA sequencing (RNA-seq) atlases were developed for both soybean (Severin et al. 2010) and common bean (O'Rurke et al. 2014). Based on RNA-seq data (Phytozome 10), genes encoding C4H are differentially expressed in six common bean and soybean tissues (Fig. 11.11). In general, the expression of the genes encoding class I C4H enzymes [*Phvul.008g247400* (*CYP73A*), *Glyma.02g236500* (*CYP73A11*), and *Glyma.14g205200* (*CYP73A90*)] compared to the class II enzymes [*Phvul.007g026000* (*CYP73A15*) and *Glyma.20g114200* (*CYP73A87*)] was higher in all tissues (flowers, pods, leaves, stems, roots, and nodules). Both common bean and soybean have two copies of genes encoding class I C4H enzymes. In both species, one of the genes (*Phvul.008g247400* and *Glyma.02g236500*) is highly expressed in all tissues. The second copy of the genes (*Glyma.14g205200* and *Phvul.006g079700*) is expressed at lower level. In soybean, *Glyma.14g205200* had approximately half of the expression of *Glyma.02g236500* in stems, roots, and nodules but very low expression in leaves, pods, and flowers. However,

Table 11.4 Protein function prediction (PFP) GO terms predicted for common bean, soybean, and *Arabidopsis* C4H proteins

Function	GO terms	Description
Molecular function	GO:0005506	Iron ion binding
	GO:0016705	Oxidoreductase activity, acting on paired donors, with incorporation or reduction of molecular oxygen
	GO:0009055	Electron carrier activity
	GO:0020037	Heme binding
Biological process	GO:0055114	Oxidation-reduction process
Cellular component	GO:0005789	Endoplasmic reticulum membrane

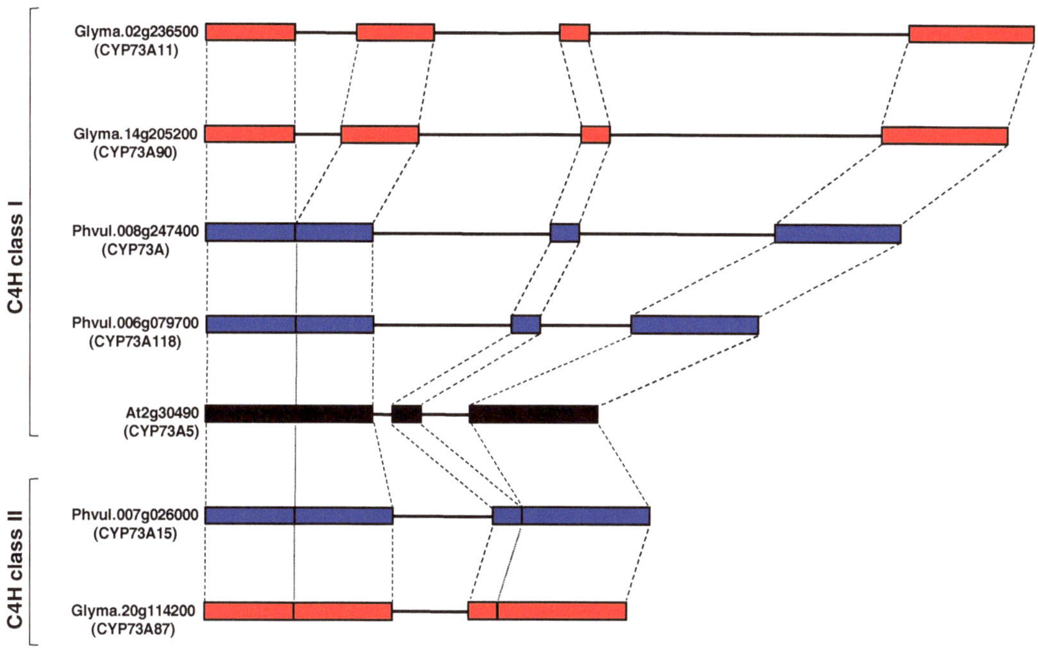

Fig. 11.10 Exon/intron structures of C4H genes in common bean, soybean, and *Arabidopsis*. Exons are represented by rectangles (common bean—blue, soybean—red, and *Arabidopsis*—black), and introns are shown as full lines. Conserved exon sequences are connected by dashed lines

Phvul.006g079700 had very low expression in all common bean tissues compared to *Phvul.008g247400* (Fig. 11.11).

Common bean C4H (*CYP73A15*) was characterized as a class II C4H enzyme, whose expression was more related to differentiation than the responses to stress (Nedelkina et al. 1999). Antisense and sense expression of cDNA coding for a truncated *CYP73A15* gene from French bean led to a reduced and delayed production of lignin in tobacco (Blee et al. 2001). Three *C4H* genes were identified in the *P. trichocarpa* genome. Two of them (*PtrC4H1* and *PtrC4H2*) were abundant in differentiating xylem, suggesting their importance in monolignol biosynthesis. Transcripts of *PtrC4H3* had little or no expression in all examined tissues (Lu et al. 2006).

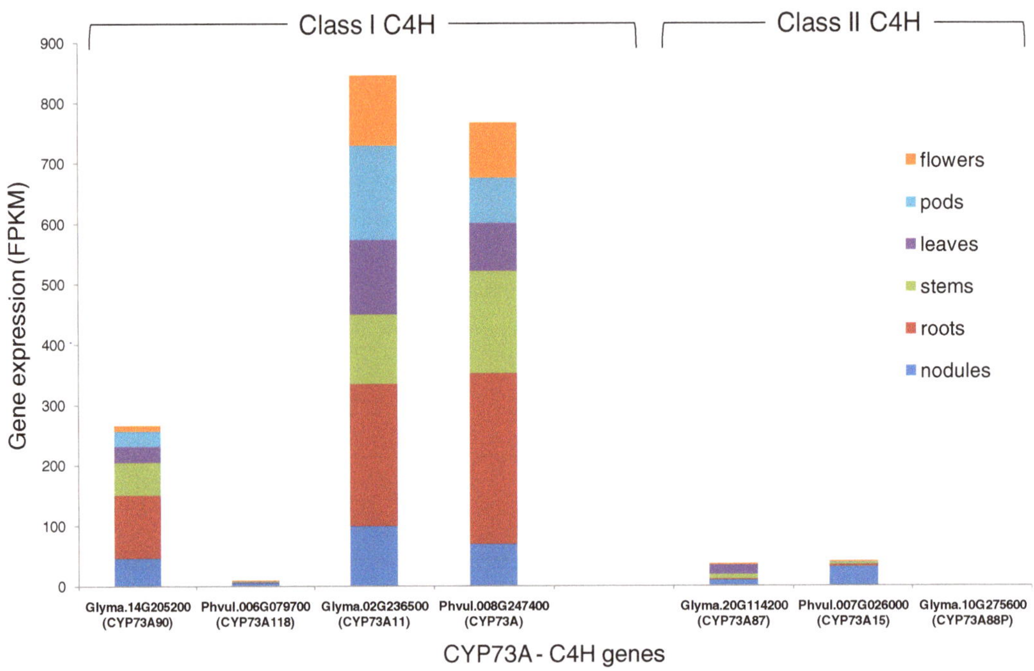

Fig. 11.11 Expression of common bean and soybean genes encoding cinnamic acid 4-hydroxylase (C4H) in six different tissues. FPKM (Fragments Per Kilobase of transcript per Million fragments mapped) data for expression levels of the genes were calculated from the RNA-seq data deposited at Phytozome 10.2 (available at http://phytozome.jgi.doe.gov/pz/portal.html)

11.4.4 *Cis*-Regulatory Regions in 5′UTRs of C4H Genes

In order to understand the functions of individual members of the *C4H* multigene families, promoters of the common bean and soybean genes were analyzed and compared to *Arabidopsis* gene (*At2g30490*) promoter, which have known functions. Promoter sequences [1 kb of 5′ regulatory sequence upstream of the coding region (1 kb 5′UTR flanking region)] of *C4H* genes were retrieved from Phytozome (10.2) and aligned in Clustal Omega at EMBL-EBI (http://www.ebi.ac.uk/Tools/msa/clustalo/) to search for possible sequence similarities among these sequences in the two C4H classes. The analysis of the 5′ regulatory regions of *C4H* genes in *Arabidopsis*, soybean, and common bean *C4H* genes revealed a moderate degree of divergence in these regions (39–60% identity). Multiple sequence alignment was sent to ClustalW2_Phylogeny to produce a phylogenetic tree, which was visualized in TreeView. Based on the 5′UTR sequences, eight *C4Hs* were split into two clusters: a three-gene class I *C4Hs* (*Phvul.008g247400, Glyma.02g236500,* and *Glyma.14g205200*) and a two-gene class II *C4Hs* (*Phvul.007g026000* and *Glyma.20g114200*) clusters. However, *Arabidopsis* (class I *At2g30490*), common bean (class I *Phvul.006g079700*), and soybean (class II pseudogene *Glyma.10g275600*) were not clearly included in any class (Fig. 11.12).

The 5′UTR sequence of *C4H* genes was analyzed for potential *cis*-acting regulatory elements using PlantCARE database (http://bioinformatics.psb.ugent.be/webtools/plantcare/html; Lescot et al. 2002). In total, 69 potential regulatory elements were identified in 5′UTR sequences of eight *C4H* genes (Fig. 11.13; Table 11.5). Twenty-six (38%) elements were present in four or more genes (Fig. 11.13, color-coded). In addition to the core TATA box and CAAT box (present in all genes), the list

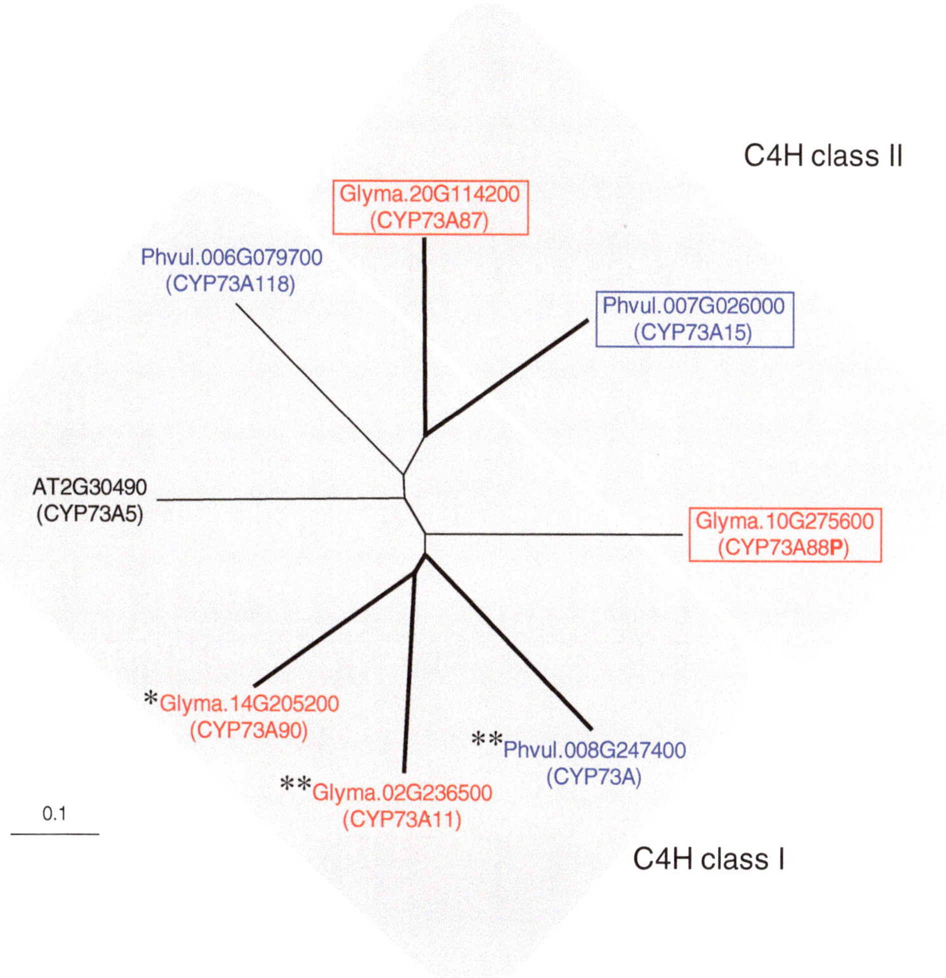

Fig. 11.12 Phylogenetic tree of 5′ upstream region (5′ UTR) sequences of the class I and class II *C4H* genes in *Arabidopsis*, common bean, and soybean. Arabidopsis sequences are labeled in black, soybean in red, and common bean are in blue; **P** at the end of the CYP name indicates pseudogene. Class II C4Hs are shown in boxes. * identifies the mostly highly expressed genes, and the number of asterisks indicates the relative levels

included a large number of light-responsive elements (27), as well as elements associated with tissue-specific expression (5), defense and stress responses (6), or hormonal responsiveness (9). A considerable number (14) of predicted regulatory elements were categorized as unknown function (Table 11.5), and two of these (AC II and unnamed_4) were present in all eight *C4H* genes.

A fraction of identified regulatory elements was specific only to class I or class II *C4H* genes (Fig. 11.13; Table 11.5). Twenty-six elements (37.7%) were present only in class I *C4H* genes. Four of these elements were identified in all five class I *C4H* genes. The CGTCA-motif and the TGACG-motif are *cis*-acting elements involved in the MeJA responsiveness, while the functions of the unnamed_1 and unnamed_3 are unknown.

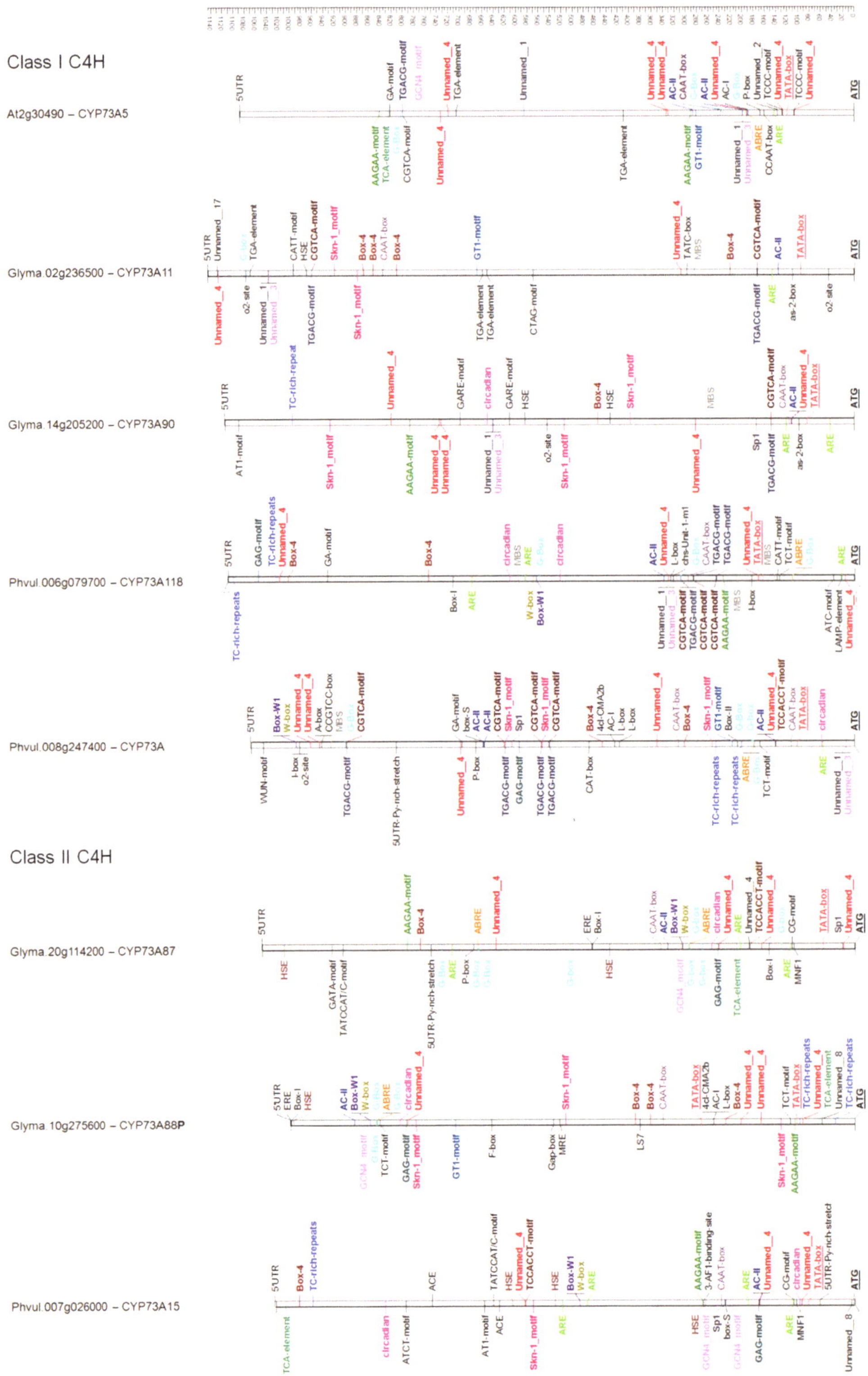

Fig. 11.13 Distribution of the putative *cis*-regulatory elements in the 5′ upstream regions (5′UTRs) in common bean, soybean, and *Arabidopsis C4H* genes, identified using PlantCARE database. The elements found in four or more genes are color-coded. Sequences and functions of elements are presented in Table 11.5

Table 11.5 Potential *cis*-acting regulatory elements identified in the 5′ regulatory sequences (5′UTR) of C4H genes in common bean, soybean, and *Arabidopsis* using PlantCARE database

Element	Sequence	Function	Present in[a]	
			Class I	Class II
TATA box	taTATAAAtc; TATAAA; ATATAA; TTTTA; TATA	Core promoter element around −30 of transcription start	At, Gm2, Gm14, Pv6, Pv8	Gm20, Gm10, Pv7
CAAT box	CAAT; CAAAT; CAATT; CCAAT; TGCCAAC; gGCAAT	Common *cis*-acting element in promoter and enhancer regions	At, Gm2, Gm14, Pv6, Pv8	Gm20, Gm10, Pv7
CAT-box	GCCACT	*cis*-acting regulatory element related to meristem expression	Pv8	
CCGTCC-box	CCGTCC	*cis*-acting regulatory element related to meristem-specific activation	Pv8	
Skn-1_motif	GTCAT	*cis*-acting regulatory element required for endosperm expression	Gm2, Gm14, Pv8	Gm10, Pv7
GCN4_motif	TGTGTCA; CAAGCCA	*cis*-regulatory element involved in endosperm expression	At2	Gm20, Gm10, Pv7
as-2-box	GATAatGATG	Involved in shoot-specific expression and light responsiveness	Gm2, Gm14	
ACE	AAAACGTTTA	*cis*-acting element involved in light responsiveness		Pv7
G-Box	CACGTT; CACGTA; TACGTG; CACGAC; CACGTC; GTACGTG; CACGTG; TACGTG; CACATGG; GACACGTAGT	*cis*-acting regulatory element involved in light responsiveness	At2, Gm2, Pv6, Pv8	Gm20, Gm10
4cl-CMA2b	TCTCACCAACC	Light-responsive element	Pv8	Gm10
Box I	TTTCAAA	Light-responsive element	Pv6	Gm20, Gm10
3-AF1 binding site	AAGAGATATTT	Light-responsive element		Pv7
GT1-motif	ATGGTGGTTGG; GGTTAA; GGTTAAT	Light-responsive element	At2, Gm2, Pv8	Gm10
MNF1	GTGCCC(A/T)(A/T)	Light-responsive element		Gm20, Pv7
Sp1	CC(G/A)CCC	Light-responsive element	Gm14, Pv8	Gm20, Pv7
MRE	AACCTAA	MYB binding site involved in light responsiveness		Gm10

(continued)

Table 11.5 (continued)

Element	Sequence	Function	Present in[a]	
			Class I	Class II
ATC-motif	AGTAATCT	Part of a conserved DNA module involved in light responsiveness	Pv6	
ATCT-motif	AATCTAATCC	Part of a conserved DNA module involved in light responsiveness		Pv7
Box 4	ATTAAT	Part of a conserved DNA module involved in light responsiveness	Gm2, Gm14, Pv6, Pv8	Gm20, Gm10, Pv7
Box II	TGGTAATAA	Part of a light-responsive element	Pv8	
CATT-motif	GCATTC	Part of a light-responsive element	Gm2, Pv6	
CG-motif	CCATGGGG	Part of a light-responsive element		Gm20, Pv7
GA-motif	AAAGATGA; ATAGATAA; AAGGAAGA	Part of a light-responsive element	At2, Pv6, Pv8	
GAG-motif	GGAGATG; AGAGATG; AGAGAGT	Part of a light-responsive element	Pv6, Pv8	Gm20, Gm10, Pv7
GATA-motif	AAGATAAGATT	Part of a light-responsive element		Gm20
I-box	TGATAATGT; GATATGG	Part of a light-responsive element	Pv6, Pv8	
L-box	TCTCACCAACC; TCTCACCTACC; CTCACCTACCAA	Part of a light-responsive element	Pv6, Pv8	Gm10
LAMP-element	CCAAAACCA	Part of a light-responsive element	Pv6	
Gap-box	CAAATGAA(A/G)A	Part of a light-responsive element		Gm10
LS7	CAGATTTATTTTTA	Part of a light-responsive element		Gm10
TCT-motif	TCTTAC	Part of a light-responsive element	Pv6, Pv8	Gm10
TCCC-motif	TCTCCCT	Part of a light-responsive element	At2	
chs-Unit 1 m1	ACCTACCACAC	Part of a light-responsive element	Pv6	
AT1-motif	AATTATTTTTTATT	Part of a light-responsive module	Gm14	Pv7
TC-rich repeat	ATTTTCTCCA; ATTTTCTTCA; GTTTTCTTAC	*cis*-acting element involved in defense and stress responsiveness	Gm14, Pv6, Pv8	Gm10, Pv7
Box-W1	TTGACC	Fungal elicitor-responsive element	Pv6, Pv8	Gm20, Gm10, Pv7

(continued)

Table 11.5 (continued)

Element	Sequence	Function	Present in[a]	
			Class I	Class II
HSE	AAAAAATTTC; AGAAAATTCG	*cis*-acting element involved in heat stress responsiveness	Gm2, Gm14	Gm20, Gm10, Pv7
ARE	TGGTTT	*cis*-acting regulatory element essential for the anaerobic induction	At2, Gm2, Gm14, Pv6, Pv8	Gm20, Pv7
WUN-motif	TCATTACGAA	Wound-responsive element	Pv8	
MBS	TAACTG; CAACTG	MYB binding site involved in drought inducibility	Gm2, Gm14, Pv6, Pv8	
TGA-element	AACGAC	Auxin-responsive element	At2, Gm2	
CGTCA-motif	CGTCA	*cis*-acting regulatory element involved in the MeJA responsiveness	At2, Gm2, Gm14, Pv6, Pv8	
TGACG-motif	TGACG	*cis*-acting regulatory element involved in the MeJA responsiveness	At2, Gm2, Gm14, Pv6, Pv8	
ERE	ATTTCAAA	Ethylene-responsive element		Gm20, Gm10
TATC-box	TATCCCA	*cis*-acting element involved in gibberellin responsiveness	Gm2	
GARE-motif	AAACAGA	Gibberellin-responsive element	Gm14	
P-box	CCTTTTG	Gibberellin-responsive element	At2, Pv8	Gm20
TCA-element	CCATCTTTTT; CAGAAAAGGA	*cis*-acting element involved in salicylic acid responsiveness	At2	Gm20, Gm10, Pv7
ABRE	TACGTG; CACGTG	*cis*-acting element involved in the abscisic acid responsiveness	At2, Pv6, Pv8	Gm20, Gm10
5UTR Py-rich stretch	TTTCTTCTCT	*cis*-acting element conferring high transcription levels	Pv8	Gm20, Pv7
O_2-site	GTTGACGTGA; GATGATGTGG; GATGACATGA; GATGATATGG	*cis*-acting regulatory element involved in zein metabolism regulation	Gm2, Gm14, Pv8	
Circadian	CAANNNNATC	*cis*-acting regulatory element involved in circadian control	Gm14, Pv6, Pv8	Gm20, Gm10, Pv7
CCAAT-box	CAACGG	MYBHv1 binding site	At2	

(continued)

Table 11.5 (continued)

Element	Sequence	Function	Present in[a]	
			Class I	Class II
W box	(T)TGAC(C/T), TTGACC	Element recognized by the family of WRKY transcription factors	Pv6, Pv8	Gm20, Gm10, Pv7
A-box	CCGTCC	*cis*-acting regulatory element	Pv8	
AAGAA-motif	gGTAAAGAAA; GAAAGAA	Unknown	At2, Gm14, Pv6	Gm20, Gm10, Pv7
TATCCAT/C-motif	TATCCAT	Unknown		Gm20, Pv7
AC-I	CCCACCTACC; TCTCACCAACC	Unknown	At2, Pv8	Gm10
AC-II	(C/T)T(T/C)(C/T)(A/C)(A/C)C(A/C)A (A/C)C(C/A)(C/A)C; TCCACCAACCCC; CCACCAACCCCC; TCACCAACCCCC; CTCACCAACCCC	Unknown	At2, Gm2, Gm14, Pv6, Pv8	Gm20, Gm10, Pv7
box S	AGCCACC	Unknown	Pv8	Pv7
F-box	CTATTCTCATT	Unknown		Gm10
CTAG-motif	ACTAGCAGAA	Unknown	Gm2	
TCCACCT-motif	TCCACCT	Unknown	Gm14, Pv8	Gm20, Pv7
Unnamed__1	GAATTTAATTAA, CGTGG	Unknown	At2, Gm2, Gm14, Pv6, Pv8	
Unnamed__2	AACCTAACCT	Unknown	At2	
Unnamed__3	CGTGG	Unknown	At2, Gm2, Gm14, Pv6, Pv8	
Unnamed__4	CTCC	Unknown	At2, Gm2, Gm14, Pv6, Pv8	Gm20, Gm10, Pv7
Unnamed__8	CATTTTTGT	Unknown		Gm10, Pv7
Unnamed__17	TAGGAGCAGCT	Unknown	Gm2	

[a]Class I: At2, *At2g30490*; Gm2, *Glyma.02g236500*; Gm14, *Glyma.14g205200*; Pv6, *Phvul.006g079700*; Pv8, *Phvul.008g247400*; class II: Gm20, *Glyma.20g114200*; Gm10, *Glyma.10g275600*; Pv7, *Phvul.007g026000*

In addition, MBS (a MYB binding site involved in drought inducibility) and O_2 site (*cis*-acting regulatory element involved in regulation of zein metabolism) were identified in the 5′UTRs of all four legume class I *C4H* genes. Eleven elements (15.9%) were unique to the class II *C4H* genes. Three of these elements were identified in both soybean (*Glyma.20g114200*) and common bean (*Phvul.007g026000*) C4H genes. The MNF1 and CG motifs are light-responsive elements, while the function of the TATCCAT/C-motif is unknown. Lu et al. (2006) reported that four divergent C4H isoforms play distinct roles in *P. trichocarpa*. The divergent upstream sequences among the two group *PtreC4H* genes suggested that the mechanisms of gene regulation might be different.

The identification of the *cis*-acting sequences regulating differential expression of *C4H* genes and transcription factors that interact with these sequences in common bean, soybean, and *Arabidopsis* could lead to an understanding of the mechanism(s) of differential regulation of these highly similar genes in these plant species.

11.4.5 Syntenic Regions Containing Common Bean C4H Genes

The availability of the complete genome sequences for numerous plant species, including soybean (Schmutz et al. 2010) and common bean (Schmutz et al. 2014), allows the organization of the individual genomes to be studied, as well as enables comparison of the genomes at the nucleotide level. The size of the common bean genome (521 Mb) is approximately half of the size of the soybean genome (978 Mb). As a result of at least two rounds of polyploidization [~59 MYA (million years ago) and ~13 MYA], the soybean genome contains significant gene duplications and redundancy (Schmutz et al. 2010). In general, for any gene in common bean, two corresponding homologous genes could potentially be found in soybean. Moreover, because of the shared synteny between the two genomes, regions homologous to regions in two soybean chromosomes were found for all 11 common bean chromosomes, with a minor marker rearrangement and/or sequence orientation (Galeano et al. 2009; McClean et al. 2010; Reinprecht et al. 2013).

Synteny analysis was performed in Plant Genome Duplication Database (PGDD, available at http://chibba.agtec.uga.edu/duplication; Lee et al. 2013) against complete genome sequences available for 47 flowering plant species. Numerous syntenic regions (26–44) with other plant species were found for common bean, soybean, and *Arabidopsis* class I *C4H*s. The blocks were of various sizes, ranging from 14 to 884 gene anchors. For example, common bean C4H on the chromosome Pv06, *CYP73A118* (*Phvul.006g079700*), was syntenic to 44 regions in 31 different plant species including two regions in soybean, poplar, pear, watermelon, rice, kale, sacred lotus, and chickpea, three regions in Chinese cabbage, and four regions in kiwifruit (data not shown). In contrast, only five syntenic blocks were identified for common bean and soybean class II *C4H*s. They were syntenic to each other and to another three legumes (*Medicago truncatula*, *Cicer arietinum*, and *Cajanus cajan*).

Several syntenic blocks containing *C4H* loci were identified among common bean, soybean, and *Arabidopsis* genomes (Table 11.6; Fig. 11.14). For example, *Phvul.006g079700* (encoding common bean class I C4H) was syntenic to other four class I *C4H*s: common bean *Phvul.008g247400*, soybean *Glyma.02g236500* and *Glyma.14g205200*, and

Table 11.6 Syntenic blocks containing *C4H* loci in genomes of common bean, soybean, and *Arabidopsis*

C4H locus (gene model) identifier[a]		Syntenic block[b]			Position within a block	Ka[c]	Ks[d]
Query	Synteny	Score	E-value	Anchors (# genes)			
Phvul.006G079700	*At2g30490*	894	6e−112	24	21	0.0	0.0
	Glyma.02g236500	1862	2e−83	51	38	0.10	1.22
	Phvul.008g247400	1537	0.0	40	20	0.08	1.05
		1789	8e−82	48	35	0.09	1.11
		539	9e−53	14	13	0.0	0.0
	Glyma.02g236500	8130	0.0	209	81	0.04	0.38
	Phvul.006g079700	1537	0.0	40	20	0.08	1.05
Phvul.007g026000	*Glyma.10g275600*	24,980	0.0	641	487	0.04	0.42
	Glyma.20g114200	21,836	2e−137	561	99	0.05	0.36
Glyma.02g236500	*Phvul.008g247400*	8130	0.0	209	81	0.04	0.38
	Phvul.006g079700	1862	2e−83	51	38	0.10	1.22
	At2g30490	1183	5e−66	31	29	0.12	0.0
Glyma.14g205200	*Phvul.006g079700*	1789	8e−82	48	35	0.09	1.11
Glyma.10g275600	*Phvul.007g026000*	24,980	0.0	641	487	0.04	0.42
	Glyma.20g114200	35,199	0.0	884	779	0.02	0.22
Glyma.20g114200	*Phvul.007g026000*	21,836	2e−137	561	99	0.05	0.36
	Glyma.10g275600	35,199	0.0	884	779	0.02	0.22
At2g30490	*Phvul.008g247400*	539	9e−53	14	13	0.0	0.0
	Phvul.006g079700	894	6e−115	24	21	0.0	0.0
	Glyma.02g236500	1183	5e−66	31	29	0.12	0.0

[a]Phytozome (https://phytozome.jgi.doe.gov/pz/portal.html)
[b]Related syntenic regions in multiple species by locus identifier were obtained from the Plant Genome Duplication Database (PGDD, available at http://chibba.agtec.uga.edu/duplication/index/locus; accessed 26 June 2015). All intra- and cross-species blocks for the query locus, graphs, and tables displayed ±200 kb region
[c]The number of non-synonymous substitutions per site (Ka)
[d]The number of synonymous substitutions per site (Ks)

Arabidopsis At2g30490. Similarly, common bean class II *C4H Phvul.007g026000* was syntenic to two soybean class II *C4Hs*: *Glyma20g.114200* and *Glyma.10g275600*. They were contained in large syntenic blocks anchored by 641 and 561 genes, respectively (Table 11.6; Fig. 11.14,). Synteny was also analyzed with SyMap v4.0 (**Syn**teny **M**apping and **A**nalysis **P**rogram; available at http://www.symapdb.org; Soderlund et al. 2011) to produce circular alignments of multiple common bean and soybean chromosomes (Fig. 11.14, right).

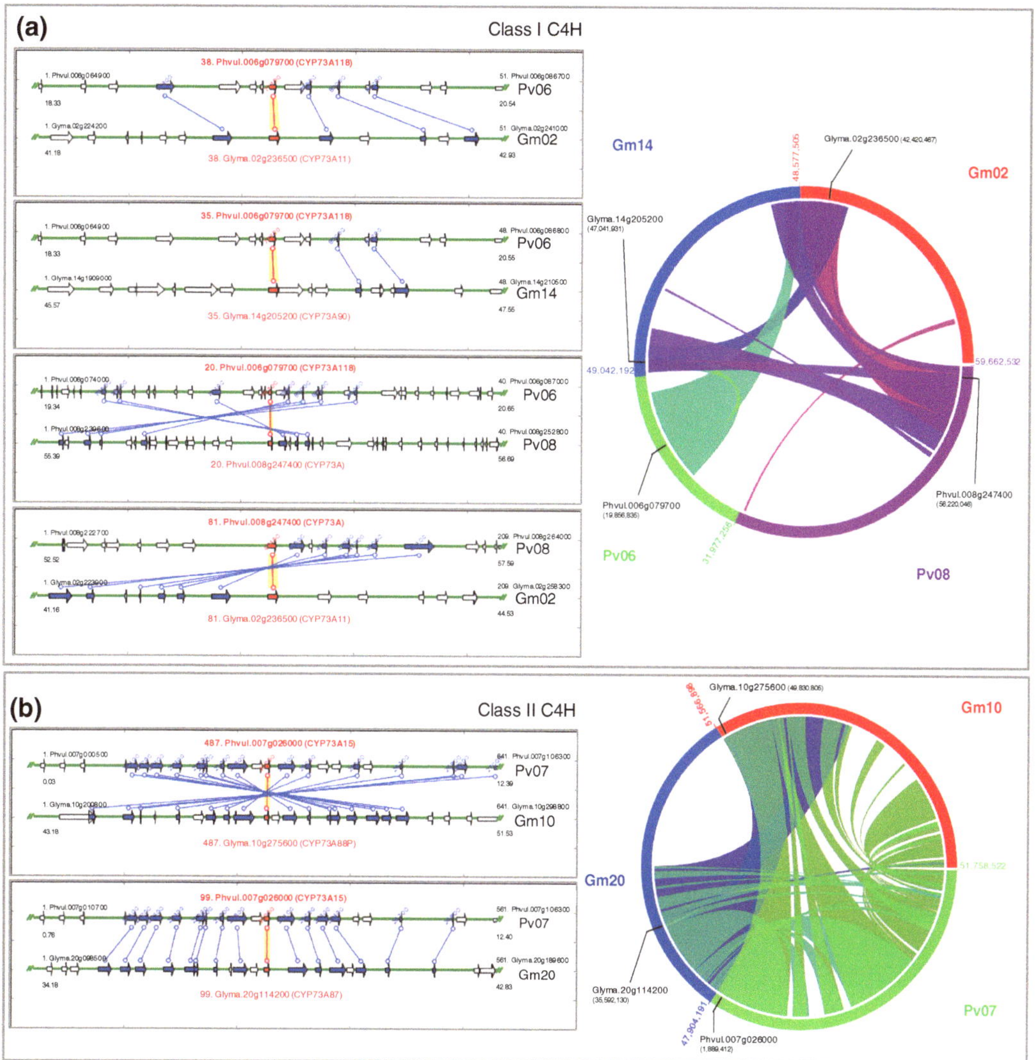

Fig. 11.14 Syntenic regions containing *C4H* loci in genomes of common bean and soybean. **a**. Class I C4H—*Phvul.006g079700* (*CYP73A118*) and *Phvul.008g247400* (*CYP73A*); **b** Class II C4H—*Phvul.007g026000* (*CYP73A15*). Left—synteny identified in Plant Genome Duplication Database. Query locus is represented by a red arrow; blue arrows are other anchor genes in the region. Right—circular alignment of common bean and soybean chromosomes containing *C4H* loci

11.4.6 Sequence Polymorphisms in C4H Genes in Common Bean

Nucleotide polymorphisms for a number of phenylpropanoid pathway genes in various plant species have been described, including *Arabidopsis* (Savolainen et al. 2000; Aguade 2001; Wright et al. 2003) and maize (Brenner et al. 2010). In the current work, sequences of three *C4H* genes in the common bean landrace G19833 (Phytozome) were BLASTed against genome sequence of cultivar OAC Rex. The structure of *C4H* genes identified in OAC Rex was predicted with the HMM-based Fgenesh gene finder (Solovyev et al. 2006; available at

Table 11.7 *C4H* gene polymorphism between common beans cultivar OAC Rex and landrace G19833

Class	*C4H* locus (gene model) identifier[a]	Polymorphism (bp difference)							
		Type[b]	5′ UTR	Exon 1	Intron 1	Exon 2	Intron 2	Exon 3	3′ UTR
I	*Phvul.006g079700* (*CYP73A118*)	SNP	1	**4**	18	0	13	**1**	49
		Del	0	0	1	0	146	0	173
		Ins	0	0	0	0	3	0	18
	Phvul.008g247400 (*CYP73A*)	SNP	27	**3**	6	0	5	**5**	11
		Del	55	0	0	0	22	0	16
		Ins	2	0	0	0	0	0	1
II	*Phvul.007g026000* (*CYP73A15*)	SNP	9	**11**	8	**3**	NA[c]	NA	8
		Del	5	0	1	0	NA	NA	0
		Ins	1	0	2	0	NA	NA	0

[a]Phytozome (http://phytozome.jgi.doe.gov/pz/portal.html)
[b]Polymorphism (SNP, single nucleotide polymorphism; del, deletion; ins, insertion) detected in OAC Rex *C4H* gene sequences [GenBank accessions: KU308554 (*Phvul.006g079700*), KU308555 (*Phvul.007g026000*), KU308556 (*Phvul.008g247400*)] compared to G19833 gene sequences
[c]NA—Not applicable
'Bold values' indicate polymorphism identified in coding (exonic) regions of the genes

http://linux1.softberry.com/berry.phtml?topic=fgenesh&group=programs&subgroup=gfind; accessed: 7 July 2015).

The C4H proteins in the two genotypes were very similar. The proteins encoded by the *Phvul.006g079700* gene in G19833 and OAC Rex were identical. A single amino acid difference was identified at position 42 between OAC Rex (I) and G19833 (V) C4H proteins encoded by the *Phvul.008g247400* gene (99.8% identity). OAC Rex and G19833 C4H proteins encoded by the *Phvul.007g026000* gene were 99.1% identical. Differences were found in five amino acids at positions 4 (V in OAC Rex, F in G19833), 7 (N in OAC Rex, K in G19833), 18 (L in OAC Rex, S in G19833), 54 (K in OAC Rex, N in G19833), and 420 (I in OAC Rex, V in G19833) (data not shown).

The *CH4* genomic sequences were also very similar between two common bean genotypes (97.2% identity for *Phvul.006g079700*, 98.5% identity for *Phvul.008g247400*, and 98.9% identity for *Phvul.007g026000*). However, by aligning the CH4 encoding sequences in the two bean genomes (G19833 and OAC Rex), polymorphism (SNPs, insertions, and deletions) was identified for all three *C4H* genes (Table 11.7; Fig. 11.15).

Although polymorphisms were detected in both the coding (one to 11 SNPs, shown in bold) and non-coding regions, the majority of the sequence differences that were identified occurred in the

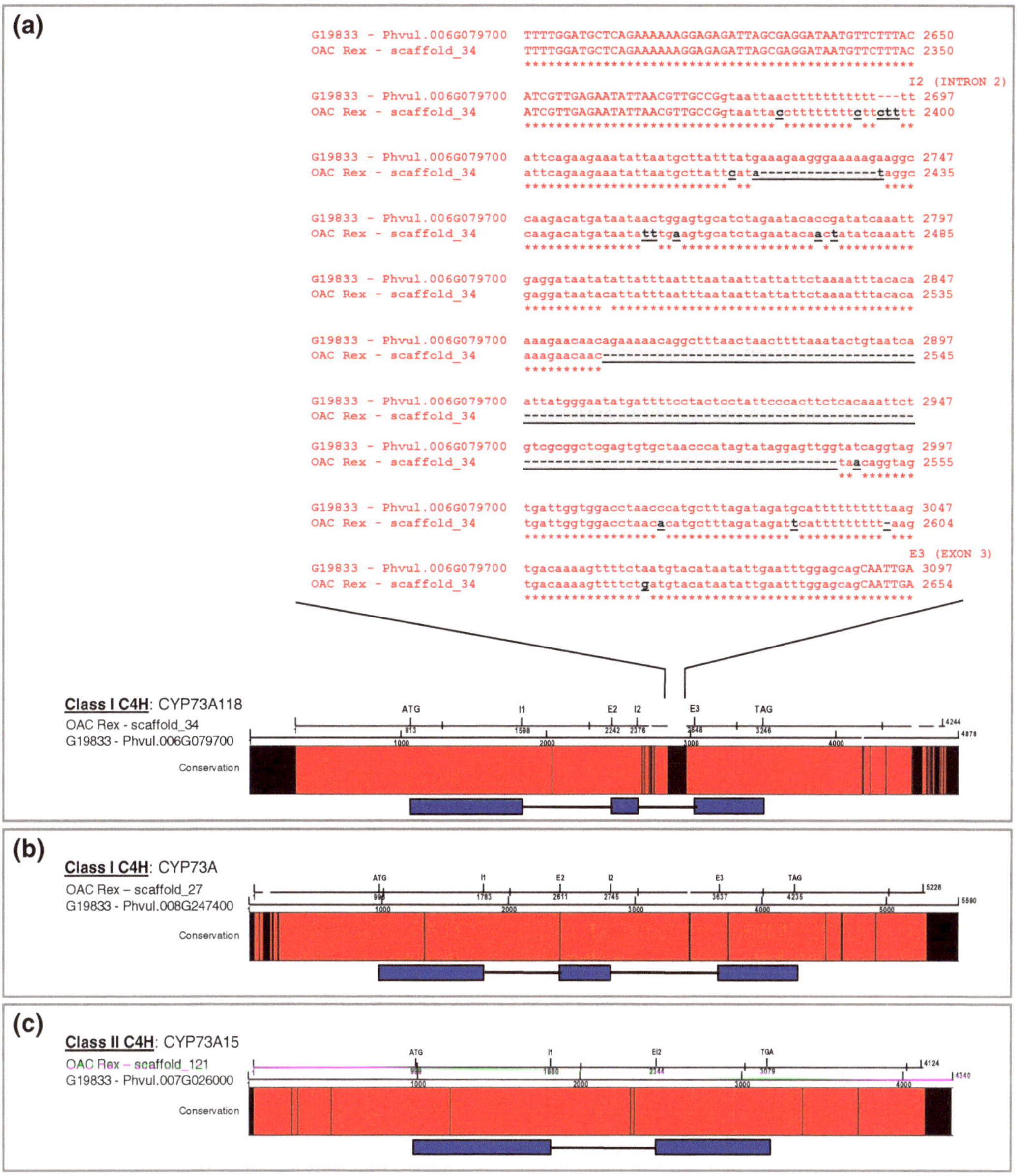

Fig. 11.15 *C4H* gene sequence polymorphisms between common bean cultivar OAC Rex (UofG) and landrace G19833 (Phytozome v10.2). **a** Class I *C4H—CYP73A118* (*Phvul.006G079700*; OAC Rex accession KU308554); in an alignment, E indicates exons (shown in capital letters) and I represents introns (shown in small letters); the sequence polymorphism in intron 2 (I2) is highlighted (shown in gray); **b** Class I *C4H—CYP73A* (*Phvul.008G247400*; OAC Rex accession KU308556); **c** Class II *C4H—CYP73A15* (*Phvul.007G026000*; OAC Rex accession KU308555)

introns and UTRs. For example, the size difference of the *Phvul.006g079700* gene (encoding class I *C4H*, *CYP73A118*) intron 2 (143 bp) in OAC Rex (272 bp) and G19833 (415 bp) can be used to develop gene-based marker(s). However, the usefulness of these polymorphisms as *C4H* gene-specific marker needs to be evaluated in additional germplasm from two common bean gene pools.

11.5 Conclusions

The availability of the whole genome sequences allowed us to identify gene families encoding major enzymes of the phenylpropanoid pathway in common bean, soybean, and *Arabidopsis*. The work focused on C4H, a cytochrome P450 that occupies an entry position in the pathway. Three genes encoding C4H proteins were identified in common bean genome compared to the four genes in soybean. The next step would be to functionally characterize these genes. The availability of the common bean genome sequence also makes it possible to identify and characterize the members of each gene family that are involved in the specific branches of the phenylpropanoid pathway. Furthermore, the identification of transcription factors that activate phenylpropanoid biosynthetic gene families could provide tools to potentially manipulate the amount of different phenylpropanoids in common bean.

References

Adlercreutz H (2007) Lignans and human health. Crit Rev Clin Lab Sci 44:483–525

Agati G, Azzarello E, Pollastri S, Tattini M (2012) Flavonoids as antioxidants in plants: location and functional significance. Plant Sci 196:67–76

Agati G, Brunetti C, Di Ferdinando M, Ferrini F, Pollastri S, Tattini M (2013) Functional roles of flavonoids in photoprotection: new evidence, lessons from the past. Plant Physiol Biochem 72:35–45

Aguade M (2001) Nucleotide sequence variation at two genes of the phenylpropanoid pathway, the *Fah1*and *F3H*genes, in *Arabidopsis thaliana*. Mol Biol Evol 18:1–9

Akada S, Dube SK (1995) Organization of soybean chalcone synthase gene clusters and characterization of a new member of the family. Plant Mol Biol 29:189–199

Akashi T, Aoki T, Ayabe S (2005) Molecular and biochemical characterization of 2-hydroxyisoflavanone dehydratase. Involvement of carboxylesterase-like proteins in leguminous isoflavone biosynthesis. Plant Physiol 137:882–891

Alber A, Ehlting J (2012) Cytochrome P450s in lignin biosynthesis. Adv Bot Res 61:113–143

Anterola AM, Lewis NG (2002) Trends in lignin modification: a comprehensive analysis of the effects of genetic manipulations/mutations on lignification and vascular integrity. Phytochemistry 61:221–294

Artigot M-P, Dayde J, Berger M (2013) Expression of flavonoid 6-hydroxylase candidate genes in normal and mutant soybean genotypes for glycitein content. Mol Biol Rep 40:4361–4369

Austin MB, Noel JP (2003) The chalcone synthase superfamily of type III polyketide synthases. Nat Prod Rep 20:79–110

Ayabe S, Akashi T (2006) Cytochrome P450s in flavonoid metabolism. Phytochem Rev 5:271–282

Bak S, Beisson F, Hamberger B, Hofer R, Paquette S, Werck-Reichhart D (2011) Cytochrome P450. Arabidopsis Book 9:e0144. doi:10.1199/tab.0144

Barakat A, Bagniewska-Zadworna A, Frost CJ, Carlson JE (2010) Phylogeny and expression profiling of *CAD and CAD-like* genes in hybrid *Populus* (*P. deltoides* × *P. nigra*): evidence from herbivore damage for subfunctionalization and functional divergence. BMC Plant Biol 10:100. doi:10.1186/1471-2229-10-100

Barker MS, Baute GJ, Liu S-L (2012) Duplications and turnover in plant genomes. In: Wendel JF, Greilhuber J, Dolezal J, Leitch IJ (eds) Plant genome diversity, vol 1. Springer-Verlag, Wien, pp 155–169

Bassard J-E, Richert L, Geerinck J, Renault H, Duval F, Ullmann P, Schmitt M, Meyer E, Mutterer J, Boerjan W, De Jaeger G, Mely Y, Goossens A, Werck-Reichhart D (2012) Protein-protein and protein-membrane associations in the lignin pathway. Plant Cell 24:4465–4482

Baxter HL, Stewart N Jr (2013) Effects of altered lignin biosynthesis on phenylpropanoid metabolism and plant stress. Biofuels 4:635–650

Bell-Lelong DA, Cusumano JC, Meyer K, Chapple C (1997) Cinnamate-4-hydroxylase expression in *Arabidopsis*. Regulation in response to development and the environment. Plant Physiol 113:729–738

Blee K, Choi JW, O'Connell AP, Jupe SC, Schuch W, Lewis NG, Bolwell GP (2001) Antisense and sense expression of cDNA coding for CYP73A15, a class II cinnamate 4-hydroxylase, leads to a delayed and reduced production of lignin in tobacco. Phytochemistry 57:1159–1166

Boerjan W, Ralp J, Baucher M (2003) Lignin biosynthesis. Annu Rev Plant Biol 54:519–546. doi:10.1146/annurev.arplant.54.031902.134938

Brenner EA, Zein I, Chen Y, Andersen JR, Wenzel G, Ouzunova M, Eder J, Darnhofer B, Frei U, Barriere Y, Lubberstedt T (2010) Polymorphisms in O-methyltransferase genes are associated with stover cell wall digestibility in European maize (*Zea mays* L.). BMC Plant Biol 10:27. doi:10.1186/1471-2229-10-27

Brunetti C, Di Ferdinando M, Fini A, Pollastri S, Tattini M (2013) Flavonoids as antioxidants and developmental regulators: relative significance in plants and humans. Int J Mol Sci 14:3540–3555. doi:10.3390/ijms14023540

Cassidy A, Hanley B, Lamuela-Raventos RM (2000) Isoflavones, lignans and stilbenes—origins, metabolism and potential importance to human health. J Sci Food Agric 80:1044–1062

Casneuf T, De Bodt S, Raes J, Maere S, Van de Peer Y (2006) Nonrandom divergence of gene expression following gene and genome duplications in the flowering plant *Arabidopsis thaliana*. Genome Biol 7:R13

Chen J, Wang L, Thompson LU (2006) Flaxseed and its components reduce metastasis after surgical excision of solid human breast tumor in nude mice. Cancer Lett 234:168–175

Cheynier V, Comte G, Davies KM, Lattanzio V, Martens S (2013) Plant phenolics: recent advances on their biosynthesis, genetics, and ecophysiology. Plant Physiol Biochem 72:1–20

Chu S, Wang J, Cheng H, Yang Q, Yu D (2014) Evolutionary study of the isoflavonoid pathway based on multiple copies analysis in soybean. BMC Genet 15:76

Costa MA, Collins RE, Anterola AM, Cochrane FC, Davin LB, Lewis NG (2003) An *in silico* assessment of gene function and organization of the phenylpropanoid pathway metabolic networks in *Arabidopsis thaliana* and limitations thereof. Phytochemistry 64:1097–1112

Crooks GE, Hon G, Chandonia JM, Brenner SE (2004) WebLogo: a sequence logo generator. Genome Res 14:1188–1190

Dhaubhadel S (2011) Regulation of isoflavonoid biosynthesis in soybean seeds. In: Ng T-B (ed) Soybean—biochemistry, chemistry and physiology. InTech, http://www.intechopen.com

Ehlting J, Hamberger B, Milliom-Rousseau R, Werck-Reichhart D (2006) Cytochrome P450 in phenolic metabolism. Phytochem Rev 5:239–270

Ehlting J, Sauveplane V, Olry A, Ginglinger J-F, Provart NJ, Werck-Reichhart D (2008) An extensive (co-)expression analysis tool for the cytochrome P450 superfamily in *Arabidopsis thaliana*. BMC Plant Biol 8:47. doi:10.1186/1471-2229-8-47

Falcone Ferreyra ML, Rius SP, Casati P (2012) Flavonoids: biosynthesis, biological functions, and biotechnological applications. Front Plant Sci 3:222. doi:10.3389/fpls.2012.00222

Ferrer JL, Austin MB, Stewart C Jr, Noel JP (2008) Structure and function of enzymes involved in the biosynthesis of phenylpropanoids. Plant Physiol Biochem 46:356–370. doi:10.1016/j.plaphy.2007.12.009

Fraser CM, Chapple C (2011) The phenylpropanoid pathway in Arabidopsis. Arabidopsis Book 9:e0152. doi:10.1199/tab.0152

Freeling M (2009) Bias in plant gene content following different sorts of duplication: tandem, whole-genome, segmental or by transposition. Annu Rev Plant Biol 60:433–453

Galeano CH, Fernandez AC, Gomez M, Blair MW (2009) Single strand conformation polymorphism based SNP and Indel markers for genetic mapping and synteny analysis of common bean (*Phaseolus vulgaris* L.). BMC Genom 10:629. doi:10.1186/1471-2164-10-629

Garnier J, Gibrat JF, Robson B (1996) GOR method for predicting protein secondary structure from amino acid sequence. Methods Enzymol 266:540–553

Gonzalez A, Zhao M, Leavitt JM, Lloyd AM (2008) Regulation of the anthocyanin biosynthetic pathway by the TTG1/bHLH/Myb transcriptional complex in *Arabidopsis* seedlings. Plant J 53:814–827

Goodstein DM, Shu S, Howson R, Neupane R, Hayes RD, Fazo J, Mitros T, Dirks W, Hellsten U, Putnam N, Rokhsar DS (2012) Phytozome: a comparative platform for green plant genomics. Nucleic Acids Res 40:D1178–D1186

Gou J-Y, Felippes FF, Liu C-J, Weigel D, Wang J-W (2011) Negative regulation of anthocyanin biosynthesis in *Arabidopsis* by a miR156-targeted SPL transcription factor. Plant Cell 23:1512–1522

Goujon T, Sibout R, Eudes A, MacKay J, Jouanin L (2003) Genes involved in the biosynthesis of lignin precursors in *Arabidopsis thaliana*. Plant Physiol Biochem 41:677–687

Graham SE, Peterson JA (1999) How similar are P450s and what can their differences teach us? Arch Biochem Biophys 369:24–29

Graham T, Graham M, Yu O (2008) Genomics of secondary metabolism in soybean. In: Stacey G (ed) Genetics and genomics of soybean. Springer Science+Business Media, LLC, Berlin, pp 211–241

Grotewold E (2005) Plant metabolic diversity: a regulatory perspective. Trends Plant Sci 10:57–62

Guttikonda SK, Trupti J, Bisht NC, Chen H, An Y-QC, Pandey S, Xu D, Yu O (2010) Whole genome co-expression analysis of soybean cytochrome P450 genes identifies nodulation-specific P450 monooxygenases. BMC Plant Biol 12:243

Hahlbrock K, Scheel D (1989) Physiology and molecular biology of phenylpropanoid metabolism. Annu Rev Plant Physiol Plant Mol Biol 40:347–369

Hamberger B, Bak S (2013) Plant P450s as versatile drivers for evolution of species-specific chemical diversity. Philos Trans R Soc B 368:20120426

Hamberger B, Ellis M, Friedmann M, de Azevedo Souza C, Barbazuk B, Douglas CJ (2007) Genome-wide analyses of phenylpropanoid-related genes in *Populus trichocarpa*, *Arabidopsis thaliana*, and *Oryza sativa*: the *Populus* lignin toolbox and conservation and diversification of angiosperm gene families. Can J Bot 85:1182–1201

Hanada K, Sawada Y, Kuromori T, Klausnitzer R, Saito K, Toyoda T, Shinozaki K, Li W-H, Hirai MY (2011) Functional compensation of primary and secondary metabolites by duplicate genes in *Arabidopsis thaliana*. Mol Biol Evol 28:377–382

Hao Z, Mohnen D (2014) A review of xylan and lignin biosynthesis: foundation for studying Arabidopsis *irregular xylem* mutants with pleiotropic phenotypes. Crit Rev Biochem Mol Biol 49:212–241

Harakava R (2005) Genes encoding enzymes of the lignin biosynthesis pathway in *Eucalyptus*. Genet Mol Biol 28:601–607

Hartmann U, Sagasser M, Mehrtens F, Stracke R, Weisshaar B (2005) Differential combinatorial interactions of *cis*-acting elements recognized by R2R3-MYB, BZIP, and BHLH factors control light-responsive and tissue-specific activation of phenylpropanoid biosynthesis genes. Plant Mol Biol 57:155–171

Hawkins T, Chitale M, Luban S, Kihara D (2009) PFP: automated prediction of Gene Ontology functional annotations with confidence scores using protein sequence data. Proteins 74:566–582

Hrazdina G, Wagner GJ (1985) Metabolic pathways as enzyme complexes: evidence for the synthesis of phenylpropanoids and flavonoids on membrane associated enzyme complexes. Arch Biochem Biophys 237:88–100

Humphreys JM, Hemm MR, Chapple C (1999) New routes for lignin biosynthesis defined by biochemical characterization of recombinant ferulate 5-hydroxylase, a multifunctional cytochrome P450-dependent monooxy genase. Proc Natl Acad Sci U S A 96:10045–10050

Humphreys JM, Chapple C (2002) Rewriting the lignin roadmap. Curr Opin Plant Biol 5:224–229

Jung W, Yu O, Lau SM, O'Keefe DP, Odell J, Fader G, McGonigle B (2000) Identification and expression of isoflavone synthase, the key enzyme for biosynthesis of isoflavones in legumes. Nat Biotechnol 18:208–212

Kelley LA, Mezulis S, Yates CM, Wass MN, Sternberg MJE (2015) The Phyre2 web portal for protein modeling, prediction and analysis. Nat Protoc 10:45–858. doi:10.1038/nprot.2015.053

Koes R, Verweij W, Quattrocchio F (2005) Flavonoids: a colorful model for the regulation and evolution of biochemical pathways. Trends Plant Sci 10:236–242

Krogh A, Larsson B, von Heijne G, Sonnhammer EL (2001) Predicting transmembrane protein topology with a hidden Markov model: application to complete genomes. J Mol Biol 305:567–580. doi:10.1006/jmbi.2000.4315

Kumar MS, Chakravarthy SS, Babu PR, Rao KV, Reddy VD (2015) Classification of cytochrome P450s in common bean (*Phaseolus vulgaris* L.). Plant Syst Evol 301:211–216

Labeeuw L, Martone LT, Boucher Y, Case RJ (2015) Ancient origin of the biosynthesis of lignin precursors. Biol Direct 10:23. doi:10.1186/s13062-015-0052-y

Lamesch P, Berardini TZ, Li D, Swarbreck D, Wilks C, Sasidharan R, Muller R, Dreher K, Alexander DL, Garcia-Hernandez M, Karthikeyan AS, Lee CH, Nelson WD, Ploetz L, Singh S, Wensel A, Huala E (2012) The Arabidopsis Information Resource (TAIR): improved gene annotation and new tools. Nucleic Acids Res 40:D1202–D1210

Latunde-Dada AO, Cabello-Hurtado F, Czittrich N, Didierjean L, Schopfer C, Hertkorni N, Werck-Reichhart D, Ebel J (2001) Flavonoid 6-hydroxylase from soybean (*Glycine max* L.), a novel plant P-450 monooxygenase. J Biol Chem 276:1688–1695

Lee TH, Tang H, Wang X, Paterson AH (2013) PGDD: a database of gene and genome duplication in plants. Nucleic Acids Res. doi:10.1093/nar/gks1104

Lescot M, Déhais P, Moreau Y, De Moor B, Rouzé P, Rombauts S (2002) PlantCARE: a database of plant *cis*-acting regulatory elements and a portal to tools for *in silico* analysis of promoter sequences. Nucleic Acids Res 30:325–327

Li S (2014) Transcriptional control of flavonoid biosynthesis—fine-tuning of the MYB-bHLH-WD40 (MBW) complex. Plant Sig Behav 9:e27522

Lozovaya VV, Lygin AV, Zernova OV, Ulanov AV, Li S, Hartman GL, Widholm JM (2007) Modification of phenolic metabolism in soybean hairy roots through down regulation of chalcone synthase or isoflavone synthase. Planta 225:665–679

Lu S, Zhou Y, Li L, Chiang VL (2006) Distinct roles of cinnamate 4-hydroxylase genes in *Populus*. Plant Cell Physiol 47:905–914

Lucheta AR, Silva-Pinhati ACO, Basílio-Palmieri AC, Berger IJ, Freitas-Astúa J, Cristofani M (2007) An *in silico* analysis of the key genes involved in flavonoid biosynthesis in *Citrus sinensis*. Genet Mol Biol 30:819–831

Lynch M, Conery JS (2000) The evolutionary fate and consequences of duplicate genes. Science 290:151–1155

Matsumura H, Watanabe S, Harada K, Senda M, Akada S, Kawasaki S, Dubouzet EG, Minaka N, Takahashi R (2005) Molecular linkage mapping and phylogeny of the chalcone synthase multigene family in soybean. Theor Appl Genet 110:1203–1209

Matsuno M, Compagnon V, Schoch GA, Schmitt M, Debayle D, Bassard J-E, Pollet B, Hehn A, Heintz D, Ullmann P, Lapierre C, Bernier F, Ehlting J, Werck-Reichhart D (2009) Evolution of a novel phenolic pathway for pollen development. Science 324:1688–1692

McClean PE, Mamidi S, McConnell M, Chikara S, Lee R (2010) Synteny mapping between common bean and soybean reveals extensive blocks of shared loci. BMC Genom 11:184. doi:10.1186/1471-2164-11-184

Miedes E, Vanholme R, Boerjan W, Molina A (2014) The role of the secondary cell wall in plant resistance to pathogens. Front Plant Sci 5:358. doi:10.3389/fpls.2014.00358

Mizutani M (2012) Impacts of diversification of cytochrome P450 on plant metabolism. Biol Pharm Bull 35:824–832

Mizutani M, Ohta D (2010) Diversification of P450 genes during land plant evolution. Annu Rev Plant Biol 61:291–315

Mizutani M, Ohta D, Sato R (1997) Isolation of a cDNA and a genomic clone encoding cinnamate 4-hydroxylase from *Arabidopsis* and its expression manner in plants. Plant Physiol 113:755–763

Mizutani M, Sato F (2011) Unusual P450 reactions in plant secondary metabolism. Arch Biochem Biophys 507:94–203

Mizutani M, Ward E, DiMaio J, Ohta D Ryals J, Sato R (1993) Molecular cloning and sequencing of a cDNA encoding mung bean cytochrome P450 ($P450_{C4H}$) possessing cinnamate 4-hydroxylase activity. Biochem Biophys Res Commun 190:875–880

Moura JC, Bonine CA, de Oliveira Fernandes Viana J, Dornelas MC, Mazzafera P (2010) Abiotic and biotic stresses and changes in the lignin content and composition in plants. J Integr Plant Biol 52:360–376. doi:10.1111/j.1744-7909.2010.00892.x

Nagano S (2014) Structural and functional diversity of cytochrome P450. In: Yamazaki H (ed) Fifty years of cytochrome P450 research. Springer, Japan, pp 95–106. doi:10.1007/978-4-431-54992-5_5

Naoumkina MA, Zhao Q, Gallego-Giraldo L, Dai X, Zhao PX, Dixon RA (2010) Genome-wide analysis of phenylpropanoid defence pathways. Mol Plant Pathol 11:829–846

Nedelkina S, Jupe SC, Blee KA, Schalk M, Werck-Reichhart D, Bolwell GP (1999) Novel characteristics and regulation of a divergent cinnamate 4-hydroxylase (CYP73A15) from French bean: engineering expression in yeast. Plant Mol Biol 39:1079–1090

Nelson DR (2006) Cytochrome P450 nomenclature, 2004. Method Mol Biol 320:1–10

Nelson DR (2009) The cytochrome P450 homepage. Hum Genomics 4:59–65

Nelson DR, Koymans L, Kamataki T, Stegeman JJ, Feyereisen R, Waxman DJ, Waterman MR, Gotoh O, Coon MJ, Estabrook RW, Gunsalus IC, Nebert DW (1996) P450 superfamily: update on new sequences, gene mapping, accession numbers and nomenclature. Pharmacogenetics 6:1–42

Nelson DR, Ming R, Alam M, Schuler MA (2008) Comparison of cytochrome P450 genes from six plant genomes. Trop Plant Biol 1:216–235

Nelson DR, Schuler MA, Paquette SM, Werck-Reichhart D, Bak S (2004) Comparative genomics of rice and *Arabidopsis*. Analysis of 727 cytochrome P450 genes and pseudogenes from a monocot and a dicot. Plant Physiol 135:756–772

Nelson DR, Werck-Reichhart D (2011) A P450-centric view of plant evolution. Plant J 66:194–211

Noel JP, Austin MB, Bomati EK (2005) Structure-function relationships in plant phenylpropanoid biosynthesis. Curr Opin Plant Biol 8:249–253

Nutzmann H-W, Osbourn A (2014) Gene clustering in plant specialized metabolism. Curr Opin Biotechnol 26:91–99

Nutzmann H-W, Osbourn A (2015) Regulation of metabolic gene clusters in *Arabidopsis thaliana*. New Phytol 205:503–510. doi:10.1111/nph.13189

O'Rourke JA, Iniguez LP, Fu F, Bucciarelli B, Miller SS, Jackson SA, McClean PE, Li J, Dai X, Zhao PX, Hernandez G, Vance CP (2014) An RNA-Seq based gene expression atlas of the common bean. BMC Genom 15:866

Osakabe K, Tsao CC, Li L, Popko JL, Umezawa T, Carraway DT, Smeltzer RH, Joshi CP, Chiang VL (1999) Coniferyl aldehyde 5-hydroxylation and methylation direct syringyl lignin biosynthesis in angiosperms. Proc Natl Acad Sci U S A 96:8955–8960

Paquette SM, Jensen K, Bak S (2009) A web-based resource for the *Arabidopsis* P450, cytochromes b5, NADPH-cytochrome P450 reductases, and family 1 glycosyltransferases (http://www.P450.kvl.dk). Phytochemistry 70:1940–1947

Parvathi K, Chen F, Guo D, Blount JW, Dixon RA (2001) Substrate preferences of *O*-methyltransferases in alfalfa suggest new pathways for 3-*O*-methylation of monolignols. Plant J 25:193–202

Paterson AH, Chapman BA, Kissinger JC, Bowers JE, Feltus FA, Estill JC (2006) Many genes and domain families have convergent fates following independent whole-genome duplication events in *Arabidopsis*, *Oryza*, *Saccharomyces* and *Tetraodon*. Trends Genet 22:597–602

Petrussa E, Braidot E, Zancani M, Peresson C, Bertolini A, Patui S, Vianello A (2013) Plant flavonoids-Biosynthesis, transport and involvement in stress responses. Int J Mol Sci 14:14950–14973. doi:10.3390/ijms140714950

Raes J, Rohde A, Christensen JH, Van de Peer Y, Boerjan W (2003) Genome-wide characterization of the lignification toolbox in *Arabidopsis*. Plant Physiol 133:1051–1071

Ralston L, Subramanian S, Matsuno M, Yu O (2005) Partial reconstruction of flavonoid and isoflavonoid biosynthesis in yeast using soybean type I and type II chalcone isomerases. Plant Physiol 137:1375–1388

Ralston L, Yu O (2006) Metabolons involving plant cytochrome P450s. Phytochem Rev 5:459–472

Ramsay NA, Glover BJ (2005) MYB-bHLH-WD40 protein complex and the evolution of cellular diversity. Trends Plant Sci 10:63–70

Reinprecht Y, Yadegari Z, Perry GE, Siddiqua M, Wright LC, McClean PE, Pauls KP (2013) *In silico* comparison of genomic regions containing genes coding for enzymes and transcription factors for the phenylpropanoid pathway in *Phaseolus vulgaris* L. and *Glycine max* L. Merr. Front Plant Sci 4:317. doi:10.3389/fpls.2013.00317

Rupasinghe S, Baudry J, Mary A, Schuler MA (2003) Common active site architecture and binding strategy of four phenylpropanoid P450s from *Arabidopsis thaliana* as revealed by molecular modeling. Protein Eng 16:721–731. doi:10.1093/protein/gzg094

Ryder TB, Hedrick SA, Bell JN, Liang XW, Clouse SD, Lamb CJ (1987) Organization and differential activation of a gene family encoding the plant defense enzyme chalcone synthase in *Phaseolus vulgaris*. Mol Gen Genet 210:219–233

Saito K, Yonekura-Sakakibara K, Nakabayashi R, Higashi Y, Yamazaki M, Tohge T, Fernie AR

(2013) The flavonoid biosynthetic pathway in *Arabidopsis*: structural and genetic diversity. Plant Physiol Biochem 72:21–34

Savolainen O, Langley CH, Lazzaro BP, Freville H (2000) Contrasting patterns of nucleotide polymorphism at the alcohol dehydrogenase locus in the outcrossing *Arabidopsis lyrata* and the selfing *Arabidopsis thaliana*. Mol Biol Evol 17:645–655

Schilmiller AL, Stout J, Weng J-K, Humphreys J, Ruegger MO, Chapple C (2009) Mutations in the *cinnamate 4-hydroxylase* gene impact metabolism, growth and development in *Arabidopsis*. Plant J 60:771–782

Schmutz J, Cannon SB, Schlueter J, Ma J, Mitros T, Nelson W, Hyten DL, Song Q, Thelen JJ, Cheng J, Xu D, Hellsten U, May GD, Yu Y, Sakurai T, Umezawa T, Bhattacharyya MK, Sandhu D, Valliyodan B, Lindquist E, Peto M, Grant D, Shu S, Goodstein D, Barry K, Futrell-Griggs M, Abernathy B, Du J, Tian Z, Zhu L, Gill N, Joshi T, Libault M, Sethuraman A, Zhang XC, Shinozaki K, Nguyen HT, Wing RA, Cregan P, Specht J, Grimwood J, Rokhsar D, Stacey G, Shoemaker RC, Jackson SA (2010) Genome sequence of the palaeopolyploid soybean. Nature 463:178–183

Schmutz J, McClean PE, Mamidi S, Wu GA, Cannon SB, Grimwood J, Jenkins J, Shu S, Song Q, Chavarro C, Torres-Torres M, Geffroy V, Moghaddam SM, Gao D, Abernathy B, Barry K, Blair M, Brick MA, Chovatia M, Gepts P, Goodstein DM, Gonzales M, Hellsten U, Hyten DL, Jia G, Kelly JD, Kudrna D, Lee R, Richard MM, Miklas PN, Osorno JM, Rodrigues J, Thareau V, Urrea CA, Wang M, Yu Y, Zhang M, Wing RA, Cregan PB, Rokhsar DS, Jackson SA (2014) A reference genome for common bean and genome-wide analysis of dual domestications. Nat Genet 46:707–713

Schoch G, Goepfert S, Morant M, Hehn A, Meyer D, Ullmann P, Werck-Reichhart D (2001) CYP98A3 from *Arabidopsis thaliana* is a 3′-hydroxylase of phenolic esters, a missing link in the phenylpropanoid pathway. J Biol Chem 276:36566–36574

Schopfer CR, Ebel J (1998) Identification of elicitor-induced cytochrome P450s of soybean (*Glycine max* L.) using differential display of mRNA. Mol Gen Genet 258:315–322

Schuler MA, Werck-Reichhart D (2003) Functional genomics of P450s. Annu Rev Plant Biol 54:629–667

Schuler MA, Duan H, Bilgin M, Ali S (2006) *Arabidopsis* cytochrome P450s through the looking glass: a window on plant biochemistry. Phytochem Rev 5:205–237

Senda M, Jumonji A, Yumoto S, Ishikawa R, Harada T, Niizeki M, Akada S (2002) Analysis of the duplicated *CHS1* gene related to the suppression of the seed coat pigmentation in yellow soybeans. Theor Appl Genet 104:1086–1091

Severin AJ, Woody JL, Bolon Y-T, Joseph B, Diers BW, Farmer AD, Muehlbauer GJ, Nelson RT, Grant D, Specht JE, Graham MA, Cannon SB, May GD, Vance CP, Shoemaker RC (2010) RNA-Seq atlas of *Glycine max*: a guide to the soybean transcriptome. BMC Plant Biol 10:160

Sewalt VJH, Ni W, Blount JW, Jung HG, Masoud SA, Howles PA, Lamb C, Dixon RA (1997) Reduced lignin content and altered lignin composition in transgenic tobacco down-regulated in expression of l-phenylalanine ammonia-lyase or cinnamate 4-hydroxylase. Plant Physiol 115:41–50

Sezutsu H, Le Goff G, Feyereisen R (2013) Origins of P450 diversity. Philos Trans R Soc B 368:20120428

Shelton D, Stranne M, Mikkelsen L, Pakseresht N, Welham T, Hiraka H, Tabata S, Sato S, Paquette S, Wang TL, Martin C, Bailey P (2012) Transcription factors of lotus: regulation of isoflavonoid biosynthesis requires coordinated changes in transcription factor activity. Plant Physiol 159:531–547

Shi R, Sun Y-H, Li Q, Heber S, Seferoff R, Chiang VL (2010) Towards a system approach for lignin biosynthesis in *Populus trichocarpa*: transcript abundance and specificity of the monolignol biosynthetic genes. Plant Cell Physiol 51:144–163

Shimamura M, Akashi T, Sakurai N, Suzuki H, Saito K, Shibata D, Ayabe S, Aoki T (2007) 2-hydroxyisoflavanone dehydratase is a critical determinant of isoflavone productivity in hairy root cultures of *Lotus japonicus*. Plant Cell Physiol 48:1652–1657

Soderlund C, Bomhoff M, Nelson WM (2011) SyMAP v3.4: a turnkey synteny system with application to plant genomes. Nucleic Acids Res 39:e68. doi:10.1093/nar/gkr123

Solovyev V, Kosarev P, Seledsov I, Vorobyev D (2006) Automatic annotation of eukaryotic genes, pseudogenes and promoters. Genome Biol 7:S10. doi:10.1186/gb-2006-7-s1-s10

Steele CL, Gijzen M, Qutob D, Dixon RA (1999) Molecular characterization of the enzyme catalyzing the aryl migration reaction of isoflavonoid biosynthesis in soybean. Arch Biochem Biophys 367:146–150

Tamura K, Stecher G, Peterson D, Filipski A, Kumar S (2013) MEGA6: molecular evolutionary genetic analysis version 6.0. Mol Biol Evol 30:2725–2829

Tanaka Y (2006) Flower colour and cytochromes P450. Phytochem Rev 5:283–291. doi:10.1007/s11101-006-9003-7

Tanaka Y, Brugliera F (2013) Flower colour and cytochromes P450. Philos Trans R Soc Lond B Biol Sci 368:20120432. doi:10.1098/rstb.2012.0432

Teutsch HG, Hasenfratz M, Lesot A, Stoltz C, Garnier J-M, Jeltsch J-M, Durst F, Werck-Reichhart D (1993) Isolation and sequence of a cDNA encoding the Jerusalem artichoke cinnamate 4-hydroxylase, a major plant cytochrome P450 involved in the general phenylpropanoid pathway. Proc Natl Acad Sci U S A 90:4102–4106

The Arabidopsis Genome Initiative (2000) Analysis of the genome sequence of the flowering plant *Arabidopsis thaliana*. Nature 408:796–815

Tohge T, Watanabe M, Hoefgen R, Fernie AR (2013) The evolution of phenylpropanoid pathway in the green lineage. Crit Rev Biochem Mol Biol 48:123–152

Tohge T, Yonekura-Sakakibara K, Niida R, Watanabe-Takahashi A, Saito K (2007) Phytochemical genomics in *Arabidopsis thaliana*: a case study for functional identification of flavonoid biosynthesis genes. Pure Appl Chem 79:811–823

Tsai C-J, Harding SA, Tschaplinski TJ, Lindroth RL, Yuan Y (2006) Genome-wide analysis of the structural genes regulating defense phenylpropanoid metabolism in *Populus*. New Phytol 172:47–62. doi:10.1111/j.1469-8137.2006.01798.x

Vanholme R, Demedts B, Morreel K, Ralph J, Boerjan W (2010) Lignin biosynthesis and structure. Plant Physiol 153:895–905

Vogt T (2010) Phenylpropanoid biosynthesis. Mol Plant 3:2–20

Wang X (2011) Structure, function, and engineering of enzymes in isoflavonoid biosynthesis. Funct Integr Genomics 11:13–22

Wang Y, Tan X, Paterson AH (2013) Different patterns of gene structure divergence following gene duplication in Arabidopsis. BMC Genom 14:652

Wang Y, Wang X, Paterson AH (2012) Genome and gene duplications and gene expression divergence: a view from plats. Ann N Y Acad Sci 1256:1–14

Weng J-K, Chapple C (2010) The origin and evolution of lignin biosynthesis. New Phytol 187:273–285

Werck-Reichhart D (1995) Cytochromes P450 in phenylpropanoid metabolism. Drug Metabol Drug Interact 12:221–243

Werck-Reichhart D, Feyereisen R (2000) Cytochrome P450: a success story. Genome Biol 1:reviews3003.1–reviews3003.9

Winkel-Shirley B (1999) Evidence for enzyme complexes in the phenylpropanoid and flavonoid pathways. Physiol Plant 107:142–149

Winkel-Shirley B (2001) Flavonoid biosynthesis. A colorful model for genetics, biochemistry, cell biology, and biotechnology. Plant Physiol 126:485–493

Wright SI, Lauga B, Charlesworth D (2003) Subdivision and haplotype structure in natural populations of *Arabidopsis lyrata*. Mol Ecol 12:1247–1263

Xiao CW (2008) Health effects of soy protein and isoflavones in humans. J Nutr 138:1244S–1249S

Xu Z, Zhang D, Hu J, Zhou X, Ye X, Reichel KL, Stewart NR, Syrenne RD, Yang X, Gao P, Shi W, Doeppke C, Sykes RW, Burris JN, Bozell JJ, Cheng Z-M, Hayes DG, Labbe N, Davis M, Stewart CN, Yuan JS (2009) Comparative genome analysis of lignin biosynthesis gene families across the plant kingdom. BMC Bioinform 10:S3

Xu W, Dubos C, Lepiniec L (2015) Transcriptional control of flavonoid biosynthesis by MYB-bHLH-WDR complexes. Trends Plant Sci 20:176–185

Xu W, Grain D, Bobet S, Le Gourrierec J, Thevenin J, Kelemen Z, Lepiniec L, Dubos C (2014) Complexity and robustness of the flavonoid transcriptional regulatory network revealed by comprehensive analyses of MYB-bHLH-WDR complexes and their targets in Arabidopsis seed. New Phytol 202:132–144. doi:10.1111/nph.12620

Yadegari Z (2013) Molecular mapping and characterization of phenylpropanoid pathway genes in common bean (*Phaseolus vulgaris* L.). Ph.D. thesis, University of Guelph, Guelph, Canada

Yi J, Derynck MR, Chen L, Dhaubhadel S (2010a) Differential expression of *CHS7* and *CHS8* genes in soybean. Planta 231:741–753

Yi J, Derynck MR, Li X, Telmer P, Marsolais F, Dhaubhadel S (2010b) A single repeat MYB transcription factor, GmMYB176, regulates *CHS8* gene expression and affects isoflavonoid biosynthesis in soybean. Plant J 62:1019–1034

Yoon J, Choi H, An G (2015) Roles of lignin biosynthesis and regulatory genes in plant development. J Integr Plant Biol 57:902–912

Zhao Q, Dixon RA (2011) Transcriptional networks for lignin biosynthesis: more complex than we thought? Trends Plant Sci 16:227–233

Zhong R, Ye Z-H (2009) Transcriptional regulation of lignin biosynthesis. Plant Signal Behav 4:1028–1034

Zubieta C, Parvathi K, Ferrer J-L, Dixon RA, Noel JP (2002) Structural basis for the modulation of lignin monomer methylation by caffeic acid/5-hydroxyferulic acid3/5-*O*-methyltransferase. Plant Cell 14:1265–1277

Phylogenomics: The Evolution of Common Bean as Seen from the Perspective of All of Its Genes

12

Salvador Capella-Gutiérrez, Anna Vlasova and Toni Gabaldón

Abstract

Phaseolus vulgaris is the most important legume species for human nourishment. However, until very recently genomics resources for this plant have been scarce, which preventing fully understanding the parallel domestications occurred at two geographical regions: Mesoamerica and Andes. The first reference genome for *P. vulgaris*, the Andean landrace G19833, was published in 2014, followed in 2016 by the Mesoamerican reference genome, the breeding line BAT93. These resources have allowed elucidating the evolutionary trajectory of *P. vulgaris* as species, and of both gene pools. First, it has been possible to confirm that the common bean has not undergone a specific whole genome duplication event similarly to the one of *Glycine max* around $\sim$12 million years ago. Second, there is a high degree of concordance between both gene pools in terms of gene content and evolutionary profiles. This includes also the pattern of specialization of gene expression profiles across different relative evolutionary ages. We confirmed the trend observed for the Mesoamerican genome: retained duplicated genes tend to specialize their expression profiles overtime. New analyses using available transcriptomic data gene co-expression networks for both gene pools have been generated and compared for this review in order to look for commonalities and differences. Genes associated to photosynthesis and to response to

S. Capella-Gutiérrez (✉) · A. Vlasova · T. Gabaldón
Bioinformatics and Genomics Programme, Centre for Genomic Regulation (CRG), Dr. Aiguader 88, 08003 Barcelona, Spain
e-mail: scapella@cnio.es

S. Capella-Gutiérrez · A. Vlasova · T. Gabaldón
Universitat Pompeu Fabra (UPF), 08003 Barcelona, Spain

S. Capella-Gutiérrez
Spanish National Bioinformatics Institute (INB), Spanish National Cancer Research Centre (CNIO), Melchor Fernandez Almagro 3, 28029 Madrid, Spain

T. Gabaldón (✉)
Institució Catalana de Recerca I Estudis Avançats (ICREA), Pg. Lluís Companys 23, 08010 Barcelona, Spain
e-mail: tgabaldon@crg.es

M. Pérez de la Vega et al. (eds.), *The Common Bean Genome*,
Compendium of Plant Genomes, https://doi.org/10.1007/978-3-319-63526-2_12

different stresses account for the largest modules of these networks, although some differences were detected which may have roles in the domestication syndrome of both gene pools. However, more sequencing data are needed to a better understanding of common bean genome function and to deepen on the domestication processes of both gene pools. It is expected that third generation sequencing technologies will play an important role in those efforts, leading to better genome assemblies and gene-sets. This will focus further efforts on improving breeding lines while keeping genetic diversity of landraces and wild accessions of *P. vulgaris*.

Keywords
Common bean · BAT93 · G19833 · Legumes · Gene duplication · Phylogenomics · Transcriptome

12.1 Introduction

Legumes are the third largest family of flowering plants, comprising over 670 genera and about 20,000 described species, of which approximately 25 are domesticated (Klitgaard and Bruneau 2003). One of the distinctive features of legumes is their ability to fix atmospheric nitrogen (N_2). Nitrogen fixation is achieved by establishing symbiotic relationships with beneficial bacteria, which are collectively known as rhizobia (Sprent 2001). This process takes place at specific root structures called nodules. Although neither nitrogen fixation nor nodules are exclusive of legumes, this system confers them a clear advantage to outcompete other plants (Doyle 1998; Sprent 2001). Many legume species have been successfully domesticated for human and animal nourishment, including, among many other species, *Phaseolus vulgaris* (common bean), *Cicer arietinum* (chickpea), *Pisum sativum* (pea), and *Lens culinaris* (lentil) (Kislev and Bar-Yosef 1988; Fuller 2007; Cannon et al. 2009). Legumes are also used as nitrogen fixers, providing up to 30% of needed nitrogen for the next crop season (O'Rourke et al. 2014b). One of the main reasons for legume domestication lies in the high protein and nitrogen content of their seeds, which currently make up to 60% of the human dietary requirements in developing countries (Vance 2001). Moreover, up to 40% of worldwide cooking oils come mainly from the legumes *Glycine max* (soybean) and *Arachis hypogaea* (peanut) (Vance 2001). In this review, we focus on common bean, for which genome sequences for two different varieties have recently become available (Schmutz et al. 2014; Vlasova et al. 2016).

Common beans are an important dietary component for more than 500 million people in developing countries (Graham 2003). Originally domesticated in the Americas, common beans are nowadays cultivated in all continents with an annual production reaching 23 million metric tons according to the 2015 report by the Food and Agriculture Organization of the United Nations (FAO). However, the origin of common bean as species and its posterior domestication has been the subject of intense debate (Kami et al. 1995; Gepts 1998; Kwak and Gepts 2009). Current consensus sets the origin of the species in Mesoamerica (Bitocchi et al. 2012) while the domestication is proposed to have occurred at least two times independently, in two different geographical regions and from two differentiated gene pools—one in Mesoamerica and another one in the Andeans (Mamidi et al. 2011, 2013). The origin of the extant *P. vulgaris* group, which includes four of the five domesticated species, *P. vulgaris*, *P. coccineus*, *P. polyanthus,* and *P. acutifolius* (the fifth cultivated species, *P. lunatus*, is part of a separate clade) can be traced back to about 2 million years ago (Mya). While the origin of the new world beans, the so-called *Phaseolus* stem clade, has been estimated between 6 and 8 Mya (Delgado-Salinas et al. 2006). The initial split of the current gene pools has been estimated to occur between 146,000 and 184,000 years ago (Schmutz et al. 2014) followed up by independent bottleneck processes with different degrees of severity which lasted about 40,000 years (Rossi et al. 2009; Bitocchi et al. 2012). Following these bottlenecks, both gene pools expanded to give rise to different landraces. Between 8,500 and 6,200 years ago, two independent and partially overlapping in time domestication processes occurred; giving rise to the two still differentiated extant varieties of common bean (Kaplan et al. 1999; Mamidi et al. 2011, 2013). While the temporal estimation of the extant gene pools split has been estimated using genomics data (Schmutz et al. 2014), the timing of both domestication processes was initially estimated using only limited loci (Mamidi et al. 2011, 2013) and archeological records (Kaplan et al. 1999). This estimation was later confirmed by genomics studies (Schmutz et al. 2014).

The constant improvement of sequencing technologies over the last two decades has provided an unprecedented availability of genomics data for almost any domain of life [http://www.genome.gov/sequencingcosts/]. Fast and affordable technologies have made it possible to fully sequence a species representative in order to get an initial reference genome. Then, RNA from different tissues and/or developmental stages are re-sequenced using alternative technologies to improve the initial reference data leading to a better understanding of the species (Bentley 2006). It is also becoming a standard approach to sequence RNA under different stress conditions in order to gain a better understanding about how organisms react to such disturbances. Finally, re-sequencing of different individuals representing different genetic backgrounds and/or populations can shed light on the recent evolution of a species at unprecedented levels of resolution and accuracy. Altogether, genomics data open the door to a multitude of studies such as the variation between and within species including the effects of species domestication, the evolution of organs and pathways, and the epigenetic modifications, among others (O'Rourke et al. 2014b). However, this genomics explosion has not fully reached the flowering plants (Fig. 12.1). The gigantic size of some plant genomes together with the high content of repetitive sequence elements has severely limited the number of fully sequenced species and individuals in plants as compared to other groups (Michael and Scott 2013). Despite vast efforts to produce high-quality reference plant genomes, including the use of different sequencing technologies, many of these genomes are still highly fragmented with *Arabidopsis thaliana* as one of the few species with a complete genome (Schneeberger et al. 2011).

Legumes are not an exception to this trend with only nine crop species with fully sequenced genomes and released for public use (Table 12.1). In some cases, wild relatives to these crops species have also been sequenced to provide a better understanding of the domestication process (Kim et al. 2010; Varshney et al. 2013a; Branca et al. 2011). Using these data, it has been possible to confirm an ancient Whole Genome Duplication event (WGD) shared by all extant legumes from the *Papilionoideae* subfamily which occurred approximately 56 Mya (Lavin et al. 2005). Furthermore, it has also been possible to detect a more recent WGD event occurring approximately 10 Mya in the lineage leading to *Glycine max* (Schmutz et al. 2010), assess the loss of genetic diversity among domesticated crops as compared to their wild relatives (Valliyodan et al. 2016; Zhou et al. 2015), as well as associate phenotypic traits with specific loci, making them potential candidates to drive further improvements for these crops (Varshney et al. 2013b; Li et al. 2016b; Zhang et al. 2016). However, much more data are needed to understand the complex history of these species. In the case of common bean,

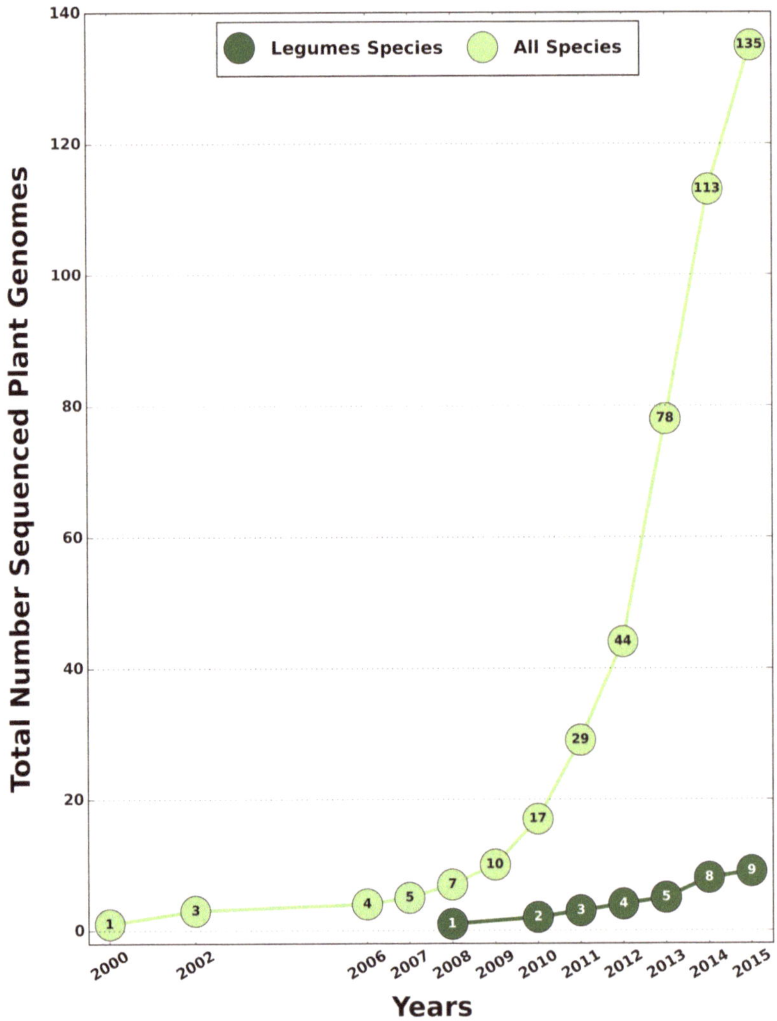

Fig. 12.1 Number of fully sequenced plant genomes. Cumulative number of plant genomes sequenced since 2000 (light green line) and cumulative number of legume genomes (dark green line). Number of sequenced plant genomes is approximately doubling each year

despite the recent publication of reference genomes for the two main domesticated varieties (Schmutz et al. 2014; Vlasova et al. 2016), genomics data from the wild relatives of both gene pools are still needed to disentangle the genotypic and phenotypic changes derived from the split and isolation of these gene pools from the parallel domestication events. Commonalities between both processes will, therefore, allow identifying common genetic signatures and making them perfect candidates for improving current crops.

12.2 Resources for Legume Genomics

In order to achieve a better understanding of common bean evolution, it is necessary to take into account the genomic context of the species. Comparison with other legumes species will allow us to understand which traits were already present in the common ancestor of legumes, and which ones are specific to each lineage and/or species. Currently, there are nine reference

Table 12.1 Overview of the reference legume genomes

Species	Chromosomes (2n)	Sequenced Genome Size (Mbp)	Number scaffolds	Number genes	GC (%)	% Repeats	Ref
Lotus japonicus v.3.0	12	394	132	39,734	36.64	34.28	Sato et al. (2008)
Cicer arietinum v.1.0	16	738	7,163	28,269	31.04	NA*	Varshney et al. (2013a)
Medicago truncatula v.4.0	16	412	~ 1,000	43,205	33.08	32.6	Young et al. (2011)
Cajanus cajan v.1.0	22	605.78	137,542	40,071	32.82	51.67	Varshney et al. (2012)
Glycine soja v.1.0	40	813–985	NA	56,655–62,048	NA	NA	Li et al. (2014)
Glycine max v.2.0	40	978.5	1190	56,044	34.76	57	Schmutz et al. (2010)
Trifolium pratense v.2.1	14	309	39,904	40,868	32.31	41.82	De Vega et al. (2015)
Vigna angularis v.1.0	22	450	3883	34,183	34.13	43.1	Yang et al. (2015)
Vigna radiata v.1.0	22	431	2748	22,427	33.16	50.1	Kang et al. (2014)
Phaseolus vulgaris G19833 *v.1.0*	22	472.5	708	27,197	34.96	45.42	Schmutz et al. (2014)
Phaseolus vulgaris BAT93 *v.1.0*	22	549.6	9,047	30,491	34.86	35	Vlasova et al. (2016)

GC content was calculated on the corresponding sequence as proportion of GC nucleotides over the entire length, excluding 'N' characters. NA: Not analyzed

sequenced genomes for legume species, with some of them with more than one accession and/or wild relatives, e.g., *Glycine soja,* the wild relative for soybean [http://legumeinfo.org/genomes] (Dash et al. 2016). Moreover, at the time of writing this chapter, there are at least twelve more genomics projects at different stages of completion, as listed in the Genomes Online database [https://gold.jgi.doe.gov] (Reddy et al. 2015). Despite the low number of fully sequenced legume species and the slow pace of data generation, it is impossible to have a stable vision of available resources as data from re-sequencing projects of different accessions and/or conditions are constantly becoming available. Table 12.1 offers an updated version of data gathered by O'Rourke and colleagues in 2014 (O'Rourke et al. 2014b).

All this data are available through resources with different objectives and scopes. For instance, the Legume Information System (LIS) [http://legumeinfo.org/] (Dash et al. 2016) is the web server for the Legume Federation Project. During more than 15 years, LIS has integrated data from crop and model legume species available at specialized sites. On this web server is possible to localize orthologous and paralogous sequences, navigate synteny maps among all sequenced legumes species, look for specific biomarkers and query for quantitative trait loci (QTL) data. It also provides information for other legumes species for which there is no available genome data. LegumeIP is another legume-specific web server [http://plantgrn.noble.org/LegumeIP] (Li et al. 2016a) which provides a platform for comparative genomics studies across legumes. It includes information for six out of the nine fully sequenced legumes species with the aim to study their evolution and assign biological function to legume genes. Importantly, it includes two outgroup species, e.g., *Arabidopsis thaliana* and *Populus trichocarpa*, which allows to differentiate common traits in legumes from other plants.

Generalist web servers such as Phytozome [https://phytozome.jgi.doe.gov] (Goodstein et al. 2012), Plaza 3.0 [http://bioinformatics.psb.ugent.be/plaza/] (Proost et al. 2014), and Ensembl Plants [http://plants.ensembl.org/] (Kersey et al. 2016) host genomics resources associated to leguminous and non-leguminous species. The general focus of these resources on plant genomics offers an unique opportunity to perform comparative genomics studies to understand which traits are specific to a given species or/and group of species, i.e., leguminous species, and which ones are more general. Data hosted at these web servers complement and extend the ones stored at specific resources such as LIS and LegumeIP. In addition, they allow comparing legume genome characteristic with those of other plant species, allowing identification of unique and shared features.

PhylomeDB [http://phylomedb.org] (Huerta-Cepas et al. 2014) is the largest public repository of phylomes, i.e., the complete collection of single-genes trees from a species in a given evolutionary context. There are more than 5,500,000 alignments and phylogenetic trees at this database with more than 250,000 of them associated to bean proteins either from the fully sequenced Mesoamerican accession BAT93 (Vlasova et al. 2016) or the Andean accession G19833 (Schmutz et al. 2014). Bean phylomes are focused mainly on legumes species with up to eight different species studied. For a better understanding about which genetic traits are legume-specific and which ones are more widespread, outgroup species such as *A. thaliana* and *S. lycopersicum* have been included. The main difference between PhylomeDB and other resources is that PhylomeDB is a gene-centric repository while others are family-centric. In a gene-centric approach, a phylogenetic tree is reconstructed for each protein-coding gene (PCGs) present in a genome in the evolutionary context set by other species of interest. In the case of family-centric approach, sequences are first clustered according to certain strategies, i.e., Best-Bidirectional Blast Hits (BBHs), and then evolutionary relationships are derived using phylogenetic trees and/or other sequences clustering methods.

12.3 Common Bean Genomics

As soon as sequencing became affordable, and despite the complexity of generating a reference genome for a plant, the complete genome for common bean was generated by the scientific community in 2014. By the time the first reference genome was released, over 15 mapping populations and few linkage maps to study the genealogy and genetic diversity of common bean were available, including the core map of BAT93 x Jalo EEP558 crossing (Nodari et al. 1993; Gepts et al. 2008). In 2005, Ramirez and colleagues sequenced and analyzed ~21,000 expressed sequence tag (ESTs) derived mainly from the Mesoamerican accession Negro Jamapa, five different cDNA libraries, and complemented with data from the Andean accession G19033, and one cDNA library from leaf samples (Ramírez et al. 2005). Using this dataset, the authors identified about 8,000 unique genes and this collection became the reference dataset for subsequent bean functional genomics studies. Moreover, different genomic libraries, mainly Bacterial Artificial Chromosomes (BACs), were constructed for a number of bean varieties, including landraces and wild lines (Vanhouten and MacKenzie 1999; Yu 2000; Yu et al. 2006; Kami et al. 2006; Grisi et al. 2007; Gepts et al. 2008; Blair et al. 2014). In fact, Schlueter and colleagues published in 2008 the first genome-wide physical map of the common bean based on BAC-end sequencing (Schlueter et al. 2008). BAC libraries were derived from the G19833 Andean common bean landrace and completed with sequences from the cultivar line BAT7. Authors were able to complete around 9.5% of the genome, which was sufficient to provide an overview of the overall genome organization, to estimate the proportion of repetitive and genic regions, and to perform some functional analyses based on transferred Gene Ontology (GO) functional terms. It is important to highlight that many of the bean accessions used in these initial analyses were later selected for full genome sequencing. In this way, it has been possible to confirm and further explore previous findings at genome-wide scales.

In 2014, an international team led by the Joint Genome Institute in US published the first reference genome for common bean. Scientists sequenced the Andean landrace G19833 (Schmutz et al. 2014). Additionally, four different gene pools were sequenced: 60 wild and 100 domesticated accessions from Mesoamerican and Andean gene pools. Using this data, scientists could confirm previous findings based on just few loci (Bitocchi et al. 2013) that suggested parallel domestication processes of the common bean in two geographically isolated regions: Mesoamerica and the Andean. According to their results, wild gene pools diverged approximately ~165,000 years ago with an asymmetric gene flow from the Mesoamerican wild relatives toward the Andean ones (Papa and Gepts 2003; Papa et al. 2005; Blair et al. 2012). While the wild Andean population experienced a severe bottleneck, which lasted for ~76,000 years, there is no evidence of a similar effect for the wild Mesoamerican population. Then, both wild populations, in particular the Andean one, have expanded and diversified until nowadays (Blair et al. 2012). Approximately 10,000 years ago, parallel domestication processes started in each of these two diverged populations, resulting in what we know today as cultivated populations and landraces (Gepts 1998; Chacón et al. 2005). As it is the case in other crops, these domestication processes have been accompanied with profound morphological changes together with a significant genetic diversity reduction.

To detect genomic regions and genes potentially associated to the domestication syndrome, authors compared four different gene pools: wild and domesticated Mesoamerican, and wild and domesticated Andean accessions (Schmutz et al. 2014). Authors compared accessions from the same geographical origin to detect regions with decreased genetic diversity, which may be the result of past selective sweeps. For this, the ratio of the observed genetic diversity across gene pools to the expected population differentiation patterns was assessed in genomic windows. This analysis identified 930 genomic windows accounting for ~74 Mb in the Mesoamerican populations and 750 windows, accounting for roughly 60 Mb, for Andean populations. Furthermore, using the same genetic pools and principles, they identified 1,835 and 745 genes in the Mesoamerican and Andean pools that may have been involved in the domestication process, as suggested by their patterns of low genetic diversity and high differentiation. In fact, when looking at the functional annotation, these genes were enriched in functions related to seed and leaf size. Remarkably, only 10% of the genomic windows and candidate genes are shared by the two gene pools. This highlights that it is possible to develop similar domestication-related phenotypes following different genotypic changes.

In 2016, an international team led by Iberoamerican researchers published the genome sequence the Mesoamerican accession BAT93, a breeding line, which became the Mesoamerican common bean reference genome and provided a counterpart for the Andean accession (Vlasova et al. 2016). While authors working on the Andean genome focused on an extensive comparison with Soybean, the teams working on the Mesoamerican genome included comparisons with several leguminous and non-leguminous plant species, which enable a first global understanding of the evolutionary trajectories of common bean genes. It was confirmed that, contrary to Soybean (Schmutz et al. 2010), none of the common bean gene pools have gone through a specific WGD. Conversely, these comparisons confirmed the existence of an ancient WGD shared by all legume species about 56 Mya (Lavin et al. 2005).

Transcriptomics data from various tissues and developmental stages, including plants under biotic and abiotic stress conditions, were taken into account for the analysis of the

Mesoamerican accession, which allowed authors to better understand different physiological phenomena. First, it allows to make accurate annotation of different genomic elements such as protein-coding genes with multiples isoforms, and long and small non-coding RNAs. Second, it was possible to understand the transcriptional changes along the plant developmental stages, which revealed more changes during the vegetative stage as compared to the reproductive one. Clustering based on the expression patterns allowed to recapitulate tissue types with a clear distinction between the root and aerial samples for PCGs, or between reproductive (seeds, pods) and the rest of the tissues for long non-coding RNAs (lncRNAs). Similar to results observed in animals, PCGs have higher levels of sequence conservation than lncRNAs across multiple plant species and higher expression levels. Only few lncRNAs were conserved in all plant genomes used for the analysis, and they had higher level of expression compare to the whole set of lncRNAs.

When studying in more detail the evolutionary trajectories of PCGs together with their expression profiles, authors uncovered interesting patterns. First, they found out that older genes and genes without paralogs, i.e., single-copy genes, tend to have a more widespread expression across a larger number of tissues and developmental stages, and a larger number of co-expression partners, as compared to younger and duplicated genes, which, in turn, tend to have more narrow and specialized expression patterns. In fact, genes associated to old duplication events are less correlated and tend to have a more complementary expression profiles than those ones associated to more recent events. Second, the number of paralogs, independently of their relative duplication age, is an important factor driving gene expression specificity, as gene expression tends to increase with the number of detected paralogs until a certain plateau is reached. Third, there are a higher number of genes identified as legume-specific among those associated to root development and flowering activities, as compared to those associated to the development of other plant organs. This is consistent with major innovations in these two tissues, including the symbiosis with rhizobia, and suggests an involvement of gene duplication in providing the substrate for the origin of novel functions in this clade.

When comparing both sequencing efforts from the technical point of view (Table 12.2), both genomes were sequenced using a combination of 454-Roche, Illumina, and Sanger platforms. The Andean accession (G19833) is 521.1 Mb while its Mesoamerican counterpart (BAT93) is 549.6 Mb long which is closer to the previously estimated genome size (587 Mb) (Arumuganathan and Earle 1991). Despite the highly fragmented assembly for the Mesoamerican accession, 21,600 BAT93 PCGs were mapped to 21,604 PCGs in the G19833 genome which ensures a high degree of concordance in terms of the gene space between both resources. Direct comparison of these two assemblies with the aim of study structural variations and rearrangements is very difficult due to the assembly fragmentation. Therefore, it will be only possible when newly improved assemblies will be available.

A significant milestone was achieved with the publication of the Mesoamerican reference genome for common bean. However, more sequencing efforts are needed to understand in greater detail the biology of this important crop. On the one hand, to fully identify the genetic bases of the two parallel domestication processes, reference sequences for the wild relatives of the Mesoamerican and Andean gene pools are needed. These efforts will allow researchers to disentangle the genomics landscape of the common ancestor from each center of domestication. It will contribute to identify genes associated to the observed phenotypic changes and establish better breeding strategies and conservation policies. On the other hand, sequencing wild relatives to common bean will help to elucidate which genotypic and phenotypic features of species at this specific *Phaseolus* clade made them prone to domestication. In fact, common bean belongs to one of the two main clades that include four of the five known domesticated species (Delgado-Salinas et al. 2006). Finally, a third generation of massive sequencing technologies, i.e., Pacific

Table 12.2 Genome assembly comparison of both common bean reference genomes

Reference	Schmutz et al. (2014)	Vlasova et al. (2016)
Accession	Andean landrace, G19833	Mesoamerican breeding line, BAT93
Data-type	Genomics	Genomics and transcriptomics atlas
Assembly size	473 Mb	549.6 Mb
Scaffolds total	708	9,047
Contigs total	41,391	59,332
Assembly N50	5 Mb	355 kb
%Ns	9.3	34.96
Chromosomally anchored size (%)	468.2 Mb (98%)	450.8 Mb (81%)
Repeat elements	214.6 Mb (45.4%)	177.2 Mb (32.2%)
Number of protein-coding genes	27,197	30,491
Number of transcripts	31,688	66,634
Number of small RNAs	–	2,523
Number of long non-coding RNAs	–	1,033
Number of resistance-associated genes	376 genes belonging to the NBS-LRR class	852 genes with 234 ones belonging to the NBS-LRR class
Other	Annotated centromere regions	Functional annotation of protein-coding genes

Biosciences, and Oxford Nanopore sequencers, promises to bring sequencing efforts to an even higher level. Long sequencing reads will allow a better characterization of highly repetitive genomic regions, and, potentially, assemble these sequences into full chromosomes. Counting with better assemblies will contribute to a better definition of common bean gene content, i.e., there is roughly a difference of 2,500 genes between Andean and Mesoamerican reference genomes, and therefore, producing refined results, which will confirm or not previous findings. This technology will center efforts on understanding the common bean biology rather than in the technical differences across studies.

12.4 Common Bean Transcriptomics Resources at Public Repositories

The reference genomes for the common bean have been complemented with the publication of different transcriptomes (Table 12.3). Transcriptomes, which comprise quantitative catalogs of expressed transcripts in a given tissue (or set of tissues) and physiological conditions, have broad applications ranging from the identification of protein-coding genes, exons, introns, alternative isoforms, and detection of long and small RNAs, for the development of molecular markers with different purposes. In addition, they are instrumental for providing a first approximation of how tissues and plants respond to different stimuli, including stress and infection. This coupled to the affordable prices of the technology makes transcriptomes the ideal approach to understand the plant gene regulation to different environmental conditions and/or developmental stages. Reference genome sequences are needed to better understand transcriptome results. When this is not possible, results should be taken with caution because the absence of genes is not an indication of gene losses. In the particular case of common bean, it is also important to consider which gene set is used as reference because using different references may produce slightly biased results.

Table 12.3 General overview of the transcriptome studies already published for common bean

Accessions	Tissues	Technology	Scope	Ref
Negro Jamapa 81, G19833	Roots, Nodules, Pods, Leaves	Sanger Sequencing	Transcriptome dataset (ESTs)	Ramírez et al. (2005)
Sierra, BAT93	Leaves, Flowers, Roots, Pods	454 Genome Sequencer-FLX	Transcriptome dataset (ESTs)	Kalavacharla et al. (2011)
G19833, DOR364	Root, Leaf	BigDye (Applied Biosystems), di-deoxy-based Sanger sequencing	Gene-based markers collection (ESTs)	Blair et al. (2011c)
BAT 477, G19833	Roots, Seeds, Leaves, Stem, Shoots, Flowers, Small pods	ABI 3730 × from Applied Biosystems	Drought stress ESTs	Blair et al. (2011b)
DOR364	Roots	Amersham Hybond N + nylon membranes	Phosphorus adaptation (ESTs)	Blair et al. (2011a)
SARC1, SMARC1 N-PN1	Developing seeds	CustomArray 90 K	Developing seeds analysis	Liao et al. (2012)
Negro Jamapa 81	Roots	CombiMatrix microarray platform, custom array 90 K array	Transcriptional factors (TF) characterization under stress conditions	Aparicio-Fabre et al. (2013)
BAT93	Seeds	454 Genome Sequencer-FLX	Profile developing seeds transcriptome	Liao et al. (2013)
Ispir	Roots, leaves	Illumina HiSeq 2000	Salt tolerance study	Hiz et al. (2014)
Long 22-0579, Naihua	Leaves	Illumina Hiseq 2500	Responses to drought stress	Wu et al. (2014)
Negro Jamapa	Leaves, Stem, Flower, Pods, Seeds, Roots, Nodules	Illumina HiSeq 1000	RNA-seq-based expression atlas	O'Rourke et al. (2014a)
21 samples: Mesoamerican (10 wild and 8 domesticated); Andean (1 wild and 2 domesticated)	Leaves	Illumina HiSeq 1000	Comparison of the domesticated and wild accessions	Bellucci et al. (2014)
Voyager, Albion	Developing pods	Illumina Genome Analyzer II (GA II)	Differences in Zinc (Zn) content	Astudillo-Reyes et al. (2015)
NAG12	Roots	Illumina HiSeq 2500	Interaction with *Rhizobium etli*	Dalla Via et al. (2015)

(continued)

Table 12.3 (continued)

Accessions	Tissues	Technology	Scope	Ref
PI533561 GTS-900	Roots	Ion Torrent PGM	Soybean cyst nematode resistance	Jain et al. (2016)
Sierra	Leaves	Illumina HiSeq 2500	Stress response to rust caused by *Uromyces appendiculatus*	Ayyappan et al. (2015)
Stringless green refugee	Leaves	Illumina HiSeq 2000	Time-course experiment of Bean common mosaic virus (BCMV) infection	Martin et al. (2016)
IPR Saracura	Roots	Illumina MiSeq	Resistance response to the root-knot nematode *Meloidogyne incognita*	Santini et al. (2016)
BAT93	Leaves, Roots, Pods, Seeds, Stem, Axial meristem, Flowers, Flower buds	Illumina HiSeq 2000	RNA-seq-based expression atlas	Vlasova et al. 2016)

Many of these studies are focused on the plant response to specific stresses using different technologies such as RNA-Seq and microarrays

Vlasova et al. (2016) were the first ones to publish the genomic sequence of a Mesoamerican accession for common bean and then generated an extensive transcriptomics atlas from the same biological sample (Vlasova et al. 2016). The atlas comprises 27 samples covering eight organs at different developmental stages allowing to identify changes in expression of genes in a temporal (across developmental stages), and spatial (in different plant tissues) manner. Moreover, this data was central to understand how bean genes tend to specialize their expression patterns overtime depending on the number of retained paralogs, and their role as central hubs in co-expression networks.

A similar effort was done by O'Rourke and colleagues taking as reference the Andean genome for the common bean. However, in this case, the sequencing was done using the Mesoamerican accession Negro Jamapa (O'Rourke et al. 2014a). This atlas is composed of 24 samples covering seven tissues and developmental stages. Authors focused on how different nitrogen sources impact the gene expression patterns of root nodules, and how that influences seed formation overtime.

Bellucci et al. (2014) published another large-scale transcriptomics study aiming to understand the differences between wild and domesticated Mesoamerican accessions using samples obtained from leaves. Rather than mapping transcripts to a reference genome, authors generated a de novo reference transcriptome, using for that purpose the results of four highly divergent accessions sequenced at much higher read coverage. Following this strategy, authors expected to capture all gene expression variability across Mesoamerican accessions. Authors found a profound loss of genetic diversity in domesticated accessions as compared to wild ones. This loss of genetic diversity deeply affected the gene expression patterns with most of the changes leading to down-regulation of transcripts. These changes may be associated to loss-of-function mutations as previously described in other species (Olson 1999).

In contrast to these large-scale efforts, many of the recent transcriptomics analyses have focused on gene expression changes under a number of stress conditions. One such study has characterized which transcripts change their expression between normal and drought conditions (Wu et al. 2014). Identifying genes associated to a higher drought-tolerance is crucial given the current unpredictable weather conditions in a context of climate change. Another study has characterized gene expression changes upon exposure to different salinity levels (Hiz et al. 2014; Wu et al. 2014). Salt is one of the main abiotic stress factors which limit crop yields, and, therefore, identifying which genes can confer a higher tolerance to high salt concentrations can be key for breeding programs aiming to introduce the crop in new environments. Finally, transcriptomics studies can be also used to understand differences among close relative accessions from the same gene pools. For instance, Astudillo-Reyes et al. (2015) studied the differences in gene expression patterns of two Mesoamerican accessions with different Zn content in seeds. Beans are an important source of dietary Zn for humans and understanding how to maximize the seed content of this nutrient is important for developing countries where access to non-plant Zn sources is limited.

12.5 Large-Scale Phylogenomics Analyses of *P. Vulgaris*

To deepen our understanding of common bean genome evolution, we reconstructed two additional phylomes, i.e., the complete collection of common bean gene phylogenetic trees, using PCGs derived from both reference genomes. These two collections of gene trees complement the ones generated by Vlasova et al (2016). In this case, we zoomed into the evolution of common bean genes in the context of an expanded dataset of fully-sequenced legumes. We included in our analyses all legumes species sequenced so far with the sole exception of *Trifolium pratense* (Table 12.1). Specifically, we have reconstructed 28,124 trees for the BAT93 accession which represents a phylogenetic tree for 92.5% of the predicted PCGs. In the case of G19833, 26,377 trees were obtained which covers 97.2% of the predicted PCGs. We used the previously described pipeline to generate these single-gene trees (Vlasova et al. 2016). In fact, when considering previous efforts, there are 136,866 common bean phylogenetic trees stored in PhylomeDB [http://phylomedb.org] (Huerta-Cepas et al. 2014).

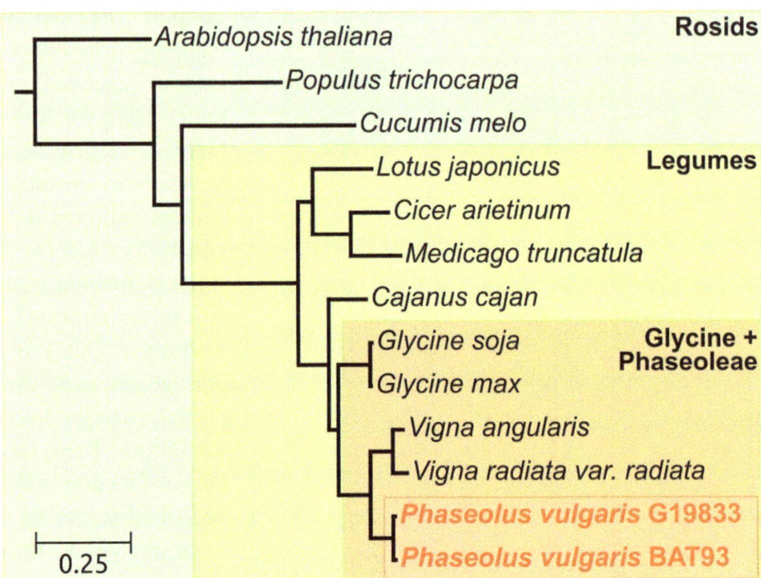

Fig. 12.2 Common bean species phylogeny. Maximum-likelihood tree reconstructed using a concatenation of 271 sets of widespread one-to-one orthologous genes. This tree offers an evolutionary context centered in legume species to study the evolution of common bean

To provide an evolutionary framework for downstream analyses, we reconstructed a species phylogeny using two complementary approaches: (i) a super-matrix reconstruction analysis of 271 sets of widespread one-to-one orthologous genes present in at least 10 out of the 13 species used in this work; and (ii) a super-tree reconstruction using 54,501 single-gene trees from the two already mentioned phylomes. Both analyses yielded an identical topology (Fig. 12.2), which was used to define five relative ages as the lineages predating the divergence of different species and/or group of them: (1) *Phaseolus*-specific including specific events for each of the studied accessions; (2) basal to *Phaseolus*; (3) basal to legume species which groups all Phaseoleae and *Glycine* species; (4) basal to all Rosids species used in this study. It is important to mention that some relative ages, e.g., basal to Phaseoleae, were collapsed into a more ancestral one given the low number of detected events.

We scanned all gene trees to detect and date duplication events following the already defined ages, delineate orthology and paralogy relationships, and assign a relative age to each PCG. Then, we used this data to perform an extensive comparison between both common bean accessions to measure how similar results are when using each accession (Fig. 12.3).

First, we compare the positional conservation, called synteny, for the 20,617 BAT93 PCGs uniquely mapping to 20,618 PCGs in the Andean genome. We used mainly one to one orthologs between both accessions to establish the correspondence between each loci in the two genomes. This was completed by homologous genes located at the same syntenic blocks. Second, we projected the number of paralogs detected for each gene placed at the linkage groups (Mesoamerican accession) and pseudo-chromosomes (Andean accession). When looking at the image, it seems that both gene-sets follow the same distribution for the number of paralogs. However, there are statistically significant differences between both distributions (p-value = 1.3873e−132 for the Kolmogorov-Smirnov test on two samples) with a clear tendency of the Mesoamerican PCGs for having just one paralog (Fig. 12.4a). Third, we analyzed the distribution of relative ages for both gene-sets. As shown in Fig. 12.4b, there is an enrichment of *Phaseolus*-specific genes for the Mesoamerican accession (BAT93) compared with the Andean one. This may be caused by two independent factors: (1) the highly fragmented assembly for BAT93 might cause over-predicting genes, since genes may fall into more than one contig. (2) G19833 *ab initio* gene-set was filtered to keep only expressed genes and/or with homologs in other species; therefore, it is possible than some of the youngest genes were filtered out. Finally, when looking at the gene-density distribution for both reference genomes, it could be appreciated the following patterns: (1) centromere regions tend to be depleted of PCGs as expected given the low recombination rate and the highly repetitive sequences present there. (2) Conversely, pseudo-chromosomes (G19833) and linkage groups (BAT93) tend to accumulate most of the genes toward their ends. (3) It is possible to observe an irregular gene-density distribution for the Mesoamerican PCGs possibly due to the highly fragmented assembly of the genome. Future efforts using long-reads can help to confirm this pattern and/or to improve the gene calls for this genome.

We combined newly generated phylogenetic information with RNA-seq data to study the temporal and spatial patterns of expression for duplicated genes across different evolutionary periods. Gene expression matrices, raw read counts and normalized reads per kilobase per million mapped reads (RPKMs), for accessions BAT93 and Negro Jamapa were obtained from previously published studies (O'Rourke et al. 2014a; Vlasova et al. 2016). For downstream analyses, we only considered genes with at least 1 RPKM in at least one tissue. Applying this threshold, 21,472 and 25,260 genes were selected for the BAT93 and G19833/Negro Jamapa accessions, respectively. Negro Jamapa is a Mesoamerican accession which was used by O'Rourke et al. (2014a) to generate the transcriptomic atlas of the Andean reference genome. Although both accessions belong to the same species, given the evolutionary history of these

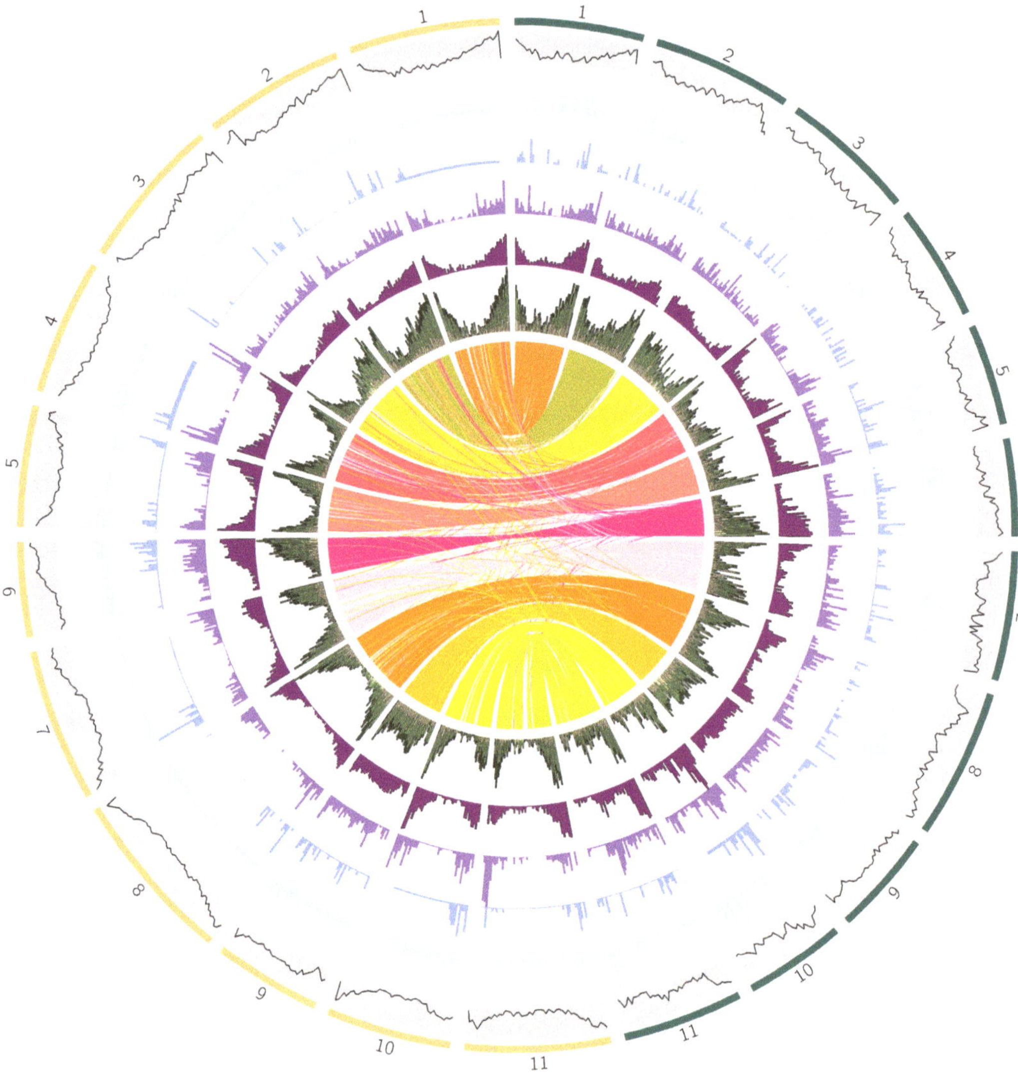

Fig. 12.3 Extensive comparison between the Andean (left) and Mesoamerican (right) reference gene-sets for common bean. The first inner track shows the gene correspondence between genes placed at linkage groups (BAT93) and pseudo-chromosomes (G19833). The second inner track shows the detected number of paralogs for each gene. There are three possible colors for the histogram: light green for one paralog, green for two to five paralogs; and dark green when more than five paralogs have been detected. Following inner tracks represent the number of genes of a given age: basal to rosids (third), basal to legumes (forth), basal to *Glycine* and Phaseoleae (fifth), and *Phaseolus*-specific (sixth). In this case, basal to *Phaseolus vulgaris* was collapsed into a more ancestral age given the few detected cases. Seven inner tracks represent the gene density across each linkage group/pseudo-chromosome, and the outer track represents pseudo-chromosome (Andean) and linkage group (Mesoamerican) ideograms for both accessions

two geographically distinct gene pools, this approximation might introduce technical and/or biological noise in downstream analyses. As it was previously shown with 18 *A. thaliana* natural accessions, about one-third of PCGs have non-synonymous mutations in at least one accession. This implies that reference gene annotation should be used for all cases with

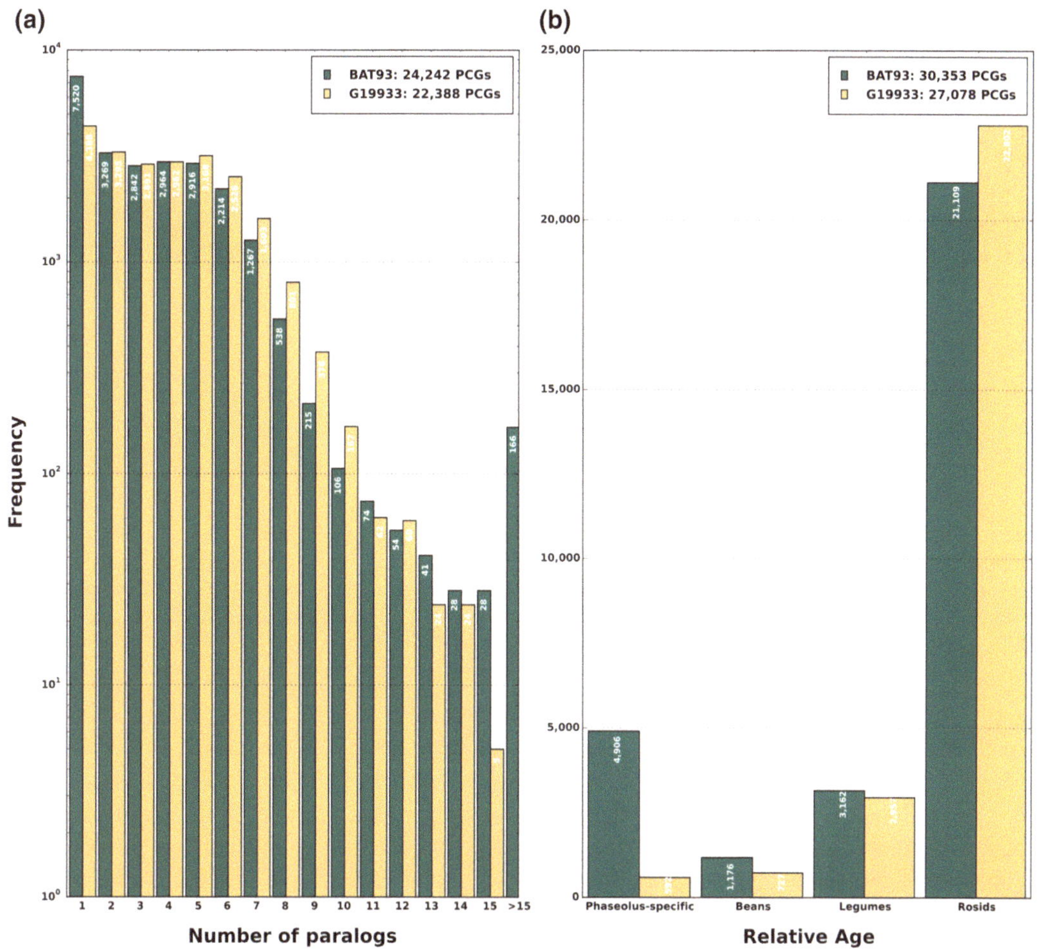

Fig. 12.4 Distributions for number of paralogs and relative ages for both reference gene-sets. **a** Both gene-sets share the same pattern regarding the number of paralogs counts. However, both distribution are statistically significant different as the Mesoamerican (BAT93) gene-set tends to accumulate many PCGs with just one paralog. **b** If the youngest relative age is not considered, both gene-set shown a similar pattern. However, there is enrichment for *Phaseolus*-specific genes in the Mesoamerican (BAT93) PCG set

caution, and there is a real need to predict de novo reference gene-set for assemblies coming from highly variable accessions (Gan et al. 2011). For the RNA-seq data, it is possible that reads from accessions different of the reference one could not be aligned correctly, which leads to wrong estimates of particular gene expression levels. In addition to difficulties in measuring true gene expression levels between accessions, there are many technical factors, i.e., different software and normalization methods (Wagner et al. 2012; Engström et al. 2013; SEQC/MAQC-III Consortium 2014; Teng et al. 2016) which can introduce non-biological differences. RNA-seq analyses often use RPKM measurement (Mortazavi et al. 2008) that is a normalized value which takes into account gene lengths and total number of reads generated in the library. However, as shown by Teng et al. (2016), the total number of transcripts in the library depends on the transcripts size distribution in the sample. It may happen that few long transcripts capture the majority of reads hence altering the overall RPKM distribution. Another important factor is the length of the reads because it may also influence gene expression

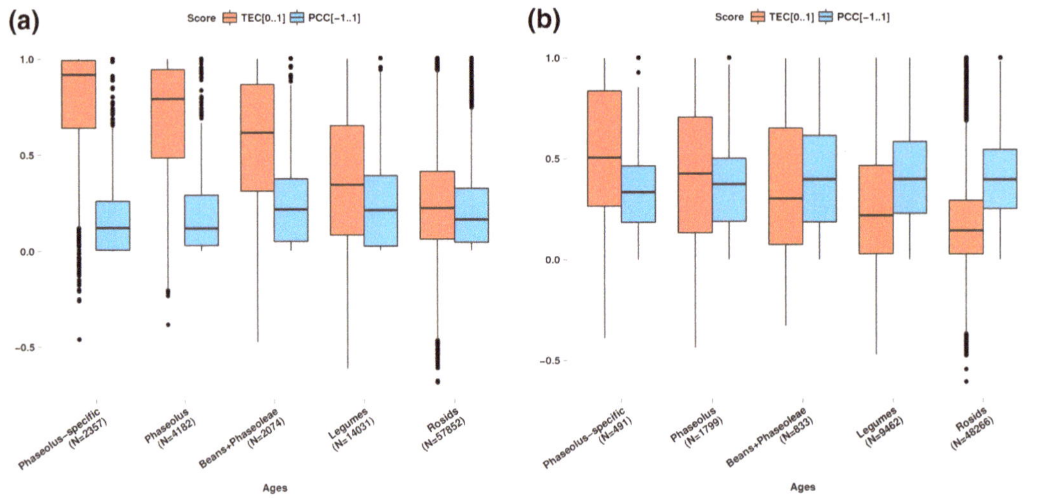

Fig. 12.5 Gene expression complementary measures. Andean and Mesoamerican gene pool results are shown in panels **a** and **b**, respectively. The boxplots correspond to the distributions of Pearson correlation coefficient (PCC, orange) and tissue expression complementarity score (TEC, blue) in the group of proteins associated to particular duplication events. Average PCC and TEC scores were computed for proteins mapping to duplication events occurring at five relative ages—*Phaseolus*-specific, *Phaseolus*, Beans-Phaseoleae, Legumes, and Rosids. Only PCGs with at least one RPKM in at least one sample were considered for these analyses

quantifications. The shorter the fragments, the larger the number of multiple mappings which may lead to counting multiple times the same fragment (Chhangawala et al. 2015).

As previously reported, Vlasova et al (2016) measured the level of tissue expression complementarity when duplications for the BAT93 PCGs were considered. We repeated the measurements for both accessions using a legume-centered evolutionary framework. As seen in Fig. 12.5, average Pearson correlation coefficient (PCC) and tissue expression complementarity (TEC) scores follow the same trend for both gene pools. However, TEC variation among ages is much more prominent than any other measure for any gene pool. This result can be partially explained due to the technical differences between both studies. Data by O'Rourke et al. (2014a) contains relatively short, 36 bp Illumina single-end reads sequenced at ~25 million reads per library; as compared with data by Vlasova et al (2016) which are longer, 100 bp paired-end reads ~50 million reads per library. Thus, same RPKM threshold applied to these datasets will lead to different representation of expression levels for the same gene. Moreover, this approach will either capture only moderate to highly expressed genes in selected tissue or only lowly expressed genes. Despite, potential technical noise, these results confirm previous findings obtained in the Mesoamerican genome (Vlasova et al. 2016) and other species (Lynch and Force 2000; Prince and Pickett 2002) where partitioning of the gene expression in a temporal and/or spatial manner plays an important role in the initial retention of duplicated genes. As time occurs, ancestral gene expression tends to diversify across duplicated genes rendering them indispensable. Further events may lead to acquire novel expression patterns and/or functions which will contribute to increase the divergence of expression patterns across duplicated genes.

To complement the gene co-expression network generated for the Mesoamerican gene pool, we reconstructed the corresponding one for the Andean accession using the transcriptomics data from Negro Jamapa samples (Fig. 12.6). As previously described (Vlasova et al. 2016), this network was constructed using the Graphical Lasso algorithm as implemented in R (Friedman et al. 2008). Initial dataset was filtered out to remove outlier genes which capture the majority

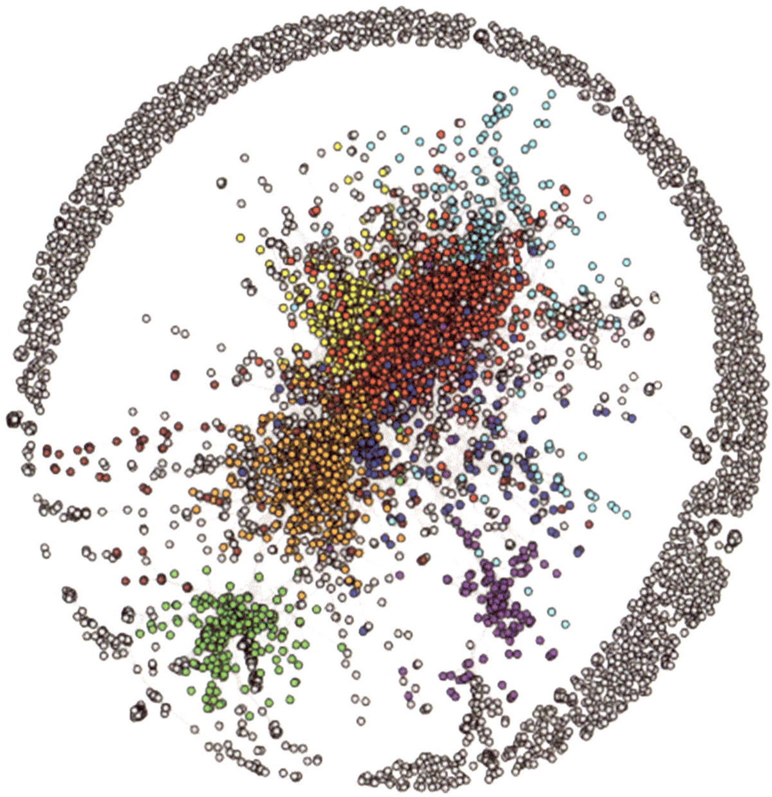

Fig. 12.6 Gene co-expression network for the Andean gene pool. This network complements the previously generated one for the Mesoamerican gene pool (Vlasova et al. 2016). To facilitate the identification of gene clusters, the ten biggest clusters have been colored. Table 12.4 contains their putative main functions

Table 12.4 Putative main functions of the biggest co-expression modules from the Andean gene pool

Module ID	Module color	Module size	Module main function
1	Red	1,628	Oxidation reduction
2	Orange	1,246	Photosynthesis
3	Yellow	559	Defense response
4	Green	431	Regulation of biosynthesis
5	Cyan	220	Lignin, coumarin metabolism
6	Blue	209	Response to stimulus
7	Purple	143	DNA repair response, cell cycle processes
8	Pink	116	Cell wall organization
9	Brown	110	Translation, methylation
10	Beige	110	Riboflavin metabolism

These functions were assigned based on the results of GO terms enrichment analysis performed for these modules

of the expression in one tissue, and not matched RNA-seq samples between these two studies—nodules samples from Negro Jamapa and axial meristems from BAT93. Only the top 10,000 genes with the highest coefficient variation were considered for this analysis. Table 12.5 compares the main features of the already published co-expression gene network in Vlasova et al. (2016) and the one generated in this review. Fast-greedy community algorithm (Clauset et al.

Table 12.5 Gene co-expression network main features for both gene pools

Accession	BAT93	G19833/Negro Jamapa
Number of nodes (genes)	4,292	7,479
Number of edges (connections)	50,034	160,882
Number of edges per node. min–max	1–198	1–247
Number of edges per node. average ± standard deviation	15 ± 26	26 ± 40
Graph density	0.0054	0.0057
Graph transitivity	0.4616	0.4902
Graph modularity	0.5866	0.6455
Number of modules	372	544
Number of modules With >100 nodes	9	11

Only modules with at least one connection per gene were considered to compute these values

2004) was used to divide this network into modules. We then analyzed the functional enrichments, in term of GO terms, for the biggest modules.

When comparing both gene co-expression networks (Table 12.5), we observed a larger number of connected genes for the Andean accession as compared to the Mesoamerican one, which implies a much higher number of connections in the former. In this context, a connection represents similar gene expression patterns among several conditions, i.e., tissues and/or developmental stages, shared by any pair of genes. However, when looking at relative metrics like graph density, transitivity, and modularity, the two networks are largely similar. Differences are even less evident when looking at the largest modules in terms of co-expressing genes, for instance, the largest module and second largest module for the Mesoamerican and Andean co-expression networks, respectively, are enriched in genes associated to photosynthesis. Other large modules are related to the cell cycle and wall organization. Many of the biggest modules in both networks are enriched in genes associated to stress and defense responses, which requires a coordinated response among the implied genes. When looking at differences, there is a large module for the Mesoamerican co-expression gene network which is associated to the flavonoids biosynthesis pathway, which produces one of the most common secondary metabolite sets in plants (Schijlen et al. 2004). Besides the importance of this pathway in different biological processes like pigmentation and/or UV-scavenging, it is attractive for their potential role in human health for its antioxidant activity. Regarding the Andean co-expression network, one of the largest modules is enriched in genes associated to the riboflavin metabolic activity which is involved in the synthesis of tetrapyrrole. It has been shown for *A. thaliana* that deficiencies on the riboflavin activity impacts on the crucial processes such as photosynthesis and respiration via a decay of tetrapyrrole activity (Tanaka and Tanaka 2007; Hedtke et al. 2012).

Phylogenomics offers the possibility of study how species evolve overtime at gene and systems level. Moreover, when evolutionary information is combined with other data, i.e., transcriptomics data, it is possible to have a more dynamic view of the species. In the case of common bean, the availability of two reference genomes allows to study whether these two independently domesticated lines have similar evolutionary trajectories and how similar the gene expression landscapes are. Despite technical differences, both accessions have similar evolutionary and expression patterns as observed when comparing both PCG sets, gene expression specialization overtime and gene co-expression modules. We expect that future sequencing efforts will help to overcome technical noise allowing a finer study of similarities and differences between independent domestication events. This, in turn, will accelerate current breeding efforts to improve current domesticated lines and keep genetic diversity among wild and landrace accessions.

12.6 Exploring the Evolution of Common Bean Protein-Coding Genes Through PhylomeDB

To provide a glimpse of the usefulness of the available phylogenetic resources for the common bean, we describe how a user could obtain evolutionary information on a gene of interest. Although evolutionary aspects are often neglected in favor of purely functional ones when analyzing a gene sequence, we consider that this negatively impacts the biological understanding of the gene under consideration. Every gene sequence is the result of an evolutionary process which, through the role of natural selection, is linked to the functional role of that gene. Processes such as gene duplication and gene loss, accelerated the rate of evolution and functional shift of duplicated genes leave its footprints on gene sequences and, therefore, the phylogenetic trees derived from them. In addition, resolving what the orthology and/or paralogy relationships are among homologs genes is fundamental. These relationships allow assessing whether a function may be conserved across species (Gabaldón and Koonin 2013). This information is especially relevant for crop species to support breeding programs aiming to improve crops.

It is possible to gather information about any common bean PCG from the Andean (G19833) and Mesoamerican (BAT93) accessions by introducing its sequence identifier (ID) at PhylomeDB [http://phylomedb.org] (Huerta-Cepas et al. 2014). If available, a phylogenetic tree will be then shown to the user. It is important to mention that phylogenetic trees are always associated to a phylome, i.e., the complete collection of gene phylogenies of a given proteome. A phylome defines the species of interest, which is used as starting point for any homology search, and a set of species which defines the evolutionary framework of the study. There are five publically available common bean phylomes in PhylomeDB covering different evolutionary contexts. One of them is focused on the relationship of BAT93 PCGs with PCGs from 13 legume and non-legume species. Two other phylomes which include PCGs from both common bean accessions used the same set of species. These phylomes were used to infer one-to-one orthology relationships between both common bean gene-sets and establish the correspondence between them, among other studies. The two newly generated phylomes, for this work, are centered in legume species allowing tune-fine studies regarding evolutionary relationships for common bean genes.

When a phylogenetic tree is found, the top panel shows to which phylome is associated the input ID. It also indicates whether the sequence has been used as starting point (seed) or not, the evolutionary models used for reconstructing maximum-likelihood trees, and whether the input sequence has been used in other phylogenetic trees in PhylomeDB (collateral trees). The second top panel offers a number of options to customize displayed tree features, i.e., sequences identifiers provided by external databases such as UniProt; to search for additional sequence identifiers; to download the current image using different formats (PNG or SVG); to download an OrthoXML file containing the different orthology and paralogy relationships inferred from the current tree; to download all data associated to the tree, i.e., raw and clean alignments, set of sequences, etc.; and to visualize the multiple sequence alignment used to generate the visualized tree.

The central panel shows the phylogenetic tree found for the input sequence identifier. By default, it shows evolutionary blocks, species names, branch supports (which is shown if different from 1.0), PFAM domains (if available),

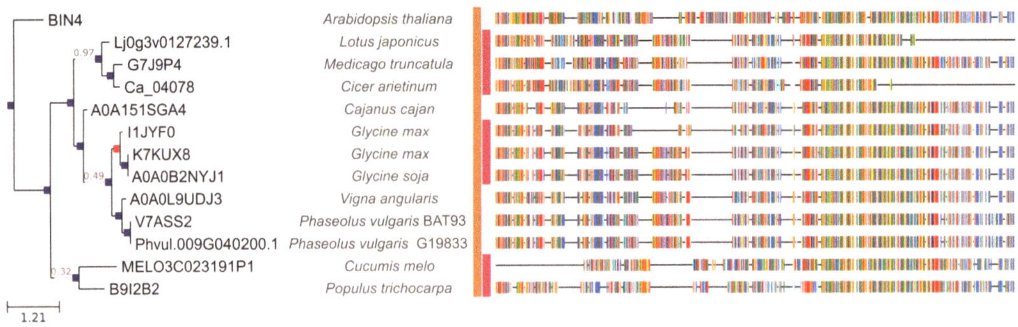

Fig. 12.7 Single-gene tree for *Phvul.009G040200* from the Andean reference genome. This phylogenetic tree is stored at PhylomeDB.org where can be queried at any time using the gene name. This figure shows the relationship of this gene with its Mesoamerican counterpart and other legume and non-legume sequences. Colors located at the right side of each sequence reflect their amino-acid composition and the presence of specific PFAM domains

and leave names with a preference for UniProt, Ensembl, and Genome project identifiers over PhylomeDB ones. Any feature can be adjusted and change using the second top panel. Figure 12.7 shows the phylogenetic tree for *Phvul.009G040200*, a PCG for the Andean reference genome which is ortholog to *BIN4*, an *A. thaliana* gene involved in cell expansion and final plant size regulation (Breuer et al. 2007). The Mesoamerican bean ortholog (*V7ASS2*) has been detected as a putative gene involved in domestication (Schmutz et al. 2014). As seen in the Fig. 12.7, it is possible to detect speciation events (blue squares) and duplication events (red squares) which recapitulate the phylogeny of the species used for the phylome. This tree also confirms one-to-one orthology relationships between the Andean gene and genes at other legume species. Moreover, it is possible to visualize that the domain architecture is somehow conserved across the entire gene family.

However, there are much more complex scenarios where the number of sequences included in the trees is limited to 150. For instance, the Mesoamerican gene *PHASIBEAM10F022389* belongs to a large multigene family with an intricate evolutionary story (Fig. 12.8). This gene is also associated to the domestication of the Mesoamerican accession while its Andean ortholog is not (*Phvul.01G038800*) associated to this syndrome (Schmutz et al. 2014). Orthologous genes at *A. thaliana* have been shown to encode cytokinin oxidase and dehydrogenase proteins which regulate a pathway associated to seed size and weight by degrading active cytokinin (Van Daele et al. 2012). To have a better insight about functional annotation at external sources, it is possible to access directly from the tree. Users can use this functionality by right-clicking at any sequence name at the tree as shown in Fig. 12.8 for *CKX7* from *A. thaliana*. Using this menu, users can also set a new root for the current tree and collapse part of it (bottom part at Fig. 12.8) for a better visualization of evolutionary relationships at the tree. Finally, users can get further orthology and paralogy predictions using MetaPhOrs which consider information stored across several public databases like TreeFAM, eggNOG, OrthoMCL, among others (Pryszcz et al. 2011). MetaPhOrs inform users about the reliability of orthology and paralogy relationships detected at the current tree which may be crucial to manually transfer functional information across orthologs sequences.

PhylomeDB is an invaluable resource to understand the evolutionary trajectory of common bean genes from the Mesoamerican and Andean accessions. Phylogenetic trees are at the base of accurate functional predictions and can assist researchers in elucidating which genes have appeared or duplicated at specific evolutionary ages. Moreover, additional information such as PFAM domains and links to external databases can accelerate the characterization of

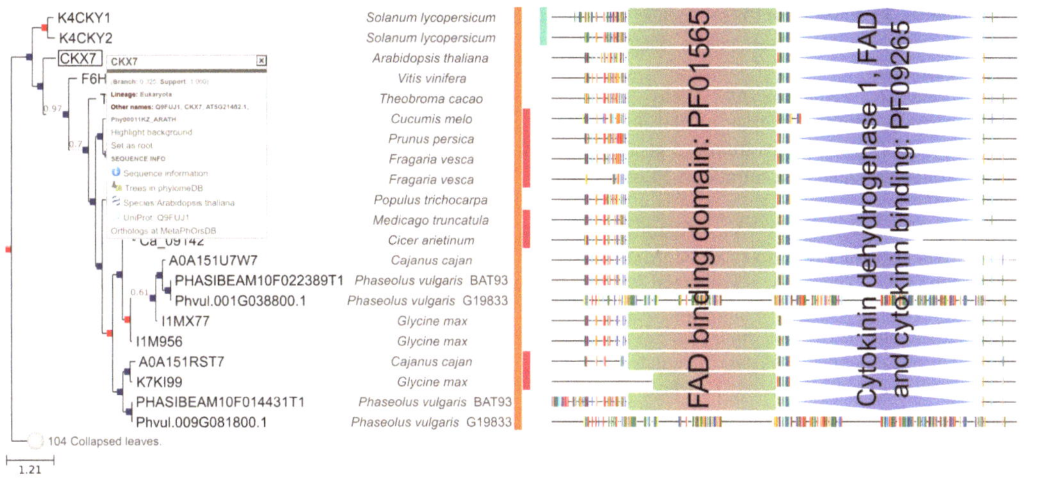

Fig. 12.8 Phylogenetic tree for *Phvul.001G038800*. A similar tree can be found querying for *PHASIBEAM10F022389* in PhylomeDB.org. Considering the tree complex relationships among 150 sequences from 13 species, some of the nodes have been collapsed—bottom part of the figure—to center our attention in the gene of interest. Information about sequences domain architecture is provided as well as how to use external information to gather additional information, i.e., from UniProt

candidate genes associated to domestication, stress response, and/or other traits of interest. Undoubtedly, PhylomeDB can play a role in breeding programs by aggregating information from different sources and make it available to the bean community.

12.7 Conclusions

Common bean is the most important grain legume for its crucial role in the nourishment of more than 500 million people in developing countries. Improving this crop in the context of changing climate conditions requires a deep understanding of the genetic basis of its domestication and the plant response to biotic and abiotic stress conditions. The recent release of the Mesoamerican reference genome (Vlasova et al. 2016) paves the way to elucidating the evolutionary trajectories of the species, and of the two domesticated gene pools. Released data complements the available one for the Andean gene pool (Schmutz et al. 2014) and will allow studying commonalities and differences between parallel domestication processes. After the initial analyses, it is possible to suggest that most gene family expansions predated the split of both domesticated lineages which may have facilitated later adaptations leading to domesticated accessions. Together with reference genomes, there is a plethora of other genomics resources such as the transcriptomic atlases which allow deepening the understanding of gene expression changes across different developmental stages and/or tissues. Many of these resources also allow understanding changes in gene expression under abiotic and biotic stress conditions assisting in the identification of key genes related to plant responses to different environments. All these studies, made available through public databases, are fundamental to develop strategies for preserving and improving common bean in the context of changing climate conditions and increase of population sizes in developing countries.

References

Aparicio-Fabre R, Guillén G, Loredo M, Arellano J, Valdés-López O, Ramírez M, Iñiguez LP, Panzeri D, Castiglioni B, Cremonesi P et al (2013) Common bean (*Phaseolus vulgaris* L.) PvTIFY orchestrates global changes in transcript profile response to jasmonate and phosphorus deficiency. BMC Plant Biol 13:26

Arumuganathan K, Earle ED (1991) Nuclear DNA content of some important plant species. Plant Mol Biol Rep 9:208–218

Astudillo-Reyes C, Fernandez AC, Cichy KA (2015) Transcriptome characterization of developing bean (*Phaseolus vulgaris* L.) pods from two genotypes with contrasting seed Zinc concentrations. PLoS ONE 10: e0137157

Ayyappan V, Kalavacharla V, Thimmapuram J, Bhide KP, Sripathi VR, Smolinski TG, Manoharan M, Thurston Y, Todd A, Kingham B (2015) Genome-wide profiling of histone modifications (H3K9me2 and H4K12ac) and gene expression in rust (*Uromyces appendiculatus*) inoculated common bean (*Phaseolus vulgaris* L.). PLoS ONE 10: e0132176

Bellucci E, Bitocchi E, Ferrarini A, Benazzo A, Biagetti E, Klie S, Minio A, Rau D, Rodriguez M, Panziera A et al (2014) Decreased nucleotide and expression diversity and modified coexpression patterns characterize domestication in the common bean. Plant Cell 26:1901–1912

Bentley DR (2006) Whole-genome re-sequencing. Curr Opin Genet Dev 16:545–552

Bitocchi E, Bellucci E, Giardini A, Rau D, Rodriguez M, Biagetti E, Santilocchi R, Spagnoletti Zeuli P, Gioia T, Logozzo G et al (2013) Molecular analysis of the parallel domestication of the common bean (*Phaseolus vulgaris*) in Mesoamerica and the Andes. New Phytol 197:300–313

Bitocchi E, Nanni L, Bellucci E, Rossi M, Giardini A, Zeuli PS, Logozzo G, Stougaard J, McClean P, Attene G et al (2012) Mesoamerican origin of the common bean (*Phaseolus vulgaris* L.) is revealed by sequence data. Proc Natl Acad Sci USA 109:E788–E796

Blair MW, Córdoba JM, Muñoz C, Yuyó DK (2014) BAC-end microsatellites from intra and inter-genic regions of the common bean genome and their correlation with cytogenetic features. PLoS ONE 9: e101873

Blair MW, Fernandez AC, Fabio P, Muñoz-Torres MC, Kapu NS, Kathleen B, Lynch JP (2011a) Parallel sequencing of expressed sequence tags from two complementary DNA libraries for high and low Phosphorus adaptation in common beans. Plant Genome 4:204–217

Blair MW, Fernandez AC, Ishitani M, Moreta D, Seki M, Ayling S, Shinozaki K (2011b) Construction and EST sequencing of full-length, drought stress cDNA libraries for common beans (*Phaseolus vulgaris* L.). BMC Plant Biol 11:171

Blair MW, Hurtado N, Chavarro CM, Muñoz-Torres MC, Giraldo MC, Pedraza F, Tomkins J, Wing R (2011c) Gene-based SSR markers for common bean (*Phaseolus vulgaris* L.) derived from root and leaf tissue ESTs: an integration of the BMc series. BMC Plant Biol 11:50

Blair MW, Soler A, Cortés AJ (2012) Diversification and population structure in common beans (*Phaseolus vulgaris* L.). PLoS ONE 7:e49488

Branca A, Paape TD, Zhou P, Briskine R, Farmer AD, Mudge J, Bharti AK, Woodward JE, May GD, Gentzbittel L et al (2011) Whole-genome nucleotide diversity, recombination, and linkage disequilibrium in the model legume Medicago truncatula. Proc Natl Acad Sci USA 108:E864–E870

Breuer C, Stacey NJ, West CE, Zhao Y, Chory J, Tsukaya H, Azumi Y, Maxwell A, Roberts K, Sugimoto-Shirasu K (2007) BIN4, a novel component of the plant DNA topoisomerase VI complex, is required for endoreduplication in *Arabidopsis*. Plant Cell 19:3655–3668

Cannon SB, May GD, Jackson SA (2009) Three sequenced legume genomes and many crop species: rich opportunities for translational genomics. Plant Physiol 151:970–977

Chacón S, Pickersgill B, Debouck DG (2005) Domestication patterns in common bean (*Phaselous vulgaris L.*) and the origin of the Mesoamerican and Andean cultivated races. Theor Appl Genet 110:432–444. doi:10.1007/s00122-004-1842-2

Chhangawala S, Rudy G, Mason CE, Rosenfeld JA (2015) The impact of read length on quantification of differentially expressed genes and splice junction detection. Genome Biol 16:131

Clauset A, Newman ME, Moore C (2004) Finding community structure in very large networks. Phys Rev E Stat Nonlin Soft Matter Phys 70

Dalla Via V, Narduzzi C, Aguilar OM, Zanetti ME, Blanco FA (2015) Changes in the common bean transcriptome in response to secreted and surface signal molecules of *Rhizobium etli*. Plant Physiol 169:1356–1370

Dash S, Campbell JD, Cannon EKS, Cleary AM, Huang W, Kalberer SR, Karingula V, Rice AG, Singh J, Umale PE et al (2016) Legume information system (LegumeInfo.org): a key component of a set of federated data resources for the legume family. Nucleic Acids Res 44:D1181–D1188

Delgado-Salinas A, Alfonso D-S, Ryan B, Matt L (2006) Phylogeny of the genus *Phaseolus* (Leguminosae): a recent diversification in an ancient landscape. Syst Bot 31:779–791

De Vega JJ, Ayling S, Hegarty M, Kudrna D, Goicoechea JL, Ergon A, Rognli OA, Jones C, Swain M, Geurts R, et al. (2015) Red clover (*Trifolium pratense* L.) draft genome provides a platform for trait improvement. Sci Rep 5:17394

Doyle JJ (1998) Phylogenetic perspectives on nodulation: evolving views of plants and symbiotic bacteria. Trends Plant Sci 3:473–478

Engström PG, Steijger T, Sipos B, Grant GR, Kahles A, Rätsch G, Goldman N, Hubbard TJ, Harrow J, Guigó R et al (2013) Systematic evaluation of spliced alignment programs for RNA-seq data. Nat Methods 10:1185–1191

Friedman J, Hastie T, Tibshirani R (2008) Sparse inverse covariance estimation with the graphical lasso. Biostatistics 9:432–441

Fuller DQ (2007) Contrasting patterns in crop domestication and domestication rates: recent archaeobotanical insights from the Old World. Ann Bot 100:903–924

Gabaldón T, Koonin EV (2013) Functional and evolutionary implications of gene orthology. Nat Rev Genet 14:360–366

Gan X, Stegle O, Behr J, Steffen JG, Drewe P, Hildebrand KL, Lyngsoe R, Schultheiss SJ, Osborne EJ, Sreedharan VT et al (2011) Multiple reference genomes and transcriptomes for *Arabidopsis thaliana*. Nature 477:419–423

Gepts P (1998) Origin and evolution of common bean: past events and recent trends. HortScience 33:1124–1130

Gepts P, Aragão FJL, de Barros E, Blair MW, Brondani R, Broughton W, Galasso I, Hernández G, Kami J, Lariguet P et al (2008) Genomics of *Phaseolus* beans, a major source of dietary protein and micronutrients in the tropics. In: Moore PH, Ming R (eds) Genomics of tropical crop plants, vol 1. Plant genetics and genomics: crops and models. Springer, New York, NY, pp 113–143

Goodstein DM, Shu S, Howson R, Neupane R, Hayes RD, Fazo J (2012) Phytozome: a comparative platform for green plant genomics. Nucleic Acids Res 40:D1178–D1186. doi:10.1093/nar/gkr944

Graham PH (2003) Legumes: importance and constraints to greater use. Plant Physiol 131:872–877

Grisi MCM, Blair MW, Gepts P, Brondani C, Pereira PAA, Brondani RPV (2007) Genetic mapping of a new set of microsatellite markers in a reference common bean (*Phaseolus vulgaris*) population BAT93 x Jalo EEP558. Genet Mol Res 6:691–706

Hedtke B, Alawady A, Albacete A, Kobayashi K, Melzer M, Roitsch T, Masuda T, Grimm B (2012) Deficiency in riboflavin biosynthesis affects tetrapyrrole biosynthesis in etiolated *Arabidopsis* tissue. Plant Mol Biol 78:77–93

Hiz MC, Canher B, Niron H, Turet M (2014) Transcriptome analysis of salt tolerant common bean (*Phaseolus vulgaris* L.) under saline conditions. PLoS ONE 9:e92598

Huerta-Cepas J, Capella-Gutiérrez S, Pryszcz LP, Marcet-Houben M, Gabaldón T (2014) PhylomeDB v4: zooming into the plurality of evolutionary histories of a genome. Nucleic Acids Res 42:D897–D902

Jain S, Chittem K, Brueggeman R, Osorno JM, Richards J, Nelson BD Jr (2016) Comparative transcriptome analysis of resistant and susceptible common bean genotypes in response to soybean cyst nematode Infection. PLoS ONE 11:e0159338

Kalavacharla V, Liu Z, Meyers BC, Thimmapuram J, Melmaiee K (2011) Identification and analysis of common bean (*Phaseolus vulgaris* L.) transcriptomes by massively parallel pyrosequencing. BMC Plant Biol 11:135

Kami J, Poncet V, Geffroy V, Gepts P (2006) Development of four phylogenetically-arrayed BAC libraries and sequence of the *APA* locus in *Phaseolus vulgaris*. Theor Appl Genet 112:987–998

Kami J, Velásquez VB, Debouck DG, Gepts P (1995) Identification of presumed ancestral DNA sequences of phaseolin in *Phaseolus vulgaris*. Proc Natl Acad Sci USA 92:1101–1104

Kang YJ, Kim SK, Kim MY, Lestari P, Kim KH, Ha B-K, Jun TH, Hwang WJ, Lee T, Lee J et al (2014) Genome sequence of mungbean and insights into evolution within *Vigna* species. Nat Commun 5:5443

Kaplan L, Lawrence K, Lynch TF (1999) *Phaseolus* (Fabaceae) in Archaeology: AMS. Econ Bot 53: 261–272

Kersey PJ, Allen JE, Armean I, Boddu S, Bolt BJ, Carvalho-Silva D, Christensen M, Davis P, Falin LJ, Grabmueller C et al (2016) Ensembl Genomes 2016: more genomes, more complexity. Nucleic Acids Res 44:D574–D580

Kim MY, Lee S, Van K, Kim T-H, Jeong S-C, Choi I-Y, Kim D-S, Lee Y-S, Park D, Ma J et al (2010) Whole-genome sequencing and intensive analysis of the undomesticated soybean (*Glycine soja* Sieb. and Zucc.) genome. Proc Natl Acad Sci USA 107:22032–22037

Kislev ME, Bar-Yosef O (1988) The legumes: the earliest domesticated plants in the Near East? Curr Anthropol 29:175–179

Klitgaard BB, Bruneau A (2003) Advances in legume systematics: higher level systematics. Royal Botanic Gardens, Kew, UK

Kwak M, Gepts P (2009) Structure of genetic diversity in the two major gene pools of common bean (*Phaseolus vulgaris* L., Fabaceae). Theor Appl Genet 118: 979–992

Lavin M, Herendeen PS, Wojciechowski MF (2005) Evolutionary rates analysis of Leguminosae implicates a rapid diversification of lineages during the tertiary. Syst Biol 54:575–594

Liao D, Cram D, Sharpe AG, Marsolais F (2013) Transcriptome profiling identifies candidate genes associated with the accumulation of distinct sulfur γ-glutamyl dipeptides in *Phaseolus vulgaris* and *Vigna mungo* seeds. Front Plant Sci 4:60

Liao D, Pajak A, Karcz SR, Chapman BP, Sharpe AG, Austin RS, Datla R, Dhaubhadel S, Marsolais F (2012) Transcripts of sulphur metabolic genes are co-ordinately regulated in developing seeds of common bean lacking phaseolin and major lectins. J Exp Bot 63:6283–6295

Li J, Dai X, Zhuang Z, Zhao PX (2016a) LegumeIP 2.0—a platform for the study of gene function and genome evolution in legumes. Nucleic Acids Res 44:D1189–D1194

Li M-W, Xin D, Gao Y, Li K-P, Fan K, Muñoz NB, Yung W-S, Lam H-M (2016b) Using genomic information to improve soybean adaptability to climate change. J Exp Bot (online first). doi:10.1093/jxb/erw348

Li Y-H, Zhou G, Ma J, Jiang W, Jin L-G, Zhang Z, Guo Y, Zhang J, Sui Y, Zheng L et al (2014) De novo assembly of soybean wild relatives for pan-genome analysis of diversity and agronomic traits. Nat Biotechnol 32:1045–1052

Lynch M, Force A (2000) The probability of duplicate gene preservation by subfunctionalization. Genetics 154:459–473

Mamidi S, Rossi M, Moghaddam SM, Annam D, Lee R, Papa R, McClean PE (2013) Demographic factors shaped diversity in the two gene pools of wild common bean *Phaseolus vulgaris* L. Heredity 110:267–276

Mamidi S, Sujan M, Monica R, Deepti A, Samira M, Rian L, Roberto P, Phillip M (2011) Investigation of the domestication of common bean (*Phaseolus vulgaris*) using multilocus sequence data. Funct Plant Biol 38:953

Martin K, Singh J, Hill JH, Whitham SA, Cannon SB (2016) Dynamic transcriptome profiling of Bean Common Mosaic Virus (BCMV) infection in common bean (*Phaseolus vulgaris* L.). BMC Genom 17:613

Michael TP, Scott J (2013) The first 50 plant genomes. Plant Genome 6:2. doi:10.3835/plantgenome2013.03.0001in

Mortazavi A, Williams BA, McCue K, Schaeffer L, Wold B (2008) Mapping and quantifying mammalian transcriptomes by RNA-Seq. Nat Methods 5:621–628

Nodari RO, Tsail SM, Gilbertson RL, Gepts P (1993) Towards an integrated linkage map of common bean 2. Development of an RFLP-based linkage map. Theor Appl Genet 85:513–520

Olson MV (1999) When less is more: gene loss as an engine of evolutionary change. Am J Hum Genet 64:18–23

O'Rourke JA, Iniguez LP, Fu F, Bucciarelli B, Miller SS, Jackson SA, McClean PE, Li J, Dai X, Zhao PX et al (2014a) An RNA-Seq based gene expression atlas of the common bean. BMC Genom 15:866

O'Rourke JA, Bolon Y-T, Bucciarelli B, Vance CP (2014b) Legume genomics: understanding biology through DNA and RNA sequencing. Ann Bot 113:1107–1120

Papa R, Acosta J, Delgado-Salinas A, Gepts P (2005) A genome-wide analysis of differentiation between wild and domesticated *Phaseolus vulgaris* from Mesoamerica. Theor Appl Genet 111:1147–1158

Papa R, Gepts P (2003) Asymmetry of gene flow and differential geographical structure of molecular diversity in wild and domesticated common bean (*Phaseolus vulgaris* L.) from Mesoamerica. Theor Appl Genet 106:239–250

Prince VE, Pickett FB (2002) Splitting pairs: the diverging fates of duplicated genes. Nat Rev Genet 3:827–837

Proost S, Van Bel M, Vaneechoutte D, Van de Peer Y, Inze D, Mueller-Roeber B, Vandepoele K (2014) PLAZA 3.0: an access point for plant comparative genomics. Nucleic Acids Res 43:D974–D981

Pryszcz LP, Huerta-Cepas J, Gabaldón T (2011) MetaPhOrs: orthology and paralogy predictions from multiple phylogenetic evidence using a consistency-based confidence score. Nucleic Acids Res 39:e32

Ramírez M, Graham MA, Blanco-López L, Silvente S, Medrano-Soto A, Blair MW, Hernández G, Vance CP, Lara M (2005) Sequencing and analysis of common bean ESTs. Building a foundation for functional genomics. Plant Physiol 137:1211–1227

Reddy TBK, Thomas AD, Stamatis D, Bertsch J, Isbandi M, Jansson J, Mallajosyula J, Pagani I, Lobos EA, Kyrpides NC (2015) The Genomes OnLine Database (GOLD) v. 5: a metadata management system based on a four level (meta)genome project classification. Nucleic Acids Res 43:D1099–D1106

Rossi M, Bitocchi E, Bellucci E, Nanni L, Rau D, Attene G, Papa R (2009) Linkage disequilibrium and population structure in wild and domesticated populations of *Phaseolus vulgaris* L. Evol Appl 2:504–522

Santini L, Munhoz C de F, Bonfim MF Jr, Brandão MM, Inomoto MM, Vieira MLC (2016) Host transcriptional profiling at early and later stages of the compatible interaction between *Phaseolus vulgaris* and *Meloidogyne incognita*. Phytopathology 106:282–294

Sato S, Nakamura Y, Kaneko T, Asamizu E, Kato T, Nakao M, Sasamoto S, Watanabe A, Ono A, Kawashima K et al (2008) Genome structure of the legume, *Lotus japonicus*. DNA Res 15:227–239

Schijlen EGWM, Ric de Vos CH, van Tunen AJ, Bovy AG (2004) Modification of flavonoid biosynthesis in crop plants. Phytochemistry 65:2631–2648

Schlueter JA, Goicoechea JL, Collura K, Gill N, Lin J-Y, Yu Y, Kudrna D, Zuccolo A, Vallejos CE, Muñoz--Torres M et al (2008) BAC-end sequence analysis and a draft physical map of the common bean (*Phaseolus vulgaris* L.) genome. Trop Plant Biol 1:40–48

Schmutz J, Cannon SB, Schlueter J, Ma J, Mitros T, Nelson W, Hyten DL, Song Q, Thelen JJ, Cheng J et al (2010) Genome sequence of the palaeopolyploid soybean. Nature 463:178–183

Schmutz J, McClean PE, Mamidi S, Wu GA, Cannon SB, Grimwood J, Jenkins J, Shu S, Song Q, Chavarro C et al (2014) A reference genome for common bean and genome-wide analysis of dual domestications. Nat Genet 46:707–713

Schneeberger K, Ossowski S, Ott F, Klein JD, Wang X, Lanz C, Smith LM, Cao J, Fitz J, Warthmann N et al (2011) Reference-guided assembly of four diverse Arabidopsis thaliana genomes. Proc Natl Acad Sci USA 108:10249–10254

Sprent JI (2001) Nodulation in Legumes. Royal Botanic Gardens, Kew, UK

SEQC/MAQC-III Consortium (2014) A comprehensive assessment of RNA-seq accuracy, reproducibility and information content by the Sequencing Quality Control Consortium. Nat Biotechnol 32:903–914

Tanaka R, Tanaka A (2007) Tetrapyrrole biosynthesis in higher plants. Annu Rev Plant Biol 58:321–346

Teng M, Love MI, Davis CA, Djebali S, Dobin A, Graveley BR, Li S, Mason CE, Olson S, Pervouchine D et al (2016) A benchmark for RNA-seq quantification pipelines. Genome Biol 17:74

Valliyodan B, Qiu Dan, Patil G, Zeng P, Huang J, Dai L, Chen C, Li Y, Joshi T, Song L et al (2016) Landscape of genomic diversity and trait discovery in soybean. Sci Rep 6:23598

Vance CP (2001) Symbiotic Nitrogen fixation and Phosphorus acquisition. Plant nutrition in a world of declining renewable resources. Plant Physiol 127: 390–397

Van Daele I, Gonzalez N, Vercauteren I, de Smet L, Inzé D, Roldán-Ruiz I, Vuylsteke M (2012) A comparative study of seed yield parameters in *Arabidopsis thaliana* mutants and transgenics. Plant Biotechnol J 10: 488–500

Vanhouten W, MacKenzie S (1999) Construction and characterization of a common bean bacterial artificial chromosome library. Plant Mol Biol 40:977–983

Varshney RK, Chen W, Li Y, Bharti AK, Saxena RK, Schlueter JA, Donoghue MTA, Azam S, Fan G, Whaley AM et al (2012) Draft genome sequence of pigeon pea (*Cajanus cajan*), an orphan legume crop of resource-poor farmers. Nat Biotechnol 30:83–89

Varshney RK, Chi S, Saxena RK, Sarwar A, Sheng Y, Sharpe AG, Steven C, Jongmin B, Rosen BD, Bunyamin T'an, et al. (2013a) Draft genome sequence of chickpea (*Cicer arietinum*) provides a resource for trait improvement. Nat Biotechnol 31:240–246

Varshney RK, Mohan SM, Gaur PM, Gangarao NVPR, Pandey MK, Bohra A, Sawargaonkar SL, Chitikineni A, Kimurto PK, Janila P et al (2013b) Achievements and prospects of genomics-assisted breeding in three legume crops of the semi-arid tropics. Biotechnol Adv 31:1120–1134

Vlasova A, Capella-Gutiérrez S, Rendón-Anaya M, Hernández-Oñate M, Minoche AE, Erb I, Câmara F, Prieto-Barja P, Corvelo A, Sanseverino W et al (2016) Genome and transcriptome analysis of the Mesoamerican common bean and the role of gene duplications in establishing tissue and temporal specialization of genes. Genome Biol 17:32

Wagner GP, Kin K, Lynch VJ (2012) Measurement of mRNA abundance using RNA-seq data: RPKM measure is inconsistent among samples. Theory Biosci 131:281–285

Wu J, Wang L, Li L, Wang S (2014) De novo assembly of the common bean transcriptome using short reads for the discovery of drought-responsive genes. PLoS ONE 9:e109262

Yang K, Kai Y, Zhixi T, Chunhai C, Longhai L, Bo Z, Zhuo W, Lili Y, Yisong L, Yudong S et al (2015) Genome sequencing of adzuki bean (*Vigna angularis*) provides insight into high starch and low fat accumulation and domestication. Proc Natl Acad Sci USA 112:13213–13218

Young ND, Debellé F, Oldroyd GED, Geurts R, Cannon SB, Udvardi MK, Benedito VA, Mayer KFX, Gouzy J, Schoof H et al (2011) The *Medicago* genome provides insight into the evolution of rhizobial symbioses. Nature 480:520–524

Yu K (2000) Integration of simple sequence repeat (SSR) markers into a molecular linkage map of common bean (*Phaseolus vulgaris* L.). J Hered 91:429–434

Yu K, Haffner M, Park SJ (2006) Construction and characterization of a common bean BAC library. Annual Report-Bean. http://naldc.nal.usda.gov/download/IND43805348/PDF

Zhang H, Li C, Davis EL, Wang J, Griffin JD, Kofsky J, Song B-H (2016) Genome-wide association study of resistance to soybean cyst nematode (*Heterodera glycines*) HG Type 2.5.7 in wild soybean (*Glycine soja*). Front Plant Sci 7:1214

Zhou Z, Jiang Y, Wang Z, Gou Z, Lyu J, Li W, Yu Y, Shu L, Zhao Y, Ma Y et al (2015) Resequencing 302 wild and cultivated accessions identifies genes related to domestication and improvement in soybean. Nat Biotechnol 33:408–414

Prospects: The Importance of Common Bean as a Model Crop

13

Marcelino Pérez de la Vega, Marta Santalla and Frédéric Marsolais

Abstract

Common bean is the most important grain legume crop for food consumption worldwide and has a role in sustainable agriculture as a main source of proteins and nutrients. This book provides insights into some of the key achievements made in the study of common bean, as well as a timely overview of topics that are pertinent for future developments in legume genomics. At the conclusion of this volume, it is as important to take a look back so as to put a forward view in proper perspective.

Keywords

Genome · Common bean · Sequencing · Breeding

We would like to open this summary and further perspectives chapter with details both of the process to breeding the performance of the common bean crop and of the outcome of the genome sequencing effort as an indispensable element for discovery that will continue to be improved and that will serve researchers for years to come.

Common bean (*Phaseolus vulgaris* L.) is an important grain legume crop at global scale. According to Singh (1999), common bean is the third most important food legume after soybean (*Glycine max* L. Merr.) and peanut (*Arachis hypogea* L.). The principal products derived from this crop are dry beans, shell beans, and snap beans. Snap bean includes varieties and cultivars of common bean selected for succulent immature pods with reduced fiber. Snap bean is one of the forms to designate this vegetable, but the term French bean, green bean, garden bean, or haricot are also used. While dry bean market classes are

M. Pérez de la Vega (✉)
Area de Genética, Departamento de Biología Molecular, Universidad de León, 24071 León, Spain
e-mail: m.perez.delavega@unileon.es

M. Santalla
Grupo de Biología de Agrosistemas, Misión Biológica de Galicia-CSIC, PO. Box 28, 36080 Pontevedra, Spain

F. Marsolais
Department of Biology, University of Western Ontario, London, ON, Canada

F. Marsolais
Genomics and Biotechnology, London Research and Development Centre, Agriculture and Agri-Food Canada, London, ON, Canada

M. Pérez de la Vega et al. (eds.), *The Common Bean Genome*,
Compendium of Plant Genomes, https://doi.org/10.1007/978-3-319-63526-2_13

based primarily on seed characteristics, snap bean market classes are based on pod characteristics and plant architecture. Fewer and less distinct market classes of snap bean exist compared to dry bean, but given their importance as a vegetable, breeders have developed many cultivars. At small scale and subsistence farming, it is possible that the same bean variety is consumed as green pods or dry seeds. This characteristic is not exceptional among other legume grain crops. For instance, pea is also consumed as dry seeds, green seeds, and green pods, and faba bean is consumed as green or dry seeds too. Likewise, common bean is also frequently eaten as canned or frozen products, contributing to a well-developed food transformation industry in several countries, although soybean primary and derived food products are far more numerous. Canned beans can be found as fully cooked products, generally as stews with meat and vegetables, or semi-cooked to eat in salads. The consumption of common bean in the form of dry seeds or green pods is relevant for breeding and for distinct market classes' production.

Chapter 1 reviews the economic importance and relevance to biological research of common bean. One of the major problems of the world common bean production is the main source of data; in fact, the FAO production statistics include several species under the term "dry beans" (Singh 1999). FAO data collectors disregard taxonomic data, probably since many farmers do. For instance, in some areas of Spain, farmers include as "alubias" (common bean) not only *P. vulgaris* but also *P. coccineus* and even *Vigna* spp. Thus, at least six species, some belonging to entirely different genera and domesticated in different continents, are considered "dry beans" (Singh 1999). According to Singh (1999), the approximate figures for the world area harvested and world production of common bean in 1997 were 14.3 million of hectares and 11.6 million of tons, respectively, while the corresponding figures in FAOSTAT data were 25.5 millon of hectares and 26.3 million of tons, respectively. America is the largest producer continent of common bean, and Brazil is the largest producer and consumer country in the world (Singh 1999). As for dry bean, *Phaseolus* snap beans are confounded with other immature green pods of other legumes such as *Vigna unguiculata*, consumed mainly in some Asian countries (Myers and Baggett 1999).

The genus *Phaseolus* is included in the tribe Phaseoleae (although this tribe does not appear to be monophyletic) together with more than 20 crop species such as soybean, several *Vigna* species (cowpea, mung bean, adzuki bean, etc.), and pigeon pea (*Cajanus cajan*), designated as warm-season food legumes. Close to the Phaseoleae is placed the numerous genus *Lupinus*, containing several fodder crops, and the less numerous *Arachis* where the main food crop peanut is included. More phylogenetically distant from the Phaseoleae are the tribes where the cool-season food legumes (pea, lentil, faba bean, grasspea, chickpea, etc.) and other close-related cool-season fodder legumes (*Trifolium, Medicago*) are included. These tribes are included in the inverted repeat loss clade (IRLC) that is characterized by the lack of the inverted repeat sequence present in the chloroplast genome of most plant species.

The genus *Phaseolus* includes a large and non-well precise number of species (Debouck 1999; Delgado-Salinas et al. 2006); among these species, five were domesticated. The domesticated species, all of American origin, are namely tepary bean (*P. acutifolius* A. Gray), runner bean (*P. coccineus* L.), Lima bean (*P. lunatus* L.), and year-long bean (*P. dumosus* Macfad.) in addition to common bean (Gepts and Debouck 1991; Debouck 1999; Singh 1999). Thus, the information gained from the common bean genome will be useful not only to this crop species, but it will speed the research and breeding of the other *Phaseolus* crops. Common bean is largely the most grown crop within the genus; it occupies more than 90% of the surface sown to *Phaseolus* species in the world. The production of *P. dumosus* and *P. acutifolius* is largely restricted to some American countries, while the other two minor species are also sown in Africa, Asia, and Europe (Singh 1999). The origin and domestication of snap bean are not as clear as the dry bean. Bean genetics suggest that snap beans are derived from domesticated dry beans because less genetic changes would be required to derive snap bean from domesticated beans than from wild beans, and the first cultivars of snap beans would be originated from the Andean gene pool (Myers and Baggett 1999).

Another peculiarity of common bean is that among crop species it is one of the few well documented examples of multiple domestications. Most of the domesticated crop species were domesticated in a single event, that is, in a relatively small and localized area (Zohary 1999). If the wild progenitor was taken into cultivation only once, the crop is a product of a single domestication event and a monophyletic evolution. Alternatively, the wild progenitor could have been introduced into cultivation several times and in different places. If this is the case, one is faced with multiple events of domestication and with a polyphyletic evolution of the crop (Zohary 1999). It is well known that common beans were domesticated at least in two places, perhaps three; thus, there are two main gene pools, the Andean and the Mesoamerican, and that their evolution was polyphyletic. These two gene pools most likely evolved independently until Spanish and Portuguese first introduced both of them in Europe and in Africa. Then, when the different gene pools hybridized, a new semi-unconscious or semi-unintentional selection started outside from the respective centers of origins. Moreover, since other species of the genus have also been domesticated, several different processes of domestication implicating some genes in common among species and some different genes between species represent one of the best examples to analyze in depth the various genetic changes involved in the process of crop domestication. The importance in itself of the process of domestication and how this evolutionary process modeled the genetic structure of cultigens is indisputable. Chapter 2 provides a comprehensive up-to-date review of the evolution of all the domesticated *Phaseolus* species. A different approach to evolutionary processes between species is to analyze the relative position of genes and repetitive sequences in genomes, synteny. Physical maps of *Phaseolus* species have revealed complete macrosynteny among the investigated species with only a few breaks in collinearity. However, the variation in the repetitive fraction of the genome is much greater than expected. These aspects are reviewed in Chap. 3.

A reductive thinking, adopted by some researchers, is that genomics is replacing and displacing genetics. Certainly it is not the case, genomics and genetics are complementary and genomics makes little sense if it is not from the point of view of genetics. In this context, this sentence from Dr. Noel Ellis is clearly relevant: "Genetic maps are not made redundant by genome sequence: they describe the distribution of alleles in populations and this is essential for connecting genotype and phenotype. In order to exploit mutant populations and genomic tools we need robust genetic analysis. The efficient application of genomics to breeding requires the sharing of both genotypic and phenotypic data: a lesson to be learnt from the rapid progress in genomics that was driven by data sharing. The time is ripe for a re-emphasis on the use of genetics to dissect and understand traits in terms of their discrete determinants" (Ellis 2014).

Genetic map setting up, that is, the linear order of genes in linkage groups that must correspond to chromosomes of the haploid complement, is one of the goals of genetics. Moreover, genetic maps are also essential in breeding crops since fine genetic mapping allows the identification of genetic markers linked to qualitative and quantitative genes or QTLs and hence to carry out marker-assisted selection (MAS). In turn, genomic and transcriptomic data are nowadays basic in obtaining thousands and thousands of good quality genetic markers, such as SNPs or SSRs, allowing fine mapping and help in MAS. Chapter 4 gives a deep review of the progressive development of common bean genetics maps, the state-of-the-art in the genomic era, and the application of these maps in breeding for disease resistance, yield, precocity, and plant architecture.

A sequenced genome is not the final goal. Although the amount of relevant information gained with the whole sequence is outstanding, the number of genes, the amount and classes of mobile genetic elements, transposons and retrotransposons, the average number and length of exons and introns, the GO functional distribution of genes, etc., it is only part of the importance of this information. The genome is a step in the development of useful information to breeding in such amount and quality not reachable by alternative ways. After the first high-coverage genome, genomes from different cultivars, accessions, or wild relatives are usually sequenced at a lower coverage. The comparison between genome sequences or even to nucleotide sequences in public databases (genomic, ESTs, TSAs, etc.) affords thousands to millions of SNPs, SSRs, and other molecular markers useful in MAS and other breeding techniques. In common bean, representative genomes from the two main germplasm sources have been sequenced, and this knowledge will help in the developing of breeding programs which include both germplasm sources. Since the current common bean cultigen is the result of different events of domestication, and hence original gene pools, and the cultigen has a large history of introductions and movements between continents and countries; a large set of cultivars and varieties should be included in a program of resequencing. Snap beans are one of the first instances necessary to complete the genomic information of the crop. An advantage of the common bean for a genomic resequencing project is that its genome size is relatively small, according to the classical review on plant species genome size by Arumuganathan and Earle (1991). Although the 1C estimations are higher than the assembled genomes (637 Mb versus 521 Mb for common bean, for instance), the estimations are accurate considering that many repetitive DNA sequences, such as retrotransposons, are not included in the assembled scaffolds. In fact, common bean is closer to other small-genome crop legumes such as chickpea (738 Mb) or the tetraploid soya bean (1115 Mb) than important crop legumes such as pea and lentil which have genomes close to the 4000 Mb. The importance of uncovering intra- and inter-species introgression events that could indicate gene transfer and the relevance of wild germplam in further breeding process is described in Chap. 5. This chapter highlights that resequencing efforts for both wild and domesticated genotypes will play a major role in the future of this crop because it will allow the genetic dissection of the characters involved in the evolution and adaptation to different environments and their further use in breeding programs.

The review by Michael and Jackson (2013) revised the first 50 plant genomes published between 2000 (*A. thaliana*) and 2014. Actually, the list included 55 genome sequences, but several species were represented by two or three different genome sources (e.g., *O. sativa*). The list included five legume species genomes: the model species *Medicago truncatula* and *Lotus japonicus*, and the crops *Glycine max*, *Cicer arietinum* (Kabuli and Desi varieties), and *Cajanus cajan*. Now the two genomes from common bean, two genomes of *Arachis* ssp., and one from *Trifolium pratense, Vigna angularis, V. radiata and Lupinus angustifolius* have expanded the list of sequenced legume genomes, and others, such as *Pisum sativum* and *Lens culinaris,* will be added soon (http://legumeinfo.org/, Oct. 2016). Although *Phaseolus coccineus* appears in the list "Species Access and Data" in the old Legume Information System homepage (http://comparative-legumes.org/), to date there is no available genome data. The available nucleotide sequence resources in the NCBI (https://www.ncbi.nlm.nih.gov/) under the heading of *P. vulgaris* are 161,691 as genomic and cDNA clones, 563,161 as genome survey sequences (GSSs), 202,679 DNA and RNA sequences, 153,337 as expressed sequence tags (ESTs), and 35,269 genes (Oct. 2016). The numbers of resources of *P. coccineus* are 368 DNA and RNA sequences and 391,164 ESTs. For *P. acutifolius,* the respective figures are 157 and 54,936. For *P. lunatus,* the proportion inverts 35 ESTs and 2250 RNA nucleotide sequences. Finally, the information contained under the headings *P. dumosus* or *P. polyanthus* is limited to a few nucleotide sequences. The information on the genus *Phaseolus* has greatly increased in the last months; currently, there is information on two genome assemblies and six genome sequencing projects are announced, in addition to 563,161 GSSs, 243,784 nucleotide sequences, 606,191 ESTs, and 77,199 genes. In turn, Chap. 6 is dedicated to understanding the importance of genome sequencing in common bean breeding.

Since the ability to genetically engineer plants was established, researchers have modified a great number of plant species to satisfy agricultural, horticultural, industrial, medicinal, or veterinary requirements. Almost thirty years after the first approaches to the genetic modification of pulse crops, it is possible to transform many grain legumes. However, *Phaseolus vulgaris* still lacks some practical tools for genomic research, such as routine genetic transformation. Its recalcitrance toward in vitro regeneration and rooting significantly hampers the possibilities of using this technology for genetic improvement. Thus, an efficient and reproducible system for regeneration of a whole plant is desired. Although noticeable progress has been made, the rate of recovery of transgenic lines is still low (Hnatuszko-Konka et al. 2014). Genetic engineering is important since genetic transformation is a powerful tool to gain valuable information on gene expression and function and to ascertain the true role of candidate genes and determining key genes in a particular process of improvement. A different dimension is the use of transformed cultivars of common bean in food production due to the bans to grow transgenic crops in some countries (mostly European countries) and the rejection of these crops by part of the consumers.

Current transformation in plants can be directed to the nuclear genome and to the plastid genome. Plastid genome transformation highly increases the number of the copies of the allospecific gene in relation to nuclear transformation. Therefore, a deep knowledge of the common bean and related species chloroplast genome and their polymorphisms is a valuable start point in designing chloroplast transformation techniques. Chapter 7 affords a complete review on common bean organelle genomes.

Breeding crops in the twenty-first century face new, or at least reinforced, challenges such as organic farming or the harvesting of healthier and/or biofortified products. In developed countries the demand of "ecological" crops from organic farming is increasing, but there is a global demand of crops with more bioactive components, with increased amounts of oligoelements, vitamins, better composition of amino acids, etc. This book affords reviews on "better quality" bean crops, and how genomics information can help to reach these breeding objectives. The content of sulfur amino acid is relatively low in common bean seeds, a general characteristic of legume seeds. This limiting amount of methionine and cysteine is a main constrain in seed nutritional value. Chapter 8 focuses on the genomics on a key component of the nutritional value of common bean, phytic acid. Phytic acid is the major form of phosphorus stored in the seed, but it binds trace elements and reduces their intestinal absorption, leading, under certain dietary circumstances, to oligoelement deficiencies. Thus, phytic acid is a key component in the plant phosphorous metabolism and a limiting factor in iron biofortification of common bean. In this case, the genomic knowledge of the genes implicated in the metabolic routes and their expression in different plant organs and developmental stages can help in breeding common bean with phytic acid levels that, without restraining the phosphorous availability to the plant development, allows a better iron and zinc intestinal absorption. A particularly relevant family of genes is the NB-LRR (nucleotide-binding leucine rich-repeat) family because the coded proteins are implicated in the disease resistance mechanisms. Chapter 9 reviews the current knowledge of this gene family in common bean and describes some unusual features in their clustering pattern within the genome. Likewise, the chapter reviews the genomic distribution of a particular satellite DNA, *khipu*, and its colocation with NB-LRR sequences. Chapter 10 gives information of the different families of seed storage proteins, their relative richness in these sulfur amino acids and on diverse common bean genotypes with different gene copy numbers, and how genomic information opens a path to breed new cultivars with a better seed amino acid composition. Chapter 11 explores the roles of the phenylpropanoid pathway gene families in various processes such as lignin or anthocyanin biosynthesis.

Other examples of legume species being massively sequenced to generate a large number of user data in plant breeding are chickpea (*Cicer arietinum*) and pigeonpea (*Cajanus cajan*). These two crops are playing a vital role in ensuring the nutritional food security in several countries of Asia and sub-Saharan Africa. With an objective to "explore this huge genetic diversity available to address the issue of low productivity and bottlenecks" associated with narrow genetic diversity, >1000 pulse genomes including 554 chickpea (reference set, elite varieties, and parents of several mapping populations) and 526 pigeonpea (reference set, hybrid parental lines, wild species accessions, and parents of mapping populations) genomes were sequenced at 5× to 13× coverage. Detailed analysis provided comprehensive data on diversity features, gene loss, domestication, and selection sweep. Multilocation phenotyping data for high priority traits for breeding is being assembled to undertake genome-wide association studies. In brief, this initiative is expected to identify superior/novel alleles associated with the traits of interest for enhancing crop productivity of these pulse crops (Varshney 2014). Another important dimension is the comparative genomics. It helps in advancing the information gained from one species genome to speed the deduced structure and function of new species genomes, in particular in phylogenetically related species. The list of the legume genomes so far sequenced, or near to be sequenced, indicates that only cool-season legumes (or Mediterranean) and warm-season legumes are represented. The comparative genomics must afford clues on the genetic difference which allow the cool-season legumes to tolerate frost temperatures during the vegetative period, while the warm-season species are very susceptible or totally intolerant (opening a possible way to manipulate this response); the symbiotic relationships between the plants and the corresponding nitrogen-fixing bacteria; and other biologically important questions. Chapter 12 insists on how the publication of genomes representative of both genetic pools, Andean and Mesoamerican, has not only corroborated the main hypotheses about the bean domestication events but has given a finer grained picture of the process. But as these later authors indicate: ".... more sequencing data are needed for a better understanding of common bean genome function and to deepen on the domestication processes of both gene pools. It is expected that third generation sequencing technologies will play an important role in those efforts, ...". While the genome efforts will continue to improve this crop, the work described in this volume represents a turning point upon which future investigations in common bean and broader plant genome evolution and biological inquiry has and will continue to be built.

References

Arumuganathan K, Earle ED (1991) Nuclear DNA content of some important plant species. Plant Mol Biol Report 9:208–218

Debouck DG (1999) Diversity in *Phaseolus* species in relation to the common bean. In: Singh SP (ed) Common bean improvement in the twenty-first century. Kluwer Academic Publishers, Dordrecht, The Netherlands, pp 25–52

Delgado-Salinas A, Bibler R, Lavin M (2006) Phylogeny of the genus *Phaseolus* (Leguminosae): A recent diversification in an ancient landscape. Syst Bot 31:779–791

Ellis TH (2014) Shifting the focus: populations, genetics and genomics. In: Sixth international food legume research conferences & seventh international conference on legume genetics and genomics program and abstract book. Saskatoon, Saskatchewan, Canada, 7–11 July 2014, p 27

Gepts P, Debouck DG (1991) Origin, domestication, and evolution of the common bean (*Phaseolus vulgaris* L.). In: van Schoonhoven A, Voyset O (eds) Common beans: research for crop improvement, C.A.B. Int., Wallingford, U.K. and C.I.A.T., Cali, Colombia, pp 7–53

Hnatuszko-Konka K, Kowalczyk T, Gerszberg A, Wiktorek-Smagur A, Kononowicz AK (2014) *Phaseolus vulgaris*—Recalcitrant potential. Biotechnol Adv 32:1205–1215

Michael TP, Jackson S (2013) The first 50 plant genomes. Plant Genome 6. doi:10.3835/plantgenome2013.03.0001in

Myers JR, Baggett JR (1999) Improvement of snap bean. In: Singh SP (ed) Common bean improvement in the twenty-first century. Kluwer Academic Publishers, Dordrecht, The Netherlands, pp 289–329

Singh SP (1999) Production and utilization. In: Common bean improvement in the twenty-first century. Kluwer Academic Publishers, Dordrecht, The Netherlands, pp 1–24

Varshney RK (2014) The 1000 pulse genome sequencing initiative: Connecting genes to traits. In: Sixth international food legume research conferences & seventh international conference on legume genetics and genomics program and abstract book. Saskatoon, Saskatchewan, Canada, 7–11 July 2014, p 23

Zohary D (1999) Monophyletic versus polyphyletic origin of the crops on which agriculture was founded in the Near East. Genet Resour Crop Evol 46:133–142

MIX
Papier aus verantwortungsvollen Quellen
Paper from responsible sources
FSC® C105338

If you have any concerns about our products,
you can contact us on
ProductSafety@springernature.com

In case Publisher is established outside the EU,
the EU authorized representative is:
Springer Nature Customer Service Center GmbH
Europaplatz 3, 69115 Heidelberg, Germany

Printed by Libri Plureos GmbH
in Hamburg, Germany